Lecture Notes in Civil Engineering

Volume 368

Lecture Notes in Civil Engineering (LNCE) publishes the latest developments in Civil Engineering—quickly, informally and in top quality. Though original research reported in proceedings and post-proceedings represents the core of LNCE, edited volumes of exceptionally high quality and interest may also be considered for publication. Volumes published in LNCE embrace all aspects and subfields of, as well as new challenges in, Civil Engineering. Topics in the series include:

- Construction and Structural Mechanics
- Building Materials
- Concrete, Steel and Timber Structures
- Geotechnical Engineering
- Earthquake Engineering
- Coastal Engineering
- Ocean and Offshore Engineering; Ships and Floating Structures
- Hydraulics, Hydrology and Water Resources Engineering
- Environmental Engineering and Sustainability
- Structural Health and Monitoring
- Surveying and Geographical Information Systems
- Indoor Environments
- Transportation and Traffic
- Risk Analysis
- Safety and Security

To submit a proposal or request further information, please contact the appropriate Springer Editor:

- Pierpaolo Riva at pierpaolo.riva@springer.com (Europe and Americas);
- Swati Meherishi at swati.meherishi@springer.com (Asia—except China, Australia, and New Zealand);
- Wayne Hu at wayne.hu@springer.com (China).

All books in the series now indexed by Scopus and EI Compendex database!

Scott Arthur · Masato Saitoh · Asiful Hoque
Editors

Proceedings of the 6th International Conference on Advances in Civil Engineering

ICACE 2022

Springer

Editors
Scott Arthur
School of Energy, Geoscience,
Infrastructure and Society
Heriot-Watt University
Edinburgh, Scotland

Asiful Hoque
Department of Civil Engineering
Chittagong University of Engineering
and Technology
Chittagong, Bangladesh

Masato Saitoh
Department of Civil and Environmental
Engineering
Graduate School of Science
and Engineering
Saitama University
Saitama, Japan

ISSN 2366-2557 ISSN 2366-2565 (electronic)
Lecture Notes in Civil Engineering
ISBN 978-981-99-3825-4 ISBN 978-981-99-3826-1 (eBook)
https://doi.org/10.1007/978-981-99-3826-1

This Springer imprint is published by the registered company Springer Nature Singapore Pte Ltd.
The registered company address is: 152 Beach Road, #21-01/04 Gateway East, Singapore 189721, Singapore

Contents

Understanding the Sedimentation Process of the Maeinee River for Developing a Future Drainage Plan Using Mathemetical Model 1
S. M. R. Newaz and M. A. Islam

GIS and Remote Sensing-Based Agricultural and Meteorological Drought Assessment in Rajshahi Division 13
M. Ferdousi and A. Narzis

Adsorption of Chloride Ions from Aqueous Solution Using Activated Carbon Obtained by Chemical Activation of Jute Stick 25
S. Tarannum, M. N. Abir, A. F. Mita, B. K. Banik, M. A. Hoque, and M. I. Kabir

Assessment of Carbon Footprint for the Bricks Manufactured in Bangladesh 35
Pritom Bhowmik Akash, Md. Jahedul Islam, and Md. Reaz Akter Mullick

Investigation on Pre-cracked RC Slab Strengthening in Flexure Region Under Patch Loading by CFRP 47
S. M. Z. Islam, M. M. Islam, M. Ahamed, M. M. H. Rakib, M. R. Hasan, and M. M. Rahman

Performance Evaluation of Different Intercity Trains of Bangladesh Based on Punctuality 59
Md. Rifat Hossain Bhuiyan, Shah Aymaan Ibtihal, Md. Tanvirul Islam, Shanjidul Hassan Shakif, and Kashfia Nahrin Nokshi

Flood Susceptibility Assessment in Southwest Coastal Region of Bangladesh Using an AHP-GIS Based Approach 73
M. Z. Hossain and S. K. Adhikary

Identifying Waterlogging Risk Zones in an Urban Area of Bangladesh Using AHP-Fuzzy Rule Based Approach 87
M. A. Azad, S. K. Adhikary, and M. M. Rana

Microplastics in River Sediments Around the Dhaka City: A Case Study for Occurrence and Quantification 101
M. R. Karim, M. T. Islam, M. S. Islam, H. T. Oishi, and Z. Tasnim

Uncertainty Based Assessment of Drought Using Standardized Precipitation Index-A Case Study 115
M. F. Rabby and S. K. Adhikary

Spatial Distribution of Heavy Metals and Identifying Factors that Influence Soil Properties at Rajbandh Landfill Site in Khulna 127
Ashif Adnan Khan Aoyon and Islam M. Rafizul

Use of Water Treatment Plant Sludge (WTPS) as a Cementitious Material in Concrete 141
M. H. Masum, I. J. Ruva, A. Tahsin, and J. Ferdous

A MCDM Based Approach to Prioritizing National Highways for Road Safety Improvements 155
Md. Rifat Hossain Bhuiyan, Md. Asif Raihan, and Moinul Hossain

Predicting Future Land Use and Land Cover Changes and Their Effects on Land Surface Temperature in Chattogram City of Bangladesh 167
T. Das, M. I. Islam, and D. R. Raja

Assessment of Air Quality and Noise Level in Chattogram City, Bangladesh 177
M. J. Alam, M. A. Aziz, A. Haque, and M. H. Masum

Scenario of Medical Waste Management in Bangladesh During Covid-19 189
K. N. Mim, Md. Z. Alam, and N. Nowshin

Factors Influencing Pedestrian Crashes in Dhaka City: A Multiple Correspondence Analysis Approach 201
S. A. Iqra, A. S. Huq, and S. H. Iqra

Modeling of Non-stationary Adsorption Behavior of Binary Heavy Metals to Porous Media 213
F. T. Jahara and M. Kikumoto

Morphological Analysis of Bishkhali River Using Multi-temporal Satellite Images and Historical Cross-Sectional Data 225
I. Sarkar and M. A. Rahman

User Perception and Assessment of Existing Public Toilets in Jessore, Bangladesh 237
S. A. S. Rafee and M. M. Rahman

Water Treatment and Scaling in Distribution Pipes—The Impact of Silica and Phosphate on Calcium Carbonate Precipitation 249
T. Bhattacharjee, S. Tarannum, and M. A. Ali

Effect of Overweight Vehicles on the Design of PC Girder Bridges 259
R. S. Tomal and K. M. Amanat

Effect of the Inclination Angle of Hooked Steel Fiber on the Flexural Behavior of Steel Fiber Reinforced Concrete 273
J. A. Mahmud, A. Islam, and M. S. Hosen

Analyzing Historical Temperature Variation in Dhaka, Bangladesh in Climate Change Perspective 285
A. Rudaiba and R. Mamtaz

BIM and CFD Based Simulation Approach in Reducing Thermal Energy Demands of a Residential Building: A Case Study in Bangladesh 297
S. A. Islam, M. T. Hassan, and M. R. Awall

Assessing the Fuel Potential of Fecal Sludge and Coal Through Co-combustion 311
R. P. Saha, T. S. Raaz, and A. B. M. Badruzzaman

On the Method to Evaluate the Post-earthquake State of Collapsed Structures and Its Applications to Numerical Analyses 323
Zongmu Liu and M. Saitoh

Relationship Between Rate of Corrosion of TMT Bars and Quality of Electrolyte in Electrochemical Corrosion Process 337
M. N. Bari and M. Mohshin

Performance of Rice Husk Ash Concrete in Acidic Environment 347
Md. S. Islam, Md. A. Hasan, Md. M. Islam, and S. S. Das

Shape Effects on Shear Behavior of Superdisk Systems Studied by Level Set Discrete Element Method 359
Haoran Jiang, Reid Kawamoto, and Takashi Matsushima

Morphological Evolution of Andesite Railway Ballast Particle Under Rotating Drum Abrasion Test 373
O. C. Debanath, T. Matsushima, T. Ijichi, and M. Miwa

Geopolymer-Based Building Blocks Using Construction and Demolition Waste 387
Md. Zubair, Md. Humayun Mahmood, and G. M. Sadiqul Islam

Burst Pressure Prediction of Dent-Cracked Steel Pipeline Subjected to Tensile Force 401
B. C. Mondal, A. Farzana, and C. K. Debnath

Heavy Metal Pollution Assesment in Urban Road Network; a Case Study of Chattogram 415
S. Deb, R. Prodhan, and S. K. Pal

Effect of Loading Directionality on the Horizontal Stiffness of Unbonded Scrap Tire Rubber Pad Isolator 429
M. B. Zisan and A. Igarashi

Optimization of Particle Packing Density of Industrial Wastes Incorporated Concrete 439
A. R. Ziad and G. M. Sadiqul Islam

Comparative Study on Riverbank Erosion and Accretion Using HEC-RAS and Landsat Imagery 453
B. K. Nath and A. Akter

Bangladeshi Drivers' Perception on Vehicle Safety Features 463
A. Huq, S. Iqra, and Z. Ikram

About the Editors

Dr. Scott Arthur is currently a professor at the Institute for Infrastructure and Environment, Heriot-Watt University, UK. He began his research career in 1992 working under the supervision of Prof. Richard Ashley on a high profile EPSRC funded project investigating sediment and pollutant transport within combined sewer networks. Following this work in Dundee, in 1996, he moved to a Post-doctoral Research Associate position at Heriot-Watt University and subsequently (August 2000) appointed within Heriot-Watt University to a lecturing position. Since that time, his research has been supported by EPSRC, the European Union, the Scottish Government, the Royal Academy of Engineering, SNIFFER and CREW. Professor Arthur research comprises work on all aspects of urban drainage systems from roof systems to urban watercourse and large sewer networks. He has published more than 60 international papers in indexed journals and has been involved in several research projects funded by Scottish Government and European Union.

Dr. Masato Saitoh is currently a professor at the Graduate School of Science and Engineering, Saitama University, Japan. After obtaining a Doctor of Engineering (D.Engg.) from Saitama University in 2001, Dr. Saitoh has moved to academia as an Assistant Professor at Saitama University. Earlier, he was working as a researcher at the Railway Technical Research Institute (RTRI) in Japan. His primary research interests include the dynamic response of the soil-pile system, foundation engineering. He has published more than 30 international papers in top-notch journals along with 13 national journals. Professor Saitoh also holds 13 patents including international PCT subsidized by the Japan Science and Technology Agency (JST) on his developed research works and lead a significant number of research projects funded by the Japanese Government (JSPS). In 2017, Prof. Saitoh received best lecturer award from Saitama University for his outstanding teaching contribution. He also received the best paper award from the Japanese Geotechnical Society in 2003 and best speaker award of JSCE conference in 1997 and 1999. At present, he is an editorial board member of JSCE Journal H-division and reviewer of several reputed journals.

Dr. Asiful Hoque is currently a professor and Head at the Department of Civil Engineering, Chittagong University of Engineering and Technology (CUET) in Bangladesh. He earned his Ph.D. from Hokkaido University, Japan in 2012. He has 20 years of experience at leading Bangladeshi academic institution and abroad through research activities and teaching students from various social and cultural backgrounds. His primary research interest focuses on environmental engineering, water treatment, wastewater treatment, air pollution, environmental impact assessment. Professor Asiful Hoque has published 16 international journals and more than 16 conference proceedings both in Bangladesh and abroad. He has been involved in few research projects, e.g. Assessment of Environmental Pollution in Ship breaking Yard, Assessment of Air Pollution in Chittagong City, Bangladesh funded by University Grant Commission, Bangladesh, assessment of Rainwater for the use of portable purpose funded by Wateraid Bangladesh. He is a member of many professional organizations like ASCE, Earthquake Society Bangladesh, IEB. In addition, he serves the academic community as a reviewer of distinguished journals.

Understanding the Sedimentation Process of the Maeinee River for Developing a Future Drainage Plan Using Mathemetical Model

S. M. R. Newaz and M. A. Islam

Abstract Mathematical model is widely used tool for planning and design of water resources project [1]. One-dimensional morphological modeling provides a scenario of sediment transport phenomena and morphological behavior in-terms of trend in long term bed level changes. Since two- and three-dimensional model provides better results but large extent of data requirement where one-dimensional morphological model can be applied for the general understanding of river system in long run river morpho-dynamics [2]. In this study 1D morphological model is developed for Maeinee River, which is one of the major rivers that feed Kaptai Lake. And Kaptai Lake is the reservoir of Karnafuli Hydroelectric Power station (230 MW). The model is calibrated with observed suspended sediment concentration by IWM. The key finding of the study is that the Maeinee River system shows that the perfect equilibrium has not yet been obtained, rather a dynamic equilibrium has been observed. This study shows upstream reaches of Maeinee River are likely to be under erosion process whereas downstream reaches mostly under deposition process in the bed. the results should be taken as indicative rather considering its quantitative outputs. Better result could be obtained if comparison between more approach could be analyzed. Lack of adequate data is one of the main problems faced during this study. The results are useful in investigating the response of a river system to large changes in the flow regime.

Keywords Transport sediment · MIKE 11 · Mathematical modelling · Drainage plan · Maeinee river

---

S. M. R. Newaz
Flood Management Division, Institute of Water Modelling, BD Dhaka, Bangladesh
e-mail: smz@iwmbd.org

M. A. Islam (✉)
Research and Development Unit, Institute of Water Modelling, BD Dhaka, Bangladesh
e-mail: asf@iwmbd.org

S. Arthur et al. (eds.), *Proceedings of the 6th International Conference on Advances in Civil Engineering*, Lecture Notes in Civil Engineering 368,
https://doi.org/10.1007/978-981-99-3826-1_1

1 Introduction

The objective of the morphological study is to understand the sedimentation process of the main drainage system for developing a sustainable drainage improvement plan. River morphology largely depends on the longitudinal slope, channel conveyance, off take of the sediment load, sediment transport rate, mean diameter of the sediment, particle settling velocity. For developing the sediment transport model required sediment data have been collected from secondary sources like BWDB, BMD. The analysis of sediment data, development of sediment model and assessment the sedimentation process of the rivers have been described in the methodology part. The combined flow of Maeinee and Kasalang river falls to Kaptai lake which fulfill water demand of the Kaptai Hydro Power Plant (230 MW). Therefore, Maeinee river play an important role in the operation of Kaptai Hydro Power plant. It is also important for supplying irrigation water in the surrounding area. There are cultivable lands on both the banks. The land is medium high land and subject to occasional flash flood during the rainy season. But the flood water usually recedes within a short time. Farmers are not able to grow rice due to scarcity of irrigation water [3]. The Maeinee river is flashy and carries sand (coarse & fine mixed) during flush flood. It is observed that the sand is mostly deposited in the river bed. By using sediment transport module of MIKE 11, total sedimentation process of the river can be observed. One dimensional morphological model needs comparatively less data and time to simulate though it has limitation of overlooking bend scour. But for long term simulation it is convenient to get sediment transport and bed level change as an indicative solution for the planning process [4]. Thus, it is helpful for decision maker and planners for the long-term planning relevant to river restoration. The study area and overall methodology are shown in Figs. 1 and 2.

2 Study Area

The Maeinee River is located at Dighinala Upazila of Khagrachhari district. The river is originated from the hilly area of Tripura, India and rolls down to Dighinala Upazila of Khagrachhari district of Bangladesh. The Maeinee River is about 109 km long and width varies from 72 m to 45 m. The average depth of the river is about 1 m. The river flows to Rangamati district at Babushara UP and falls to Kasalang river at Kheder mara UP of Baghai Chhari UZ. The Maeinee river is a perennial river, however, during the dry months the flow comes down to as low as 2.82 m^3/s (field measurement). It is perennial and flashy in nature. Three temporary gauge stations have been established by IWM under a project on Maeinee at km 0.0 near Beltola, at km 21.40 near Dighinala Steel Bridge & at km 73.35 near Maeinee Mukh. The average bed slope of the River is 41 cm per km within the project area. It is mentioned earlier that the Maeinee river is flashy and carries sand (coarse & fine mixed) during flash flood. Map of the study area is shown in Fig. 1.

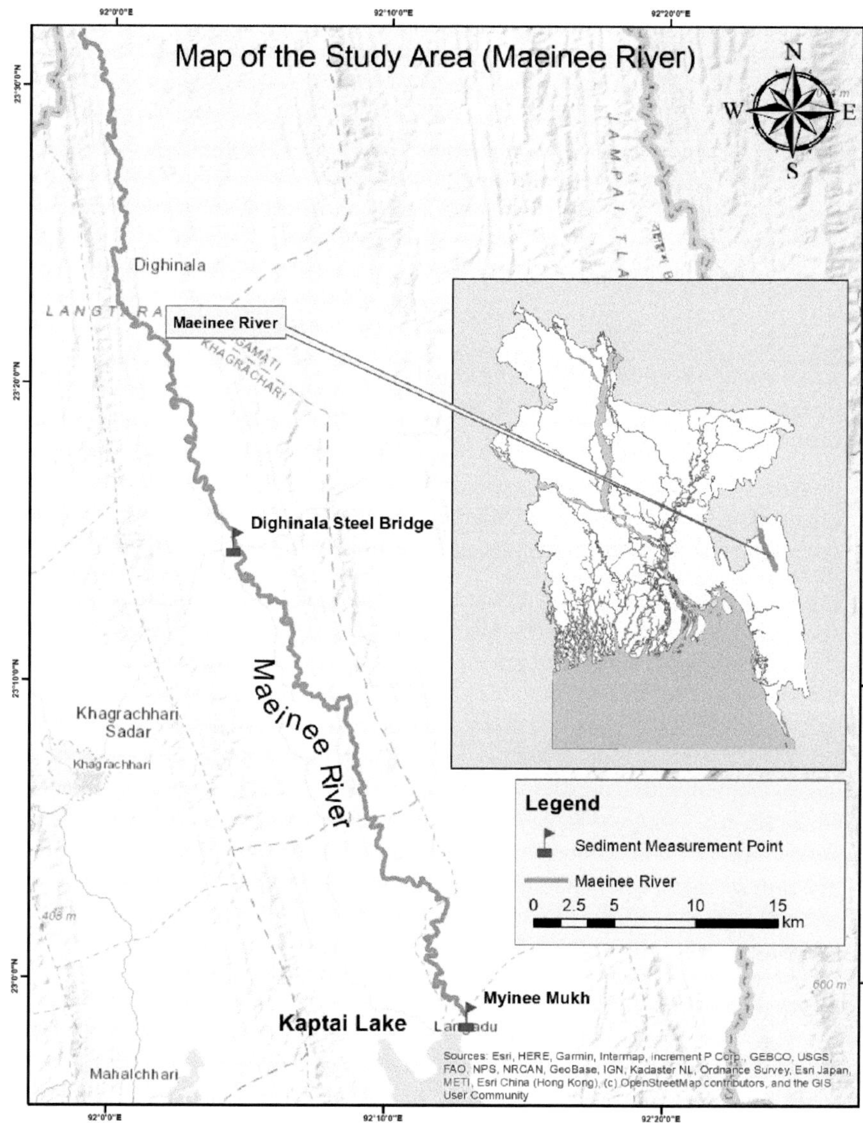

Fig. 1 Map of the study area

3 Methodology

Relevant publications and project reports have been reviewed to finalize this methodology. A one-dimensional morphological model has been developed coupling with hydrodynamic simulation using secondary hydrological data and recently collected topographic data. In this study, the non- cohesive sediment transport and morphology

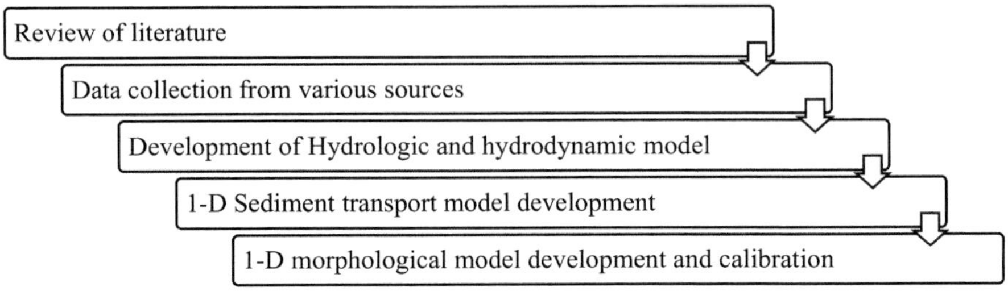

Fig. 2 Methodology flowchart of the study

module of mike 11 is used. By using this module, we have the understanding of sediment transport phenomenon and morphological behavior. The computation of sediment transport rate depends on the development of 3 modules:

- Computation of rainfall-runoff using MIKE 11-NAM
- Computation of hydrodynamic behavior using MIKE 11-HD
- Computation of sediment transport and bed level change using MIKE-11 NST

The hydrological and hydrodynamic Model input data have been collected from various source like IWM, BWDB. Historical data from 1992 to 2019 have been used to computation of hydrodynamic model and calibrated for year 2019. Development and validation of Eastern Hilly Regional Model (EHRM) have been carried out regularly by IWM. Our model network system has been developed on the basis of Eastern Hilly Regional Model (EHRM), where an upstream boundary is the Dighinala Maeinee catchment runoff and downstream boundary is Water level of Maeinee mukh is used. Sediment samples data have been collected by IWM under a study and analysis project of IWM. Simulation of the sediment transport model is carried out for the period of Jan 2019 to September 2019 to assess the sediment during monsoon season. To get the agreement with the model simulated and observed sediment concentration, it is required to calibrate the model.

4 Hydrological Model Development

The hydrological model data is collected from IWM feasibility study model on the area. The hydrological model covers 11 sub-catchments located both in India and Bangladesh. Total area of these 11 sub-catchments is 10810 sq.km. The catchment area covers mainly hilly terrain with heavy vegetation. The catchments generate flash flood flow during any rainfall event in the upstream hill area. The sub-catchment delineation is needed to compute the runoff generated in the study area. The delineation has been done from the land level data collected from the secondary source (SRTM) using MIKE Hydro Basin software (a water management tool of DHI) (Fig. 3).

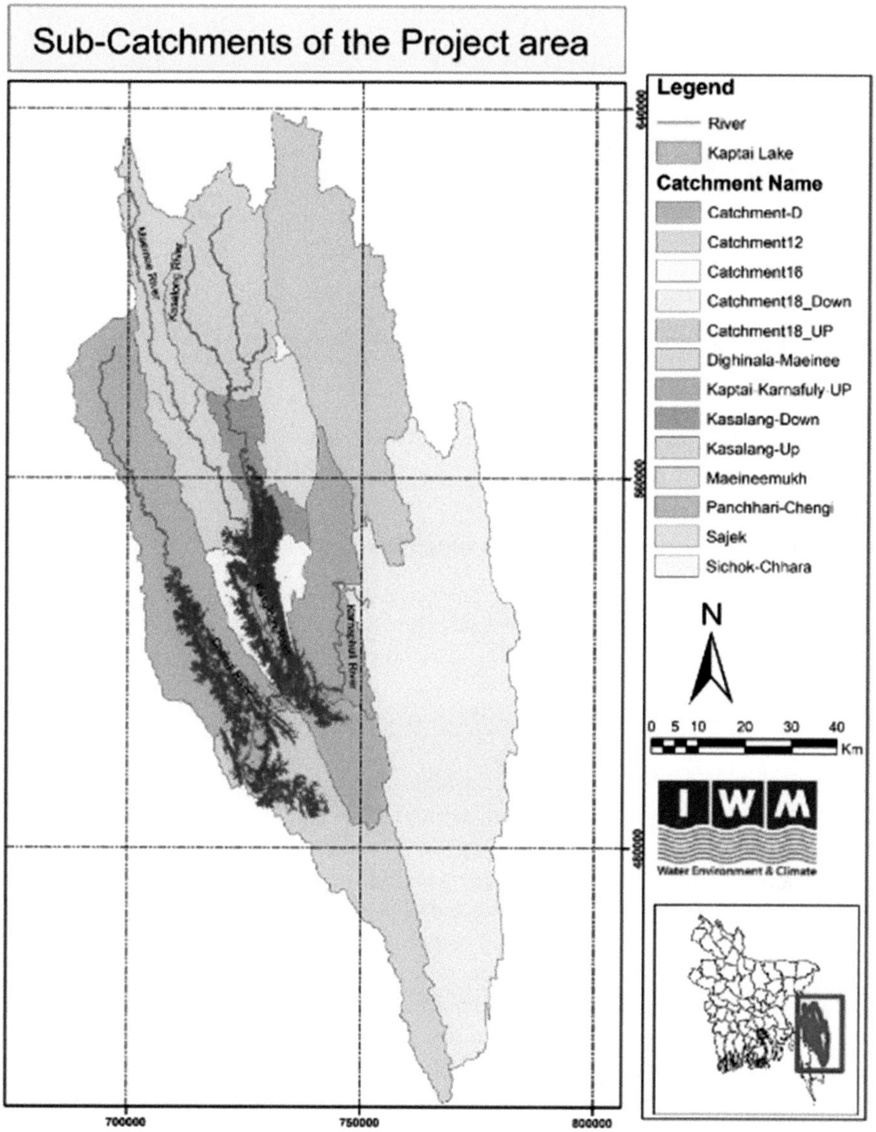

Fig. 3 Sub catchments of the project area (*Source* IWM)

The rainfall-runoff model has been calibrated comparing with observed flow measured near the outlet of the "Dighinala-Maeinee Sub-Catchment". The comparison plot of the computed runoff with measured discharge is shown in the following Fig. 4.

From the comparison, it has been observed good agreement between the field measured discharge and model simulated runoff (discharge).

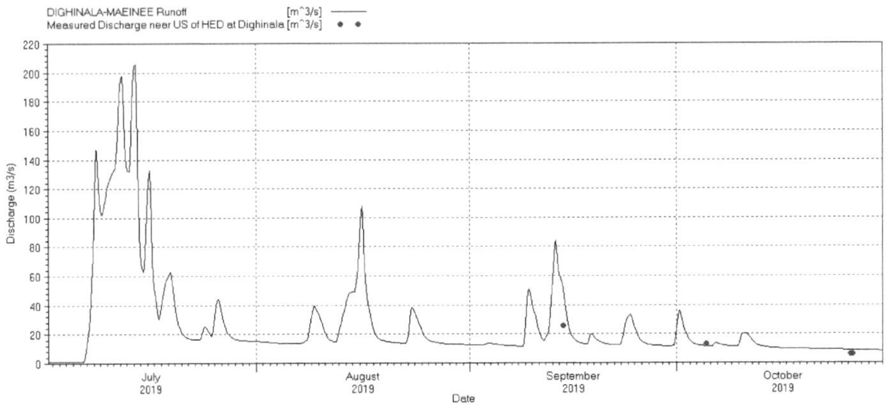

Fig. 4 Comparison of simulated runoff with observed discharge (Q) at the outlet of the Dighinala-Maeinee Sub-Catchment. (*Source* IWM)

5 Hydrodynamic Model Development

The project model of the study has been developed for area covering the entire Maeinee River of Dighinala Upazila. The project model comprises of lump information of study area and it has been utilized for simulation of hydraulics in Maeinee River. Figure 5 shows schematized River network and cross section setup of the project model.

For our project we used Dighinala-Maeinee catchment runoff taken from the hydrological model as upstream boundary and water level at Maeinee much as downstream boundary. Maeinee mukh water level is collected from calibrated hydrodynamic model used by IWM for adjacent project. The project has been calibrated comparing simulated water level with the observed water level at location Beltola and Dighinala. Comparison plot of the simulated and observed water level at location Beltola and Dighinala are shown respectively in the Figs. 6 and 7.

From the above figures we can say that model predicted result is quite close with the observed result. The observed data was collected from IWM.

6 Sediment Transport Modelling

Bed samples were collected from Maeinee River using Van Veen grab sampler. Laboratory analysis is carried out in IWM Laboratory. The model has been run using a number of grain sizes (D50) representing grain size fractions of individual reaches of Maeinee River system shown in flowing Fig. 8.

The non-cohesive sediment transport module of MIKE 11 [5] has been developed and used to evaluate the sediment transport rate and accumulated sediment transport data. The model run was carried out using Van Rijn sediment transport formula.

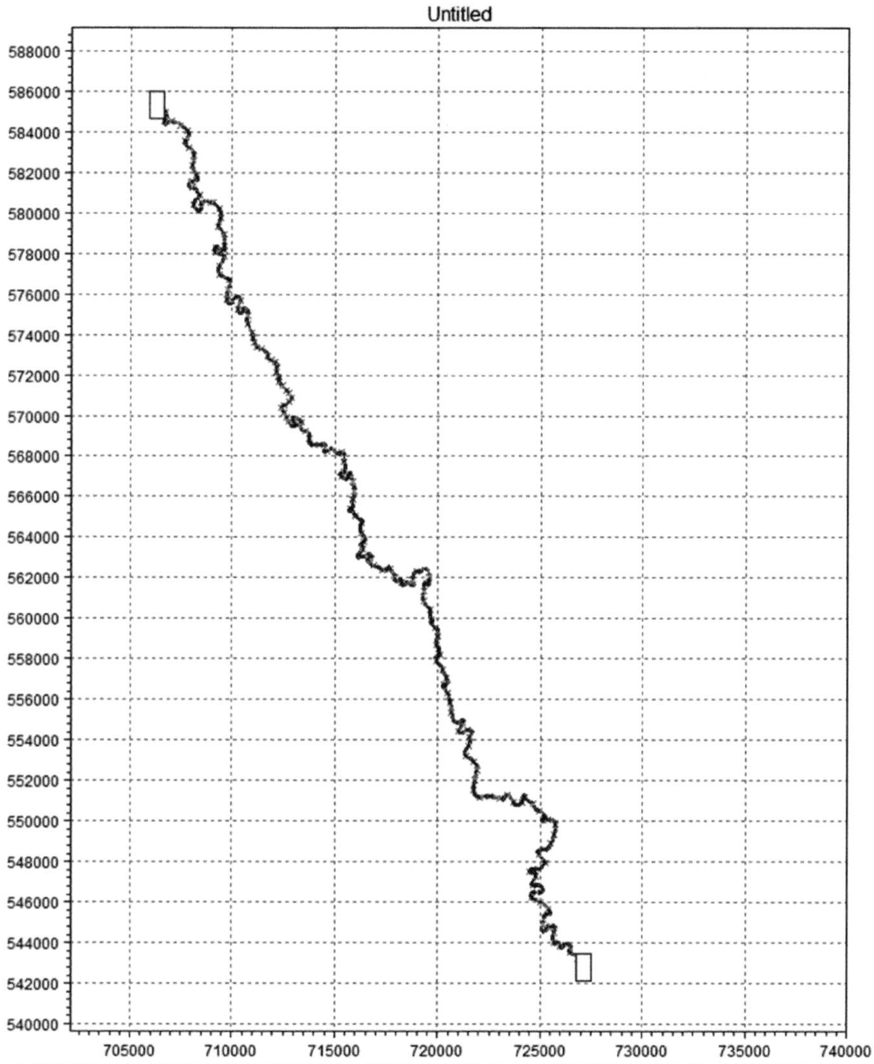

Fig. 5 Model network of Maeinee river

The model run was carried out for the long period of 28 years from 1992 to 2019 based on the availability of boundary data. The sediment transport calculations are based on the results of the HD model. There is no feedback from the sediment transport calculations to the HD model. Results are in the form of transport rates and accumulated volumes (per meter width) of deposition or erosion (Fig. 9).

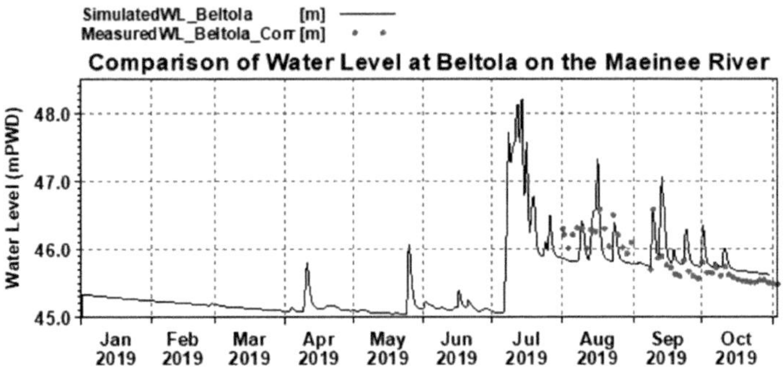

Fig. 6 Water level comparison at Beltola on Maeinee river

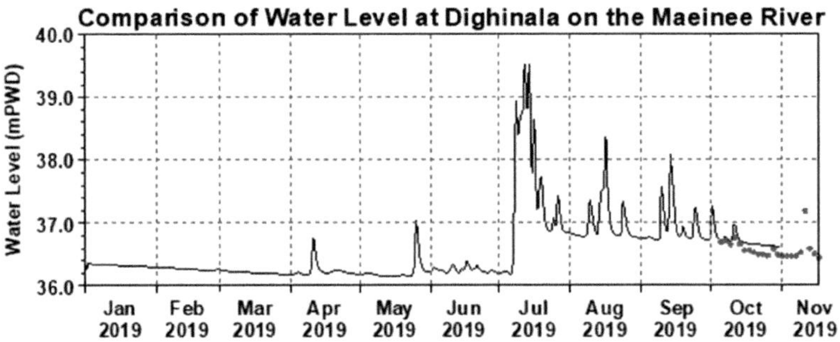

Fig. 7 Water level comparison at Dighinala Steel bridge on Maeinee river

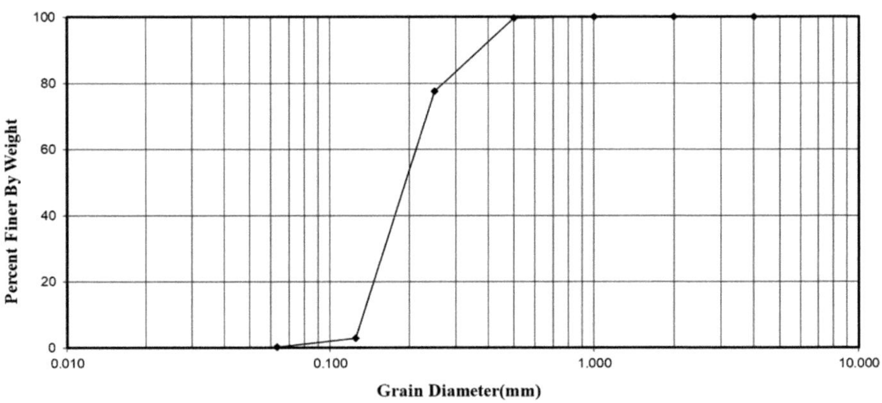

Fig. 8 The median grain size distribution (d50) of bed material samples (*Source* IWM)

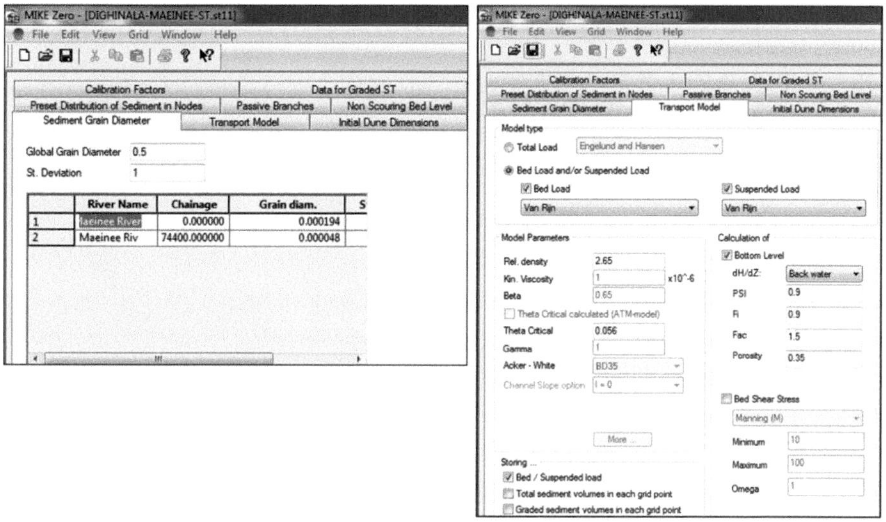

Fig. 9 Parameter setup and approach adopted for the sediment transport model

7 Sediment Transport and Morphological Modelling

A 28 years (1992 to 2019) simulation was made using the existing topography. The main objective was to observe the long-term development of bed levels. So, the model results were saved every day though the bed levels were updated every time step of model simulation. To get the agreement with the model simulated and observed sediment concentration, it is required to calibrate the model. We could get to compare our result only in one location due to lack of observed data. In this purpose, the sediment model was calibrated near Dighinala steel bride on Maeinee River. We collected the observed data from Feasibility Study report of IWM. Suspended sediment samples were collected from Maeinee using pump sampling at selected locations during flow measurements. Laboratory analysis of sediment samples have been carried out to obtain total concentration as well as concentration of wash load (silt) and of bed material load (sand) (Fig. 10).

8 Result and Discussion

The key finding of the study is that the Maeinee River system shows that the perfect equilibrium has not yet been obtained, rather a dynamic equilibrium has been observed. The trend of bed level changes (Sedimentation and erosion pattern) along the Maeinee river system is shown in the Fig. 11. From the figure it is clear that the upper reach of the river system shows mainly erosion pattern whereas the lower part of the reach shows mainly deposition pattern. For the first 30 km of the reach

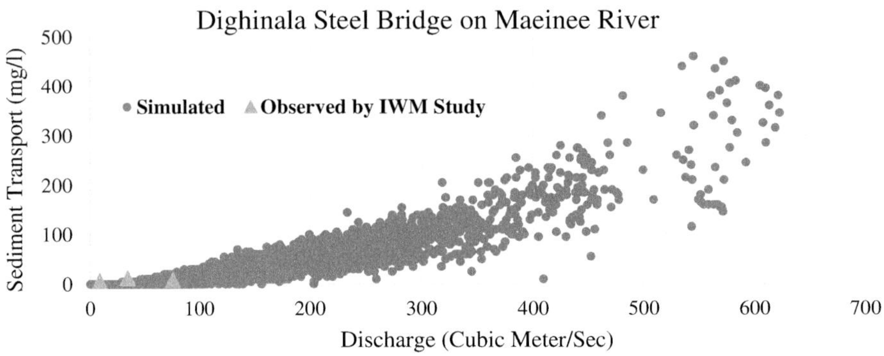

Fig. 10 Plot showing agreement of measured and model simulated concentration at Dighinala steel bridge on Maeinee river

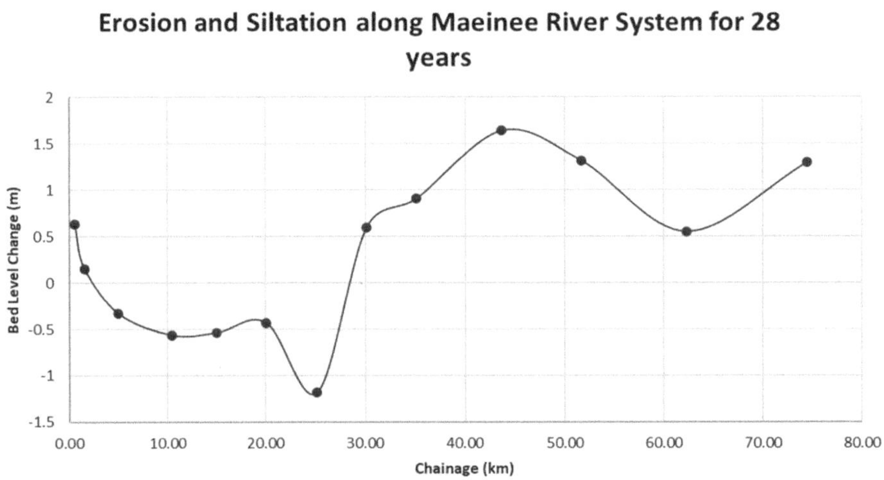

Fig. 11 Bed level changes along the Maeinee river system

shows erosion, where the range of bed level change varies from −1.181 to 0.594 m. From 31 km to up to 74.4 km shows mainly deposition pattern, where the range of bed level change varies from 0.55 to 1.65 m.

9 Conclusions and Recommendations

Erosion tendency is observed in the upper reaches of the river system whereas deposition tendency is observed in the lower reaches. Better Hydrodynamic model results can be achieved in case of adequate data. Due to lack of enough sediment data the model is calibrated only in one location. More sediment transport calculation approach can be adopted to find the best possible methods. It should be mentioned that the one-dimensional morphological models generally have limitations and is developed and calibrated with limited sediment data and information, thus the results should be taken as indicative rather considering its quantitative outputs. Nevertheless, the results are useful in investigating the response of a river system to large changes in the flow regime. It is, therefore, believed that the present development of the morphological model would be useful in planning activities such as assessment of trend of sensitive morphological features, any intervention made in the river.

Acknowledgements The authors wish to acknowledge the support of Institute of Water Modelling (IWM). It was not possible to accomplish the study without the software and data supports from IWM.

References

1. Chollet JP, Cunge JA (1980) Simulation of unsteady flow in alluvial streams. Appl Math Modell 4:234–244
2. Kafi MAH, Masud MS, Magumdar TK, Pk S, Hossain MS (2018) Assessment of long-term evolution of morphodynamics of the Kalni-Kushiyara river system in Bangladesh using 1-dimensional morphological model. J Civil Eng (IEB) 46(1):21–30
3. FAP6 (1994) Flood action plan, Northeast regional water management project, specialist study, river sedimentation and morphology, final report. SNC Lavalin International Northwest Hydraulic Consultants, Flood Plan Coordination Organization, Bangladesh Water Development Board
4. Nones M, Silvio GD, Guerrero M, Pastorello L (2015) 1-D long term simulation of the Paraná River morphodynamics in the light of climate variability
5. DHI (2014) A modelling system for rivers and channels. Danish Hydraulic Institute (DHI), Reference Manual

GIS and Remote Sensing-Based Agricultural and Meteorological Drought Assessment in Rajshahi Division

M. Ferdousi and A. Narzis

Abstract Drought in the northwest region of Bangladesh is mostly influenced by several climatic parameters such as rainfall, evapotranspiration, temperature, etc. To adapt the unfavorable climatic conditions, it is vital to use alternating cropping patterns and land use, even during the rainy season. From this context, the present study aims at monitoring agricultural and meteorological drought patterns during Kharif II (mid-July to mid-November) season in the Rajshahi division. Satellite images (Landsat and Sentinel-2A) were collected at two years intervals to monitor agricultural drought within the study area using the Normalized Difference Vegetation Index (NDVI). Then the values were categorized into six different classes: extreme dry, dry, moderate, wet, and extremely wet soil conditions that governed the crop growth and indicate agricultural drought severity. To assess the meteorological drought condition Standardized Precipitation Index (SPI) is calculated using the SPI tool. Calculated SPI values were then divided into six different classes from no drought to exceptional drought. The coupling effects of rainfall and vegetative conditions were then determined using NDVI anomalies and plotted against SPI values. The month of September is noticed to be the most critical crop period as the NDVI values are at the peak representing the fullest crop growth. It is also observed that Naogaon, Natore, Bogra, and Sirajganj experienced dry, hot to extremely dry soil moisture conditions indicating agricultural drought almost every year from 2010–2020. The negative NDVI values in 2010, 2012, and 2014 suggest that there was a drought in those years because there was less available rainfall. Both indices show that crop production is possible when rainfall is available. In 2018, there was less rainfall but a positive vegetation index suggesting irrigation/water storage. Future studies may combine crop yield data with the findings from this study to better understand drought in Rajshahi.

Keywords Agricultural drought · Meteorological drought · NDVI · SPI · Satellite images

M. Ferdousi · A. Narzis (✉)
Department of Environmental, Water Resources, and Coastal Engineering (EWCE), Military Institute of Science and Technology (MIST), Dhaka, Bangladesh
e-mail: narzis@ewce.mist.ac.bd

S. Arthur et al. (eds.), *Proceedings of the 6th International Conference on Advances in Civil Engineering*, Lecture Notes in Civil Engineering 368,
https://doi.org/10.1007/978-981-99-3826-1_2

1 Introduction

Drought is a complicated process that can last for a long period. It is usually measured in months or even years and occurs when the moisture availability at a certain location falls far short of what is climatically predicted or adequate [1]. Drought should not be confused with related concepts such as aridity, in which low rainfall is a permanent feature of the climatic environment, seasonality, in which water scarcity is a normal part of the annual climatic regime, or desertification, in which an arid region shifts due to poor management [2] The southern and south-eastern portions of the Rajshahi division, as well as the southern portion of the Rangpur division, all faced a severe drought that caused agricultural damage, particularly to rice and jute, according to earlier studies. These studies were done using MODIS and Landsat data for years before 2010. Using data from Landsat and MODIS, M. Aziz created a drought severity SPI map for the entirety of Bangladesh in 2012 [3]. For the entirety of Bangladesh, M. Mondal et al. created an SPI map between 1981 and 2010[4]. Since the sentinel-2A product was launched in 2014, there has been no such research using Sentinel-2A images to detect drought in the north-western region of Bangladesh. NDVI from LANDSAT images and precipitation data collected from ground stations were used for the identification and extraction of drought risk in these areas on a large scale. The main objective was to find the correlation between precipitation and NDVI to estimate how smoothly and effectively the drought-prone regions can be calculated [5].

54 rivers traverse both Bangladesh and India; these are referred to as trans-boundary rivers. Most of these rivers enter the nation from India and Myanmar and travel through the north and southwest. The upstream withdrawal of water for industrial, domestic, and international water management structures causes trans-boundary rivers' natural flow to be disrupted. As a result, the Tista, Punarbhaba, and Ganges (at Farakka) experience an interruption in their regular water flow.

Bangladesh is a low-lying country located downstream and faces the consequences of these changes upstream. These restrict the groundwater level from adequately filling and prevent surface water from being available in north-western locations and cause a water shortage [6]. With an average of just 100 cm from June to October, the northwest experiences the least amount of seasonal precipitation nationwide [7]. Compared to the rest of the region, these areas have seen consistently high temperatures and dry weather [8]. Large amounts of groundwater are mostly harvested in this area for irrigation purposes. The amount of surface water has been drastically decreasing. The study's goals are to identify meteorological drought using Standardized Precipitation Index (SPI) and remote sensing methods, to identify agricultural drought using NDVI (Normalized Difference Vegetation Index), and to assess drought in the north-western region by combining agricultural and meteorological drought.

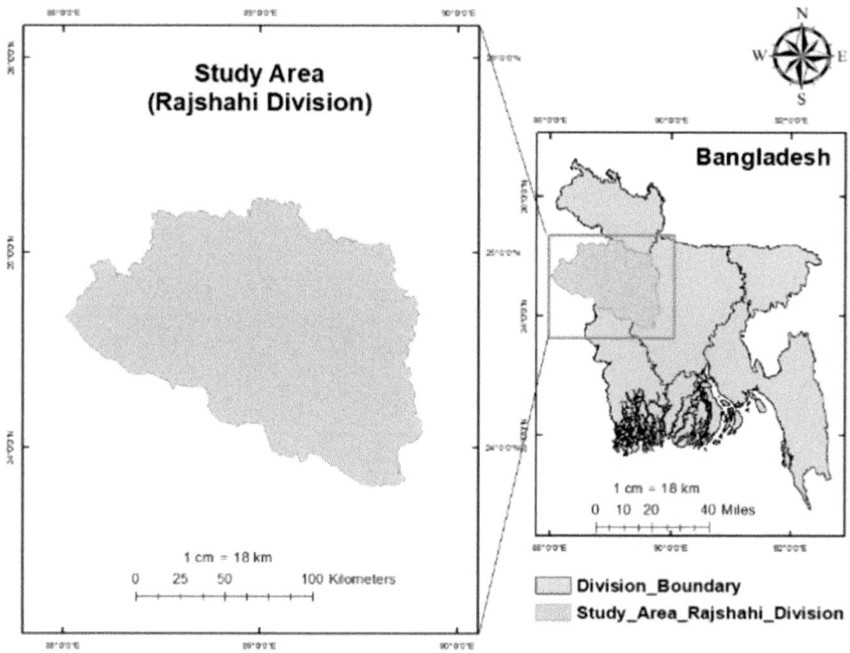

Fig. 1 Location of study area (Rajshahi division)

1.1 Study Area

The Rajshahi division is situated in western Bangladesh (Fig. 1). It covers an area of 18,174.40 square kilometers and is located between 23°48' and 25°16' north latitudes and 88°01' and 89°48' east longitudes [9]. The country is entirely flat plains, with a few swaths of forest remaining. The Padma River forms the international boundary with India to the west, and the Jamuna River forms the international border with Bangladesh to the east. These rivers meet in the Pabna district, at the southeastern tip of the Rajshahi division [10].

2 Materials and Methods

2.1 Data Acquisition

The majority of the data came from two sources: available satellite images from USGS and rainfall data collected from BMD (Table 1).

Table 1 Collected data and their sources

Data type	Source	Period
Remote sensing products (satellite images: Sentinel 2A, Landsat 4–8)	USGS Global Visualization Viewer (GloVis)	2010–2020 (2 years interval)
Meteorological data (monthly rainfall) of six rainfall stations (Rajshahi division: Pabna, Sirajganj, Natore, Naogaon, Nawabganj, Jaypurhat)	Bangladesh Meteorological Department (BMD)	2010–2020 (2 years interval)

The processing of data and analysis was carried out using the Standardized Precipitation Index (SPI) tool and ArcGIS 10.8.

2.2 *Data Processing and Analysis*

For the years 2010, 2012, 2014, 2016, 2018, and 2020 correlation and regression approaches were employed to see if there was a link between NDVI and rainfall patterns in the N-W area of Bangladesh. For agricultural areas in the northwest region, district-level NDVI anomalies were linked to SPI and crop yield. Finally, the agricultural and meteorological droughts were integrated to determine the total drought for the research region (Fig. 2).

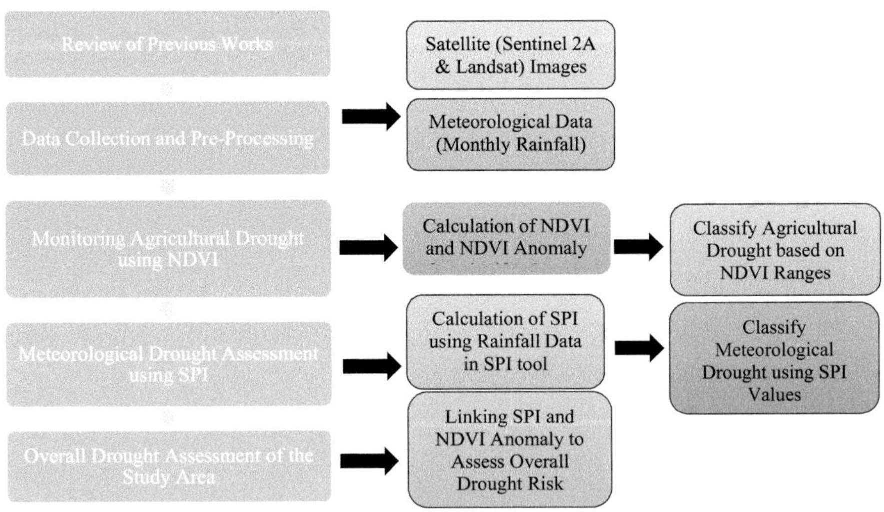

Fig. 2 Methodological framework of the study area

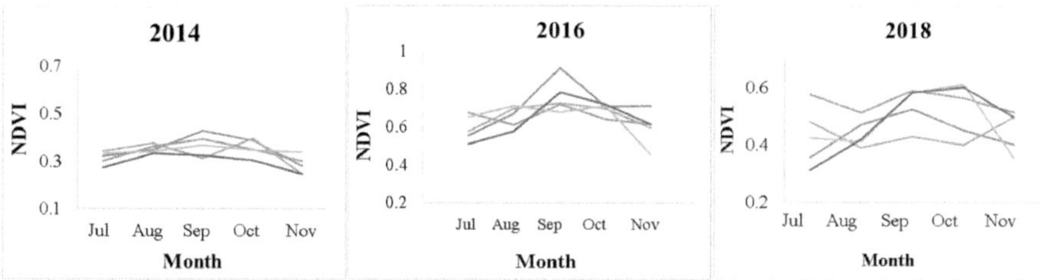

Fig. 3 NDVI value for the selected coordinates in the study area

2.2.1 Preprocessing of Satellite Data

Earth Explorer or USGS Global Visualization Viewer (GloVis) is be used to retrieve Sentinel-2 data and Landsat 7–8 imagery. Only a portion of the available ESA acquisitions is represented in the USGS Sentinel-2 collection. To meet the stated goals, data was gathered from several sources. All of the data was examined using ArcGIS. The processing includes several steps like-

1. Mosaic dataset
2. Cloud cover removal
3. Clipping to extent
4. Computation of NDVI

2.2.2 Post-Processing of Satellite Data to Derive Vegetation Indices and NDVI Calculation

To derive the seasonal pattern of NDVI for the selected years, firstly, the average NDVI for each year was computed by using Eq. (1).

$$\text{Average NDVIy} = (\text{NDVI7} + \text{NDVI8} + \cdots + \text{NDVI11})/5 \tag{1}$$

where, NDVIy is the NDVI for y year, and NDVI7, NDVI8……….. NDVI11 refers to the NDVI of specific months during that year. Equation [2] was used to get the mean NDVI for 6 years.

$$\text{Mean NDVI} = (\text{Average NDVIy1} + \text{Average NDVIy2} + \cdots + \text{Average NDVIy6})/6 \tag{2}$$

where Average NDVIy1……….. Average NDVIy9 is the 6-year yearly average NDVI value. The maximum NDVI of 2010, 2012, 2014, 2016, 2018, and 2020 was used to calculate the NDVI anomaly using Eq. (3).

$$\text{NDVI max i} = (\text{NDVI 1, NDVI 2, } \ldots \text{NDVI n}) \tag{3}$$

Table 2 Classification of NDVI [12]

NDVI range	Drought
<0	Extreme dry
0–0.2	Dry
0.2–0.4	Moderate
0.4–0.6	Wet
≥0.6	Extremely wet

where the maximum NDVI in the ith year is NDVI max i and NDVI n is the difference between the NDVI in month 1 and the NDVI in month n of the ith year. The data was a monthly composite, but it was nearly cloud-free and editable. To reduce the effects of cloud contamination, the maximum NDVI for the growing season has been established each year. By using the highest value compositing procedure, NDVI composites are less affected by cloud contamination, shifting sun zenith angles, and surface topography [11]. After the NDVI max was calculated, the mean NDVI max values over the past six years were determined by averaging these NDVI max photos. Equation [4] was used for computing the NDVI anomaly.

$$\text{Anomaly NDVI i} = (\text{NDVI max i} - \text{mean NDVI max})/(\text{mean NDVI max}) * 100 \tag{4}$$

where Anomaly NDVI i denotes an NDVI anomaly in the ith year, NDVI max denotes the maximum NDVI, and mean NDVI max denotes the average of maximum NDVI across the study period. Finally, images are subset as district-wise for 16 districts in the N-W area to compute district-wise NDVI anomaly for all 6 images.

NDVI values are divided into five categories to classify NDVI results using Table 2. The categorization of NDVI values is done to distinguish between vegetated and non-vegetated areas, and it is also used to assess dry and wet areas. Negative NDVI values indicate below-normal vegetation conditions, implying a drought situation such as severe or extreme drought. Reclassification of drought severity of Kharif season is done for each year which indicates the locations of wet and dry areas [5].

2.3 SPI Calculation

SPI is calculated using Eq. (5):

$$SPI = \frac{(Xi - Xm)}{o'} \tag{5}$$

The station's average monthly rainfall is X_i, it's mean is X_m, and its standard deviation is ό. Six rainfall stations' monthly precipitation data are used as input to the SPI software for the chosen years. Input data are arranged in the three-column

Table 3 Drought Severity Classification (modified) [5]

Category	Description	Standardized precipitation index (SPI)
D0	No drought	−0.50 and above
D1	Abnormally dry	−0.51–0.70
D2	Moderate drought	−0.71–1.10
D3	Severe drought	−1.11−1.50
D4	Extreme drought	−1.51−1.99
D5	Exceptional drought	−2 or less

format is yyyy mm pppp. Where pppp = precipitation, mm = month, and yyyy = year.

Few observations were made during SPI calculation using the SPI tool. (i) The precipitation quantities entered contain decimals after being multiplied by 100. Actual precipitation should only have two decimal places because the real input precipitation data should be integers (after being multiplied by 100). (ii) The input file does not contain enough values. There should be a minimum number of readings for precipitation each month. However, if only one year's worth of data is submitted, an error will be found, and nothing will be calculated. It is unclear how many is the minimum.

For this study, only monthly rainfall data for the crop growing season from 2010, 2012, 2014, 2016, 2018, and 2020 were used to calculate at each of the six rainfall stations (July–November). The cutoff point established by the U.S. Drought Mitigation Centre for determining the severity of the meteorological drought has been implemented. The drought severity classification chart (Table 3) category column has been changed to account for the SPI map reclassification [5].

2.4 Developing a Linkage Between NDVI Anomaly and SPI Value

Divisional NDVI anomalies in the N-W region have been linked to divisional SPI. Linear regression between NDVI anomaly has been computed for the entire region to assess the impact of rainfall on the crop, which leads to agricultural drought. Positive SPI indicates that water has been accessible to plants in just the correct amount, resulting in a positive NDVI value, while negative SPI indicates that water has not been available to plants in just the right amount, resulting in a negative NDVI value. The coupling effects of precipitation and vegetation conditions were then determined by creating a graph to show the relationships between SPI and NDVI anomaly [5].

3 Results and Discussions

3.1 *Crop Seasons in the N-W Region of Bangladesh*

Five random coordinates were selected from the study area. Monthly NDVI for five random coordinates of 2010, 2012, 2014, 2016, 2018, and 2020 were entered into an excel spreadsheet, and crop cultivation trends were created (Fig. 2). The Kharif-2 season begins in July and ends in October/November. According to the graph, there is one peak NDVI value throughout the Kharif-2 season. The duration of a crop season is calculated using the start and endpoints of a peak. The peak is at its highest in September, with a few exceptions in October. As a result, September is a critical month for Bangladesh's northwestern region.

3.2 *NDVI and Agricultural Drought*

From + 1.0 to −1.0, the NDVI scale is used. In places of a dead plant or gravel, NDVI values are often relatively low. NDVI values between 0.2 and 0.4 typically correspond to areas with sparse vegetation; moderate vegetation ranges between 0.4 and 0.6, and anything above 0.6 indicates the highest possible density of green leaves. We estimated the variation of NDVI throughout the Kharif season for the years 2010, 2012, 2014, 2016, 2018, and 2020 (Fig. 4).

NDVI values were highest in September which indicates dense vegetation in 2012, 2016, 2018, and 2020. The NDVI values range from −0.83 to 0.99, −0.70 to 0.99, −0.25 to 0.66, −0.32 to 0.90, −0.28 to 0.91, and −0.28 to 0.99. The average high value of NDVI is 0.86 in six years. Wet area increased in 2018 and 2020 which falls between the NDVI range of 0.4 and 0.6.

According to our research, the Rajshahi division experienced severe drought in 2010, 2012, 2014, and 2018. No drought or moderate drought was experienced in 2016 and 2020 (Fig. 5).

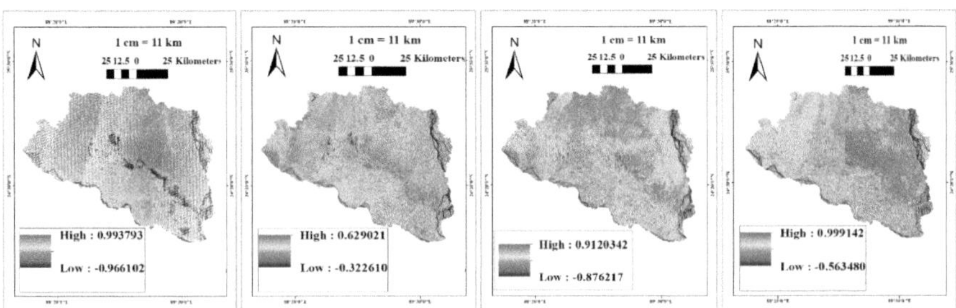

Fig. 4 Range of NDVI values of 2012, 2016, 2018, and 2020 for September (Monsoon season)

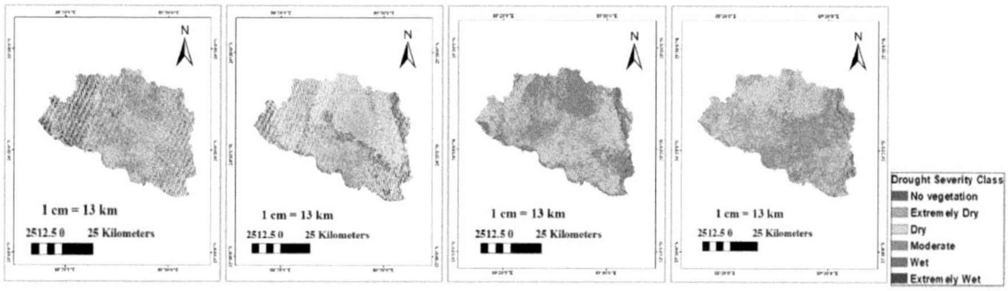

Fig. 5 Classification of Agricultural drought of 2010, 2012, 2014 and 2018

3.3 SPI and Meteorological Drought

In this study, 1-month SPI was calculated for 6 rainfall stations using monthly rainfall data of 6 years only for the crop growing season (July–November). Monthly rainfall data from six rainfall stations are utilized as input to the SPI program for the selected years.

The SPI value of the entire region was determined by interpolation (Inverse Distance Weighted method) using ArcGIS. If the SPI value is low, it indicates less rainfall in that area. In the year 2010, there was a severe drought in Nawabgonj in August, a severe drought in Pabna and Sirajgonj in September, and no drought in October. There was a severe drought in August 2012 in Pabna, Sirajganj, and Nawabganj. Pabna experienced an abnormal drought in September and a moderate drought in October. In 2014, Pabna and Nawabganj witnessed an abnormal drought in September and a severe drought in the Naogaon district. It can be seen that there was almost no drought in 2016 according to our research. In August of 2018, parts of Sirajgonj and Pabna were hit by severe drought, while Nawabgonj was hit by severe drought in September. In October of the same year, a moderate drought struck almost the whole Rajshahi division. In August 2020, Nawabgonj faced an abnormally dry spell. 2016 is a wet year, while the others are dry years.

3.3.1 SPI and Drought Severity

Based on the interpolated SPI values in September, a threshold from the modified SPI table (Table 3) for the N-W region was computed for each of the 6 years, and meteorological drought risk was thus classified as slight, moderate, or severe. Slight drought has a value of > -0.71, moderate drought has a value of > -1.11, and severe drought has a value of -1.11 (Fig. 6). A severe meteorological drought was witnessed on a large scale in 2010.

2020 and 2016 is the wet year and 2010 is a dry year (Fig. 6). Many research has shown that short-term (1 month) and long-term (3 and 6 months) drought indicators can be used in agriculture and hydrology [13].

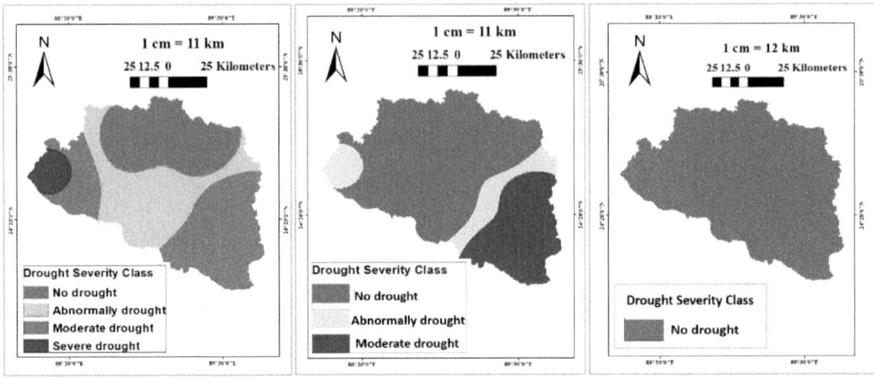

Fig. 6 Classification of Meteorological drought for 2010 (Dry year) **a** August **b** September **c** October

3.3.2 Relationships Among SPI, NDVI Anomaly

For the entire N-W region, NDVI anomaly and SPI have been computed, and it shows that when SPI is positive, NDVI anomaly is also positive, indicating that NDVI anomaly and SPI have a linear correlation (Fig. 7). Because SPI represents scarcity of water or excess, a positive SPI indicates that just the right amount of water was available to plants, resulting in a positive NDVI. And a negative SPI refers to the deficit of water resulting in negative NDVI. Even though the SPI value was negative in 2018, the NDVI value was positive. The reason for this is that wetlands increased in 2018, and soil moisture content was high. It is seen that when the NDVI anomaly is −13.7%, the SPI value is −0.97, and when the NDVI value is near −3.4%, the SPI value is −0.55. These correspondingly refer to the years 2010, 2012, and 2014. NDVI anomaly and SPI both are positive in the years 2016 and 2020 indicating there was almost no drought in these years.

Fig. 7 Relationship of September 1 month SPI with NDVI anomaly

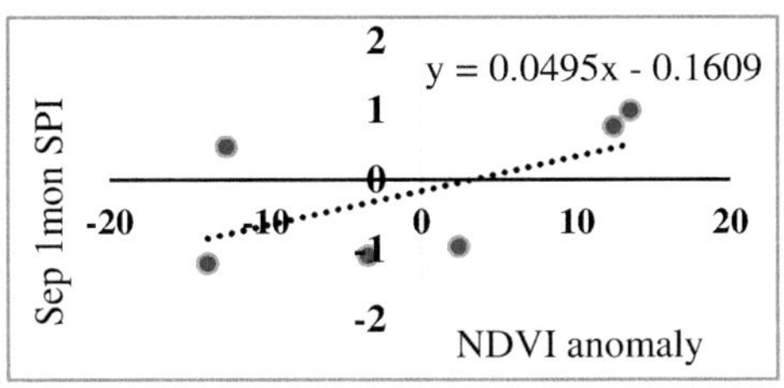

4 Conclusions

Using GIS and remote sensing, it became possible to calculate both the agricultural and meteorological drought using the NDVI and SPI index.

Kharif II season drought has been evaluated in this study. In 2010 and 2012, Nawabganj, Sirajganj, and Pabna experienced severe meteorological drought (SPI ranges from −1.4 to −0.04). The southern part of the Rajshahi division was affected by an abnormal drought during 2012 (SPI ranges from −1.4 to −0.55). The majority of the Rajshahi district was hit by mild drought in October, while there was no drought in other months of 2014. Pabna, Nawabganj, and parts of Sirajganj were hit by severe drought in 2018 (SPI ranges from −1.3 to 1.1), while other parts experienced moderate and abnormal drought. Although the meteorological drought must have happened geographically in certain locations, it has not altered drought severity due to irrigation projects, effective land management practices, and other considerations.

The NDVI is a standard remote sensing metric for agriculture that captures how more near-infrared radiation is reflected than visible red light. If there is drought resolved with supplementary irrigation, we can not determine those areas facing agricultural drought. The NDVI and precipitation indicator has a close connection where moisture is a major constraint for plant development. Agricultural drought is evident in some areas, although analysis for these areas does not represent meteorological drought because there is lack of rainfall stations in those areas. For instance, severe agricultural drought was seen in some places of Rajshahi, Naogaon, and Natore, but these areas experienced moderate meteorological drought. In this study, drought risks were estimated for 6 years (2010, 2012, 2014, 2016, 2018, 2020), but a minimum of 20 years of data are required to evaluate drought accurately. The SPI values from the six rainfall stations do not accurately represent the rainfall conditions in the entire N-W region due to the shortage of rainfall stations. So, it is suggested to use more stations to detect meteorological droughts. Landsat-4, 5, 7, 8, and Sentinal-2 images have been used in this research. Recently in 2021, Landsat-9 has been launched which has a higher resolution. Future researchers can use Landsat-9 for further research. For this study, the Kharif-II season was used. All cropping patterns, including those for the Rabi, Kharif-I, and Kharif-II seasons, are recommended for future studies. For a more accurate outcome, this study might be expanded by using the information on crop yields with NDVI and SPI indices.

References

1. Alley WM (1985) The palmer drought severity index as a measure of hydrologic drought. JAWRA J Am Water Resour Assoc 21:105–114. https://doi.org/10.1111/J.1752-1688.1985.TB05357.X
2. White DH, Walcott JJ (2009) The role of seasonal indices in monitoring and assessing agricultural and other droughts: a review. Crop Pasture Sci 60:599–616. https://doi.org/10.1071/CP08378
3. Aziz MdA, Hossain ABMZ, Moniruzzaman Md, Ahmed R, Zahan T, Azim S, Qayum MdA, Al Mamun MdA, Kader MdA, Rahman NMdF, Aziz MdA, Hossain ABMZ, Moniruzzaman Md, Ahmed R, Zahan T, Azim S, Qayum MdA, Al Mamun MdA, Kader MdA, Rahman NMdF (2022) Mapping of agricultural drought in bangladesh using geographic information system (GIS). ESE 6:657–667https://doi.org/10.1007/S41748-021-00231-8
4. Mondol MAH, Ara I, Das SC (2017) Meteorological drought index mapping in Bangladesh using standardized precipitation index during 1981–2010. Adv Meteorol 2017.https://doi.org/10.1155/2017/4642060
5. Murad H (2010) Agricultural and meteorological drought assessment using remote sensing and GIS in North - West region of Bangladesh
6. Trans-boundary Rivers—Banglapedia. https://en.banglapedia.org/index.php/Trans-boundary_Rivers. Accessed 4 Sep 2022
7. Garrity DP, Oldeman LR, Morris RA, Lenka D (1986) Rainfed lowland rice ecosystems: characterization and distribution (South Asia; Southeast Asia). https://api.semanticscholar.org/CorpusID:135354161
8. Hassan MS, Syed M-U-I (2013) Drought vulnerability assessment in the high barind tract of Bangladesh using MODIS NDVI and Land Surface Temperature (LST) imageries. Int J Sci Res 4:2319–7064
9. Rajshahi Division—Banglapedia. https://en.banglapedia.org/index.php/Rajshahi_Division. Accessed 4 Sep 2022
10. Rajshahi Division—Wikipedia. https://en.wikipedia.org/wiki/Rajshahi_Division. Accessed 4 Sep 2022
11. Anyamba A, Tucker CJ (2005) Analysis of Sahelian vegetation dynamics using NOAA-AVHRR NDVI data from 1981–2003. J Arid Environ 63:596–614. https://doi.org/10.1016/J.JARIDENV.2005.03.007
12. Hill G, Tadesse S, Atnafu T (2011) Using satellite images for drought monitoring: a knowledge discovery approach. J Strateg Innov Sustain 7:135–153
13. Hayes MJ, Svoboda MD, Wilhite DA (2000) Chapter 12 Monitoring drought using the standardized precipitation index. Drought Mitigation Center Faculty Publications. 70. http://digitalcommons.unl.edu/droughtfacpub/70

Adsorption of Chloride Ions from Aqueous Solution Using Activated Carbon Obtained by Chemical Activation of Jute Stick

S. Tarannum, M. N. Abir, A. F. Mita, B. K. Banik, M. A. Hoque, and M. I. Kabir

Abstract The coastal belt of Bangladesh is severely impacted by salinity intrusion. One way to address this problem is to use efficient and low-cost materials as adsorbents to remove saline ions from water. Activated carbon (AC) is the most efficient adsorbent in water treatment among those materials. This study demonstrates the effectiveness of activated carbon, derived from jute stick (JS), as an adsorbent and its adsorption performance in removing salinity. Both column and batch adsorption tests were carried out to evaluate the performance of AC on the synthesized saline water. As part of the column adsorption process, five experimental filters with a continuous flow were set up for five different initial concentrations of chloride ions. The best two AC samples obtained from the batch adsorption process (out of 32 samples), each having triplicates, were selected for further analysis. In batch adsorption, the removal efficiency of activated carbon was evaluated for chloride concentrations ranging from 500–900 mg/L. It was found that AC impregnated with $ZnCl_2$ (1:1) at 600°C and $FeSO_4$ (1:1) at 700°C showed the maximum removal efficiency of 48% for optimum dosing of 4.4 g/L and 46% for optimum dosing of 4.8 g/L respectively. As for the column adsorption process with continuous flow having the same initial chloride concentration as the batch experiment, AC impregnated with $ZnCl_2$ (1:1) at 600°C and $FeSO_4$ (1:1) at 700°C showed removal efficiencies of 58.13% and 56.91% when subjected to initial chloride ion concentration of 500 mg/L.

S. Tarannum · M. N. Abir (✉) · A. F. Mita · B. K. Banik · M. A. Hoque · M. I. Kabir
Department of Civil and Environmental Engineering, Shahjalal University of Science and Technology, Sylhet, Bangladesh

A. F. Mita
e-mail: mita-cee@sust.edu

B. K. Banik
e-mail: bijit-cee@sust.edu

M. A. Hoque
e-mail: ahoque-cee@sust.edu

M. I. Kabir
e-mail: imran.kabir-cee@sust.edu

S. Arthur et al. (eds.), *Proceedings of the 6th International Conference on Advances in Civil Engineering*, Lecture Notes in Civil Engineering 368,
https://doi.org/10.1007/978-981-99-3826-1_3

Keywords Salinity · Activated carbon · Jute stick · Low-cost · Adsorption

1 Introduction

Bangladesh has an abundance of water, but almost one-fourth of the population lacks access to clean water for drinking. People in the coastal region are at risk of saline water due to climate change, rising sea levels, population growth, and decreasing freshwater supply. Direct medical issues, including hypertension and preeclampsia, skin conditions, acute respiratory infections, diarrheal disorders, and the transmission of mosquito-borne diseases, are all linked to an increase in water salinity [1]. According to a World Health Organization (WHO) study, the concentration of chloride in the main or second aquifer in most coastal areas ranges from 103 to 12,433 parts per million (ppm) during the dry months and 34 to 11,366 ppm during the wet season. Sodium chloride might taste salty; however, calcium or magnesium chloride is usually identifiable by taste up to 1000 mg/L [2]. Reverse osmosis, electrodialysis, ion exchange technology, and thermal technology are examples of conventional desalination methods that are effective in removing salt from drinking water but are unsuitable for rural communities [3].

Bangladesh is a country with a large agricultural sector and produces a lot of agricultural trash every year, especially jute sticks, which are made by separating the fibers [4]. Jute stick, a typical agricultural by product in most Asian countries, was used as a precursor for chemical and physical activation methods to make activated carbon [5]. Activated carbon can be used to eliminate chlorine [6]. Activated carbons are effective adsorbents for purifying drinking water and wastewater as well as extracting hazardous components from gases [7]. Adsorption by activated carbon is the most popular method due to its highly developed interior surface area and porosity, as well as its straightforward design and simplicity of use [8]. The fundamental goal of this work is to develop workable desalination methods utilizing adsorbents made from materials that are widely available locally to reduce the scarcity of safe drinking water. Moreover, the performance of utilized adsorbents is determined by two parameters in the adsorption process: maximum adsorption capacity and maximum elimination of contaminants; to examine the adsorbent for various parameters utilizing both batch and column adsorption techniques.

2 Materials and Methods

2.1 *Sample Preparation*

Producing activated carbon from jute stick with high cellulose and lignin content consists of steps like washing, pretreatment, impregnation, activation, and acid leaching to get carbons with micropores and high specific surface area. The steps of processing have been described accordingly.

2.2 *Washing of Raw JS*

At first, the jute sticks are washed thoroughly with water to remove dirt and other contaminants. Then they are oven-dried at about 110 °C for 24 h to eliminate excess water content.

2.3 *Pre-treatment of JS*

The JS are ground with a mixer grinder and sieved through ASTM #30 sieve to achieve a consistent particle size of 0.6 mm, which was later used to produce granular activated carbon.

In this study, about 20 g of granular jute stick is added with 660 ml of 0.25 M, 0.5 M and 1 M sodium hydroxide solution. A portion of the sample has not been mixed with sodium hydroxide to further investigate the effect of base leaching on the produced AC. According to studies done on similar lignocellulosic materials, this solution provides a minimum amount of NaOH to produce silicate with $SiO_2/NaOH$ ratio = 1 [9]. JS powder is subjected to immersion in NaOH solution for 1 h with constant stirring on a magnetic stirrer hotplate. After the base solution is drained, JS powder is rinsed with deionized water until the pH reaches 7. Since the Na_2SiO_3 formed is soluble in water, it can be easily removed by water washing. After washing, the leached jute stick is oven-dried at 110 °C for 24 h. to remove excess water.

2.4 *Impregnation of JS for Activation*

To activate the leached JS, two different samples have been prepared by mixing 200 ml of 10% (w/w) $ZnCl_2$, and 10% (w/w) $FeSO_4$ as activating agents with 20 g base leached JS in each case. Impregnation is carried out at 70 °C in a thermostatic water bath for all two samples until excess water gets evaporated.

2.5 Carbonization of Impregnated JS

For carbonization, the impregnated JS is inserted into a stainless-steel tubular reactor 6 inches in length and 2 inches in diameter. After placing JS into the reactor, an inert atmosphere is created using nitrogen gas. Then the reactors are placed into the furnace where carbonization is performed at 300 °C, 400 °C, 500 °C, 600 °C and 700 °C for 1 h.

2.6 Acid Washing of AC

For acid washing, the AC is immersed in a 3 M HCl solution. Then the slurry is magnetically stirred for 30 min at ambient temperature. After the acidic solution is drained, the samples are washed with hot distilled water several times until neutralization. The washed samples are dried at 100 °C for 24 h. and stored in zip-locked bags for further use.

At the end of this step, thirty types of activated carbons are obtained, characterized by three concentrations of the base (0.25 M, 0.5 M, and 1 M), two activating agents ($ZnCl_2$ and $FeSO_4$), and five activation temperatures (300 °C, 400 °C, 500 °C, 600 °C and 700 °C). These samples are then used as adsorbents in the preparation of filter media.

2.7 Adsorption Process

As a part of the experimental work on a laboratory scale, two types of adsorption techniques have been adopted:

- Column adsorption with continuous flow
- Batch adsorption in suspension

2.8 Experimental Setup for Column Adsorption

For the removal of salinity, the performance of produced AC as an adsorbent is checked on an experimental basis. Five different filtration units have been established for column adsorption, having chloride concentrations of 500, 600, 700,800, and 900 mg/L respectively. Each of the filters is set up using a total of 10 types of AC samples containing triplicate for each to conduct the filtration process in laboratory scale. The filtration is carried out for 21 days to check the efficiencies of the adsorbents.

2.9 Experimental Setup for Batch Adsorption

The adsorption capacities of 32 types of activated carbons are checked in suspension for dosing of 2 g/L, 4 g/L, 6 g/L and 8 g/L. Among them, the top two activated carbons obtained from using different activating agents have been selected for further analysis. Suspension batches are made by mixing the required amount of sample with 1L saline solution of initial chloride concentrations of 500 mg/L, 600 mg/L, 700 mg/L, 800 mg/L, and 900 mg/L ensuring a contact time of 180 min.

Figure 1 shows the flow chart of the significant steps regarding activated carbon production and the adsorption process.

3 Results and Discussions

The quality of the JS derived activated carbon is influenced by pre-carbonization conditions, type of activating agent, impregnation ratio, inert atmosphere, activation temperature, and activation time. As such, these factors need to be carefully controlled to obtain activated carbon with the desired adsorption properties.

In this section, the adsorption capacities of different adsorbents have been summarized. Also, the effect of activating agent, concentration of base, different concentrations of ion, and temperature have been discussed. In addition to this, the effect of dosing and effect of temperature on suspension have also been described.

Table 1 represents the notation and salinity removal efficiency of derived ACs using column and batch adsorption. From Table 1, it has been found that AC with

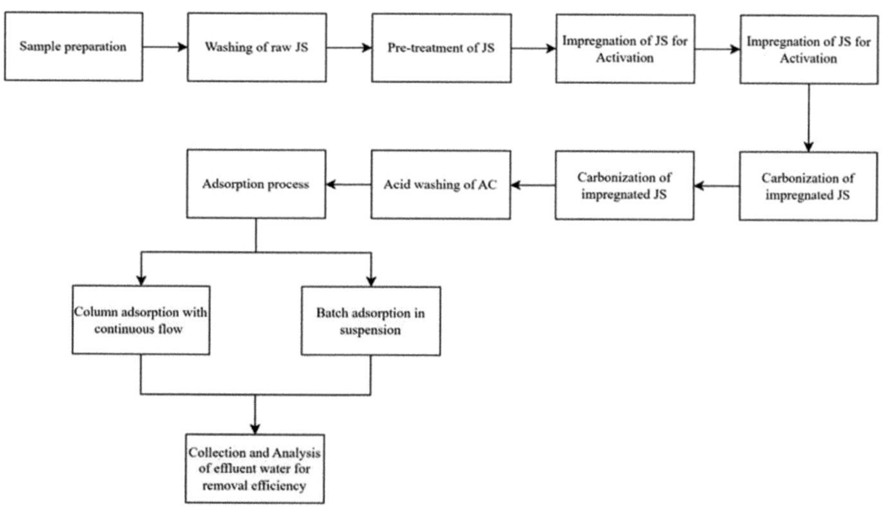

Fig. 1 Overview of the study

Table 1 Removal efficiencies of derived activated carbons

Notation	Initial Salinity (g/L)	Final salinity (g/L)		Removal efficiency (%)	
		Column adsorption	Batch adsorption	Column adsorption	Batch adsorption
FS(7)-5	0.82	0.35	0.44	56.91	46.34
FS(7)-6	0.99	0.44	0.53	55.89	46.46
FS(7)-7	1.15	0.50	0.62	56.52	46.09
FS(7)-8	1.32	0.60	0.72	54.80	45.45
FS(7)-9	1.48	0.68	0.81	54.05	45.27
ZC(6)-5	0.82	0.34	0.42	58.13	48.78
ZC(6)-6	0.99	0.44	0.51	55.89	48.48
ZC(6)-7	1.15	0.51	0.60	55.94	47.83
ZC(6)-8	1.32	0.60	0.71	54.80	46.21
ZC(6)-9	1.48	0.69	0.80	53.83	45.95

FS(7)-5, FS(7)-6, FS(7)-7, FS(7)-8, and FS(7)-9 represents AC with $FeSO_4$ (1:1) at 700 °C for chloride concentration of 500, 600, 700, 800, and 900 mg/L respectively.
ZC(6)-5, ZC(6)-6, ZC(6)-7, ZC(6)-8, and ZC(6)-9 represents AC with $ZnCl_2$ (1:1) at 600 °C for chloride concentration of 500, 600, 700, 800, and 900 mg/L respectively.

$ZnCl_2$ (1:1) at 600 °C has the best salinity removal efficiency (58.13%). Hence the most efficient activating agent is $ZnCl_2$ (1:1) for producing AC in Column adsorption.

3.1 *Effect of Temperature for Different Dosing in Suspension*

The samples carbonized at five different temperatures (300 °C, 400 °C, 500 °C, 600 °C, and 700 °C) show different thermal responses. $ZnCl_2$ (1:1) and $FeSO_4$ (1:1) produces AC most efficiently at 600 °C and 700 °C respectively. Figure 2 exhibits that both ZC(6) and FS(7) have the best salinity removal efficiency at a dosing of 4 g/L. When the temperature is lower than 700 °C, the active reaction increases with the increase of temperature and a large number of pores are formed, so the specific surface area increases. However, when the temperature rises higher than 600 °C, the destruction of pores plays a principal role to decrease the specific surface area [10]. Despite using three samples for each case having a standard deviation value close to 0.01, in Fig. 2, FS(7) shows an increasing trend for 4 g/L and 6 g/L dosing, which is irregular. Further experiments are required using higher activation temperatures to evaluate the effect on removal efficiency for temperatures greater than 700 °C.

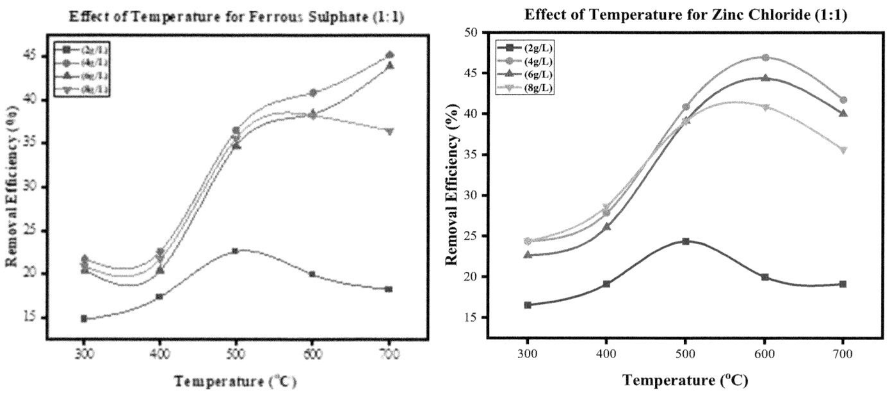

Fig. 2 Comparison of the effect of activation temperatures

3.2 Effect of Dosing for $FeSO_4$ (1:1)

The best adsorption capacity of AC impregnated with $FeSO_4$ (1:1) at 700 °C has been found for a dosing of 4 g/L.

From Fig. 3, In case of 2 g/L dosing, the amount of adsorbent is not sufficient to remove the saline ion in a greater proportion. Thus, the removal efficiency becomes lower. Conversely, for a dosing of 8 g/L, the amount of adsorbent is more than required which causes the agglomeration of adsorbent particles. This results the decreased removal efficiency of AC with $FeSO_4$ (1:1). From Fig. 3, the optimum dosing is found to be 4.8 g/L.

Fig. 3 Relationship between removal efficiencies and dosing for $FeSO_4$

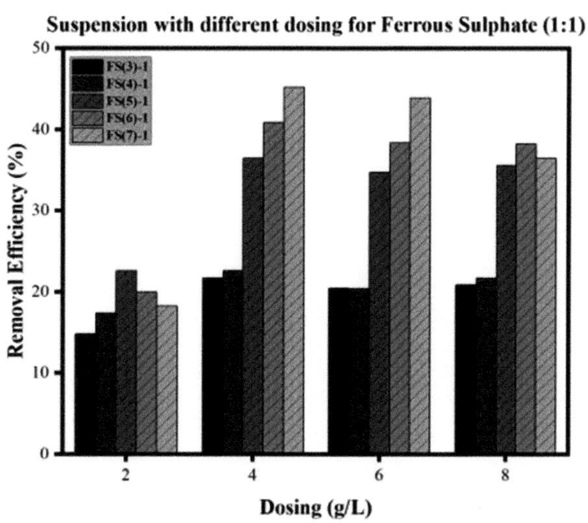

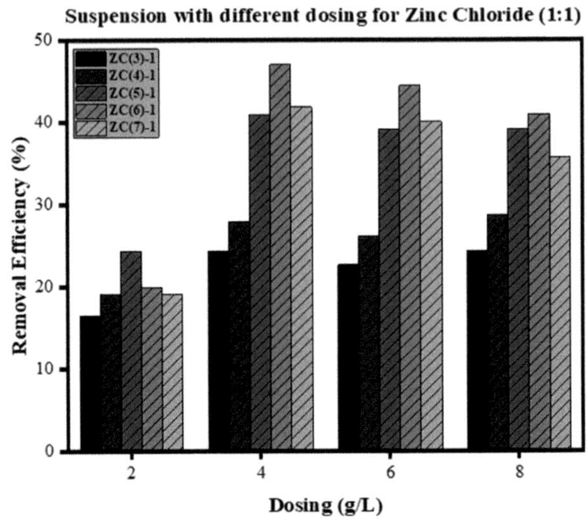

Fig. 4 Relationship between removal efficiencies and dosing for $ZnCl_2$

3.3 *Effect of Dosing for $ZnCl_2$ (1:1)*

The best adsorption capacity of AC impregnated with $ZnCl_2$ (1:1) at 600 °C has been found for a dosing of 4 g/L.

From Fig. 4, In case of 2 g/L dosing, the amount of adsorbent is not sufficient to remove the saline ion in a greater proportion. Thus, the removal efficiency becomes lower. Conversely, for a dosing of 8 g/L, the amount of adsorbent is more than required which causes the agglomeration of adsorbent particles. This results in the decreased removal efficiency of AC with $ZnCl_2$ (1:1). From Fig. 4, the optimum dosing is found to be 4.4 g/L.

3.4 *Analysis of Carbon Yield*

The yield of AC with different activating agents at three temperatures are calculated by Eq. (1).

$$Carbonyield(\%) = (Massofactivatedcarbon/Massofdriedricehusk) \times 100 \quad (1)$$

For two chemical activating agents and three base concentrations, the yields of carbon decreases with the increase of reaction temperature as shown in Fig. 5. By increasing the reaction temperature, the gasification becomes severe, and ultimately leads to a reduction in the yield.

Fig. 5 Effect of temperature on carbon yield

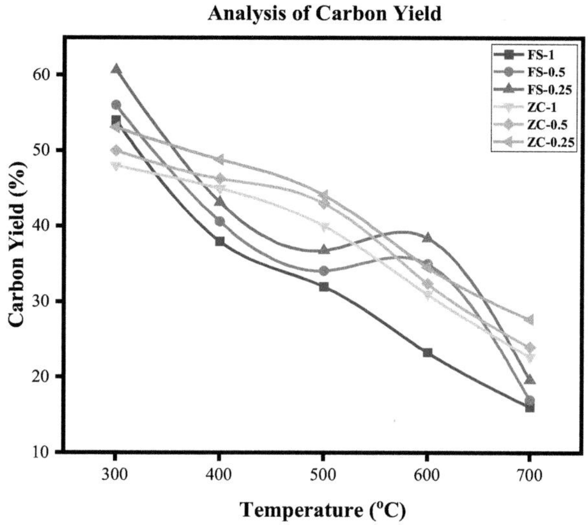

Figure 5 also shows that the carbon yield for 0.25 M base concentration is higher than 0.5 base concentration. Similarly, 0.5 M base concentration has a higher value of yield compared to 1 M base concentration. This is because the lignin tissue is removed after base-leaching of the JS, and this causes a decrease in the carbon yield.

4 Conclusion

The porous structure of AC can be optimized with the alteration of the activation parameters such as the type of activation (physical and chemical), choice of activating agent, activation temperature, impregnation ratio, and activation sequence. The optimization of the activation parameters yields the desired physicochemical properties and consequently the highest adsorption capacity. AC impregnated with $ZnCl_2$ (1:1) at 600 °C has the best salinity removal efficiency (58.13%) for 500 mg/L in column adsorption process with continuous flow. With increasing concentrations of saline water, removal efficiency slightly decreases. AC impregnated with $ZnCl_2$ (1:1) at 600 °C shows the maximum removal efficiency of 48% can be obtained for an optimum dosing of 4.4 g/L in suspension. 1 M NaOH is the best for base leaching of granular JS among the concentration of 0.25 M, 0.5 M, and 1 M NaOH.

AC impregnated with $ZnCl_2$ and $FeSO_4$ represents the best adsorption capacity at 600 °C and 700 °C among the five activation temperatures.

Acknowledgements The authors acknowledge the University Research Centre (URC), Shahjalal University of Science and Technology, Sylhet-3114, Bangladesh, for providing financial assistance to conduct the study.

References

1. Akib Jabed M, Paul A, Nath TK (2020) Peoples' perception of the water salinity impacts on human health: a case study in south-eastern coastal region of Bangladesh. Expo Health 12:41–50. https://doi.org/10.1007/S12403-018-0283-0
2. Kumar M, Puri A (2012) A review of permissible limits of drinking water. Indian J Occup Environ Med 16:40–44. https://doi.org/10.4103/0019-5278.99696
3. Younos T, Tulou KE, Overview of techniques overview of desalination techniques
4. Nitu IP, Islam MN, Ashaduzzaman M, Amin MK, Shams MI (2020) Optimization of processing parameters for the manufacturing of jute stick binderless particleboard. J Wood Sci 66:65. https://doi.org/10.1186/s10086-020-01913-z
5. Asadullah M, Kabir MS, Ahmed MB, Razak NA, Rasid NSA, Aezzira A (2013) Role of microporosity and surface functionality of activated carbon in methylene blue dye removal from water. Korean J Chem Eng 30:2228–2234
6. Asadullah M, Rahman MA, Motin MA, Sultan MB (2006) Preparation and adsorption studies of high specific surface area activated carbons obtained from the chemical activation of jute stick. Adsorpt Sci Technol 24:761–770
7. Marsh H, Reinoso FR (2006) Activated carbon. Elsevier
8. Bhatnagar A, Sillanpää M (2017) Removal of natural organic matter (NOM) and its constituents from water by adsorption—a review. Chemosphere 166:497–510. https://doi.org/10.1016/J.CHEMOSPHERE.2016.09.098
9. Rungrodnimitchai S, Phokhanusai W, Sungkhaho N (2009) Preparation of silica gel from rice husk ash using microwave heating
10. Lan X, Jiang X, Song Y, Jing X, Xing X (2019) The effect of activation temperature on structure and properties of blue coke-based activated carbon by CO_2 activation. Green Process Synth 8:837–845. https://doi.org/10.1515/GPS-2019-0054

Assessment of Carbon Footprint for the Bricks Manufactured in Bangladesh

Pritom Bhowmik Akash, Md. Jahedul Islam, and Md. Reaz Akter Mullick

Abstract To meet the ever-increasing demand for construction materials for various development projects and for the housing of growing population, the brick production trend is always upward. However, emission from brick industries is a significant concern, particularly in a developing country like Bangladesh. Carbon Footprint (CF) is an important and widely used environmental indicator defined as the equivalent carbon dioxide emitted into the environment from the direct or indirect activities linked with the manufacturing process of a product. This study reports a detailed evaluation of CF of brick technologies and compares comprehensive carbon efficiency of different types of kilns currently operating in Bangladesh from the perspective of Carbon Footprint analysis. A field survey of six different types of brick kilns was conducted to collect information regarding fuel, electricity, soil, and water used in various operations of the brick manufacturing process. The assessment was carried out by the Intergovernmental Panel on Climate Change (IPCC) guidelines. Among the four different technologies, two Zigzag kilns showed the lowest CF values, 217.561 gm CO_2-eq/kg fired brick and 218.798 gm CO_2-eq/kg fired brick, accordingly, for the production of 100,000 bricks. In comparison, the two Fixed-Chimney Kilns (FCK) obtained a higher value of 291.096 gm CO_2-eq/kg and 265.547 gm CO_2-eq/kg. Despite being the most efficient technology, Tunnel Kiln showed a higher CF value than Hybrid Hoffman Kiln (HHK). Their CF values are 284.170 gm CO_2-eq/kg fired brick and 266.795 gm CO_2-eq/kg fired brick, respectively. This study also identified HHK with the lowest CF value of 203.768 gm CO_2-eq/kg if fired with natural gas. Being the first carbon footprint study of the brick industry of Bangladesh as per the authors' knowledge, this paper aims to help policymakers taking appropriate

P. B. Akash · Md. J. Islam · Md. R. A. Mullick (✉)
Department of Civil Engineering, CUET, Chattogram, Bangladesh
e-mail: reazmullick@cuet.ac.bd

P. B. Akash
e-mail: u1601118@student.cuet.ac.bd

Md. J. Islam
e-mail: u1601074@student.cuet.ac.bd

S. Arthur et al. (eds.), *Proceedings of the 6th International Conference on Advances in Civil Engineering*, Lecture Notes in Civil Engineering 368,
https://doi.org/10.1007/978-981-99-3826-1_4

policy interventions to progress specifically in the implementation of the Sustainable Development Goal 10, i.e., take urgent action to combat climate change and its impacts.

Keywords Brick kiln · Carbon footprint · CO_2 emissions · Chattogram–Bangladesh

1 Introduction

Human activities are bringing on a rise of 0.2 °C of warming every ten years, thus making Global Warming one of the most globally critical issues [1, 2]. As CO_2 accounts for about 79% of all human-caused greenhouse gas emissions [3], measuring the gas while producing various goods and assessing different products for their contributions to climate change has now become a burning necessity. Carbon footprinting is the primary method for doing this. Carbon Footprint (CF) is a significant environmental indicator to compare the greenhouse gas emission rate from different goods or activities. Usually, it is defined as the amount of equivalent carbon dioxide emitted into the environment over a timeframe or the entire life span of the direct or indirect activities of a single person, organization, or community. Carbon dioxide (CO_2), methane (CH_4), nitrous oxide(N_2O), and various fluorinated gases are the vital greenhouse gases released into the atmosphere as a consequence of human activities, thus contributing the most to the carbon footprint [3, 4]. Even though CO_2 has the least capacity to absorb heat in the atmosphere, it is released in such large amounts that all greenhouse gases are converted to CO_2 equivalent (CO_2-eq) to make the total footprint of all gases easier to quantify [5].

Bangladesh has some of the worst air quality globally, and brick kilns are a significant contributor to deteriorating air quality [6]. The brick-making industry is a critical component of Bangladesh's severe air pollution crisis [7]. Improvements in living conditions, continuous technological progress, and growing demand for new infrastructure have contributed to a globally substantial rise in construction material consumption in recent years [8]. Being a developing country, such continuous increment in Bangladesh's housing industry has forced its construction industry to depend on bricks as the primary construction material and thus making the country world's fourth largest brick manufacturer with an annual production of 23 billion bricks [6, 9]. So, the brick industry of Bangladesh is highly accountable for the country's total CO_2 emission.

Although there are a few estimates of carbon emissions from various brick kilns used in Bangladesh, as per author's knowledge, no precise estimate of Carbon Footprint analysis from any kiln in the country is available. Imran et al. [10] estimated CO_2 emissions from the coal/natural gas combustion in the kilns of four different technologies (Bull's Trench Kiln, Fixed Chimney Kiln, Zigzag Kiln, and Hoffman Kiln) in Bangladesh. However, over a while, there has been a change in the existing brick technologies in the country. As of June 2017, the country has 6744 brick kilns, where Fixed-Chimney Kiln (FCK), Zigzag Kiln, Hybrid Hoffman Kiln (HHK), and Tunnel Kiln are the leading four technologies holding the country's 99.91% of all types of brick kilns [9]. These four types of brick technologies are different from one another in terms of kiln structure, type of fuel used, production capacity, and many more. As a result, an approach has been taken to cover this gap by measuring the CF of different types of brick technologies currently operating in Bangladesh and comparing their carbon efficiency according to their type from the perspective of Carbon Footprint analysis. Moreover, the study also tries to identify and compare the major sources that have the greatest impact on the carbon footprint as a whole. For collecting data regarding all the site activities, a field survey was conducted in the six kilns of four different types of brick technologies in Chittagong. The calculations of CF due to various activities in brick kilns were done using guidelines provided by IPCC & other related resources. However, being the pioneering study on estimating carbon footprint of brick industry of Bangladesh this paper aims to help policymakers suggesting appropriate policy interventions to progress specifically in the implementation of the Sustainable Development Goal 10, i.e., take urgent action to combat climate change and its impacts.

2 Materials and Methods

The study region Chattogram located in southern Bangladesh, has all the main types of brick kilns currently available in the country. Hence, six brick kilns of the main four brick technologies of Bangladesh shown in Fig. 1 are designated as: Wood powered FCK (F-1), Coal powered FCK (F-2), Zigzag kiln with electric brick molding machine (Z-1), Zigzag kiln with no brick molding machine (Z-2), Tunnel kiln (T-1), Hybrid Hoffman Kiln (H-1). Before the monsoon season, a field survey was carried out between March and May of 2021. For all six kiln sites, data on the activities involved in the brick manufacturing process, different unit operations, the size and weight of bricks, and the amounts of fuel consumed in various activities are gathered during the field survey.

The "Cradle to Gate" technique estimates the current study's carbon footprint, considering all unit operations and activities until the burned clay brick is prepared for transmission to its final destination from the kiln site [11]. For this study's carbon footprint assessment, Fig. 2a depicts several activities in the brick manufacturing process inside a specific system boundary. Clay bricks are either dried under sun or dried by reusing temperature of firing chamber. So, this is not included in the system boundary. Depending on the data availability for an activity, Tier—I or Tier—II approach has been followed while selecting the emission factors and calculation formulas. For the ease of calculation and data representation, the activities are classified into four sources of CF depicted in the Fig. 2b.

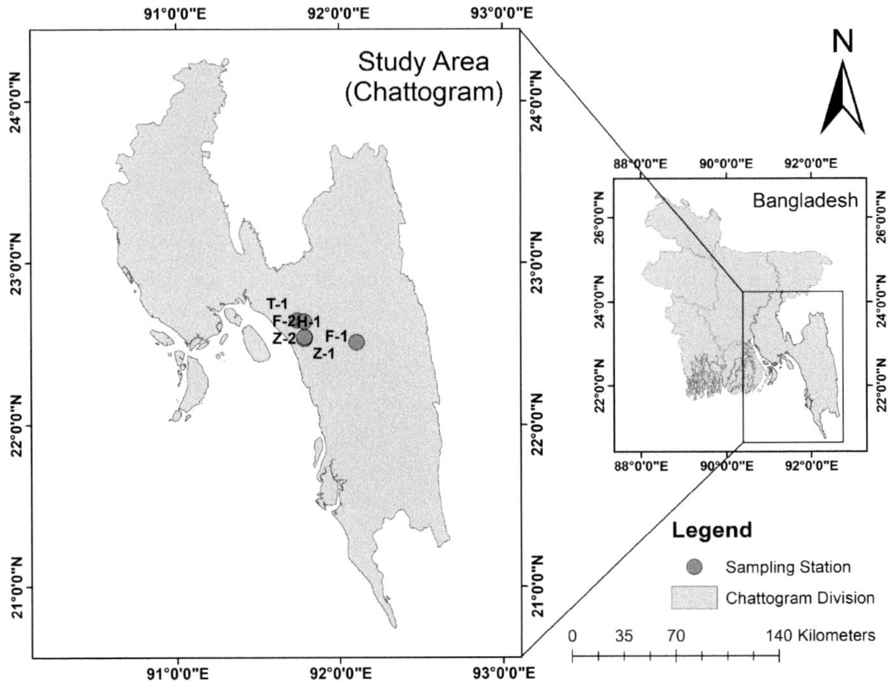

Fig. 1 Location of sampling sites (Chattogram–Bangladesh)

2.1 *Emission from Coal/ Natural Gas Combustion and Transportation*

This study used the following Eq. (1) given by the IPCC (1995) to determine the CO_2 emissions from coal/ natural gas combustion and transportation (except Shipping of Coal) depending on the parameters such as energy usage, emission factors of carbon, and the fraction of oxidized carbon in the fuel.

$$CE_i^t = \sum CE_{ij}^t = \sum E_{ij}^t \times EF_j \times \left(1 - CS_{jt}\right) \times O_j \times M \tag{1}$$

where, $CE_i{}^t$ stands for total CO_2 emissions in year t (in tons, t), $\sum CE_{ij}{}^t$ for total CO_2 emissions of the I sector in year t (in tons, t), $\sum E_{ij}{}^t$ for total energy consumption in year t (in tons, t), EF_j for carbon emission factor of the j fuel (in tons/ton of energy consumed), CS_{jt} for a fraction of jth fuel that is not oxidized as raw materials in year t, O_j denotes the percentage of carbon according to fuel type that has been burned, and M the molecular weight ratio of carbon dioxide to carbon (44/12) [12, 13].

As all the kilns are using Indonesian coal, the net calorific value (NCV) of 25.75 TJ/kt was used [14]. Whilst NCV of diesel is 0.00003594 TJ/L [15]. Table 1 shows the fraction of carbon that has been burned and the carbon emission factors. Fuel was

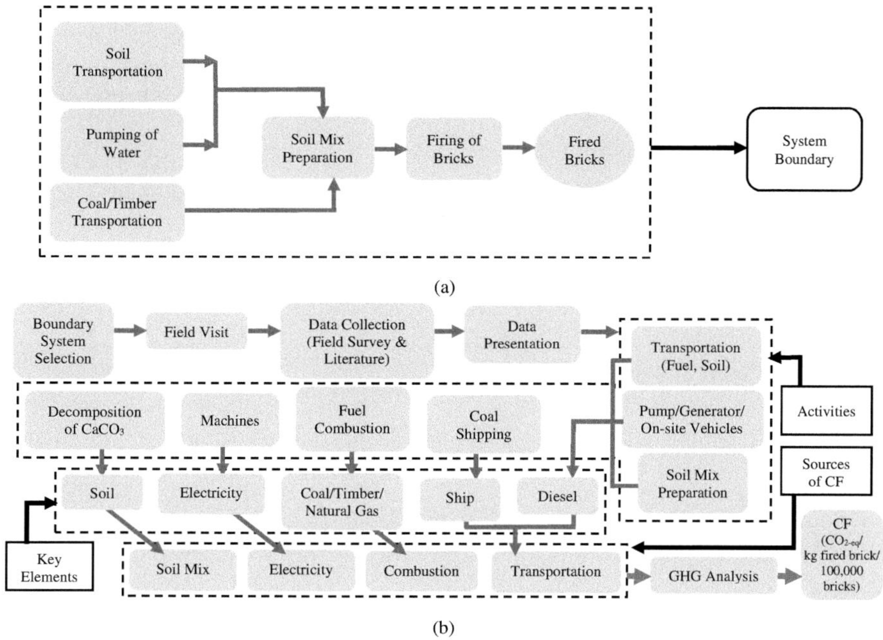

Fig. 2 Schematic flowchart for carbon footprint analysis **a** system boundary **b** research framework

taken out of the total amount of paper's overall energy usage because it was a raw material for producing the end product [13]. Therefore, this analysis also accounted for the complete oxidation of fuels. Although HHK is usually powered by natural gas [9] but H-1 uses coal to fire brick due to delay in getting approval and arrangement of gas supply. Therefore, for a typical gas powered HHK H-1(G), CO_2 emission from gas combustion is calculated using Eq. (1). Here the energy consumption, $\sum E_{ij}{}^{t}$ is assumed to be equivalent to the energy consumption due to coal firing in H-1.

Table 1 Carbon emission factor and fraction of carbon oxidized[a]

Fuel	Carbon emission factor, EF_j (tC/TJ)	Fraction of carbon oxidized, O_j
Coal	25.8	0.98
Diesel	20.2	0.98
Natural gas	15.3	0.995

[a] [12, 13]

2.2 Emission from Timber Combustion

Equation [2] is used to determine the carbon emission produced by the combustion of timber in kilns [16]:

$$\text{Carbon emission} = \text{Quantity} \times \text{NCV} \times \text{CO}_2\text{emission factor} \quad (2)$$

where, NCV of wood is 0.015 TJ/ton [17].

2.3 Emission from Electricity

The following Eq. (3) is used to find the CO_2 emission from electricity consumption [16]:

$$\text{Carbon emission} = \text{Electricity consumption} \times \text{Grid emission factor} \quad (3)$$

2.4 Emission from Soil Mix

Kulkarni & Rao (2016) [11] assumed the clamps' firing temperature is to be over 840ºC, which causes $CaCO_3$ of brick-making soil to break down into CaO and CO_2. Hence, the current study estimated the emission from this sector is according to Eq. (4) [11]. Soil samples were collected in a triplicate manner from each site and percentage of $CaCO_3$ estimated via titration method [18].

$$CO_{2(soil)} = \frac{M_{CO_2}}{M_{CaCO_3}} \times \%_{CaCO_{3(soil)}} \quad (4)$$

where, $CO_{2(Soil)}$ stands for CO_2 emission from soil mix, M_{CO2} and M_{CaCO3} stands for molecular weight of CO_2 and $CaCO_3$.

2.5 Emission from Shipping of Coal

As per field survey, all the sites use Indonesian coal. Hence shipment of coal from Indonesia to Chittagong port is calculated by Eqs. (5) and (6) [19].

$$G_W = F \times g_w \quad (5)$$

Table 2 Emission factors of various fuels for CO_2, CH_4, N_2O

Fuel	CO_2	CH_4	N_2O
Coal		0.001 tCH_4/TJ[a]	0.0015 tN_2O/TJ[a]
Timber	112 tCO_2/TJ[a]	0.03 tCH_4/TJ[a]	0.004 tN_2O/TJ[a]
Diesel		0.003 tCH_4/TJ[a]	0.0006 tN_2O/TJ[a]
Electricity	0.67 tCO_2/MWh[b]	0.00001095 tCH_4/MWh[c]	0.000008283 tN_2O/MWh[c]
Natural gas		0.001 tCH_4/TJ[a]	0.0001 tN_2O/TJ[a]

[a] [16]; [b] [23]; [c] [24]

$$F = W \times D \times E \tag{7}$$

where, G_W stands for Well-to-wheels GHG emissions CO_2 equivalents, F for energy consumption, g_w for emission factor, W for weight, D for distance, and E for specific energy consumption.

For bulk ships carrying coal in Asia E is 0.0014 kg/tkm [19, 20]. The shipping distance of coal from Indonesia to Bangladesh is 2306 nautical miles [21].

2.6 *Calculation of Carbon Footprint*

For the calculation of CH_4 and N_2O emission from each sector Eq. (7) is used [16].

$$\text{Emission}_{\text{GHG, fuel}} = \text{Fuel Consumption} \times \text{Emission Factor}_{\text{GHG, fuel}} \tag{7}$$

The estimated CH_4 and N_2O emission for each kiln are taken into their CO_2 equivalent (CO_2-eq) value applying global warming potential value of CH_4 as 86 and N_2O as 268 [22]. Hence the Carbon footprint of a site expressed as gm CO_2-eq per kg of fired brick for the production of 100,000 bricks is the summation of all CO_2 and its equivalent emission from each sector. Following Table 2 depicts the emission factors of different fuels for CO_2, CH_4, and N_2O.

3 Results and Discussions

Figure 3a displays that CO_2 emissions are much higher than N_2O and CH_4. Due to the higher global warming potential value, the average N_2O emission of all kilns is larger than CH_4. F-1 showed a higher emission of all three gases resulting in its highest position in CF analysis depicted in Fig. 4. Although F-1 showed the highest value of CF, but coal-powered FCK (F-2) holds the fourth highest position in the CF analysis. This is because of the higher emission factor of timber than coal. The usage of electric machines may affect the CF values. It is observed that electric machines are used in the updated technologies like T-1 and H-1 for transporting bricks and excavated soil from one operation site to another inside the brickfield, molding of soil, and chimney fan, whereas conventional brick technologies like F-1 and F-2 use labor for these operations. Thus use of labor in different activities lessens the energy input and thus results in a lower CF value since human energy infusions into various operations are mentioned by BSI (2011) to be excluded from the boundary system [11, 25]. This is why modern technologies Tunnel kiln(T-1) and HHK (H-1) ranked second and third, respectively. Another reason for the higher value of T-1 is the lower brick production, as the kiln was not being operated at full capacity due to some maintenance works. However, the gas-powered HHK showed the lowest CF value of all. From Fig. 3b, it can be stated that firing in the kiln is the primary source of CF. As the carbon emission factor of coal is almost 1.67 times higher than natural gas, this is why H-1(G) yielded a much lower value than H-1. In the two zigzag kilns, usage of both labor and machines was seen. Also, a smaller distance of soil transport was reported during the field survey. The lower height of the chimney might be another reason why the zigzag kilns showed the lowest values than others. Z-1 having an electric brick molding machine, shows a higher CF value than Z-2 with no brick molding machine.

Combustion of Coal/Timber for firing brick is the most responsible source of CO_2 emission, followed by Soil Mix, Transportation, and Electricity, illustrated in Fig. 3b. The emission of CO_2 due to soil mix plays a significant role in the overall CF of the brick kiln. Only FCKs use local soil and bring soil from a smaller distance of 0.5 km. This lessens the diesel consumption to transport soil. Hence emission from transportation in F-1 and F-2 is much lower than in the others. The results of this study are in line with the previous worldwide studies. For example, Kulkarni & Rao (2016) [11] reported CF of a UK tunnel kiln having as 234.24 gm CO_2/kg of fired brick which is a bit lower than the CF value of T-1, which is 284.17 gm CO_2/kg of

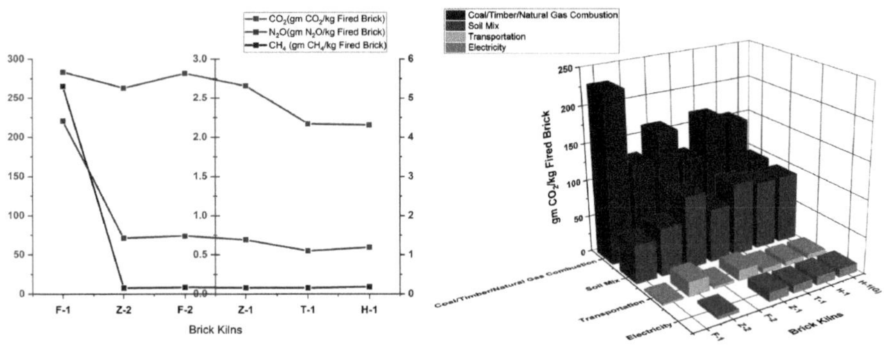

(a) (b)

Fig. 3 Emission from brick kilns for production of 100,000 bricks **a** CO_2, N_2O, CH_4 emission **b** CO_2 emission from the sources of CF

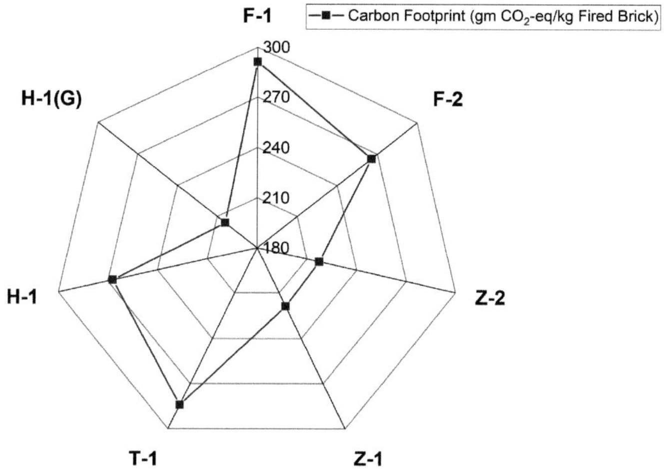

Fig. 4 CF of brick kilns (for production of 100,000 bricks)

fired brick found in this study. Whilst, tunnel kilns of the USA and Canada have CF of 232.25 gm CO_2/kg of fired brick (powered by natural gas) and 338.19 gm CO_2/kg of fired brick (powered by light-fuel oil) [11]. A previous study based on the CO_2 emission from energy consumption in the firing chamber also designated Zigzag kilns as the best kiln in Bangladesh for its both energy and carbon efficiency [26]. The study also showed the gas-powered Hoffmann kiln as the least carbon-emitting kiln in the country. The current study also identified the HHK with the lowest CF value if fired with natural gas. The average CF of the six sites estimated in the paper is 257.33 gm CO_2/kg of fired brick, slightly greater but in line with the average CF of brick in the UK is 239.4 gm CO_2/kg of fired brick [27].

In an announcement in 2019, the Ministry of Environment, Forest and Climate Change took an initiative to construct all public construction projects by replacing clay bricks with concrete blocks and thus aimed for the gradual phasing out of bricks in governmental construction projects by 2025 [28]. For the production of every unit volume of concrete blocks, a CF value of 290 kg CO_2-eq has been evaluated [29], which is higher than the average CF value of bricks estimated in this paper. Another study defined hollow cement block (HCV) as the most harmful building material, showing an elevated CF value of HCV than bricks [30]. However, replacing Portland cement in concrete with fly ash results in a net reduction in GHG emissions as 0.006 CO_2-eq kg/ ton of fly ash is recorded only to collect the fly ash from the power plant station [31]. Although fly ash is designated as an industrial waste of thermal power plants, the production of total ash is roughly 10% of the amount of coal fired in the plant, of which 80% is estimated to be fly ash [32]. Therefore, the life cycle carbon analysis of fly ash should be studied before highlighting fly ash bricks as the best alternative to clay bricks from the carbon footprint perspective. The study relates to the fifth theme of Bangladesh Climate Change Strategy and Action Plan 2009 (BCCSAP 2009), which is "Mitigation of Low Carbon Development." Concerning

CO_2 emissions, natural gas is the greenest energy source [33]. The current study also highlighted Hoffman Kilns as the least CF brick technology if fueled by gas, so using natural gas in firing bricks can be associated with the first program of the fifth theme of BCCSAP (2009). However, firing coal in the kiln accounts for higher CF values. Hence using clean coal technology in this sector can be linked with the third program of the theme. Lastly, using solar energy in the brickfield for various operations can serve both the fourth and the tenth programs of the fifth theme of BCCSAP (2009).

4 Conclusion

According to this baseline estimate, two Zigzag kilns showed the lowest CF values among the four different technologies. In comparison, wood-powered FCK showed the highest of all, whereas coal-powered FCK showed a lower value. Hybrid Hoffman Kiln holding a third-highest position, showed a lower CF value than Tunnel Kiln. However, Hybrid Hoffman Kiln would be the best choice in terms of CF analysis if natural gas is fired instead of coal. Additionally, it is estimated that the average carbon footprint of bricks made in Bangladeshi kilns is 257.33 g CO_2-eq/kg of burnt brick, which is comparable to values for the US, Canada, and the UK, but somewhat higher than the average CF of brick in the UK. The primary source of CO_2 emissions has been discovered to be the burning of coal or wood, or natural gas, followed by soil composition, transportation, and power. Although from CF analysis, it can be recommended that using bricks produced from gas-powered HHK and Zigzag kilns is sustainable. Nevertheless, further studies should be carried out to compare the quality and economic analysis of the bricks of the Zigzag with the modern brick-making technologies, i.e., Tunnel and Hybrid Hoffman kilns. The study is a pioneering study of this kind in Bangladesh to analyze CF of different types of brickfields in Bangladesh. The limited sample size can hardly projectile the overall scenario of the country; however, the study covered all the types of kilns that are in practice in Bangladesh. Therefore, further research, primarily based on the CF from brickfields across the country, can be done focusing on implementing a larger sample size to have a more accurate outcome from statistical analysis. This study can guide decision-makers to adopt the right policies to advance specifically in the application and realization of SDG 10.

Acknowledgements The authors would like to express their gratitude to the Department of Environment, Chattogram Division, for their logistics support during the field survey, and the Department of Civil Engineering, CUET, for supervision and laboratory facilities regarding the study.

References

1. KZ (Canada/G Fernando Aragón-Durand (Mexico)), Cramer W (France/Germany), Humphreys S (UK/ Ireland), Kainuma M (Japan), Kala J (Australia), Mahowald N (USA), Mulugetta Y (UK/Ethiopia), Perez R (Philippines), Wairiu M (Solomon Isl) (2018) Framing and context in: global warming of 1.5 °C. An IPCC special report on the impacts of global warming of 1.5 °C above pre-industrial levels and related global greenhouse gas emission pathways, in the context of strengthening the global response to the t. https://www.ipcc.ch/site/assets/uploads/sites/2/2019/05/SR15_Chapter1_Low_Res.pdf
2. Azad AK, Iqbal A (2006) Impact of brick kilning on the environment : a study on Khulna region of Bangladesh. 7(1):13–18. https://ku.ac.bd/uploads/kustudies/Vol7/N1/13-18.pdf
3. US EPA (2021) Overview of greenhouse gases|US EPA. https://www.epa.gov/ghgemissions/overview-greenhouse-gases. Accessed 6 May 2021
4. Strutt J, Wilson S, Shorney-Darby H, Shaw A, Byers A (2008) Assessing the carbon footprint of water production. J/Am Water Work Assoc 100(6):80–91. https://doi.org/10.1002/j.1551-8833.2008.tb09654.x
5. Shrestha E, Ahmad S, Johnson W, Batista JR (2012) The carbon footprint of water management policy options. Energy Policy 42:201–212. https://doi.org/10.1016/j.enpol.2011.11.074
6. Saha CK, Hosain J (2016) Impact of brick kilning industry in peri-urban Bangladesh. Int J Environ Stud 73(4):491–501. https://doi.org/10.1080/00207233.2016.1179014
7. Haque MI, Nahar K, Kabir MH, Salam A (2018) Particulate black carbon and gaseous emission from brick kilns in Greater Dhaka region, Bangladesh. Air Qual Atmos Heal 11(8):925–935. https://doi.org/10.1007/s11869-018-0596-y
8. Valdes H, Vilches J, Felmer G, Hurtado M, Figueroa J (2020) Artisan brick kilns: State-of-the-art and future trends. Sustain 12(18):1–19. https://doi.org/10.3390/su12187724
9. DoE (2017) National strategy for sustainable brick production in Bangladesh. http://ccacoalition.org/sites/default/files/resources/2017_strategy-brick-production-bangladesh.pdf?fbclid=IwAR1WT1w24zIBqJFagPqZ6SfMj_8KzQjhCq8BJNfPzOkBgW-JIjkdkJjvAjY
10. Imran M, Baten M, Nahar B, Morshed N (2014) Carbon dioxide emission from brickfields around Bangladesh. Int J Agric Res Innov Technol 4(2):70–75. https://doi.org/10.3329/ijarit.v4i2.22653
11. Kulkarni NG, Rao AB (2016) Carbon footprint of solid clay bricks fired in clamps of India. J Clean Prod 135:1396–1406. https://doi.org/10.1016/j.jclepro.2016.06.152
12. IPCC (1995) Greenhouse gas inventory: IPCC guidelines for national greenhouse gas inventories. United Kingdom Meteorological Office, Bracknell, UK
13. Mitra N et al (2020) Assessing energy-based CO2 emission and workers' health risks at the shipbreaking industries in Bangladesh. Environment 7(5):35. https://doi.org/10.3390/ENVIRONMENTS7050035
14. IPCC (1996) Chapter 1 energy revised 1996 IPCC guidelines for national green house gas inventories: reference manual, pp 01–20. https://www.ipcc-nggip.iges.or.jp/public/gl/guidelin/ch1ref1.pdf. Accessed 9 July 2022
15. Staffel I (2011) The energy and fuel data sheet. Clavert Energy Res Group 11. http://www.claverton-energy.com/the-energy-and-fuel-data-sheet.html
16. Darío Gómez AR et al (2006) Chapter 2: stationary combustion 2006 IPCC guidelines for national greenhouse gas inventories. https://www.ipcc-nggip.iges.or.jp/public/2006gl/pdf/2_Volume2/V2_2_Ch2_Stationary_Combustion.pdf. Accessed 09 Jul 2022
17. IPCC (1996) Chapter 1 energy revised 1996 IPCC guidelines for national green house gas inventories: reference manual, no 1990, pp 41–60. https://www.ipcc-nggip.iges.or.jp/public/gl/invs6.html
18. Bashour II, Sayegh AH (1990) Methods of analysis for soils of arid and semi-arid regions. Food and Agriculture Organization. United Nations Rome 2007:1–128
19. Schmied W, Knörr M (2012) Calculating GHG emissions for freight forwarding and logistics services in accordance with EN 16258, pp 1–63. https://www.clecat.org/media/CLECAT_Guide_on_Calculating_GHG_emissions_for_freight_forwarding_and_logistics_services.pdf. Accessed 22 Oct 2019

20. Witte J (2010) Noise from moored ships. 39th Int Congr Noise Control Eng 2010, INTER-NOISE 2010 5:3389–3398
21. Sea route & distance—ports.com (2022). http://ports.com/sea-route/#/?a=4141&b=4065&c=PortofTanjungPriokIndonesia&d=PortofChittagongBangladesh. Accessed 10 Jul 2022
22. Takemu T, Zhang H, Myhre G, Shindell D, Bréon F-M, Collins W, Fuglestvedt J, Huang J, Koch D, Lamarquec J-F, Lee D, Mendoza B, Nakajima T, Robock A, Stephens G (2013) Anthropogenic and natural radiative forcing. https://www.ipcc.ch/site/assets/uploads/2018/02/WG1AR5_Chapter08_FINAL.pdf
23. DOE (2022) Grid-Emission-Factor (GEF)-of-Bangladesh. Department of Environment, Bangladesh. http://www.doe.gov.bd/site/notices/059ddf35-53d3-49a7-8ce6-175320cd59f1/Grid-Emission-Factor(GEF)-of-Bangladesh. Accessed 9 Jul 2022
24. US EPA (2014) Emission factors for greenhouse gas inventories. pp 1–5. papers2://publication/uuid/FDF7F83F-129C-44B6-AD7E-9620A9C9ED08
25. BSI (2011) PAS 2050:2011 Specification for the assessment of the life cycle greenhouse gas emissions of goods and services. British Standards Institution, London, pp 1–45
26. Saadat AHM (2008) Estimation of carbon dioxide emission from Brick Kilns in Bangladesh. pp. 2–6
27. BDA (2014) BDA 2014. http://www.brick.org.uk/2014/03/lucideonproduces-%0Ageneric-carbon-footprint-report-for-the-uk-clay-brickmanufacturing/
28. Zaman MA (2022) Use of concrete blocks in public construction still low. The Daily Star
29. Van Gijlswijk RN, Pascale S, De Vos SE, Urbano G (2015) Carbon footprint of concrete based on secondary materials. Heron 60(1–2):113–139
30. Udawattha C, Halwatura R (2017) Life cycle cost of different walling material used for affordable housing in tropics. Case Stud Constr Mater 7:15–29. https://doi.org/10.1016/j.cscm.2017.04.005
31. O'Brien KR, Ménaché J, O'Moore LM (2009) Impact of fly ash content and fly ash transportation distance on embodied greenhouse gas emissions and water consumption in concrete. Int J Life Cycle Assess 147, 14(7):621–629. https://doi.org/10.1007/S11367-009-0105-5
32. Tamim MM, Dhar A, Hossain MS (2013) Fly ash in Bangladesh—an overview. Int J Sci Eng Res 4(Ŝ):809–812. http://www.ijser.org/paper/Fly-ash-in-Bangladesh-An-Overview.html
33. MoEFCC (2009) Bangladesh climate change and action plan 2009. In: Ministry of environment, forest and climate change. Government of the People's Republic of Bangladesh, Dhaka, Bangladesh, p 76

Investigation on Pre-cracked RC Slab Strengthening in Flexure Region Under Patch Loading by CFRP

S. M. Z. Islam, M. M. Islam, M. Ahamed, M. M. H. Rakib, M. R. Hasan, and M. M. Rahman

Abstract Reinforced concrete (RC) slabs may often experience deficiency due to steel corrosion, temperature effect, increased live loads, fatigue cracking by impact loading, variation in utilizations, a mistake in design and calculation, and error in construction. Carbon Fiber Reinforced Polymer (CFRP) is one of the most important, effective, potential, and advanced composite materials for retrofitting and strengthening metal structures. The objective of this research is to investigate the structural strength and behaviour of CFRP-strengthened RC slabs subjected to patch loading. A series of tests have been conducted on RC slabs which are strengthened by CFRP. Sixteen RC slabs including two reference specimens and CFRP-strengthened specimens were tested in this study. The specimens of the reinforced concrete (RC) slabs were cast with dimensions of 1050 × 700 × 75 mm. Different lengths and configurations of CFRP strips are placed along the slabs. Hydraulic contorted universal testing machine were used for patch loading. The collapse loads, collapse modes, and the load-deformation behaviour of the reference specimen and CFRP strengthen specimen are also presented in this paper. Strengthening of pre-cracked RC slabs at flexural region using CFRP provided better performance than reference slab. Based on test results, it was found that the structural performance of CFRP strengthening pre-cracked RC slab including deflection, and stiffness cracking load is more significant than reference RC slab. The flexural capacity load carrying capacity as well as serviceability improved significantly and varied for different strengthening techniques. Therefore, it can be concluded that better performance can be attained by CFRP strengthening in the flexural region damaged RC slab with the appropriate technique.

S. M. Z. Islam (✉) · M. M. Islam · M. Ahamed · M. M. H. Rakib
Department of Civil Engineering, RUET, Rajshahi, Bangladesh
e-mail: zahurul90@gmail.com

M. R. Hasan
Department of Civil Engineering, BAUET, Natore, Bangladesh

M. M. Rahman
Nutech Construction Chemicals Company Limited, BUET, Dhaka, Bangladesh

S. Arthur et al. (eds.), *Proceedings of the 6th International Conference on Advances in Civil Engineering*, Lecture Notes in Civil Engineering 368,
https://doi.org/10.1007/978-981-99-3826-1_5

Keywords Fibre reinforced polymer · Punching load · Reinforced concrete slab · Strengthening and repair

1 Introduction

Concrete is weak in tensile stress; hence RC slab may often experience flexural cracking due to poor quality of the material's workmanship, a mistake in design, and changes in uses. Retrofitting and/or strengthening can be considered to restore the structural strength and load-bearing capacity of the defected RC slab [1, 2]. Recently, strengthening and retrofitting of RC slabs have gotten extra importance for the reduction of strength and load-carrying capacity of structures due to unpredictable climatic conditions, use of sub-standard materials, impact loads, corrosion or deterioration of concrete, settlement, and poor workmanship. For retrofitting and strengthening of RC slab, many different techniques are available such as ferro-cement, cement steel plates, epoxy injection, grout, and externally adhesive bonded fiber reinforced polymers. Fiber-reinforced polymers for structural retrofitting are increasingly used in different concrete structures. There are many disadvantages to conventional slab strengthening and retrofitting techniques, namely bulky, heavyweight, laborious, complex and fatigue prone, etc. In the literature review, the different researchers applied different techniques for RC slab strengthening [3–9]. Strengthening/retrofitting is needed to retain the slab under serviceable conditions instead of newly constructed slabs and structures. Fiber-reinforced polymers for structural retrofitting are increasingly used in different concrete structures [10]. Several failure modes have been observed for RC slab bonded with FRP soffit plates in various experimental studies to date. Strut and GFRP anchor were used for strengthening of RC slab [10, 11]. Anchoring technology was proposed for one-way reinforced concrete slabs by the external fiber-reinforced polymer tendons [12]. The investigation was carried out on reinforced concrete one-way slabs strengthened by CFRP sheets in the flexural zone by a few researchers [13–16]. Different failure modes were observed for RC slab bonded with FRP soffit plates in previous various experimental studies. Smith and Kim [17] researched strengthening one-way RC slabs using cutouts by FRP composites. Research on two ways concrete slabs strengthen by FRP composite laminates by Mosllana [18]. Cracks were developed on the top surface of the slab deck and the bottom face of the slab soffit of the Bangabandhu multipurpose bridge over the Jamuna River due to shrinkage. This crack in Bangabandhu Jamuna multipurpose bridge at the slab deck was repaired by CFRP retrofitting [19]. Moreover, the Kanchpur bridge deck slab was also developed cracked, and retrofitted by FRP [20]. In the literature review, the strengthening of concrete slabs with FRP composite is described in previous research. Still, a few researches have been conducted on the effectiveness of retrofitting and strengthening RC slab failure by CFRP materials. FRP composites are advanced composite materials used as retrofit materials has gained much outstanding success in recent years. However, little research has been conducted on the application of CFRP materials for strengthening and retrofitting pre-crack RC column failure. Therefore, it is an innovative approach to study on strengthening and retrofitting of pre-crack RC slab by CFRP.

The purpose of this research is to investigate the structural strength and behaviour of CFRP-strengthened RC slabs subjected to patch loading. A series of tests have been carried out to CFRP-strengthen the RC slab by CFRP patch loading by a universal testing machine. Sixteen RC slabs including one reference slab and another nine initially cracked slabs were tested in this study with varying different lengths and configurations of CFRP strips applied. Hydraulic contorted universal testing machine were used for the patch loading system. The flexural regions of the RC slab have been strengthened and retrofitted by CFRP materials to observe the failure mode, deflection, and ultimate load-carrying capacity. CFRP materials can be applied for strengthening and retrofitting of Pre-crack RC slab effectively for better performance against flexural.

2 Material Properties

External bonded strengthening highly depends on the properties of adhesive and CFRP materials. The effective bond strength, elastic modulus, and elongation are the key mechanical properties of adhesive for strengthening structures. CFRP material is a composite material that typically consists of fibers embedded in a resin matrix. Epoxy resin is the most widely used resin for CFRP. CFRP could be more than 120 times higher in strength than concrete in terms of tensile strength. Four materials have been used to prepare the specimens such as primer and saturant, CFRP fabric, adhesive, and RC slab as shown in Fig. 1.

According to manufacturer specification, carbon CFRP fabrics Kor-CFW450 is used in this research having fiber strength of 4900 MPa, fiber stiffness of 230 GPa, an areal weight of 450 g/m^2, and fabric thickness of 0.255 mm. Primer and saturant were used having density 1.14 gm/cm^3, 1.8 gm/cm^3; pot life 30 min, 1 h 30 min, tensile strength 1350 MPa, 4875 MPa, Modulus of elasticity 99.37 GPa, 238.00 GPa, respectively. Adhesive Kor-CPA 10 Base Resin and hardener used in this research have a tensile strength of 49.8 MPa, shear strength of adhesive of 29 MPa, and pot life of 70 min. The concrete slab was cast targeted at 25 MPa concrete.

Fig. 1 Primer, CFRP fabric, and adhesive

3 Experimental Program

A series of concrete slabs were made for testing to meet the main aim of this research to find the effectiveness of CFRP strengthening of pre-crack RC slab at flexural region. A total of 16 slabs were cast and tested for this research. The targeted concrete was 25 MPa. The slab was cast using cement, sand, and reinforcement 3.75 mm as shown in Fig. 2. Reinforcement and framework for slab casting are shown in Fig. 2. Slab was 711.2 mm in width and 1050 mm long. In slab design, the BNBC code required reinforced will be provided in the tested slab. In all the slabs, the tension steel reinforcement ratio was 0.56%, within the range of minimum and maximum amounts of reinforcement ratio required by BNBC codes, which are respectively 0.20% and 4.39%. Distribution reinforcements were designed in the slabs to avoid punching shear failure. The casted slab specimen maintains strip cutting 600 and 750 mm as well as square and rectangular 300 × 300, 300 × 375 and 300 × 450 mm Pre-cracked RCC slab specimen is sown in Fig. 3. Figure 4 shows the mixing of primer and base resin and hardener adhesive. Applying primer and adhesive and cutting and then attaching are shown in Figs. 5 and 6 respectively.

The specimens were labelled as $Sl_{1050}C_0F_0$, $Sl_{1050}C_0F_1$, $Sl_{1050}C_{500}F_0$, $Sl_{1050}C_{500}F_1$, $Sl_{1050}C_{600}F_0$, $Sl_{1050}C_{600}F_1$, $Sl_{1050}C_{750}F_0$ and $Sl_{1050}C_{750}F_1$, where,

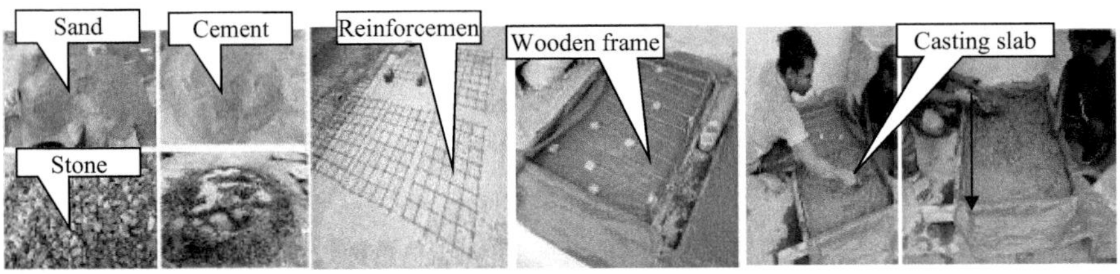

Fig. 2 Binding material (cement), fine aggregate (sand), coarse aggregate (stone chips), and mixing with water for concrete reinforcement and framework for slab casting

Fig. 3 Casted slab specimen, 600 and 750 mm pre-cracked RCC slab specimen 300 × 300, 300 × 375and 300 × 450 mm pre-cracked RC slab specimen

Fig. 4 Mixing primer and base resin and hardener adhesive

Fig. 5 Applying primer on slab specimen and applying adhesive on slab specimen

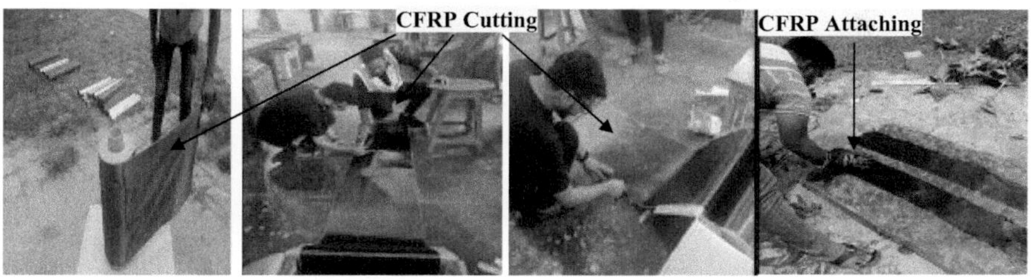

Fig. 6 Cutting and attaching CFRP with adhesive on slab specimen

Sl_{1050} indicates slab length 1050 mm, C_0 indicates reference slab no pre-crack and F_o indicate no CFRP, moreover C_{500}, C_{600}, CF_{750} indicate pre-crack strip length 500, 600, and 750 mm. Slabs also have been considered square and rectangular area CFRP strengthen such as 300 × 300, 300 × 375 and 300 × 450 mm^2 respectively as shown in Fig. 9. Square and rectangular area CFRP strengthen specimen is symbolized as $Sl_{1050}C_0F_0R$, $Sl_{1050}C_0F_1R$ $Sl_{1050}C_{300\times300}F_0$, $Sl_{1050}C_{300\times300}F_1$, $Sl_{1050}C_{300\times375}F_0$, $Sl_{1050}C_{300\times375}F_1$, $Sl_{1050}C_{300\times450}F_0$, $Sl_{1050}C_{300\times450}F_1$. In $Sl_{1050}C_0F_0R$ and $Sl_{1050}C_0F_1R$ 'R' indicate repeat test and $C_{300\times300}$, $C_{300\times375}$, and $C_{300\times450}$ indicate crack and CFRP strengthen area 300 × 300, 300 × 375 and 300 × 450 mm^2 respectively. Slab strengthening strip, square and rectangular area by CFRP as shown in Fig. 7. Schematic front, top and bottom view of test setup of the specimen as shown in Fig. 8.

In patch loading, vertical concentrated loads in a specific area were applied to the slab specimen. The vertical load was applied by patch loading a specific rectangular area (127 mm × 228.6 mm) as shown in Figs. 8 and 9. The quality of the bond between the concrete surface and adhesive interface depends on adhesive properties and surface treatment. According to the required bonding length, the primer and then adhesive was then applied uniformly on the slab surfaces. The thickness of the adhesive layer was kept uniform. For most of the strengthened RC slabs, the adhesive thickness was around 0.65 mm. Test setup, testing and failure mode of CFRP strengthen pre-crack slab are shown in Fig. 9.

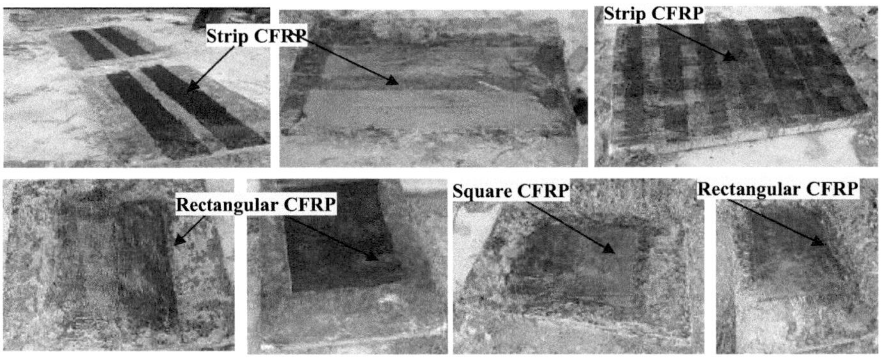

Fig. 7 Slab strengthen strip, square and rectangular area by CFRP

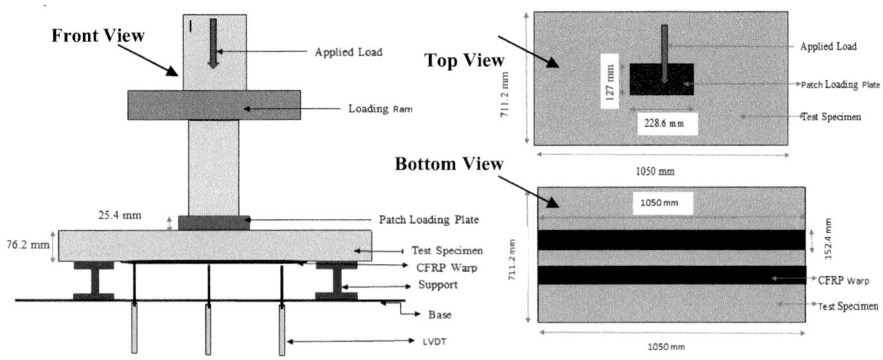

Fig. 8 Schematic front, top and bottom view of test setup of specimen

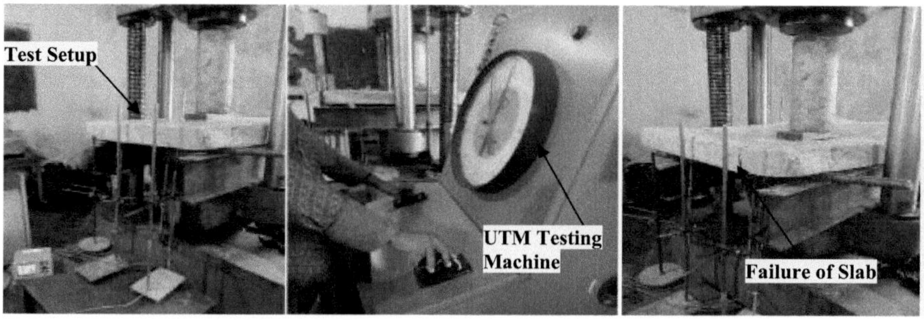

Fig. 9 Test setup and testing and failure mode of CFRP strengthen the pre-crack slab

4 Results and Discussions

A series of tests have been conducted to investigate the flexural behavior of pre-cracked RC slabs strengthened using CFRP by patch loading. Sixteen RC slabs including two reference slabs and another twelve initially cracked slabs were tested in this study with varying different CFRP lengths and strengthened areas. The failure loads, failure modes, and the load-deformation behavior of the reference slab and CFRP strengthen slab are observed in this research. The failure modes of the unstrengthened and CFRP-strengthened RC slab at the flexural region are shown in Fig. 10. At the reference slab, the first visible crack appeared at the base of the slab, directly beneath the loading point. The number of cracks started to increase more frequently after that, although their numbers eventually stabilized. Following that, the primary crack quickly spread to the top of the slab and the slab almost completely ruptured as the load increased to some extent. The entire load–deflection curve was graphed for each tested specimen, and the development of cracks was tracked. To investigate the impact of the type of strengthening material on the mechanical behavior of the slab under punching loads, the attained outcomes were analyzed and discussed. The experimental results on the reference and strip CFRP strengthening slab are given in Table 1. Test results show that the ultimate load-carrying capacity of the pre-crack slab (l × b × t-1050 × 700 × 75) was increased by 188%, 200%, 212%, and 244% compared to reference strengthen with CFRP strip lengths 500, 600, and 750 mm respectively.

The experimental results of reference and square and rectangular area CFRP strength are presented in Table 2. Test results show that the ultimate load carrying capacity of the pre-crack slab (l × b × t-1050 × 700 × 75) was increased by 164%, 177%, 195%, and 216% compared to reference strengthen with the square and rectangular area of CFRP 300 × 300, 300 × 300, 300 × 375 and 300 × 450 mm respectively. A comparison of crack and ultimate loading for strip and square-rectangular area CFRP strengthen is shown in Fig. 11 respectively. It is revealed that the initial crack was started around half of the ultimate loading.

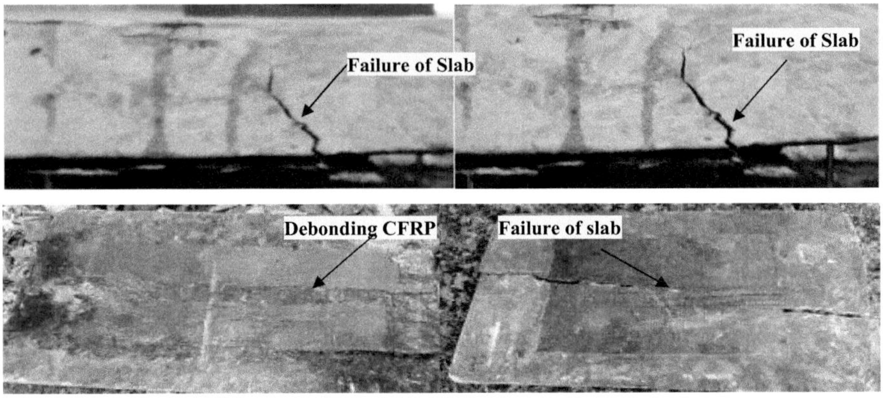

Fig. 10 Failure mode without CFRP attached slab and with CFRP attached slab

Table 1 Initial crack load, ultimate load, and enhancement of load carrying capacity by CFRP strip strengthening with respect to reference slab

Sl. No.	Specimen	Initial crack load (kN)	Ultimate load P_o and P_u (kN)	Initial load enhancement P_u/P_{u0}	Ultimate load enhancement P_u/P_{u0}
1	$Sl_{1050}C_0F_0$	13.01	25.02	–	–
2	$Sl_{1050}C_0F_1$	18.85	47.00	1.45	1.88
3	$Sl_{1050}C_{500}F_0$	12.1	22.05	–	–
4	$S1_{1050}C_{500}F_1$	19.11	44.04	1.58	2.00
5	$Sl_{1050}C_{600}F_0$	11.88	20.5	–	–
6	$Sl_{1050}C_{600}F_1$	22.09	43.60	1.86	2.12
7	$Sl_{1050}C_{750}F_0$	11.76	19.60	–	–
8	$Sl_{1050}C_{750}F_1$	23.16	47.80	1.97	2.44

Table 2 Initial crack load, ultimate load, and enhancement of load carrying capacity by CFRP square and rectangular strengthening with respect to reference slab

Sl. No.	Specimen	Initial crack load (kN)	Ultimate load P_o and P_u (kN)	Initial load enhancement P_u/P_{u0}	Ultimate load enhancement P_u/P_{u0}
1	$Sl_{1050}C_0F_0$	12.55	24.80	–	–
2	$Sl_{1050}C_0F_1$	17.31	40.67	1.38	1.64
3	$Sl_{1050}C_{300\times300}F_0$	11.98	21.96	–	–
4	$Sl_{1050}C_{300\times300}F_1$	17.13	38.86	1.43	1.77
5	$Sl_{1050}C_{300\times375}F_0$	11.35	20.05	–	–
6	$Sl_{1050}C_{300\times375}F_1$	17.36	39.08	1.53	1.95
7	$Sl_{1050}C_{300\times450}F_0$	11.02	19.25	–	–
8	$Sl_{1050}C_{300\times450}F_1$	18.18	41.58	1.65	2.16

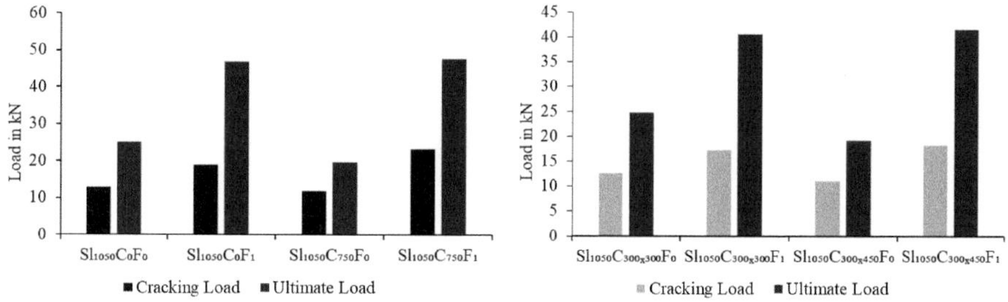

Fig. 11 Comparison of crack and ultimate loading for slab strip and rectangular strengthen by CFRP

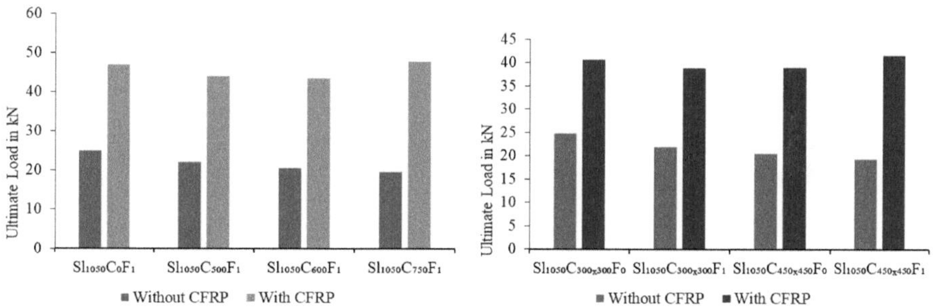

Fig. 12 Enhanced load carrying capacity for slab CFRP strip strengthening

Enhanced load carrying capacity for slab CFRP strip rectangular area strengthening is graphically represented in Fig. 12a and b respectively. The load–displacement curve of the CFRP strip and rectangular strengthening slab is shown in Fig. 13. CFRP strengthening increased the ductility and stiffness of the RC slab. Based on experimental observation, it can be revealed that by increasing CFRP length, load-carrying capacity enhancement also is increased. Deformation is also increased due to CFRP strengthening. The strength enhancement in terms of load-carrying capacity significantly varied from 188–244% and 164–216% for different CFRP strip lengths and rectangular strengthening techniques, respectively. Therefore, better performance can be attained by CFRP strengthening in the flexural region damaged RC slab with appropriate technique.

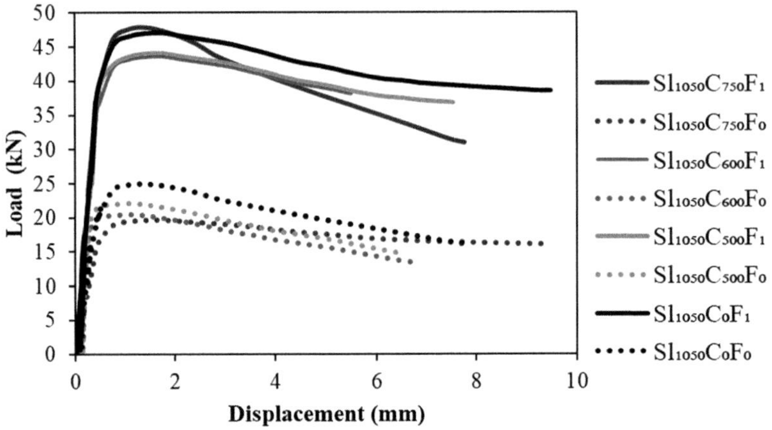

Fig. 13 Load–displacement curve of comparison for reference and CFRP strip strengthening slab

5 Conclusion

In this paper, a series of tests on strengthening the pre-crack RC slab by CFRP has been presented. Failure mode, maximum load, maximum deflection, and load-deformation behavior improvement of load-carrying capacity are also presented in this research. CFRP-wrapped sheet strengthening provided better results. CFRP strips also provide better performance techniques with an effective length of strength. Unstrengthen RC slab failed by common flexural pattern while pre-cracked CFRP strengthened RC pre-crack failed by flexural. Based on experimental results, it was found that CFRP strengthening pre-cracked RC slab provides significant performance compared to reference RC slab. The percentage increase in load-carrying capacity significantly. Based on experimental observation, it can be revealed that by increasing CFRP length, load carrying capacity enhancement also increased the value of load carrying capacity improved, and it varied from 188–244% for strip and 164–216% square and rectangular strengthening techniques, respectively. Deformation is also increased due to CFRP strengthening of the pre-crack slab. The test results revealed that externally bonded uni-directional CFRP strips strengthen and increase the flexural capacity and strength of the two-way slab. Therefore, it can be concluded that the flexural region damaged RC slab can be strengthened efficiently by CFRP with an appropriate method.

Acknowledgements The financial support (Project No.: DRE/7/RUET/489(31)/Pro/2020–2021/22) provided by the University Grants Commission of Bangladesh is gratefully acknowledged. The authors are grateful to the strength of the Materials Laboratory, Department of Civil Engineering, RUET. The authors are also gratefully acknowledging Nutech Construction Chemicals Company Limited, Dhaka, Bangladesh for supplying CFRP and adhesive materials.

References

1. Arduini M, Nanni A, Romagnolo M (2004) Performance of one-way reinforced concrete slabs with externally bonded fiber-reinforced polymer strengthening. ACI Struct J 101(2):193–201
2. Faria D, Lucio V, Ramos A (2011) Strengthening of flat slabs with post-tensioning using anchorages by bonding. Eng Struct J 33(6):2025–2043
3. Mohammed M, Salman W, Mubarak H (2019) Improving the flexural capacity of reinforced concrete one-way slabs by different techniques. J Eng Appl Sci 14(18):6769–6776
4. Zhang J, Li C, ASCE F, Nowak S, Wang S (2002) Introducing ductile strip for durability enhancement of concrete slabs. J Mater Civ Eng ASCE 14(3):253–261
5. Thanoon W, Jaafar M, Razali M, Kadir A, Noorzaei J (2005) Repair and structural performance of initially cracked reinforced concrete slabs. Constr Build Mater J ASCE 19(8):595–603
6. Foret G, Limam O (2008) Experimental and numerical analysis of RC two-way slabs strengthened with NSM CFRP rods. Constr Build Mater J ASCE 22(10):2025–2030
7. Maaddawy T, Soudki K (2008) Strengthening of reinforced concrete slabs with mechanically-anchored unbonded FRP system. Constr Build Mater J ASCE 22(4):444–455
8. Xiaohong Z, Baolin W, Peiyan H, Jinlin H (2019) Experimental study of hybrid strengthening technique using carbon fiber laminates and steel plates for reinforced concrete slabs. Constr Build Mater 210:324–337
9. Salman W (2012) Flexural behavior of bubbled reinforced concrete slabs. PhD thesis, Baghdad University, Iraq, pp 26–40
10. Moon J, Taha M, Kim J (2017) Flexural strengthening of RC slabs using a hybrid FRP-UHPC system including shear connector. Adv Mater Sci Eng 7
11. Piyong Y, Silva P, Nanni A (2003) Flexural strengthening of concrete slabs by a three-stage prestressing FRP system enhanced with the presence of GFRP anchor spikes. In: Composites in construction international conference. University of Calabria Rende (CS), Italy, pp 318–324
12. Smith S, Zhang H, Wangc Z (2013) Influence of FRP anchors on the strength and ductility of FRP-strengthened RC slabs. Constr Build Mater 49:998–1012
13. Gao D, Fang D, Tang J (2020) Flexural behavior of reinforced concrete one-way slabs strengthened via external post-tensioned FRP tendons. Eng Struct 216:110718
14. Adheem A, Ali I, Shaker M (2018) Behavior of reinforced concrete one-way slabs strengthened by CFRP sheets in flexural zone. Int J Civ Eng Technol 9(10):1872–1881
15. Adheem A, Ali I, Shaker M (2018) Flexural behavior of RC one-way slabs strengthened with fiber reinforcement cementations matrix. J Univ Babylon Eng Sci FRCM 26(10):287–298
16. Salman W, Ismael M, Ahmed W (2015) Strengthening of reinforced concrete one-way slabs using CFRP in flexural. J Eng Sci Res Technol 4(8):247–268
17. Smith S, Kim S (2009) Strengthening of one-way spanning RC slabs with cut outs using FRP composites. Constr Build Mater 23:1578–1590
18. Mosallama A, Mosalam K (2003) Strengthening of two-way concrete slabs with FRP composite laminates. Constr Build Mater 17(1):43–54
19. Khan M, Amin F, Hossain T, Kabir A (2010) Cracks in the box girders of Bangabandhu Jamuna multipurpose bridge-identification of causes based on FE analysis. In: Proceedings of the IABSE-JSCE joint conference on advances in bridge engineering-II, at BUET, Dhaka, Bangladesh, vol 2, pp 451–460
20. Rahman M, Nuruzzaman A, Haque R (2020) Rehabilitation of existing Kanchpur bridge using carbon fiber reinforced polymer laminate. In: IABSE-JSCE joint conference on advances in bridge engineering-IV, Dhaka, Bangladesh, pp 88–93

Performance Evaluation of Different Intercity Trains of Bangladesh Based on Punctuality

Md. Rifat Hossain Bhuiyan, Shah Aymaan Ibtihal, Md. Tanvirul Islam, Shanjidul Hassan Shakif, and Kashfia Nahrin Nokshi

Abstract The punctuality of train services being a critical measure of performance plays an important role in service quality and passenger satisfaction. In this study, a comprehensive scientific approach has been followed to evaluate and rank the performance of 20 major intercity trains operating under the East zone of Bangladesh Railway (BR) based on their punctuality. The railway punctuality data was obtained from the regional office of the East zone from 2015 to 2019. Afterward, the five-year data was analyzed to attain information such as on-time arrival and departure punctuality, average arrival and departure delay, and maximum arrival and departure delay. The analysis revealed how each of the intercity trains performed in the form of descriptive statistics. In the next step, FAHP (Fuzzy Analytic Hierarchy Process) was adopted to estimate the weight of six performance criteria—average departure punctuality (%), average arrival punctuality (%), average departure delay (h), average arrival delay (h), maximum departure delay (h) and maximum arrival delay (h). As a multi-criteria decision-making method, FAHP performs better than the traditional AHP method because of its ability to deal with inconsistency and vagueness while developing the pair-wise comparison matrix. In this study, the pair-wise comparison

Md. R. H. Bhuiyan (✉) · S. A. Ibtihal · S. H. Shakif
Department of Civil and Environmental Engineering, Islamic University of Technology, Gazipur, Bangladesh
e-mail: rifathossain@iut-dhaka.edu

S. A. Ibtihal
e-mail: aymaanibtihal@iut-dhaka.edu

S. H. Shakif
e-mail: shanjidulhassan@iut-dhaka.edu

Md. T. Islam
Wadsworth Department of Civil and Environmental Engineering, West Virginia University, Morgantown, USA
e-mail: mi00018@mix.wvu.edu

K. N. Nokshi
Department of Civil Engineering, International University of Business Agriculture and Technology, 4 Embankment Drive Road, Dhaka, Bangladesh
e-mail: kashfia.ce@iubat.edu

S. Arthur et al. (eds.), *Proceedings of the 6th International Conference on Advances in Civil Engineering*, Lecture Notes in Civil Engineering 368,
https://doi.org/10.1007/978-981-99-3826-1_6

matrix was developed by the co-authors along with two experts in the related field using the TFN (Triangular Fuzzy Number). The calculated fuzzy weights were defuzzified and normalized by the centroid defuzzification method to obtain a crisp numeric weight for each criterion. Once the crisp numeric weights were obtained, the weights were multiplied by the criteria values achieved from the descriptive analysis to calculate the overall punctuality score of each train. Results showed that Suborno Express achieved a maximum score of 62 followed by Bijoy Express and Turna Express which achieved 55.73 and 54.77 respectively. Trains residing at the bottom of the chart were Kalni Express, Jamuna Express, and Mohanagar Godhuli with a score of 31.73, 30.74, and 28.77 respectively. Scores of all the 20 intercity trains were calculated in a similar process and were ranked based on it. The methodology proposed in this research offers the decision-makers a tool to evaluate the performance of train services founded on scientific grounds. The findings of this research call for action on the policymakers' end to improve the overall punctuality of the trains residing at the bottom of the list and control delays to make the following trains more popular among passengers.

Keywords Punctuality · Delay · Multi-criteria decision-making method · Fuzzy AHP · Ranking

1 Introduction

The appeal of the train as a mode of transportation has been increasing since its invention, considering its ease, fuel efficiency, less travel time, and sustainability. From FY 2004–05 to FY 2013–14, Bangladesh railway experienced a significant rise (about 25 million) in passenger count [1]. Despite this heavy demand, it currently has only 2,877 route kilometers of railways connecting 44 districts [2]. The level of service provided is another cause of concern in this sector. Thus, the government has undertaken a 30-year master plan which started in 2016 to rebuild the rail transportation system [3]. The primary objective of this modernization process is to convert the railway communication system into a mass-friendly one with improved service and passenger satisfaction. Though any public organization like Bangladesh Railway (BR) must prioritize the current service quality to adjust the issues or shortcomings in the upcoming/ongoing projects, the performance of the existing trains of BR has not seen any light of measurement yet. Moreover, most of the existing literature on Bangladesh Rail Transport System either focuses on accident and safety [2, 4–6] or ongoing metro rail project [7–9].

Many factors have emergent controls on the operation quality and consumer satisfaction of a public transportation, among them robustness, reliability and punctuality are one of the most influential attributes [10]. Punctuality is the measure of having a predetermined vehicle arrive, leave, or pass at a predetermined location and time [11]. It can be denoted as the deviation between scheduled arrival/departure time (t_s) and the real arrival/departure time (t_r) [12]. Since trains of a certain route usually operate infrequently, on-time arrivals and departures play an important role in vehicle performance. Rahaman and Rahaman [13] identified, punctuality as one of the eight service quality characteristics of public transport. Chen et al. [14] considered punctuality index as on the reliability measures of public vehicles to analyze the urban bus service quality. Researchers have emphasized the punctuality of the train services since decades. Sobota and Żochowska [15] developed a transportation network based mathematical model to analyze punctuality of public transport. Grechi and Maggi [16] established a comprehensive classification of train delays based on the relationship between drivers, stimulates, and accountability. However, till now no mathematical tool has been developed to evaluate the performance of different railway transports based on punctuality.

Tomas Saaty pioneered the Analytic Hierarchy Process (AHP) approach, that has been widely utilized for decision-making for over 25 years, and that allows to approach intricate decision-making problems with substantial criteria and time periods [17]. Nevertheless, the conventional AHP appears to be insufficient and imprecise to capture the correct judgements due to the ambiguity and uncertainty on judgments of the evaluator. To address this shortcoming in the traditional AHP, triangular fuzzy number convention is incorporated to the crisp pair-wise comparison of AHP [18]. Several researchers utilized FAHP for multi-criteria decision-making, i.e., for hospital site selection, university ranking, transit system performance, and transportation corridor safety performance evaluation [19–22]. In railway system, performance assessment tool for freight transportation and rail transit operation was previously developed with FAHP application [17, 23].

This study aims to assess and score 20 intercity trains from East zone of Bangladesh Railway stressing on punctuality with descriptive statistics and Fuzzy Analytic Hierarchy Process (FAHP). East zone is of significance interest due to its geographical coverage, which covers Dhaka, Chittagong, Mymensingh, and Sylhet Division. The administrative and commercial capital of Bangladesh reside in the Dhaka and Chittagong Division, respectively. The capital of Mymensingh division is ranked first in terms of population density in Bangladesh. Chittagong and Sylhet divisions are well-known as top tourist places in Bangladesh. The approach presented in this study gives decision-makers a framework to assess the effectiveness of train services. The observation of this experiment combining with the previous studies on influencing factors of railway punctuality [24] and punctuality enhancing techniques [25] will urge policymakers to take effort to increase the overall punctuality and reduce delays of the poor performing trains.

2 Materials and Methods

Monthly punctuality data of successive five years starting from 2015 to 2019 was collected from the East zone regional office of Bangladesh Railway. The data comprises (i) no. of days operated, (ii) on-time departure in days, (iii) maximum departure delay in hours, (iv) on-time arrival in days, and (v) maximum arrival delay in hours for 20 intercity passenger trains. Five-year average departure punctuality, average departure delay, average arrival punctuality, average arrival delay, average maximum departure delay, and average maximum arrival delay were derived from the obtained data and represented as descriptive statistics. Here, departure punctuality (P_d) and arrival punctuality (P_a) denotes the percentage of on-time departure (N_d) or arrival (N_a) with respect to no. of days operated (N).

$$P_d = \frac{N_d}{\mathrm{N}} \times 100 \tag{1}$$

$$P_a = \frac{N_a}{\mathrm{N}} \times 100 \tag{2}$$

The first step of the fuzzy AHP approach is to determine the crisp relative weights of each pair of elements within the same hierarchy on a scale of 1–9. Crisp relative importance was assigned by the judgement of the co-authors and two experts from the relevant field. A 6 ×6 matrix with crisp numeric values was formed using pairwise comparison. Let M = $(m_{ij})_{k \times l}$ a crisp paise wise matrix, where if $m_{ij} = x$, then m_{ji} will satisfy $1/x$. The next step is to convert the crisp numeric values into the triangular fuzzy numbers. The notation for each crisp numeric value and its corresponding triangular fuzzy numbers are presented in Table 1.

For inverse comparison, let M = $(m_{ij})_{k \times l}$ is a pair-wise comparison matrix, where for crisp numerical value $m_{ij} = x$, the triangular fuzzy number satisfy (p, q, r). Then for $m_{ji} = 1/x$, the corresponding fuzzy number will be $(1/r,\ 1/q,\ 1/p)$.

Then the fuzzy geometric mean was estimated for each row of the matrix using the following formula. Suppose y_1 a row in the pair-wise comparison matrix M, where (p_1, q_1, r_1), (p_2, q_2, r_3), ... (p_n, q_n, r_n) are the triangular fuzzy numbers. Then the geometric mean for y_1 will be $((p_1 \times p_2 \times \cdots p_n)^{\frac{1}{4}}$, $(q_1 \times q_2 \times \cdots q_n)^{\frac{1}{4}}, (r_1 \times r_2 \times \cdots r_n)^{\frac{1}{4}})$. The geometric mean was then utilized to determine fuzzy weights for each component of the matrix. Let M a fuzzy pair-wise comparison matrix, where for rows $y_i, y_j, y_k, \ldots y_n$, the corresponding geometric mean can be denoted as $z_i, z_j, z_k, \ldots z_n$ then the fuzzy weights (w_i) for y_i will be

Table 1 Relation between crisp relative importance and triangular fuzzy relative importance

Relative importance	Crisp numeric value	Triangular fuzzy numbers
Equal importance	1	(1, 1, 1)
Moderate importance	3	(2, 3, 4)
Strong importance	5	(4, 5, 6)
Very strong importance	7	(6, 7, 8)
Extreme importance	9	(9, 9, 9)
Intermediate values	2, 4, 6, 8	(1, 2, 3); (3, 4, 5); (5, 6, 7); (7, 8, 9)
Values for inverse comparison	1/3, 1/5, 1/7, 1/9	(1/4, 1/3, 1/2); (1/6, 1/5, 1/4) (1/8, 1/7, 1/6); (1/9, 1/9, 1/9)

$(z_i \times (z_j + z_k + \ldots.. + z_n)^{-1})$. The estimated fuzzy weights were then de-fuzzified into crisp weights using Center of Area (COA) methods. Suppose for element y_i the fuzzy weight is (p_i, q_i, r_i), then crisp weight will satisfy $(\frac{p_i+q_i+r_i}{3})$.

The weight for each component was normalized by the summation of the crisp weights. The normalized weights were multiplied by the criteria values obtained from the descriptive statistics and then the multiplied values were summed up to determine the overall score of each train. More details on FAHP can be found on [26, 27].

3 Analysis and Results

3.1 Descriptive Analysis

Figure 1 reveals the performance of 20 East zone intercity trains based on average departure and arrival punctuality. Suborno Express, Mohanagar Godhuli, Turna, and Bijoy Express had high departure and arrival punctuality. Tista, Egarosindhur, and Haor Express showed satisfactory performance in arrival punctuality, but substandard performance in departure punctuality. Mohanagar Godhuli, Jamuna Express, and Kalni express were found to be the lowest performing railways considering average departure and arrival punctuality.

Some other attributes of descriptive analysis are presented in Table 2. It is seen that, Suborno Express, Tista Express, and Egarosindhur Express had the least avg. departure and arrival delay whereas the passenger of Udayan Express experienced the highest average arrival and departure delays. The average of maximum arrival and departure delays from 2015 to 19 indicated that Parabat, Udayan, and Bijoy express required much longer time than scheduled to depart or arrive at the station.

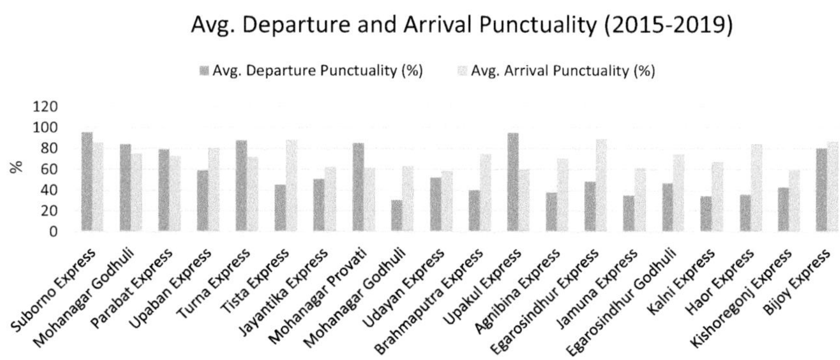

Fig. 1 Average departure and arrival punctuality of the intercity trains

Table 2 Performance of the intercity trains based on the descriptive analysis

SI	Train No.	Train name	Avg. departure punctuality (%)	Avg. arrival punctuality (%)	Avg. departure delay (h)	Avg. arrival delay (h)	Avg. max. departure delay (h)	Avg. max. arrival delay (h)
1	701	Suborno Express	95.45	85.51	0.87	1.95	4.1	6.15
2	703	Mohanagar Godhuli	84.26	75.21	1.1	2.28	4	6.17
3	709	Parabat Express	79.41	72.78	1.44	2.54	8.85	7.68
4	739	Upaban Express	58.99	80.42	1.76	2.49	5.23	6.37
5	741	Turna Express	87.88	71.74	1.41	2.95	5.48	7.95
6	707	Tista Express	45.3	88.37	0.82	1.92	1.8	4.58
7	717	Jayantika Express	50.81	62.52	2.98	3.7	6.49	7.7
8	721	Mohanagar Provati	85.24	61.49	1.55	3.02	6	6.5
9	722	Mohanagar Godhuli	30.7	63.01	2.9	3.73	6.57	8.07
10	723	Udayan Express	52.21	58.3	3.51	5.03	7.6	10.65
11	743	Brahmaputra Express	39.98	75	1.58	2.48	3.13	5.13
12	711	Upakul Express	95.08	59.62	0.75	2.64	2.8	4.97
13	735	Agnibina Express	37.81	70.56	1.74	2.81	4.63	5.83
14	737	Egarosindhur Express	48.47	89.19	0.96	1.76	2.57	5.33
15	745	Jamuna Express	35.33	61.59	1.85	2.68	3.93	4.58
16	749	Egarosindhur Godhuli	46.94	74.99	2.13	2.76	5.9	6.25
17	773	Kalni Express	34.41	67.07	2.02	3.02	5.39	6.13
18	777	Haor Express	36.01	84.58	1.46	2.11	2.9	5.07
19	781	Kishoregonj Express	42.9	60.1	1.41	2.4	2.78	4.81
20	785	Bijoy Express	80.5	86.68	2.03	3.11	8.04	7.92

3.2 FAHP for Ranking

The FAHP framework has been adopted in this study to further analyze the performance of the intercity trains. This first step into the FAHP approach requires developing a pair-wise comparison matrix which delineates priority of each evaluation criteria in comparison to another. This step requires subjective judgement from the experts. A panel of experts were involved in developing the comparison matrix for this study based on data collected through social media surveys. The crisp and fuzzy values for pair-wise comparison matrix can be seen in Table 3 and Table 4, respectively. While designing the study, avg. departure punctuality was given more importance than the avg. arrival punctuality. This hypothesis was verified through an online survey, which revealed passengers found departure delays more annoying compared to arrival delays. Thus, most and least weight was assigned for the avg. Departure Punctuality and avg. maximum arrival delay, respectively.

The crisp numeric values of the initial pair-wise comparison matrix were later converted to fuzzified pair-wise comparison matrix using the TFN (Triangular Fuzzy Number). Fuzzification reduces the uncertainty of the judgement since the numerical scores are not sharply defined. The range of values provided against linguistic variables in FAHP help curtail imprecision and vagueness making this methodology more acceptable than the traditional AHP. Table 4 shows the fuzzified pair-wise comparison matrix.

The de-fuzzified normalized scores obtained from Table 4 was utilized to determine the rank of competing trains services. From FAHP, 40.6% weightage was

Table 3 Crisp pair-wise comparison matrix

Pair-wise comparison matrix						
Criteria	Avg. departure punctuality	Avg. arrival punctuality	Avg. departure delay	Avg. arrival delay	Avg. maxm. departure delay	Avg. maxm. arrival delay
Avg. departure punctuality	1	2	4	5	7	9
Avg. arrival punctuality	1/2	1	3	4	5	7
Avg. departure delay	1/4	1/3	1	2	4	6
Avg. arrival delay	1/5	1/4	1/2	1	2	4
Avg. maxm. departure delay	1/7	1/5	1/4	1/2	1	2
Avg. maxm. arrival delay	1/9	1/7	1/6	1/4	1/2	1

Table 4 Fuzzified pair-wise comparison matrix

Fuzzified pair-wise comparison matrix							Fuzzy geometric mean	Fuzzy weights	De-fuzzified weights	De-fuzzified weights (normalized)
Criteria	Avg. departure punctuality	Avg. arrival punctuality	Avg. departure delay	Avg. arrival delay	Avg. maxm. departure delay	Avg. maxm. arrival delay				
Avg. departure punctuality	(1, 1, 1)	(1, 2, 3)	(3, 4, 5)	(4, 5, 6)	(6, 7, 8)	(9, 9, 9)	(2.94, 3.69, 4.32)	(0.267, 0.417, 0.623)	0.436	0.406
Avg. arrival punctuality	(1/3, 1/2, 1/1)	(1, 1, 1)	(2, 3, 4)	(3, 4, 5)	(4, 5, 6)	(6, 7, 8)	(1.91, 2.44, 3.14)	(0.173, 0.276, 0.453)	0.301	0.28
Avg. departure delay	(1/5, 1/4, 1/3)	(1/4, 1/3, 1/2)	(1, 1, 1)	(1, 2, 3)	(3, 4, 5)	(5, 6, 7)	(0.95, 1.26, 1.61)	(0.086, 0.142, 0.232)	0.153	0.142
Avg. arrival delay	(1/6, 1/5, 1/4)	(1/5, 1/4, 1/3)	(1/3, 1/2, 1/1)	(1, 1, 1)	(1, 2, 3)	(3, 4, 5)	(0.57, 0.76, 1.04)	(0.052, 0.086, 0.15)	0.096	0.09
Avg. maxm. departure delay	(1/8, 1/7, 1/6)	(1/6, 1/5, 1/4)	(1/5, 1/4, 1/3)	(1/3, 1/2, 1/1)	(1, 1, 1)	(1, 2, 3)	(0.33, 0.44, 0.59)	(0.03, 0.05, 0.085)	0.055	0.051
Avg. maxm. arrival delay	(1/9, 1/9, 1/9)	(1/8, 1/7, 1/6)	(1/7, 1/6, 1/5)	(1/5, 1/4, 1/3)	(1/3, 1/2, 1/1)	(1, 1, 1)	(0.23, 0.26, 0.33)	(0.021, 0.029, 0.048)	0.033	0.031
Sum									1.074	1

obtained for criteria avg. departure punctuality, and 28% weightage for avg. arrival punctuality. These two criteria were identified as the most crucial in performance evaluation. The percentage avg. arrival and departure punctuality being the indicator of an individual train's overall punctuality out of about 1560 operational days within five-year span justifies the high weightage allocation. Another key observation was high weightage allocation in departure delay criteria compared to arrival delay criteria. This is nothing but the reflection of general passengers' opinions collected through online surveys where the majority suggested that departure delay affects passenger satisfaction more than arrival delay. Table 5 shows the overall score of the intercity trains and their ranks evaluated by consideration of six mentioned evaluation criteria.

The findings from the analysis showed that Suborno express (701) ranked first among 20 intercity trains in the east zone with an overall score of 62. Suborno Express offered noticeably better performances in all the evaluation criteria in both descriptive and FAHP analysis to attain the top spot in the list. The scores of second to fifth position holder intercity trains were almost identical (55.73–54.51). Bijoy Express (785), Turna Express (741), Upakul Express (711), and Mohanagar Godhuli (703) were found to occupy second to fifth spots in the table by offering overall better performance in terms of punctuality. Most of the bottom rank trains showed poor records of departure Punctuality ranging between 30.7 and 36.01%. Mohanagar Godhuli (722) was at the bottom end of the table with an overall score of 28.77, and average departure and arrival punctuality of 30.7% and 63.01% respectively.

4 Conclusions and Policy Implications

The aim of this study was to develop an evaluation framework to rank the major intercity trains operating under East Zone of Bangladesh Railway and to identify trains residing at the bottom of the list so that actions can be taken in improving the service quality. State of an art prioritization algorithm FAHP has been applied in this study complemented by expert's participation and social media surveys. The trains were evaluated under six punctuality specific criteria named—avg. departure punctuality, avg. arrival punctuality, avg. departure delay, avg. arrival delay, max. departure delay, and max. arrival delay. The methodology proposed in this research offers the decision-makers a tool to evaluate the performance of train services founded on scientific ground. The findings of this research call for action on the policymakers' end to improve the overall punctuality of the trains residing at the bottom of the list and control delays to make the following trains more popular among the passengers.

Table 5 Score and rank of the intercity trains of East Zone of Bangladesh Railways

SI	Train No.	Train name	Avg. departure punctuality (%)	Avg. arrival punctuality (%)	Avg. departure delay (h)	Avg. arrival delay (h)	Max. departure delay (h)	Max. arrival delay (h)	Score	Rank
1	701	Suborno Express	95.45	85.51	0.87	1.95	4.1	6.15	62	1
2	703	Mohanagar Godhuli	84.26	75.21	1.1	2.28	4	6.17	54.51	5
3	709	Parabat Express	79.41	72.78	1.44	2.54	8.85	7.68	51.5	6
4	739	Upaban Express	58.99	80.42	1.76	2.49	5.23	6.37	45.53	8
5	741	Turna Express	87.88	71.74	1.41	2.95	5.48	7.95	54.77	3
6	707	Tista Express	45.3	88.37	0.82	1.92	1.8	4.58	42.61	10
7	717	Jayantika Express	50.81	62.52	2.98	3.7	6.49	7.7	36.81	13
8	721	Mohanagar Provati	85.24	61.49	1.55	3.02	6	6.5	50.83	7
9	722	Mohanagar Godhuli	30.7	63.01	2.9	3.73	6.57	8.07	28.77	20
10	723	Udayan Express	52.21	58.3	3.51	5.03	7.6	10.65	35.85	15
11	743	Brahmaputra Express	39.98	75	1.58	2.48	3.13	5.13	36.47	14
12	711	Upakul Express	95.08	59.62	0.75	2.64	2.8	4.97	54.66	4
13	735	Agnibina Express	37.81	70.56	1.74	2.81	4.63	5.83	34.19	16

(continued)

Table 5 (continued)

SI	Train No.	Train name	Avg. departure punctuality (%)	Avg. arrival punctuality (%)	Avg. departure delay (h)	Avg. arrival delay (h)	Max. departure delay (h)	Max. arrival delay (h)	Score	Rank
14	737	Egarosindhur Express	48.47	89.19	0.96	1.76	2.57	5.33	44.06	9
15	745	Jamuna Express	35.33	61.59	1.85	2.68	3.93	4.58	30.74	19
16	749	Egarosindhur Godhuli	46.94	74.99	2.13	2.76	5.9	6.25	39.01	11
17	773	Kalni Express	34.41	67.07	2.02	3.02	5.39	6.13	31.73	18
18	777	Haor Express	36.01	84.58	1.46	2.11	2.9	5.07	37.6	12
19	781	Kishoregonj Express	42.9	60.1	1.41	2.4	2.78	4.81	33.54	17
20	785	Bijoy Express	80.5	86.68	2.03	3.11	8.04	7.92	55.73	2

Acknowledgements The authors are grateful to Bangladesh Railway (East Zone) for facilitating this research by offering data.

References

1. Railway reform progress report. Asian Development Bank
2. Mohammed Ahsan H (2021) Key safety approaches in Bangladesh railway. J Eng Sci 12:57–62. https://doi.org/10.3329/jes.v12i1.53102
3. Ahsan HM, Rahman MM (2022) Development progress of railway services in Bangladesh. In: Arthur S, Saitoh M, Pal SK (eds) Advances in civil engineering. Springer, Singapore, pp 385–396
4. Probha NA, Hoque MdS (2018) A study on transport safety perspectives in Bangladesh through comparative analysis of roadway, railway and waterway accidents. In: Proceedings of the Asia-Pacific conference on intelligent medical 2018 and international conference on transportation and traffic engineering 2018 on—APCIM & ICTTE 2018. ACM Press, Beijing, China, pp 81–85
5. Azzacy MB (2012) Analysis of rail accidents due to collision and their preventive measures in Bangladesh
6. Mahfuz N, Dhali OA, Ahmed S, Nigar M (2017) Autonomous railway crack detector robot for Bangladesh: SCANOBOT. In: 2017 IEEE region 10 humanitarian technology conference (R10-HTC). IEEE, Dhaka, pp 524–527
7. Ahasan R, Kabir A, Nirzhar Y (2020) Forecasting the impact of metro rail implementation on mobility and accessibility: case of Dhaka, Bangladesh
8. Voumick D, Deb P, Sutradhar S, Salim SA, Jahan Mim KS, Tazin T, Khan MM (2021) Metro rail tracking system in Bangladesh. In: 2021 IEEE 12th Annual information technology, electronics and mobile communication conference (IEMCON). IEEE, Vancouver, BC, Canada, pp 0656–0661
9. Alam S, Chowdhury M, Siddique AB (2021) A user-friendly android application featuring smart ticketing system and destination announcement for metro rail based rapid transport system in Bangladesh. In: 2021 3rd International conference on electrical and electronic engineering (ICEEE). IEEE, Rajshahi, Bangladesh, pp 29–32
10. Parbo J, Nielsen OA, Landex A, Prato CG. Measuring robustness, reliability and punctuality within passenger railway transportation—a literature review. 4
11. Rudnicki A (1997) Measures of regularity and punctuality in public transport operation. IFAC Proc 30:661–666. https://doi.org/10.1016/S1474-6670(17)43896-1
12. Mattsson L-G (2007) Railway capacity and train delay relationships. In: Murray AT, Grubesic TH (eds) Critical infrastructure. Springer, Berlin Heidelberg, Berlin, Heidelberg, pp 129–150
13. Rahaman KR, Rahaman MA (2009) Service quality attributes affecting the satisfaction of railway passengers of selective route in southwestern part of Bangladesh. Theor Empir Res Urban Manag 4:115–125
14. Chen X, Yu L, Zhang Y, Guo J (2009) Analyzing urban bus service reliability at the stop, route, and network levels. Transp Res Part A: Policy Pract 43:722–734. https://doi.org/10.1016/j.tra.2009.07.006
15. Sobota A, Żochowska R (2008) Model of urban public transport network for the analysis of punctuality. J Achiev Mater Manuf Eng 28:4
16. Grechi D, Maggi E (2018) The importance of punctuality in rail transport investigation on the delay determinants. European Transport\Trasporti Europei
17. Blagojević A, Vesković S, Kasalica S, Gojić A, Allamani A (2020) The application of the fuzzy AHP and DEA for measuring the efficiency of freight transport railway undertakings. Oper Res Eng Sci Theory Appl 3:1–23. https://doi.org/10.31181/oresta2003001b

18. Ayağ Z, Özdemir RG (2006) A fuzzy AHP approach to evaluating machine tool alternatives. J Intell Manuf 17:179–190. https://doi.org/10.1007/s10845-005-6635-1
19. Vahidnia MH, Alesheikh AA, Alimohammadi A (2009) Hospital site selection using fuzzy AHP and its derivatives. J Environ Manag 90:3048–3056. https://doi.org/10.1016/j.jenvman.2009.04.010
20. Li X, Fan Y, Shaw J, Qi Y (2017) A fuzzy AHP approach to compare transit system performance in US urbanized areas. J Public Transp 20. https://doi.org/10.5038/2375-0901.20.2.4
21. Bhuiyan MRH, Raihan MA, Hossain M (2022) Prioritizing locations for safety improvement: an integrated disutility-based approach. In: International conference on transportation and development 2022. American Society of Civil Engineers, Reston, VA, pp 287–300. https://doi.org/10.1061/9780784484333.026
22. Aliyev R, Temizkan H, Aliyev R (2020) Fuzzy analytic hierarchy process-based multi-criteria decision making for universities ranking. Symmetry 12:1351. https://doi.org/10.3390/sym12081351
23. Liu X (2010) Fuzzy theory and AHP applied to urban rail transit operation performance. In: 2010 2nd International Asia conference on informatics in control, automation and robotics (CAR 2010). IEEE, Wuhan, China, pp 150–153
24. Olsson NOE, Haugland H (2004) Influencing factors on train punctuality—results from some Norwegian studies. Transp Policy 11:387–397. https://doi.org/10.1016/j.tranpol.2004.07.001
25. Veiseth M, Magnus Hegglund P, Wien I, Olsson NOE, Stokland Ø (2011) Development of a punctuality improvement method. TQM J 23:268–283. https://doi.org/10.1108/17542731111124334
26. Liu Y, Eckert CM, Earl C (2020) A review of fuzzy AHP methods for decision-making with subjective judgements. Expert Syst Appl 161:113738. https://doi.org/10.1016/j.eswa.2020.113738
27. Leung LC, Cao D (2000) On consistency and ranking of alternatives in fuzzy AHP. Eur J Oper Res 124:102–113. https://doi.org/10.1016/S0377-2217(99)00118-6

Flood Susceptibility Assessment in Southwest Coastal Region of Bangladesh Using an AHP-GIS Based Approach

M. Z. Hossain and S. K. Adhikary

Abstract Floods are one of the deadliest types of natural calamities that result in huge losses of lives and cause significant economic and property damages. Flood hazards are expected to persist in the future, posing severe threats to humanity. As climate change continues, the severity and frequency of floods have become a major concern in many areas of the world. Bangladesh, being located in a low-lying topography, has been facing flood disasters over the years like many other nations around the world. The country's southwest coastal region, in particular, is one of the most vulnerable to flooding. Flood management, control, and prevention measures are crucial to minimizing the likelihood of damage to agriculture, infrastructure, and other natural resources. Flood mitigation efforts and early warning systems may depend heavily on the flood vulnerability assessment. Therefore, the objective of the current study is to identify the flood susceptible zones in the six administrative districts covering Khulna, Satkhira, Pirojpur, Bagerhat, Jhalokati, and Barguna in the southwest coastal region of Bangladesh. Flood-susceptible zones are determined using a multi-criteria decision technique that has adopted an integrated geographic information system (GIS) and analytical hierarchical process (AHP)-based approach. Ten different flooding causal factors, including elevation, topographic wetness index, land use/land cover, slope, soil type, distance from the main river, surface runoff, geology, soil drainage, and soil texture, are used to identify flood susceptible zones. The flood susceptible areas with very low, low, moderate, and high risk attributes and river regions are found to be 318 km^2 (2%), 1553 km^2 (10%), 3556 km^2 (23%), 7190 km^2 (47%), and 2784 km^2 (18%), respectively. It is expected that the flood-prone zones identified by this study will be used to undertake immediate countermeasures aimed at reducing flood-related losses.

M. Z. Hossain (✉)
Department of Civil Engineering, Stamford University Bangladesh, Dhaka 1217, Bangladesh
e-mail: zahedce91@stamforduniversity.edu.bd

S. K. Adhikary
Department of Civil Engineering, Khulna University of Engineering and Technology, Khulna 9203, Bangladesh
e-mail: sajal@ce.kuet.ac.bd

S. Arthur et al. (eds.), *Proceedings of the 6th International Conference on Advances in Civil Engineering*, Lecture Notes in Civil Engineering 368,
https://doi.org/10.1007/978-981-99-3826-1_7

Keywords Flood susceptibility · AHP-GIS · Coastal region · NDVI · Multi-criteria decision technique

1 Introduction

Floods are one of the most damaging natural catastrophes among all other natural disasters. It destroys infrastructure for communication and property, and results in a major loss of human and cattle lives, and destroys agricultural products, farmland, and important properties [1]. Waterlogging and flooding are caused by sudden, broad, and continuous rainfall [2]. Future flood risks are anticipated to persist, and as climate change progresses, the severity and frequency of floods will become a problem in many regions of the world. Bangladesh, having low-lying land like many other countries of the world, has been facing flooding disasters over the years. One of Bangladesh's most seriously flood-prone regions is the southwest coastal region. Some major flood events in Bangladesh occurred in 1988, 1998, 2004, 2007, and 2014. During these occurrences, there has been a major loss of property and a fatality [3]. Floods are created by the rapid accumulation of flows from upstream and the release of runoff waters downstream as a result of extremely heavy rainfall. Discharges quickly reach a peak and then drop off just as swiftly. The breakdown of drainage management systems and major rivers are the causes of severe flood occurrences. Failure of drainage management systems and major rivers are the causes of these flood occurrences [4]. To minimize the likelihood of damage to infrastructure, agriculture, and other natural resources, flood control and prevention measures are required. Therefore, assessing flood susceptibility is crucial for future early warning systems and flood mitigation [5]. Water logging and flooding are caused by sudden, broad, and continuous rainfall. Numerous studies employ a variety of comprehensive approaches for mapping flood susceptibility, including the multi-criteria decision support approach (MCDA), the analytical hierarchy process (AHP), the weights of evidence (WoE) model, the frequency ratio (FR) model, artificial neural networks (ANN), and so on [2]. The AHP method, which has received the most attention, has been used to develop a unique framework for making decisions regarding flood susceptibility mapping. In AHP, distinct flood vulnerability criteria are prioritized according to the impact of those parameters, using various considerations. The chosen methodological framework formulates each criterion's cumulative character, which is useful for generating flood data that are subject to the spatial scale at local, regional, and national levels [6]. An effective way to give people the chance to analyze, acquire, store, combine, manipulate, retrieve, and display possible threat regions is through the use of geographic information systems (GIS) and remote sensing (RS) [7]. The objectives of the current study are to prepare maps of factors that are responsible for flood-susceptible zones as well as to detect flood-susceptible areas in the southwest coastal region of Bangladesh.

2 Description of the Study Area

This research covers a relatively larger area located in the southwest coastline of Bangladesh, which includes both urban and rural land, by considering a large number of flood-influencing factors. Six administrative districts, namely Khulna, Satkhira, Pirojpur, Bagerhat, Jhalokati, and Barguna, are selected for the current study. The location of the study area is illustrated in Fig. 1. The total area of the study area is 15402 km^2. The geographical location of the area is between 88° 52′ and 90° 23′ east longitudes and between 21° 38′ and 23° 1′ north latitudes. The study area is bounded by West Bengal on the west, India on the east, and the Bay of Bengal on the south. On the west side there are Barishal and Patuakhali districts, and on the north side there are Jessore, Narail, and Gopalganj districts. The annual rainfall of the study area ranges from 1380 to 2950 mm, with a mean of 2165 mm. Most of the rainfall occurs from May to September. The average temperature of the region varies from 15.5 to 36.2 °C. Where the warmest month is April and the coldest month is January [8]. The study area has a large portion of Sundarban, a mangrove forest that is among the biggest in the world. This is a beautiful example of how people may live in harmony with both land and aquatic flora and fauna. This forest has remained largely unmodified due to the great commitment of the government under various preservation measures. Although the recorded net decline of mangrove forests is not particularly high, the changing pattern indicates that turnover due to erosion, aggradation, reforestation, and deforestation was significantly bigger than the net change. Both human and natural elements that cause the forest's degeneration pose a threat to it, especially illnesses like top-dying and overuse of its resources [9].

3 Methodology

The development of numerous conceptual frameworks that identify numerous indications and so implicitly demonstrate the complexity of vulnerability assessment is a distinctive feature of the science of flood and water-logging vulnerability assessment [10]. The current study is based on an integrated AHP and GIS-based approach. The RS data are adopted and processed on the ArcGIS platform for flood susceptibility mapping. The process of flood susceptibility mapping starts with the collection of maps for different flood-influencing factors. Ten flood-influencing factors, namely rainfall, land use/land cover (LULC), topographic wetness index (TWI), elevation, slope, distance from the river, normalized difference vegetation index (NDVI), drainage density, distance from the road, and soil type, are used in the current study. Considering the characteristics of the study area described in Sect. 2, the current study has considered both urban and monsoon floods. The factors including rainfall, slope, drainage density, etc. are mainly responsible for causing monsoon flooding, whereas the factors including rainfall, soil type, land use/land cover, etc. are accountable for causing urban flooding. The source of the collected data is presented in Table 1. After

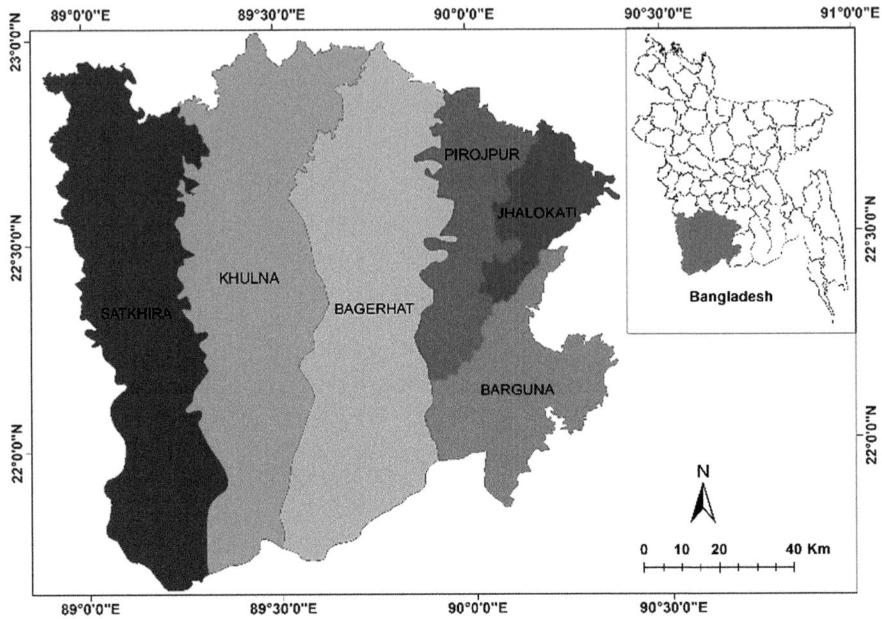

Fig. 1 Location of study area in the southwest coastal region of Bangladesh

the collection of data from different sources, a thematic map of each factor is prepared by processing the collected maps in the GIS environment. ArcGIS software is used for the processing and analysis of data and the generation of thematic layers based on the aforementioned impacting factors. The thematic map of different factors is then classified into different sub-classes depending on their effects on flood susceptibility. After the preparation of thematic maps for the factors, the AHP technique is used to assign weight to the sub-categories of the impacting factors. AHP is one of the most widely adopted multi-criteria decision analysis (MCDA) techniques. The technique is adopted by many researchers worldwide for ranking the priority-based factors [2, 6, 11–13].

In AHP, the priority or weighting of the factors is determined using a pairwise comparison matrix. A linear scale of 1–9 based on Saaty [14] is used to assign scores for building pair-wise comparison matrix. In Saaty's linear scale, 1 denotes equal importance, whereas 9 designates something extremely more important, and 1/9 indicates something extremely less important. The meaning of each score is tabulated in Table 2.

The pairwise comparison matrix of different factors is tabulated in Table 3. Based on the importance obtained from the pairwise comparison matrix (i.e., the AHP analysis), each factor's priority is determined. According to the Saaty approach [14], a perfect pairwise comparison matrix should have a consistency of less than 10%. In this experiment, the consistency is found to be 8.96%, which justifies the use of the pairwise comparison matrix. Next, the sub-categories of each factor are assigned

Table 1 Sources of flood susceptibility factors

Factors	Source	Resolution
Rainfall	Bangladesh Meteorological Department (BMD)	
DEM	US Geological Survey (USGS) Shuttle Radar Topography Mission (SRTM) data	30 m × 30 m
Elevation, slope, TWI, drainage density	Generated from the DEM	30 m × 30 m
LULC	Landsat 8 satellite image, USGS online portal	30 m × 30 m
NDVI	USGS	30 m × 30 m
Soil type, distance from main river, distance from road	Bangladesh Agricultural Research Council (BARC)	

Table 2 Saaty's linear scale

Decrease in row importance over Column				Equal	Increase in row importance over Column			
←				=				→
1/9	1/7	1/5	1/3	1	3	5	7	9

weight based on the ranking priority, which is presented in Table 4. Finally, the weighted overlay analysis of the factors is performed in the ArcGIS platform based on the assigned weights to achieve the flood susceptibility map.

Table 3 Pair-wise comparison matrix based on flood susceptibility

	TWI	EL	Slope	RF	LULC	DD	NDVI	DRi	DRo	ST
TWI	1	1	1	1	3	3	3	1	3	5
EL	1	1	1	1	2	5	4	1	3	5
Slope	1	1	1	1	3	5	3	1/2	3	3
RF	1	1	1	1	3	3	2	2	4	5
LULC	1/3	1/2	1/3	1/3	1	3	1	1/3	3	1
DD	1/3	1/5	1/5	1/3	1/3	1	3	1/3	1	3
NDVI	1/3	1/4	1/3	1/2	1	1/3	1	1/5	5	5
DRi	1	1	2	1/2	3	3	5	1	3	3
DRo	1/3	1/3	1/3	1/4	1/3	1	1/5	1/3	1	1/3
ST	1/5	1/5	1/3	1/5	1	1/3	1/5	1/3	3	1

Note EL Elevation, *TWI* Topographic Wetness Index, *LULC* Land Use/Land Cover, *RF* Rainfall, *DD* Drainage Density, *DRi* Distance from Major River, *NDVI* Normalized Difference Vegetation Index, *Dro* Distance from Road, and *ST* Soil Type

Table 4 Assigned weight to different layers

Layer	Sub class	Area (km^2)	Rank	Individual weight	Weight	Layer	Sub class	Area (km^2)	Rank	Individual weight	Weight
TWI	<8.4	6101	1	3	14	Distance from main river (m)	<2015	7589	5	15	15
	8.4–11	3193	2	6			2015–6000	5272	4	12	
	12–13	3038	3	9			6000–9500	1395	3	9	
	14–16	2574	4	12			9500–17,000	848	2	6	
	>16	495	5	14			>17,000	298	1	3	
Elevation (m)	<2	3535	5	15	15	Distance from road (m)	<12,010	11,821	5	4	4
	2–6	6235	4	12			12,010–24,021	1348	4	4	
	6–10	3882	3	9			24,021–36,031	1077	3	3	
	10–14	1473	2	6			36,031–48,042	837	2	2	
	>14	277	1	3			>48,042	319	1	1	
Slope (% rise)	<1.4	4594	5	14	14	Soil type	Alluvium	307	4	4	4
	1.4–3.8	6653	4	12			Floodplain soils	8063	3	2	
	3.8–6.6	3147	3	9			Acid sulphate soil	339	3	2	
	6.6–11.5	868	2	6			Peat (organic)	510	2	1	
	>11.5	140	1	3			Sundarbans	4316	5	4	

(continued)

Table 4 (continued)

Layer	Sub class	Area (km^2)	Rank	Individual weight	Weight	Layer	Sub class	Area (km^2)	Rank	Individual weight	Weight
Rainfall (mm)	<1510	118	1	4	16		Waterbodies	1785	5	4	
	1510–1740	2466	2	7			Urban	81	1	2	
	1740–1975	2941	3	10		NDVI	<0	2363	5	7	7
	1975–2215	4466	4	13			0–0.15	2411	4	5	
	>2215	5411	5	16			0.15–0.30	7035	3	3	
LULC	Water	36,150	5	6	6		>0.30	3593	2	2	
	Trees	56,610	2	3		Drainage density (km/ km^2)	<0.67	2669	5	5	5
	Vegetation	1498	1	2			0.67–1.34	7110	4	4	
	Crops	46,565	4	5			1.34–2.01	4569	3	3	
	Built area	9125	1	2			2.01–2.68	969	2	2	
	Bare ground	70	2	3			>2.68	84	1	1	
	Rangeland	384	2	3							

Note Here, the interpretation of Rank is 1 = very low, 2 = low, 3 = moderate, 4 = high and 5 = very high

4 Results and Discussion

4.1 *Analysis of Influenting Factors*

This study attempts to locate the flood-susceptible zone in the southwest coastal region of Bangladesh. Among so many hydrological factors based on analysis, ten major flood-influencing factors are selected. They are land use/land cover (LULC), topographic wetness index (TWI), slope, elevation, rainfall, drainage density, distance from the river, normalized difference vegetation index (NDVI), distance from the road, and soil type. The factors are then subdivided into different classes and assigned a weight based on AHP to obtain the reclassified flood susceptibility influencing factors. The area covered by each of the regions along with the weight assigned to the subclass is presented in Table 4. The spatial analysis of the factors responsible for flood susceptibility is detailed in the following sub-sections.

TWI: TWI is a wetness index that is employed to evaluate the topographic management of hydrological processes. The index is a feature of each slope and the region upstream that contributes in a direction orthogonal to the waft direction [15]. According to the AHP result, it secured the fourth position among the factors. The index is derived using DEM via several steps. The maps are divided into five subclasses. Figure 2a shows the TWI categorized map for the research area.

Elevation: The elevation is also a strong factor regarding flood susceptibility mapping. It denotes the contour of a location. The higher value indicates the upper place where the chances of water-logging are low, and the lower value indicates the lower region where the chances of water logging are higher. Therefore, it has an inverse relationship with flood susceptibility. According to the AHP result, it achieved the second position among the factors. The index is derived using DEM. Figure 2b presents the classified elevation map for the research area.

Slope: The slope is also considered as a strong factor regarding flood susceptibility mapping. It denotes elevation at a detailed level. The higher value indicates the upper place where the chances of water-logging are low, and the lower value indicates the lower region where the chances of water logging are higher. Therefore, it has an inverse relationship with flood susceptibility [16]. According to the AHP result, it achieved the fifth position among the factors. The index is derived using DEM. Figure 2c shows the classified map of slope for the research area.

Rainfall: One of the most important factors that cause flooding is rainfall. It is an indication of the availability of water. Rainfall has a proportional relationship with flood susceptibility. According to the AHP result, it achieved the first position among the factors. The collected rainfall records from BMD were used to derive the rainfall zone using IDW in the GIS domain. The maps are divided into five subclasses. Figure 2d displays the classified map of rainfall for the research area.

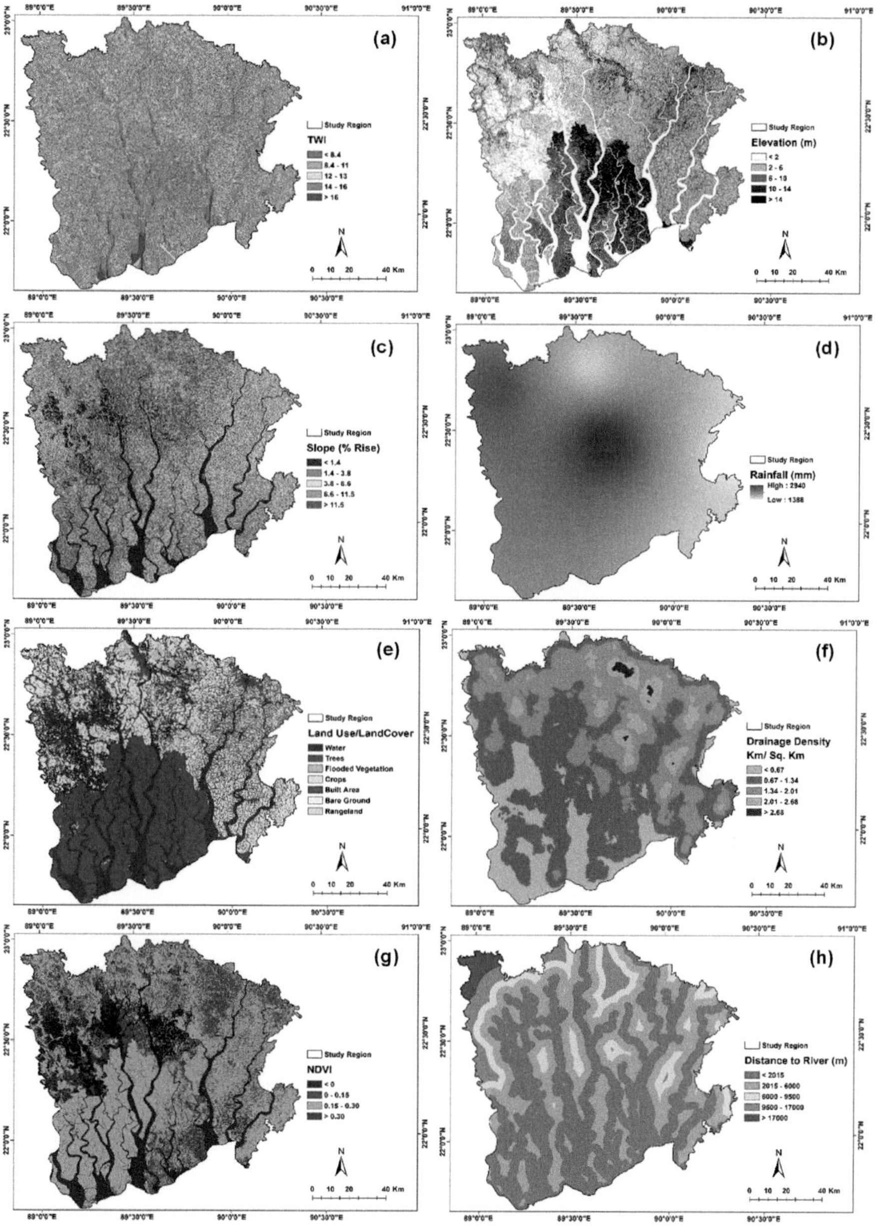

Fig. 2 Thematic maps of the study area **a** topographic wetness index (TWI), **b** elevation. Thematic maps of the study area **c** slope, **d** rainfall, **e** LULC, **f** Drainage density, **g** NDVI, **h** distance to major river, **i** distance to road, and **j** soil type

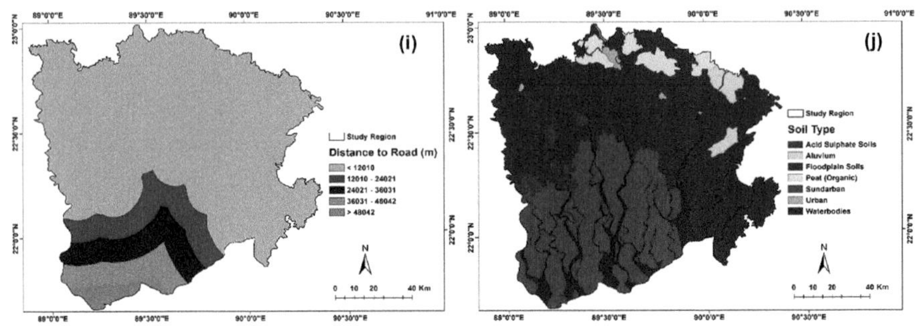

Fig. 2 (continued)

LULC: The Landsat-8 satellite image, which is obtained from the USGS Earth Explorer at 30 m resolution, is the source of the LULC map. The satellite image is used to derive a map of land use and land cover using the supervised classification method. The derived map identifies various land uses that have varying degrees of potential for groundwater recharge. According to the AHP result, it achieved the seventh position among the factors. Figure 2e shows the classified map of LULC for the research area.

Drainage Density: Another influencing factor responsible for flood susceptibility is the drainage density. It shows where the storm drainage channels are in relation to one another. Higher drainage density denotes that speed and water get the advantage in early removal from a region. Lower density means there is a lack of flow paths for water to be drained out. Therefore, it has an inverse relationship with flood susceptibility. According to the AHP result, it achieved the fifth position among the factors. The index is derived using DEM. The classified map of Fig. 2f presents the classified map of the drainage density for the research area.

NDVI: A well-known and often used index is NDVI. It is a significant vegetation index that is frequently used in studies on climatic and environmental change on a worldwide scale. The vegetation's sensitivity to the environment has been found to be a powerful defense against natural disasters, as it not only influences the ecological balance but also the climate [17]. According to the AHP result, it achieved the sixth position among the factors. The index is derived using the near and far infrared bands of the Landsat 8 satellite image. Figure 2g displays the NDVI map for the research area.

Distance to Major River: Another mighty factor used in floor susceptibility analysis is the distance from the river map. It indicates the path for water to be drained out of a region. A higher distance indicates a higher time to discharge. Hence, there is a peak of water, and a flood occurs. On the contrary, the lower distance indicates less time for water to be drained out, i.e., a lower probability of flooding. Therefore, it has an inverse relationship with flood susceptibility. According to the AHP result, it achieved the second position among the factors. The index is derived using the river map. Figure 2h presents the classified distance to the river map of the study area.

Distance to Road: A tiny factor for flood susceptibility is distance from the road map. This factor denotes the availability of a road network where many of the roads work as a dam or water barrier for a region. Therefore, it has is an inverse relationship with flood susceptibility. According to the AHP result, it achieved the tenth position among the factors. The index is derived using the road map. Figure 2i provides the classified distance to the road map for the research area.

Soil Type: Another less-influencing factor for flood susceptibility is soil type. It is commonly acknowledged that different soil types affect groundwater recharge in different ways. For instance, less clay soil results in a higher rate of recharge [18]. According to the AHP result, this factor gets the ninth position among the factors. Figure 2j shows the classified soil type for the research area.

4.2 Assessment of Flood Susceptibility

The aforementioned influencing factors were weighted using the AHP analysis and overlaidusing the weighted overlay analysis using the ArcGIS platform for the derivation of the flood susceptibility map. The map is then divided into four regions, which are categorized as very low, low, moderate, and high risk areas. Another portion that is subtracted from the map is the river region. The map of flood susceptibility is presented in Fig. 3. The area extended by each region is 318 km^2, 1553 km^2, 3556 km^2, 7190 km^2, and 2784 km^2 for very low, low, moderate, high risk, and river regions, respectively. The percentage of the total area covered by this region is 2%, 10%, 23%, 47%, and 18%, and their corresponding classifications are very low, low, moderate, high-risk, and river regions, respectively.

Furthermore, it is seen from the current study that Khulna district has a higher risk of flooding compared to other regions. The lower-risk region covers the upper part of the Bagerhat district, along with the Pirojpur and Jhalokati districts. The influence of TWI is clearly shown in the final output, where higher TWI in the upper part has a lower risk. Furthermore, the most dominating factor in the study is the elevation. The final output has a clear similarity with the different elevation regions where the relationship is proportional. The lower-risk region is shown in the moderate-slope area. The middle of the Bagerhat district has shown a higher risk, where the main responsible factor is rainfall, which is higher in this region. The LULC does not have enough significance in the final output as the majority of itis covered by the Sundarbans. Though the AHP result gives drainage density a lower ranking compared to other factors, the higher drainage density region exhibits a lower flood risk. In addition to this, an NDVI of less than 0.15 is responsible for a higher risk of flooding. Although the distance to the river is an important factor, it does not demonstrate enough influence on the flood susceptibility. It is also found that flood plain soil is responsible for causing comparatively higher flood susceptibility. Lower

Fig. 3 Flood susceptibility map of the study area

risk is seen near the areas having peat soil, but the overall influence of soil type is lower in the flood susceptible map of the study area.

5 Conclusions

The current study focuses on the quantitative assessment of flood susceptibility in the southwest coastal region of Bangladesh. The most widely used multi-criteria decision analysis technique, namely AHP integrated with GIS, is adopted in the current study using the remote sensing data and information to determine the flood-susceptible zones. Several flood-influencing factors are selected, namely elevation, topographic wetness index, rainfall, slope, drainage density, land use/land cover, distance from the river, NDVI, distance from the road, and soil type. Based on the results of the current study, the following conclusions can be drawn:

- The area covered by the region is 318 km^2, 1553 km^2, 3556 km^2, 7190 km^2, and 2784 km^2 for very low, low, moderate, high risk, and river regions, respectively. The percentage of the total area covered by this region is 2%, 10%, 23%, 47%, and 18% for the very low, low, moderate, high risk, and river regions, respectively.
- It is found from the current study that Khulna district has a higher risk of flooding compared to other regions. The lower-risk region covers the upper part of the Bagerhat district, along with the Pirojpur and Jhalokati districts.

References

1. Ahmed CF, Kranthi N (2018) Flood vulnerability assessment using geospatial techniques: Chennai, India. Indian J Sci Technol 11(6):215–223. https://doi.org/10.17485/ijst/2018/v11i6/110831
2. Natarajan L, Usha T, Gowrappan M, Kasthuri BP, Moorthy P, Chokkalingam L (2021) Flood susceptibility analysis in Chennai Corporation using frequency ratio model. J Indian Soc Remote Sens 49:1533–1543. https://doi.org/10.1007/s12524-021-01331-8
3. Annual Flood Report 2016, Flood Forecasting and Warning Center (FFWC) of Bangladesh Water Development Board (BWDB). http://www.ffwc.gov.bd/. Accessed 12 July 2022
4. Bisht DS, Chatterjee C, Kalakoti S, Upadhyay P, Sahoo M, Panda A (2016) Modeling urban floods and drainage using SWMM and MIKE URBAN: a case study. Nat Hazards 84:749–776. https://doi.org/10.1007/s11069-016-2455-1
5. Tehrany MS, Pradhan B, Jebur MN (2015) Flood susceptibility analysis and its verification using a novel ensemble support vector machine and frequency ratio method. Stoch Env Res Risk Assess 29:1149–1165. https://doi.org/10.1007/s00477-015-1021-9
6. Swain KC, Singha C, Nayak L (2020) Flood susceptibility mapping through the GIS-AHP technique using the cloud. ISPRS Int J Geoinf 9(12):720. https://doi.org/10.3390/ijgi9120720
7. Samanta S, Pal DK, Palsamanta B (2018) Flood susceptibility analysis through remote sensing, GIS and frequency ratio model. Appl Water Sci 8:66. https://doi.org/10.1007/S13201-018-0710-1
8. Climate and monthly weather forecast of Khulna, Bangladesh. https://www.weather-atlas.com/en/bangladesh/khulna-climate#rainfall. Accessed 30 Aug 2022
9. Giri C, Penga B, Zhu Z, Singh A, Tieszen LL (2007) Monitoring mangrove forest dynamics of the Sundarbans in Bangladesh and India using multi-temporal satellite data from 1973 to 2000. Estuar Coast Shelf Sci 73(1–2):91–100. https://doi.org/10.1016/j.ecss.2006.12.019
10. Veerbeek W (2018) Estimating the impacts of urban growth on future flood risk: a comparative study. CRC Press
11. Ouma YO, Tateishi R (2014) Urban flood vulnerability and risk mapping using integrated multi-parametric AHP and GIS: methodological overview and case study assessment. Water 6(6):1515–1545. https://doi.org/10.3390/w6061515
12. Nowreen S, Jalal MR, Khan MSA (2014) Historical analysis of rationalizing South West coastal polders of Bangladesh. Water Policy 16(2):264–279. https://doi.org/10.2166/wp.2013.172
13. Chattaraj D, Paul B, Sarkar S (2021) Integrated multi-parametric analytic hierarchy process (AHP) and geographic information system (GIS) based spatial modelling for flood and water logging susceptibility mapping: a case study of English Bazar Municipality of Malda, West Bengal, India. Nat Hazards Earth Syst Sci Discuss. https://doi.org/10.5194/nhess-2020-399
14. Saaty TL (1990) An exposition on the AHP in reply to the paper “remarks on the analytic hierarchy process.” Manag Sci 36(3):259–268. http://www.jstor.org/stable/2631947
15. Sørensen R, Zinko U, Seibert J (2006) On the calculation of the topographic wetness index: evaluation of different methods based on field observations. Hydrol Earth Syst Sci 10:101–112. https://doi.org/10.5194/hess-10-101-2006
16. Hammouri N, El-Naqa A, Barakat M (2012) An integrated approach to groundwater exploration using remote sensing and geographic information system. J Water Resour Prot 4(9):717–724. https://doi.org/10.4236/jwarp.2012.49081
17. Bhandari AK, Kumar A, Singh GK (2012) Feature extraction using normalized difference vegetation index (NDVI): a case study of Jabalpur City. Procedia Technol 6:612–621. https://doi.org/10.1016/j.protcy.2012.10.074
18. Thapa R, Gupta S, Guin S, Kaur H (2017) Assessment of groundwater potential zones using multi-influencing factor (MIF) and GIS: a case study from Birbhum district, West Bengal. Appl Water Sci 7:4117–4131. https://doi.org/10.1007/s13201-017-0571-z

Identifying Waterlogging Risk Zones in an Urban Area of Bangladesh Using AHP-Fuzzy Rule Based Approach

M. A. Azad, S. K. Adhikary, and M. M. Rana

Abstract Waterlogging is a very common problem during the rainy season in Bangladesh, which often causes physical, social, and economic losses in both urban and rural areas. The problem is more severe, particularly in the urban areas of the country. Khulna City Corporation (KCC), being the third-largest city of the country, is located on a low-lying area on the west bank of the Rupsha River. The city has been suffering from water logging problems due to the rapid growth of the city and urbanization, the disruption of natural drainage system, high-intensity rainfall, increased population, an unplanned urban drainage system, the siltation of nearby river beds, etc. and so on. Therefore, the aim of the current study is to explore the waterlogging situation and identify waterlogging risk zones in the KCC. At first, social, economic, and hydrological data were collected from the Bangladesh Meteorological Department (BMD), the Bangladesh Bureau of Statistics (BBS), and KCC data sources. Various hydrological data such as inundation depth, drainage density, slope, etc. have been processed from the digital elevation model (DEM) through spatial analysis in the geographic information system (GIS) platform. After data collection and processing, several indicators are identified, and a criteria weight for each indicator is obtained by the analytical hierarchy process (AHP) technique. Then a fuzzy inference system (FIS) was developed in the MATLAB platform to determine the vulnerability index of three factors, such as social, economic, and hydrological. After determining the vulnerability index of three factors, a final water logging vulnerability index was determined. Finally, the waterlogging risk zones were delineated based on the index, which was interpreted by the GIS mapping. From the analysis, it is found that about 20% of the KCC area lies in the high water logging risk zone, about 42% of the

M. A. Azad (✉) · S. K. Adhikary
Department of Civil Engineering, Khulna University of Engineering and Technology, Khulna 9203, Bangladesh
e-mail: aymanazad47@gmail.com

S. K. Adhikary
e-mail: sajal@ce.kuet.ac.bd

M. M. Rana
Department of Civil Engineering, Northern University of Business and Technology, Khulna 9100, Bangladesh

S. Arthur et al. (eds.), *Proceedings of the 6th International Conference on Advances in Civil Engineering*, Lecture Notes in Civil Engineering 368,
https://doi.org/10.1007/978-981-99-3826-1_8

KCC area is in the medium water logging risk zone, and the remaining areas cover the low water logging risk zones. Overall, this study demonstrates that the integrated AHP-fuzzy rule based approach can be effectively used for identifying waterlogging risk zones.

Keywords Water logging · AHP · Fuzzy rule · Vulnerability index · GIS · Khulna City Corporation

1 Introduction

Bangladesh is located on the extensive floodplains of the Ganges–Brahmaputra-Meghna (GBM) basin, where flooding is a natural part of life. Floods and waterlogging problems are a very common incidence in Bangladesh due to its geographical location in the lower downstream part of the GBM basin. The country is affected by flooding with either small or large ranges almost every year [1]. Due to the melt-water from the Himalayas, approximately 18% of the country is inundated annually, which results in the death of over 5,000 people and the damage of over seven million dwellings [2]. In addition to flood problems, the country has been dealing with a number of environmental concerns, including waterlogging, solid waste disposal, black smoke emissions from industries and vehicles, air and noise pollution, and water body contamination by the discharge of industrial effluents, and so on.

Bangladesh is a country of heavy rainfall and receives upstream river flows over which it has little or no control. Thus, waterlogging in urban areas is a frequent occurrence in the country, particularly during the rainy season, which has been a vital concern to city authorities. This is now a part of the life of city dwellers, and the frequency of this problem is increasing day by day. Most of the urban areas in the country have been experiencing waterlogging problems for the last few years, and even a little rainfall causes serious waterlogging in certain areas of the urban centers [3]. When a city experiences rapid urbanization, it becomes unable to manage the changing environment. Uncontrolled development in urban areas cannot satisfy sufficient drainage of excess rainwater, which eventually results in waterlogging parts of urban areas being inundated for a few hours to several days [3, 4]. In Bangladesh, waterlogging is one of the major problems in urban areas, particularly during the rainy season. With the rapid growth of population and unplanned urbanization in the country, the drainage and sewer infrastructures in Bangladesh have not been developed accordingly [3]. In our nation, waterlogging happens most frequently as a result of insufficient drain sections, outlet closures, and inappropriate maintenance of existing drainage systems. Most of these natural drains have been either filled in or have had human activities encroach upon them during the past decades. During the monsoon period, a huge amount of stormwater runoff is generated, which increases the runoff frequency. Only a small amount of that runoff can drain out successfully, and the remaining portion often causes waterlogging in urban areas [3, 5].

In order to minimize the aforementioned problems, identification of waterlogging risk areas and vulnerability assessment are crucial for developing affordable adaptation measures. Three different methodologies have been employed to quantify vulnerability assessment: historical loss data (HLD), vulnerability curve (VC), and vulnerability index (VI) [6–8]. Vulnerability indexing is the most commonly used approach. It is an effective tool that may help urban planners and policymakers prioritize waterlogging mitigation methods and raise public awareness by giving information and emphasizing hot locations for the flood risk, communities vulnerable to water congestion, and so on. Vulnerability can be analyzed in many ways [6–8]. However, the fuzzy rule-based approach [9] is one of the best tools among others and has been extensively evaluated by researchers in flood and waterlogging vulnerability studies [10]. Therefore, the fuzzy logic based approach has been selected in the current study for the waterlogging vulnerability study and identifying the waterlogging risk zones.

Urban flooding and waterlogging are very common occurrences in Khulna City, particularly during the monsoon season. In the recent years, the city has experienced a number of extreme rainfall events, which has resulted in waterlogging in some areas of Khulna City Corporation (KCC). For example, such as Royal Mohr, Dak Bangla Mohr, KD Ghosh Road, Gollamari, and Sonadanga frequently experience urban water logging problems almost every year [5, 11]. In addition to this, the city is currently in the development stage, which causes substantial changes in the land use and land cover. The combined effect of heavy rainfall in the monsoon season and urbanization induced land cover changes results in intense urban flooding in the city. Furthermore, the impact of climate change can exacerbate the situation and make the city more vulnerable to waterlogging [11]. Therefore, the objective of the current study is to study the waterlogging problems in the KCC area and delineate the waterlogging risk areas.

2 Study Area Description

Khulna is the third largest metropolitan city in Bangladesh, which is located in the lower delta of the GBM basin. The urban area of Khulna is located on the western bank of the Rupsha River, one of the KCC's most effective natural drainage outlets. The approximate location of KCC is shown in Fig. 1, which covers an area of about 51 km^2. The city is divided into 31 wards and is situated between 22° 49′ north latitude and 89° 33′ east longitude. The city has around 663,342 inhabitants [12]. The land elevation in the city is flat and ranges from 1 to 5 m above mean sea level [5, 11]. The annual average rainfall of in Khulna city is 1878 mm, and two-thirds of this rainfall occurs during the monsoon period. During heavy rainfall, it is incapable of evacuating the flow of the city, which creates water logging. Because of rapid urbanization, many drains have been closed, which puts pressure on existing drains. So, the canals (locally named "khals") should be saved anyway. Existing drains should be repaired and maintained properly. It is expected that the outcome of the current study would be supportive of the proper development and maintenance planning of drainage infrastructure in the KCC and thereby reduce the waterlogging problems.

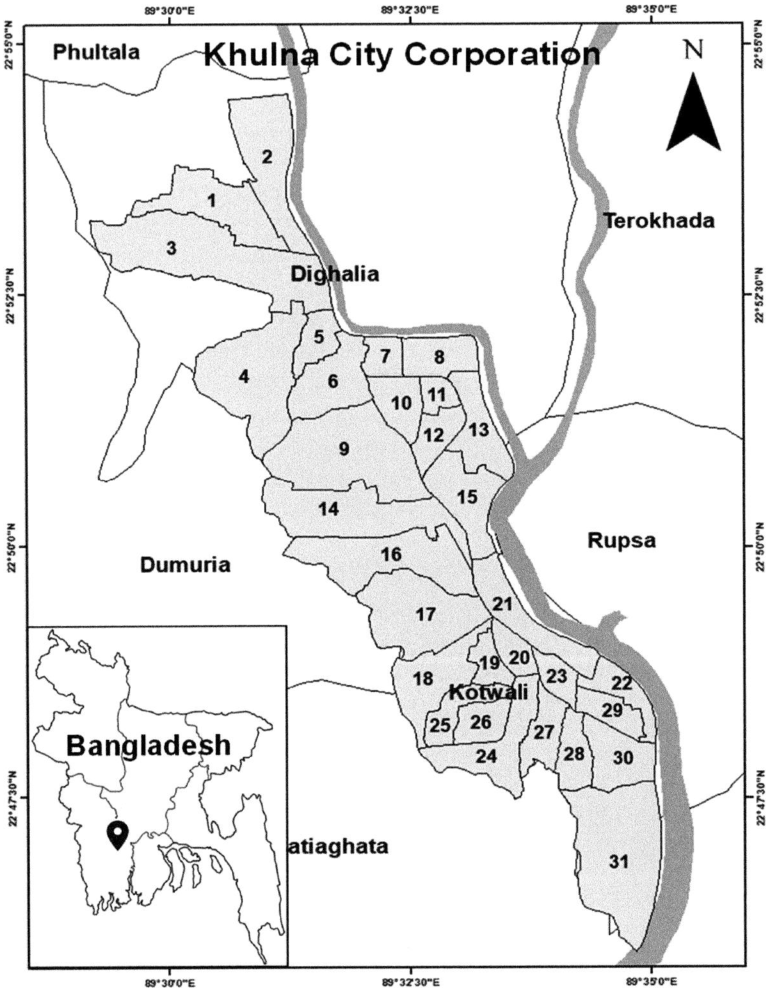

Fig. 1 The study area with different wards

3 Methodology

The methodological framework adopted in the current study is presented in Fig. 2, which is demonstrated through the use of KCC as the case study area. As can be seen from the figure, the following tasks have been undertaken in this study:

- Collection, analysis, and processing of data and selection of indicators
- Standardization of data and estimation of weighting coefficients using the analytical hierarchy process (AHP) technique
- Developing a fuzzy inference system (FIS) using the fuzzy logic toolbox in the MATLAB software platform to formulate the fuzzy logic-based water logging vulnerability index
- Identification of waterlogging risk areas in the KCC map and interpretation through spatial mapping in the ArcGIS software platform

3.1 *Data Collection and Indicators Selection*

In the current study, various hydrological, social, and economic data are collected, and suitable indicators have been selected for the case study. Data collected for each of the three components are provided in Table 1. In the case of hydrologic analysis, rainfall data, a digital elevation model (DEM), slope, inundation depth, drainage density, and soil permeability are gathered. Rainfall data are collected from the Bangladesh Meteorological Department (BMD). DEM has been processed from the SRTM-DEM data, which is freely available from the USGS website. By processing the DEM data on the ArcGIS platform, the slope, inundation depth, and drainage density data are obtained and organized ward-by-ward using the GIS-based raster analysis.

The soil permeability data are collected from the Bangladesh Agricultural Research Council (BARC). On the other hand, social and economic data have been collected from the Bangladesh Bureau of Statistics [12]. According to Balica et al. [8], an indicator is a trait that is inherent to a system and that estimates the condition of the system quantitatively. In this study, there are nineteen indicators selected based on the availability of data, which are presented in Table 2. The indicators are then clustered into three groups, such as exposure, susceptibility, and resilience.

Fig. 2 The flow chart of methodology adopted in the current study

Table 1 Collected data for each components

Hydrological	Social	Economic
• Rainfall • DEM • Slope • Inundation depth • Drainage density • Soil permeability	• Population density • Age • Disabled people • Floating people • Low income people • In house population	• Kutcha and jhupri types household • Institutional and historical buildings • Tenant • Unemployment • Electricity connection • Sanitation facility

Table 2 List of the indicators and their clustered group

Components	Exposure	Susceptibility	Resilience
Hydrological	Flooded area ratio Annual maximum rainfall No of months heavy rainfall	Slope Soil permeability	Drainage density
Social	People of 0–9 People of 60+ Disabled people Floating people	Illiteracy rate Low income people	In house population
Economical	Kutcha and jhupri household Sanitation facility Institutional and historical buildings	Tenant Unemployment	Electricity connection

3.2 Data Standardization and Weighting Estimation by AHP

The indicators have different units and measurement scales. Hence, the indicators are converted into a standard form ranging from 0 to 1 [13]. There are different methods of standardizing data. In the current study, the data were standardized using Eq. (1).

$$S = \frac{I_n}{I_{\max}} \tag{1}$$

where S is the standardized value. I_n is the value of the nth number, and $I_{\max}$ is the maximum value of each indicator obtained from all wards of KCC.

After the standardization of the data, the next step is to assign suitable weights to each of the indicators. There are several methods of weighting data, such as the Analytical Hierarchy Process (AHP), the Delphi Method, and Principal Component Analysis (PCA). In this study, the most widely used AHP method is adopted. AHP is a multi-criteria decision-making tool that is frequently used for analyzing complex decision-related problems in different fields. This is a pairwise comparison method for determining the preferred factors [14, 15]. The steps for undertaking AHP are shown in Fig. 4. As can be seen from the figure, there are a few steps to follow for the successful completion of AHP. The first step is the selection of indicators, which have been indicated earlier. The next step is to take an expert opinion survey. A scale

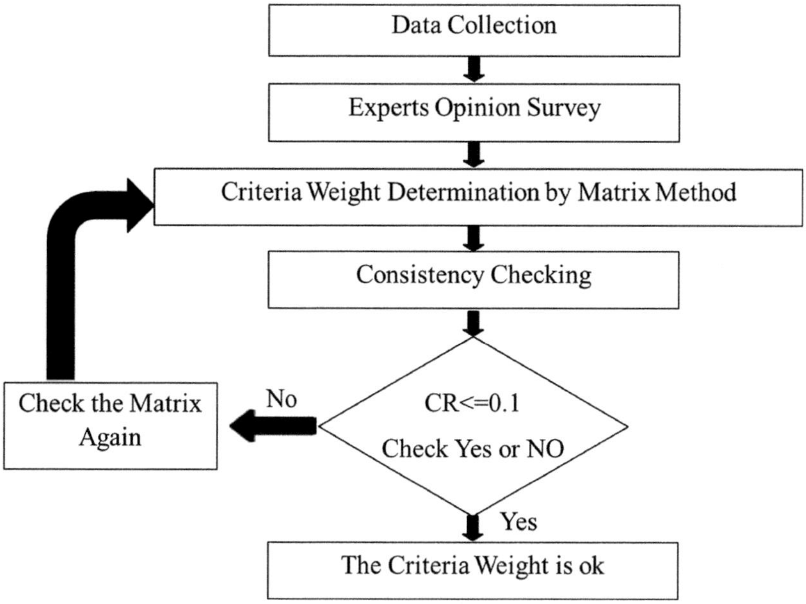

Fig. 4 Process involved in the AHP analysis technique

is used to mark the pairwise indicators. This scaling is accomplished by conducting a survey of experts in the relevant field. After the survey, a pairwise comparison matrix is formed to determine criteria weights. The final step is to check the consistency of the criteria's weight. If the consistency ratio is less than or equal to 0.10, the survey is considered consistent, and the evaluations do not require further repetitions [16]. If the consistency ratio exceeds 0.10, the decision-maker should repeat or change the analysis to improve consistency.

3.3 Development of Fuzzy Rule Based Model

In the current study, the waterlogging vulnerability index is calculated using fuzzy logic. A fuzzy rule-based model was explored by Zadeh in 1965 [6, 9]. The fuzzy method had been developed for analyzing uncertainties. In the case of the hydrological events, a model that can organize human knowledge and connect it to other data sources, such as linguistic information, is required [17, 18]. There are two types of fuzzy models, such as Mamdani FIS and Sugeno FIS. The Mamdani FIS has been selected in this study for the calculation of the waterlogging vulnerability index due to its worldwide use and acceptance. Moreover, it is simpler to comprehend than the Sugeno FIS [17]. The steps involved in the Mamdani FIS are defining fuzzy sets,

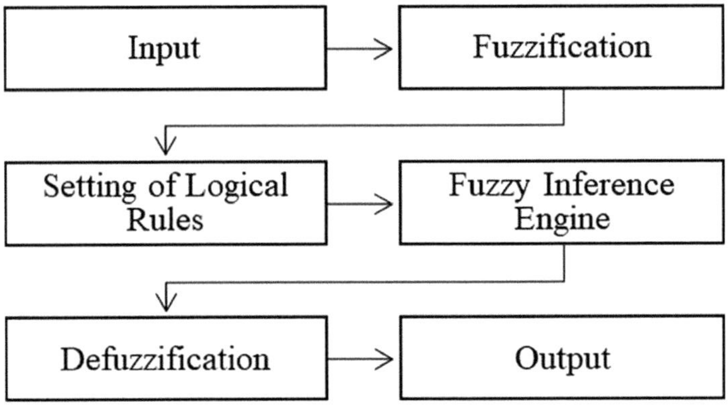

Fig. 5 Process involved with the Mamdani FIS

selecting membership functions, fuzzification logical rule setting (if–then), fuzzy inference systems, and de-fuzzification, which are shown in Fig. 5.

Fuzzy sets are sets that have degrees of membership. It indicates that the set in Boolean logic can only be defined as either 0 or 1, but in the case of fuzzy logic, the set can be any value between 0 and 1. In this study, triangular and trapezoidal membership functions are used to generate three fuzzy sets, such as low, medium, and high. This particular membership function is chosen because it gives more accurate and error-free results than other functions. In fuzzification steps, crisp input is assigned to fuzzy input according to membership functions. Since we have used if-else rules, we are familiar with Boolean logic. However, fuzzy logic follows an if–then style for its rules. For this study, it can be stated as follows: "If the inundation level is high, then the vulnerability is high." Then output is given based on rules generated by the inference engine. As indicated earlier, the data that we collected were clustered into three groups, such as exposure, susceptibility, and resilience. Three different fuzzy models were developed for these three different groups. Then a final fuzzy model is developed to determine the final water logging vulnerability index, taking the three fuzzy models into consideration, which is represented in Fig. 6.

Finally, in the defuzzification steps presented in Fig. 7, a unique value of output is shown considering all the outputs from the inference engine. The collected data have been clustered into three groups: exposure, susceptibility, and resilience. Three different fuzzy models are developed for these three indicators. As illustrated in Fig. 7, a final fuzzy model has been developed at the end to determine the water logging vulnerability index, taking the three fuzzy models into consideration.

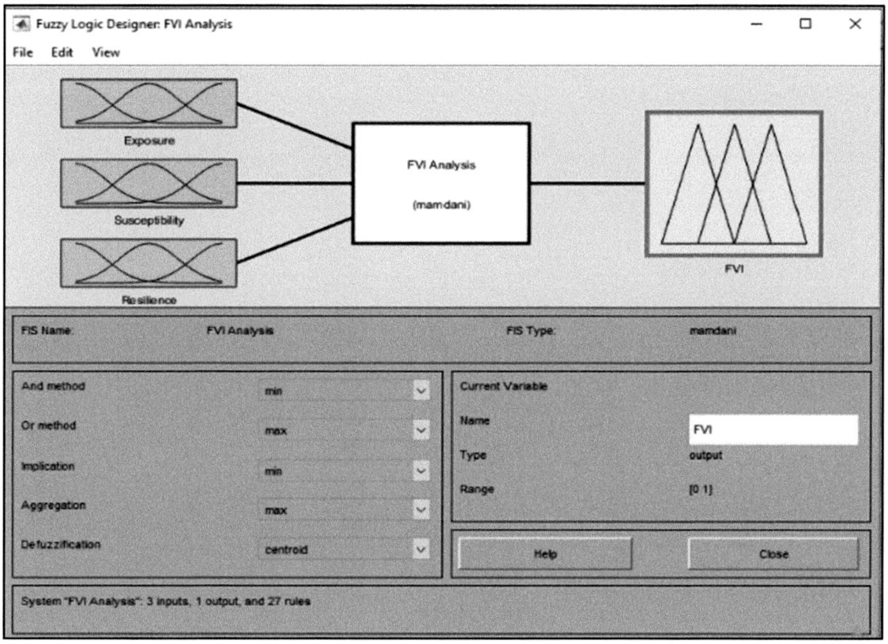

Fig. 6 Final fuzzy model considering exposure, susceptibility and resilience

4 Results and Discussion

Based on the AHP analysis, a definite weight for the criteria of the indicators has been determined. The consistency of the criteria is an important factor. We have used the pairwise comparison matrix method to determine the criteria weight. The consistency ratio of the matrix must be less than or equal to 0.1. We determined individual criteria weights for an individual expert's opinion survey. Then consistency has been checked through the matrix. Those opinions were taken into consideration, and their consistency ratio has been found to be per rules. After that, the average criteria weight of each indicator has been determined. The average maximum eigenvalue for hydrological analysis has been found to be 6.68, and the criteria weights of indicators such as the flooded area ratio, annual rainfall, number of months having heavy rainfall, slope, permeable ratio, and drainage density are 0.18, 0.15, 0.19, 0.14, 0.15, and 0.20, respectively. Likewise, for social and economic analysis, the average maximum eigenvalues are found to be 7.85, and 6.60, respectively.

The criteria weights of social indicators such as people from 0–9, people from 60+, disabled people, floating people, illiterate people, and low-income people are 0.089, 0.093, 0.266, 0.236, 0.088, and 0.177. The criteria weights of economic indicators such as kutcha and jhupri households, sanitation facilities, institutional and other households, unemployed people, non-established people, and electricity connections are found to be 0.298, 0.293, 0.112, 0.087, 0.128, and 0.083, respectively. After

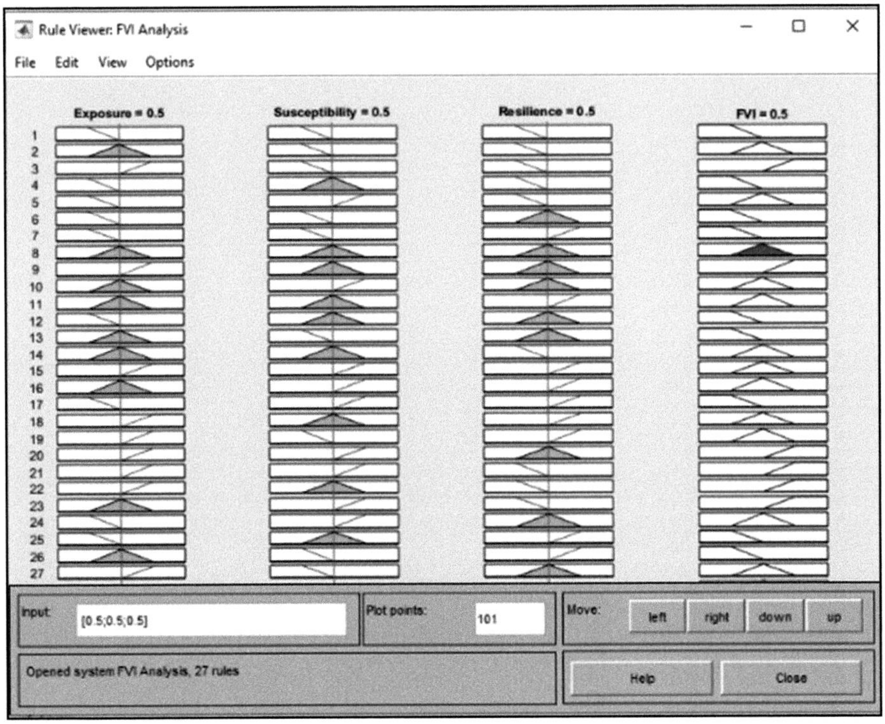

Fig. 7 Estimating water logging vulnerability index (WVI) by de-fuzzification

weighting, all the data were given as input to the fuzzy inference system, which was developed using the fuzzy toolbox in the MATLAB platform. After the calculation through fuzzy logic, all types of vulnerability indices, including SVI, EVI, and HVI, are aggregated using the arithmetic mean method in order to determine the final water logging vulnerability index. The result of the vulnerability index is then further classified into three categories, such as 'high', 'medium', and 'low'. The result is then interpreted by the spatial GIS mapping in the ArcGIS platform. The result of the analysis is presented in Fig. 8.

In the case of social vulnerability analysis, it has been found that nearly 20% of total areas are in highly vulnerable classes. On the other hand, in cases of economic vulnerability, the results range from low to medium. From hydrological analysis, 20% of total areas of the KCC have been identified as high-risk areas. Then, aggregating all the vulnerability indexes, the final vulnerability shows that 20% are in the high-water logging risk areas, 42% areas lie in the medium vulnerable class, and the remaining parts are in the low categories. Flat slopes and drainage congestion are the main factors that make them highly vulnerable to waterlogging according to hydrologic analysis. In the case of social analysis, floating people, the elderly population, and low-income people are the main reasons for vulnerability.

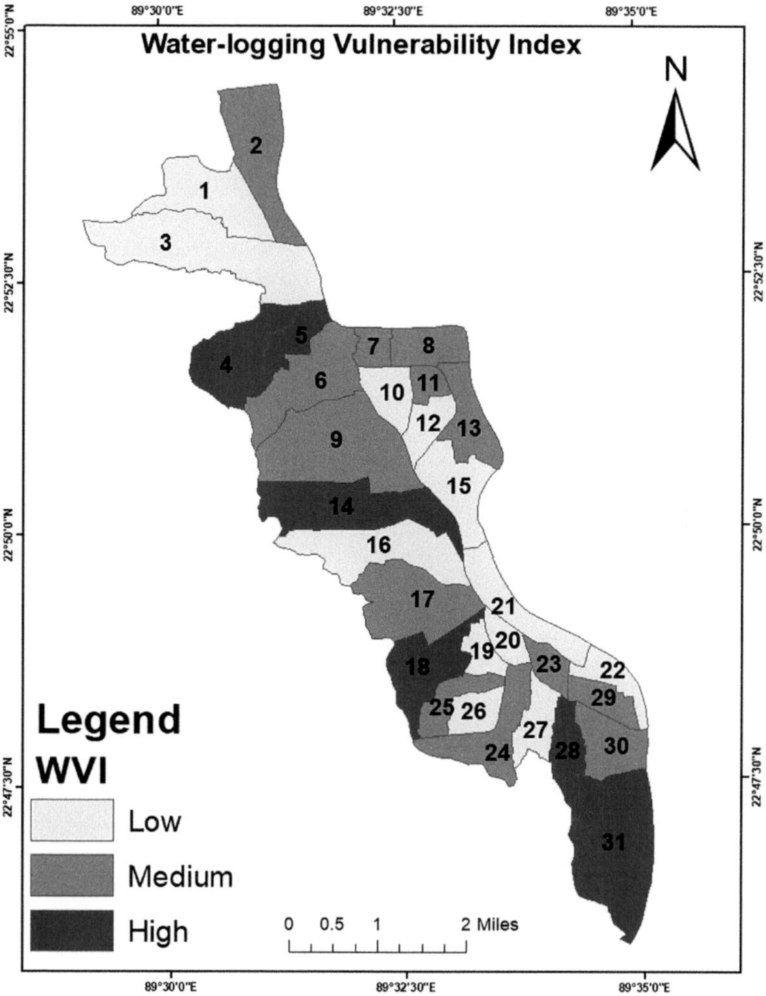

Fig. 8 Ward-wise delineation of water logging risk zones in the study area

5 Conclusions

An index-based approach is chosen in this study for exploring the water-logging vulnerability in an urban area such as the Khulna City Corporation (KCC) because it is a simple and practical tool for water management and policy makers. This study uses the AHP-fuzzy rule-based approach to assess the water logging vulnerability index (WVI) and identify the water-logging risk areas in the KCC. For the analysis, various hydrological, social, and economic data are employed. The fuzzy logic toolbox in the MATLAB platform is used to create the fuzzy model, and then the vulnerability index of each component is determined for each ward of the KCC, taking exposure, susceptibility, and resilience into account. It is found from the analysis that 6 wards, including ward nos. 4, 5, 14, 18, 28, and 31, are located in high-risk zones both socially and hydrologically. It is also found from the study that approximately 38% of the KCC area is identified as being in the low vulnerable class, and about 42% of the KCC area falls in the medium vulnerable class. Based on the findings obtained in the current study, it can be recommended that the relevant government authorities, such as KCC, Khulna Development Authority (KDA), and Bangladesh Water Development Board (BWDB), along with different national and international non-government organizations (NGO) such as WaterAID, Nobolok, and Practical Action, should undertake necessary actions and appropriate measures to maintain the undisturbed natural flows in the drainage networks.

References

1. Haque MM, Islam S, Sikder MB, Islam MS (2022) Community flood resilience assessment in Jamuna floodplain: a case study in Jamalpur District, Bangladesh. Int J Disaster Risk Reduct 72:102861. https://doi.org/10.1016/j.ijdrr.2022.102861
2. Aktar MA, Shohani K, Hasan MN, Hasan MK (2021) Flood vulnerability assessment by flood vulnerability index (FVI) method: a study on Sirajganj Sadar Upazila. Int J Disaster Risk Manag 3(1):1–13. https://doi.org/10.18485/ijdrm.2021.3.1.1
3. Anisha NF, Hossain S (2014) A case study on water logging problems in an urban area of Bangladesh and probable analytical solutions. In: Proceedings of the 2nd international conference on advances in civil engineering 2014 (ICACE 2014), 26–28 December 2014, CUET, Chittagong, Bangladesh
4. Khan MMH (2017) Analysis of causes and impact of water logging in Khulna city of Bangladesh. In: Proceedings of the international conference on mechanical, industrial and materials engineering (ICMIME 2017), 28–30 December 2017, RUET, Rajshahi, Bangladesh
5. Sarkar SK, Rahman MA, Esraz-Ul-Zannat M, Islam MF (2018) Simulation-based modeling of urban waterlogging in Khulna City. J Water and Clim Change 12(2):566–579. https://doi.org/10.2166/wcc.2020.256
6. Rashetnia S, Jahanbani H (2021) Flood vulnerability assessment using a fuzzy rule-based index in Melbourne, Australia. Sustain Water Resour Manag 7. https://doi.org/10.1007/s40899-021-00489-w
7. Balica SF, Douben N, Wright NG (2009) Flood vulnerability indices at varying spatial scales. Water Sci Technol 60(10):2571–2580. https://doi.org/10.2166/wst.2009.183
8. Balica SF, Wright NG, van der Meulen F (2012) A flood vulnerability index for coastal cities and its use in assessing climate change impacts. Nat Hazards 64:73–105. https://doi.org/10.1007/s11069-012-0234-1
9. Zadeh LA (1997) Toward a theory of fuzzy information granulation and its centrality in human reasoning and fuzzy logic. Fuzzy Sets Syst 90(2):111–127. https://doi.org/10.1016/S0165-0114(97)00077-8
10. Yazdi J, Neyshabouri SAAS (2012) Assessing flood vulnerability using a rule-based fuzzy system. Water Sci Technol 66(8):1766–1773. https://doi.org/10.2166/wst.2012.346
11. ADB (2010) Strengthening the resilience of the water sector in Khulna to climate change. Consultants' Reports, August 2010, Project Number 42469-012, The Asian Development Bank (ADB)
12. BBS (2011) Population and housing census 2011. Bangladesh Bureau of Statistics (BBS), Bangladesh

13. Wang Y, Li Z, Tang Z, Zeng G (2011) A GIS-based spatial multicriteria approach for flood risk assessment in the Dongting Lake Region, Hunan, Central China. Water Resour Manag 25(13):3465–3484. https://doi.org/10.1007/s11269-011-9866-2
14. Zou Q, Zhou J, Zhou C, Song L, Guo J (2013) Comprehensive flood risk assessment based on set pair analysis-variable fuzzy sets model and fuzzy AHP. Stoch Env Res Risk Assess 27:525–546. https://doi.org/10.1007/s00477-012-0598-5
15. Fedeski M, Gwilliam J (2007) Urban sustainability in the presence of flood and geological hazards: the development of a GIS-based vulnerability and risk assessment methodology. Landsc Urban Plan 83(1):50–61. https://doi.org/10.1016/j.landurbplan.2007.05.012
16. Saaty TL, Vargas LG (2012) Methods, concepts and applications of the analytic hierarchy process, 2nd edn. Springer, New York, USA
17. Sen Z (2010) Fuzzy logic and hydrological modeling, 1st edn. CRC Press, Taylor & Francis, London
18. Zimmermann HJ (2010) Fuzzy set theory. WIREs Comput Statist 2(3):317–332. https://doi.org/10.1002/wics.82

Microplastics in River Sediments Around the Dhaka City: A Case Study for Occurrence and Quantification

M. R. Karim, M. T. Islam, M. S. Islam, H. T. Oishi, and Z. Tasnim

Abstract Microplastics (MPs) pollution has gained global attention as an emerging environmental issue considering ubiquity, chemical inertness and adverse impact. Studies on MPs are mostly concentrated on the marine environment and there are very limited studies on MPs in freshwater river systems. To understand the distribution of MPs in urban rivers, this study has investigated the riverbed sediments in the peripheral rivers of Dhaka city. For the purposes, riverbed sediment samples were collected from 28 locations considering land use characteristics of the surrounding Buriganga, Turag and Balu Rivers. Riverbed sediments at sampling location were sampled from the top 10 cm sediment layer using an Ekman grab sampler and then placed into aluminum foil bags in order to avoid further contamination. MPs were extracted from the samples following a modified NOAA methodology. MPs were detected using Stereo Zoom Microscope and mass of microplastics were measured by using a digital balance. The mass concentrations of MPs particles were found to vary from 13.56 mg/kg to 430.65 mg/kg with an average value of 98.30 ± 70.72 and median value of 86.37 mg/kg in the river sediment. Comparatively higher MPs concentrations were found in highly urbanized and populated Buriganga river (Mean: 118.04 ± 108.46 mg/kg) than the Turag (Mean: 88.26 ± 31.04 mg/kg) and Balu River (Mean: 72.76 ± 33.35 mg/kg). This study gives an idea of the degree of MPs pollution

M. R. Karim (✉) · M. S. Islam · H. T. Oishi · Z. Tasnim
Department of Civil and Environmental Engineering, Islamic University of Technology, Gazipur, Bangladesh
e-mail: rezaulmd@iut-dhaka.edu

M. S. Islam
e-mail: saifulislam2@iut-dhaka.edu

H. T. Oishi
e-mail: humairatasnim@iut-dhaka.edu

Z. Tasnim
e-mail: zarintasnim@iut-dhaka.edu

M. T. Islam
Department of Civil and Environmental Engineering, West Virginia University, Morgantown, USA
e-mail: mi00018@mix.wvu.edu

S. Arthur et al. (eds.), *Proceedings of the 6th International Conference on Advances in Civil Engineering*, Lecture Notes in Civil Engineering 368,
https://doi.org/10.1007/978-981-99-3826-1_9

in the urban rivers surrounding of Dhaka city. This study indicated that microplastics pollution in these urban river systems could be an alarming situation and measures must be undertaken to control the MPs pollution of these rivers.

Keywords Microplastic · Freshwater · Urban river sediment · Mass concentration

1 Introduction

Plastic has become an increasingly substantial product with the advancement of technology and is being used extensively in many spheres of life [1]. Since the middle of the twentieth century, manufacturing and use of plastic have been on an upward trajectory [2]. About 5 billion tons of plastic have been manufactured globally as of 2015, and by 2050, it is predicted that at least 40 billion tons will be produced [3]. Plastic debris is prevalent due to the massive amount of production, inadequate management, as well as lack of reuse [4]. Plastic waste less than 5 mm in size is considered as microplastics (MPs). MPs are found in every place of earth including sea, surface water, sediment, soil and living creatures. Microplastics were discovered in 94 percent of all surface samples in the northeast Atlantic [5]. In the year of 2016, daily microplastic loads of American rivers ranged from 3–23 billions of particles [6].

MPs are incredibly mobile in aquatic settings due to longevity, low weight and insolubility [7]. Aquatic species from various trophic levels and diverse feeding strategies mistreat and consume them as food [7]. Marine species, seabirds, and mammals all may consume microplastics; once they amass in the food chain, they eventually find their way into the bodies of top predators like humans [8]. Even microplastics can influence the life cycle of benthic organism by amending bulk density, water holding capacity and other physical properties of the riverbed sediments [9]. Since MPs are unique in size with large surface area and stable in characteristics, microplastics serve as both breeding ground for microbes and vehicle for contaminants. Moreover, plastic products can be incorporated with various types of performance enhancing additives such as plasticizers, flame retardants and other compounds during production [10]. Microplastics may leak these compounds into the environment, posing intricate ecological hazards [8]. Various health problems including cancer may result from the leaching of these contaminants into living things tissues [11].

It is well known that MPs are commonplace in maritime environment [12]. However, ecosystems in freshwater play a significant part in the movement of microplastics. Around 70 to 80 percent of the MPs in the ocean migrated from land via river [5]. Furthermore, MPs can be redistributed, stored, processed and absorbed in the riverbed sediment [13]. MPs distribution patterns in the riverine environment may differ from those in marine environments because they are often smaller and closer to sources [14]. Although the contamination from microplastics in many freshwaters is severe or worse than other in marine waters, the scientific community has paid little attention to the MPs pollution in the freshwater system [15]. From the existing literature, Maheswaran et al. [16] identified, quantified, and characterized MPs from 14 points of Kaveri river, South India [17]. Investigated the seasonal variation in the distribution of MPs in the riverbed sediment of Liangfeng River [18]. Evaluated the MPs concentration and their correlation with biological oxygen demand (BOD) at 29 rivers of Japan [19]. Performed a study to observe distribution and traits of MPs in Chishui River water column. However, more studies on freshwater MPs at diverse part of the world is required to reduce the margin of current knowledge gap.

Rivers of Bangladesh should always be of significant interest of the researchers due to its geographic location. It is situated at the largest river delta and receives discharge from the largest sediment producer mountains, Himalayas. Considering the present economic boost and industrialization, rivers of Bangladesh should pose high risk of microplastic pollution. Surprisingly, very few experiments on microplastics, mostly focusing on beach samples [20] have been performed in Bangladesh [21]. Investigated the occurrence of MPs in the riverine environment of Bangladesh through literature review. This study aims to investigate the state of MPs abundance and concentration in the riverine freshwater systems around Dhaka city. The findings of this research can be utilized to label emergent point and non-point sources of microplastic pollutants, improve the existing database on microplastic contamination and serve as helpful sources for future studies.

2 Materials and Methods

2.1 Study Area and Selection of Sampling Points

Dhaka is one of the busiest capitals in the world with 7000 industries that release wastewater into the surrounding waterways everyday [22]. Riverbed sediment samples were collected from 28 sampling points of the riverine settings around Dhaka city, that covers three major rivers: the Buriganga (BR), Turag (TR), and Balu (BaR), and one canal- Tongi Canal (ToC). The sampling locations are distributed along 74 km of length (BR: 23, TR: 23, BaR: 15 ToC: 21) across the rivers and canals. Our study area includes three sub-basins as shown in Fig. 1 that combinedly envelops 1005.95 Km^2 (BR: 251.50 Km^2, TR and ToC: 290.56 Km^2, BaR: 463.88 Km^2). The sub-basins were developed from the Digital Elevation models (DEMs) collected from US Geological Survey (USGS) using Arc Hydro tools. Since Turag River and Tongi canal share the same sub-basin, TR and ToC was considered as a single entity in this study.

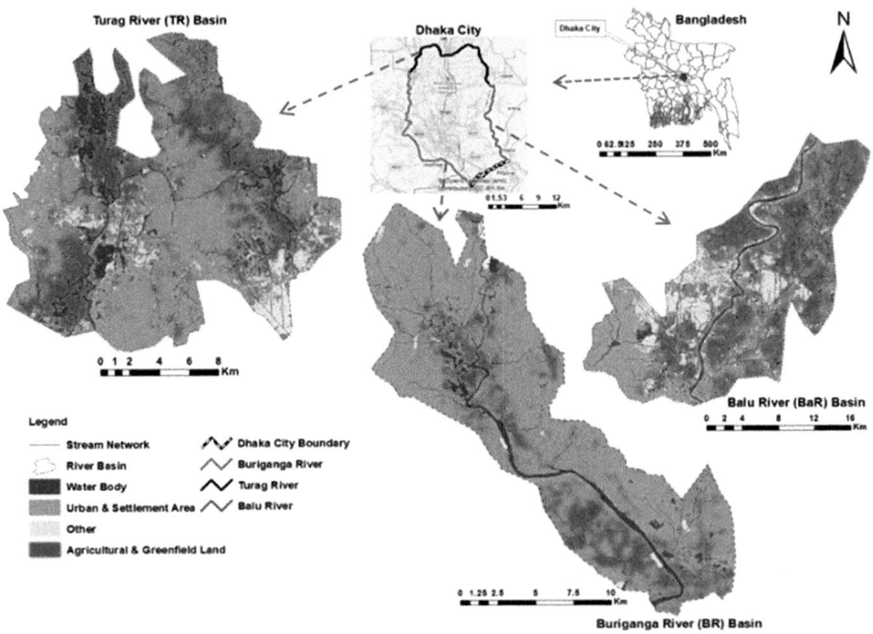

Fig. 1 Study area and corresponding river basins

Selection of sampling points were done focusing on the land-use patterns and pollution sources rather than length or catchment area. Thus, only 3 sampling points were chosen from Balu River despite being the largest basin among the three rivers. Buriganga and Turag (TR and ToC) had 11 and 14 sampling stations, respectfully. Land-use features and pollution sources information were extracted from Google Earth and USGS (Fig. 1). Figure 1 reveals that Buriganga river basin are mostly urbanized and influenced by point sources, whereas Balu river basin has complete inverse features in terms of land use characteristics. Turag river basin comprises both point (urban) and non-point (agricultural and green fields) sources. More details on the study area can be found in Table 1.

2.2 *Sample Collection*

About 1.0 kg of sediment was collected from each sampling point from a depth of 10 cm. Sediments were picked from the center of the river using an Ekman grab sampling tool with a dimension of 15 cm in each direction. Samples were placed into aluminum foil bags for storage and transfer to lab for subsequent details analysis.

Table 1 GPS Coordinates and key features of the sampling stations

SI. No	River	Assigned ID	GPS coordinates	Key features/Remarks	Source type
01	Buriganga	BR1	23.664859° N, 90.452726° E	Downstream of Pagla STP (Commercial and Industrial Area)	Point
02	Buriganga	BR2	23.679284° N, 90.439381° E	Upstream of Pagla STP (Commercial and Industrial Area)	Point
03	Buriganga	BR3	23.690331° N, 90.425369° E	Faridabadh Residential Area, Buriganga Bridge near Fatullah, Postogola Govt. Modern Flour Mill	Point
04	Buriganga	BR4	23.699379° N, 90.417263° E	Down Stream of Sadar Ghat, Sluice gate near Buriganga River, FarashGanj Bridge	Point
05	Buriganga	BR5	23.709795° N, 90.401527° E	Commercial area, Sawari Ghat, Babu Bazar, Sir Salimullah Medical College	Point
06	Buriganga	BR6	23.710026° N, 90.390533° E	Downstream of Sultanganj Residential and Commercial area, Kamrangi char	Point
07	Buriganga	BR7	23.740877° N, 90.351158° E	Boshila Residential Area, Bangladesh eye trust Hospital	Point
08	Buriganga	BR8	23.751012° N, 90.330164° E	Boshila Residential Area, Brick Firm	Point
09	Buriganga	BR9	23.769175° N, 90.344659° E	Residential Area: Baitul Aman Housing Society, Sunibir Housing	Point
10	Buriganga	BR10	23.777804° N, 90.337171° E	Mixed Area: Downstream of Gabtoli Sweeper Colony, BIWTA Landing station	Both
11	Buriganga	BR11	23.783858° N, 90.335702° E	Mixed Area: Gabtoli Cattle Market, Amin Bazar landing station, Gabtoli Bridge (Mix Zone Area)	Point
12	Turag	TR1	23.786274° N, 90.338190° E	Mixed Area: Golaptak mix zone Area, Boro Bazar, Boro Bazar Ghat	Both

(continued)

Table 1 (continued)

SI. No	River	Assigned ID	GPS coordinates	Key features/Remarks	Source type
13	Turag	TR2	23.799784° N, 90.343166° E	Residential & Homestead Plants area: Turag City, Bangladesh national Zoo, BIWTA, Landing Station, Diabari Boat Yard	Both
14	Turag	TR3	23.826958° N, 90.342968° E	Residential and Planted Garden: Eastern Housing, Botanical Garden, Tamanna Family Park, S4 Sluice Gate	Both
15	Turag	TR4	23.854818° N, 90.341898° E	Downstream of Rajuk Residential area (Effect of Ashulia Industrial Area)	Both
16	Turag	TR5	23.890476° N, 90.359335° E	Ashulia Ferry Ghat, Ashulia Landing station, Ashulia Bus Stop (Effect of Ashulia Industrial Area)	Both
17	Turag	TR6	23.893522° N, 90.362811° E	Industrial and Residential Area (Jamaldia, Tongi)	Both
18	Turag	TR7	23.898066° N, 90.383805° E	Industrial and residential Area, kathaldia Ghat, Greenland hospital	Both
19	Turag	TR8	23.880292° N, 90.393299° E	Industrial, Residential and Hospital Area: Abdullah Sluice Gate, Shaheed Mansur Ali Medical college & Hospital, Tongi Bishwa Ejtema Mydan, Near Uttara Sector 11	Point
20	Turag	TR9	23.881708° N, 90.405556° E	Mixed Zone: Tongi Bridge, Sawdagar Stone Mill, Arichpur	Point
21	Turag	TR10	23.886367° N, 90.416720° E	Effect of Industrial Area, Tongi Nodi Bondor	Point
22	Turag	TR11	23.898398° N, 90.435431° E	Effect of Industrial Area, Radix Garments	Both

(continued)

Table 1 (continued)

SI. No	River	Assigned ID	GPS coordinates	Key features/Remarks	Source type
23	Turag	TR12	23.861912° N, 90.474911° E	Agricultural land & Open Plot for Future Development	Non-Point
24	Turag	TR13	23.837315° N, 90.477250° E	Effect due to construction work of 300 ft Purbachal Road: Boalia Bridge, Balu River, Purbachal Express Highway	Both
25	Turag	TR14	23.796113° N, 90.481048° E	Beraaid Residential Area, Agar Para Mosjid, A K H Rahmatullah Stadium,	Both
26	Balu	BaR1	23.762079° N, 90.482599° E	Open Area for Future Development, Rampura Khal,	Both
27	Balu	BaR2	23.727584° N, 90.500133° E	Mixed: Demra Residential Area, Chanpara Bus Stand, Chanpara Bazar	Point
28	Balu	BaR3	23.718460° N, 90.499591° E	Confluence of Shitalakshya and Balu River, Karim Jute Mill	Both

2.3 *Sample Processing, Laboratory Analysis, and Visual Inspection*

Modified NOAA methods was adopted for the laboratory analysis of the sediment samples [23]. Necessary adjustment in the standard NOAA methods was made based on procedures described by [24–26]. In short, sediment samples were first oven dried to a constant weight at 90 ° C and then the particles outside the range of 0.3–5 mm in size were discarded through the customized sieves. Organic particles were digested at 60 ° C through wet peroxidation (WPO) technique. Wet Peroxidation was performed using Fenton's reagent, a mixture of 20-mL $FeSO_4.7H_2O$ solution and 20 mL of 30% H_2O_2 solution. Two steps density separation with aqueous Zinc Chloride ($ZnCl_2$) (density 1.5 $g.cm^{-3}$) solution was employed to eliminate clay and other inorganic particles. First step of density separation preceded WPO, whereas the final step succeeded. Extracted particles were visually verified under an OPTIKA SLX-3 Stereo Zoom microscope at numerous magnifications. Protocols set by NOAA [23] and literature reviewed by [27] were followed for the quality control measures during the entire experiment.

3 Results and Discussions

Total MPs count was found 4190 particles from 28 samples. The abundance of microplastics varies from 46 to 534 items per kg (nos./kg) of dry sediment with a mean value of 149.64 ± 83.7 nos./kg. Figure 2 shows the mean, median, and standard deviation of MPs count of the three rivers. Buriganga (165.45 ± 127.87) and Turag River (142.43 ± 37.32) possessed the highest and lowest microplastic abundance, respectfully. Buriganga river mostly receives water from point sources (drainage outfalls, industrial disposal, etc. (Table 1). Thus, our study findings advocate the hypothesis that point sources contribute much to the MPs pollution in contrast to non-point sources, which is consistent with the previous literature [28, 29].

The mass concentration of MP particles ranged from 13.56–430.65 mg/kg with a mean value of 106.52 ± 73.17 mg/kg. Figure 3 shows that Buriganga river (127.13 ± 106.85 mg/kg) also has the highest mass concentration of MPs, followed by Turag river (97.55 ± 43.24 mg/kg) and Balu river (72.76 ± 33.35 mg/kg).

Highest number and mass concentration of MPs (534 nos./Kg, 430.65 mg/kg) were found at sampling point BR4 (Table 2), the downstream of Buriganga river. Sluice gate near BR4 can be one of the possible causes of high occurrence of MPs pollution at this station. Lowest MPs abundance and concentration was found at TR12 (46 nos./kg, 13.56 mg/kg) as the adjacent land of TR12 is mostly empty and grassland. However, BR1 (112 nos./kg; 55.38 mg/kg), TR9 (158 nos./kg; 63.71 mg/kg), and BaR2 (120 nos./kg, 63.27 mg/kg) had higher particles count but lower mass concentration as opposed to BR2 (94 nos./kg; 82.74 mg/kg), BR8 (64 nos./kg; 155.27 mg/kg), and TR14 (90 nos./kg; 203.23 mg/kg) (Table 2). Apart from abundance, mass concentration is largely dependent on the density of the MPs particles; thus, this type of observation is justifiable (Fig. 3).

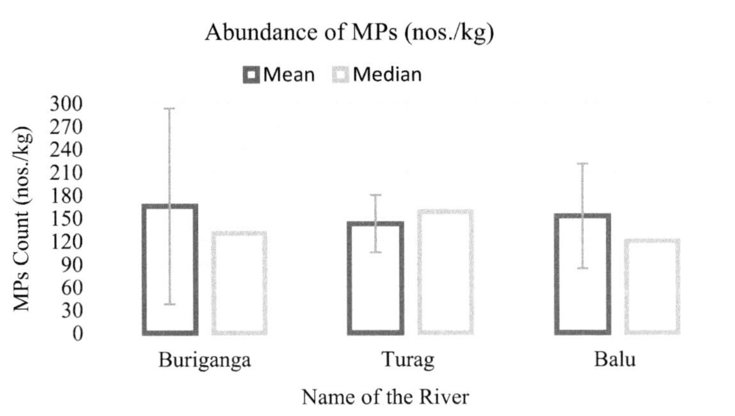

Fig. 2 Abundance of microplastics (nos./kg)

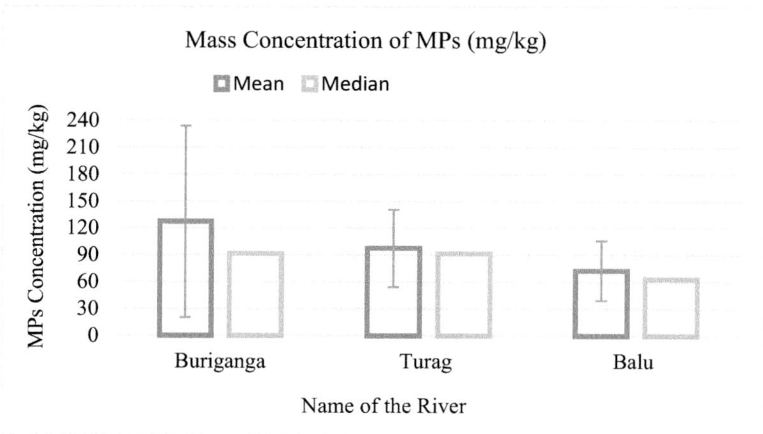

Fig. 3 :Mass concentration of MPs (mg/kg)

The outcomes of this research were compared with the published data of microplastics occurrences in the freshwater sediments, the comparison is provided in Table 2. It reveals that all the three-rivers in this study possessed higher MPs abundance than the Karnaphuli river in Bangladesh. However, Amazon rivers (range: 417–817 n/kg), St. Lawrence River (range: 65–7562 n/kg; mean: 832 $\pm 150n/kg$ and Tisza River (mean: 3177 $\pm 170n/kg$) experienced significantly severe microplastics pollution than the rivers of this study (Table 3).

Table 2 The abundance and mass concentrations of microplastics (MPs) in different sediment sampling points

Locations	Abundance (nos./kg)	Mass concentration (mg/kg)	Locations	Abundance (nos./kg)	Mass concentration (mg/kg)	Locations	Abundance (nos./kg)	Mass concentration (mg/kg)
BR1	112	55.38	BR11	122	59.82	TR9	158	63.71
BR2	94	82.74	TR1	178	107.69	TR10	172	130.29
BR3	122	132.5	TR2	166	92.37	TR11	138	87.39
BR4	534	430.65	TR3	142	84.31	TR12	46	13.56
BR5	214	156.85	TR4	160	107.27	TR13	104	56.34
BR6	160	85.34	TR5	158	135.93	TR14	90	203.23
BR7	130	93.02	TR6	154	90.49	BaR1	60	45.19
BR8	64	155.27	TR7	166	104.83	BaR2	120	63.27
BR9	130	76.31	TR8	162	88.29	BaR3	196	109.83
BR10	138	70.56						

Table 3 Global microplastics pollution in comparison with this study

Study areas	Abundance (nos./kg) (d.w.)	Concentration (mg/kg)	References
Buriganga River, Bangaldesh	65–534^r, 165.45 ± 127.87^m	55.38–430.65^r	This study
Turag River, Bangladesh	46–178^r, 142.43 ± 37.32^m	13.56–203.23^r	This study
Balu River, Bangladesh	60–196^r, 152.33 ± 68.16^m	45.19–109.83^r	This study
Karnaphuli River, Bangladesh	22.29–59.5^r		[21]
Brisbane River, Australia	10–520^r	0.18–129.20^r	[30]
Ayaragi River, Japan	24–608^r	3.94–282.45^r	[25]
Yongfeng River, China	05–72^r, 26 ± 23^m	0.5–16.75^r, 5.17 ± 16.75^r	[31]
Daliao River, China	100–467^r, 237 ± 129^m		[32]
Amazon Rivers, Brazil	417–8178^r		[33]
Antuã River, Portugal	18–629^r	2.6–71.7^r	[34]
Atoyac River Basin, Mexico	33.33–400^r		[35]
Yushan River, China	30–70^r, 44 ± 18^m	3.5–53^r, 30.5 ± 23^m	[36]
St. Lawrence River, North America	65–7562^r, 832 ± 150^m		[37]
Tisza River, Central Europe	3177 ± 1970^m		[38]

r- indicates range (Min to Max).
m- indicates mean value.

4 Conclusion

This study findings reveal that the peripheral rivers of Dhaka city have been contaminated by MPs to a larger extend, highest abundance of MPs was found in Buriganga river followed by Turag and Balu river. The peripheral rivers' ecosystem is also impacted by MPs from anthropogenic point and non-point sources of the rivers' catchment. Most of the plastic wastes disposed into the open environment in Dhaka city is ultimately entered into water bodies, which are disintegrated by weathering and photolytic actions into MPs of various sizes and types and retains in riverbed sediment. Disposal of untreated industrial wastewater also caused MPs in the receiving water bodies. Thus, the legislative authorities of Bangladesh should implement strict regulatory measures for the uncontrolled disposal of domestic and industrial wastes and wastewater, enforcing new policy and measures for the growing amount of plastic wastes management, reuse and recycling and protect the water resources from increasing plastic and MPs pollution for environmental sustainability. More researches are required on the occurrence, ecotoxicity, food chain transfer and fate of microplastics in river environment. This research findings will provide useful insight to the policymakers to take protective measures and source control against the microplastic pollution of the urban rivers in Dhaka city.

Acknowledgements (if any): This study was supported by the Seed Grand of Islamic University of Technology. The authors also acknowledge the Department of Civil and Environmental Engineering, IUT for providing with the necessary laboratory supports for this study.

References

1. Di Mauro R, Kupchik MJ, Benfield MC (2017) Abundant plankton-sized microplastic particles in shelf waters of the northern Gulf of Mexico. Environ Pollut 230:798–809. https://doi.org/10.1016/j.envpol.2017.07.030
2. Kabir AHME, Sekine M, Imai T, Yamamoto K, Kanno A, Higuchi T (2021) Assessing small-scale freshwater microplastics pollution, land-use, source-to-sink conduits, and pollution risks: Perspectives from Japanese rivers polluted with microplastics. Sci Total Environ 768:144655. https://doi.org/10.1016/j.scitotenv.2020.144655
3. Zalasiewicz J, Waters CN, Ivar do Sul JA, Corcoran PL, Barnosky AD, Cearreta A, Edgeworth M, Gałuszka A, Jeandel C, Leinfelder R, McNeill JR, Steffen W, Summerhayes C, Wagreich M, Williams M, Wolfe AP, Yonan Y, (2016) The geological cycle of plastics and their use as a stratigraphic indicator of the Anthropocene. Anthropocene 13:4–17. https://doi.org/10.1016/j.ancene.2016.01.002
4. Browne MA, Galloway TS, Thompson RC (2010) Spatial patterns of plastic debris along estuarine shorelines. Environ Sci Technol 44:3404–3409. https://doi.org/10.1021/es903784e
5. Eriksen M, Lebreton LCM, Carson HS, Thiel M, Moore CJ, Borerro JC, Galgani F, Ryan PG, Reisser J (2014) Plastic pollution in the world's oceans: more than 5 trillion plastic pieces weighing over 250,000 tons afloat at Sea. PLoS ONE 9:e111913. https://doi.org/10.1371/journal.pone.0111913
6. Baldwin AK, Corsi SR, Mason SA (2016) Plastic debris in 29 great lakes tributaries: relations to watershed attributes and hydrology. Environ Sci Technol 50:10377–10385. https://doi.org/10.1021/acs.est.6b02917
7. Holland ER, Mallory ML, Shutler D (2016) Plastics and other anthropogenic debris in freshwater birds from Canada. Sci Total Environ 571:251–258. https://doi.org/10.1016/j.scitotenv.2016.07.158
8. Jiang C, Yin L, Li Z, Wen X, Luo X, Hu S, Yang H, Long Y, Deng B, Huang L, Liu Y (2019) Microplastic pollution in the rivers of the Tibet Plateau. Environ Pollut 249:91–98. https://doi.org/10.1016/j.envpol.2019.03.022
9. Adomat Y, Grischek T (2021) Sampling and processing methods of microplastics in river sediments—A review. Sci Total Environ 758:143691. https://doi.org/10.1016/j.scitotenv.2020.143691

10. Yan M, Nie H, Xu K, He Y, Hu Y, Huang Y, Wang J (2019) Microplastic abundance, distribution and composition in the Pearl River along Guangzhou city and Pearl River estuary, China. Chemosphere 217:879–886. https://doi.org/10.1016/j.chemosphere.2018.11.093
11. Nobre CR, Santana MFM, Maluf A, Cortez FS, Cesar A, Pereira CDS, Turra A (2015) Assessment of microplastic toxicity to embryonic development of the sea urchin Lytechinus variegatus (Echinodermata: Echinoidea). Mar Pollut Bull 92:99–104. https://doi.org/10.1016/j.marpolbul.2014.12.050
12. Avio CG, Gorbi S, Regoli F (2017) Plastics and microplastics in the oceans: From emerging pollutants to emerged threat. Mar Environ Res 128:2–11. https://doi.org/10.1016/j.marenvres.2016.05.012
13. Matsuguma Y, Takada H, Kumata H, Kanke H, Sakurai S, Suzuki T, Itoh M, Okazaki Y, Boonyatumanond R, Zakaria MP, Weerts S, Newman B (2017) Microplastics in sediment cores from Asia and Africa as indicators of temporal trends in plastic pollution. Arch Environ Contam Toxicol 73:230–239. https://doi.org/10.1007/s00244-017-0414-9
14. Eerkes-Medrano D, Thompson RC, Aldridge DC (2015) Microplastics in freshwater systems: A review of the emerging threats, identification of knowledge gaps and prioritisation of research needs. Water Res 75:63–82. https://doi.org/10.1016/j.watres.2015.02.012
15. Dris R, Gasperi J, Rocher V, Saad M, Renault N, Tassin B (2015) Microplastic contamination in an urban area: a case study in greater paris. Environ Chem 12:592. https://doi.org/10.1071/EN14167
16. Maheswaran B, Karmegam N, Al-Ansari M, Subbaiya R, Al-Humaid L, Sebastin Raj J, Govarthanan M (2022) Assessment, characterization, and quantification of microplastics from river sediments. Chemosphere 298:134268. https://doi.org/10.1016/j.chemosphere.2022.134268
17. Xia F, Yao Q, Zhang J, Wang D (2021) Effects of seasonal variation and resuspension on microplastics in river sediments. Environ Pollut 286:117403. https://doi.org/10.1016/j.envpol.2021.117403
18. Kataoka T, Nihei Y, Kudou K, Hinata H (2019) Assessment of the sources and inflow processes of microplastics in the river environments of Japan. Environ Pollut 244:958–965. https://doi.org/10.1016/j.envpol.2018.10.111
19. Li J, Ouyang Z, Liu P, Zhao X, Wu R, Zhang C, Lin C, Li Y, Guo X (2021) Distribution and characteristics of microplastics in the basin of Chishui River in Renhuai. China. Science of The Total Environment 773:145591. https://doi.org/10.1016/j.scitotenv.2021.145591
20. Tajwar M, Yousuf Gazi Md, Saha SK (2022) Characterization and spatial abundance of microplastics in the coastal regions of cox's bazar, bangladesh: an integration of field, laboratory, and GIS techniques. Soil Sediment Contam: Int J 31:57–80. https://doi.org/10.1080/15320383.2021.1910622
21. Rakib MRJ, Hossain MB, Kumar R, Ullah MA, Al Nahian S, Rima NN, Choudhury TR, Liba SI, Yu J, Khandaker MU, Sulieman A, Sayed MM (2022) Spatial distribution and risk assessments due to the microplastics pollution in sediments of Karnaphuli River Estuary. Bangladesh. Sci Rep 12:8581. https://doi.org/10.1038/s41598-022-12296-0
22. Roy S, Sowgat T, Mondal J (2019) City Profile: Dhaka, Bangladesh. Environ Urban ASIA 10:216–232. https://doi.org/10.1177/0975425319859126
23. Mausra B, Foster a laboratory methods for the analysis of microplastics in the marine environment. 39
24. Frias JPGL, Pagter E, Nash R, O'Connor I, Carretero O, Filgueiras A, Viñas L, J. Gago, Antunes JC, Bessa F, Sobral P, Goruppi A, Tirelli V, Pedrotti ML, Suaria G, Aliani S, Lopes C, Raimundo J, Caetano M, Palazzo L, Lucia GAD, Camedda A, Muniategui S, Grueiro G, Fernandez V, Andrade J, Dris R, Laforsch C, Scholtz-Bottcher B, Gerdts G (2018) Standardised protocol for monitoring microplastics in sediments. https://doi.org/10.13140/RG.2.2.36256.89601/1
25. Kabir AHME, Sekine M, Imai T, Yamamoto K, Kanno A, Higuchi T (2022) Microplastics in the sediments of small-scale Japanese rivers: Abundance and distribution, characterization, sources-to-sink, and ecological risks. Sci Total Environ 812:152590. https://doi.org/10.1016/j.scitotenv.2021.152590

26. Li J, Liu H, Paul Chen J (2018) Microplastics in freshwater systems: A review on occurrence, environmental effects, and methods for microplastics detection. Water Res 137:362–374. https://doi.org/10.1016/j.watres.2017.12.056
27. Yang L, Zhang Y, Kang S, Wang Z, Wu C (2021) Microplastics in freshwater sediment: A review on methods, occurrence, and sources. Sci Total Environ 754:141948. https://doi.org/10.1016/j.scitotenv.2020.141948
28. Dikareva N, Simon KS (2019) Microplastic pollution in streams spanning an urbanisation gradient. Environ Pollut 250:292–299. https://doi.org/10.1016/j.envpol.2019.03.105
29. Tibbetts J, Krause S, Lynch I, Sambrook Smith GH (2018) Abundance, distribution, and drivers of microplastic contamination in urban river environments. Water 10:1597. https://doi.org/10.3390/w10111597
30. He B, Goonetilleke A, Ayoko GA, Rintoul L (2020) Abundance, distribution patterns, and identification of microplastics in Brisbane River sediments. Australia Sci Total Environ 700:134467. https://doi.org/10.1016/j.scitotenv.2019.134467
31. Rao Z, Niu S, Zhan N, Wang X, Song X (2020) Microplastics in Sediments of River Yongfeng from Maanshan City, Anhui Province, China. Bull Environ Contam Toxicol 104:166–172. https://doi.org/10.1007/s00128-019-02771-2
32. Xu Q, Xing R, Sun M, Gao Y, An L (2020) Microplastics in sediments from an interconnected river-estuary region. Sci Total Environ 729:139025. https://doi.org/10.1016/j.scitotenv.2020.139025
33. Gerolin CR, Pupim FN, Sawakuchi AO, Grohmann CH, Labuto G, Semensatto D (2020) Microplastics in sediments from Amazon rivers. Brazil Sci Total Environ 749:141604. https://doi.org/10.1016/j.scitotenv.2020.141604
34. Rodrigues MO, Abrantes N, Gonçalves FJM, Nogueira H, Marques JC, Gonçalves AMM (2018) Spatial and temporal distribution of microplastics in water and sediments of a freshwater system (Antuã River, Portugal). Sci Total Environ 633:1549–1559. https://doi.org/10.1016/j.scitotenv.2018.03.233
35. Shruti VC, Jonathan MP, Rodriguez-Espinosa PF, Rodríguez-González F (2019) Microplastics in freshwater sediments of Atoyac River basin, Puebla City, Mexico. Sci Total Environ 654:154–163. https://doi.org/10.1016/j.scitotenv.2018.11.054
36. Niu S, Wang X, Rao Z, Zhan N (2021) Microplastics Present in Sediments of Yushan River: A Case Study for Urban Tributary of the Yangtze River. Soil Sediment Contam: Int J 30:314–330. https://doi.org/10.1080/15320383.2020.1841731
37. Crew A, Gregory-Eaves I, Ricciardi A (2020) Distribution, abundance, and diversity of microplastics in the upper St. Lawrence River Environ Pollut 260:113994. https://doi.org/10.1016/j.envpol.2020.113994
38. Kiss T, Fórián S, Szatmári G, Sipos G (2021) Spatial distribution of microplastics in the fluvial sediments of a transboundary river – A case study of the Tisza River in Central Europe. Sci Total Environ 785:147306. https://doi.org/10.1016/j.scitotenv.2021.147306

Uncertainty Based Assessment of Drought Using Standardized Precipitation Index-A Case Study

M. F. Rabby and S. K. Adhikary

Abstract Drought is a recurrent extreme hydrological event occurring almost every year in Bangladesh. Particularly, the northwest region of the country has been severely affected by the frequent occurrence of droughts. Standardized Precipitation Index (SPI) is the most widely used index for meteorological drought assessment because of its modest data requirements and ease of use. However, SPI's reliability in drought assessment has been questioned due to the presence of uncertainty in its computational process. Although past studies have identified various sources of uncertainty, only a few have explored their influence on drought calculation. Therefore, the aim of this study is to identify the sources of uncertainty in SPI based drought assessment and analyze their effects on drought characteristics. The current study is demonstrated through two selected climate stations, namely Bogra and Ishurdi, located in the northwest region of Bangladesh. Two probability distribution functions, such as gamma and log-normal distributions, are used to calculate SPI over 6-, 12-, and 24-month time scales. The maximum likelihood method is adopted to find parameters for the distribution functions, and Run's theory is employed to determine the drought characteristics. The results indicate that the uncertainty in the SPI calculation is greater at smaller time scales and usually decreases with the increase in time scale and length of data series. It is also evident from the results that the standard gamma distribution performs similarly to the log-normal distribution at larger time scales. This demonstrates that the log-normal distribution can be adopted as a viable alternative to the standard gamma distribution for drought assessment in the study area. The results also show that the uncertainty greatly affects extreme droughts; the higher the SPI value, the greater the chances of uncertainty. It is also found that the sources of uncertainty greatly impact the characteristics and categorization of the

---

M. F. Rabby (✉)
Department of Civil Engineering, Bangladesh Army University of Science and Technology (BAUST), Saidpur 5310, Bangladesh
e-mail: fazzlece@baust.edu.bd

S. K. Adhikary
Department of Civil Engineering, Khulna University of Engineering and Technology (KUET), Khulna 9203, Bangladesh
e-mail: sajal@ce.kuet.ac.bd

S. Arthur et al. (eds.), *Proceedings of the 6th International Conference on Advances in Civil Engineering*, Lecture Notes in Civil Engineering 368,
https://doi.org/10.1007/978-981-99-3826-1_10

drought. The current study thus concludes that a correct and reliable assessment of drought using SPI can be achieved by considering diverse sources of uncertainty and their impacts on the drought assessment.

Keywords Uncertainty · SPI · Drought · Maximum likelihood · Gamma distribution · Log-normal distribution

1 Introduction

Drought is a naturally occurring extreme hydrological event that is mainly caused by the scarcity of precipitation. Due to the high spatial and temporal variability of precipitation caused by climate change, drought has become an important concern for researchers all over the world. Drought is occurring recurrently almost every year in Bangladesh. Considering the country's unique geographical location, dense population, low income, and uncontrolled urbanization, as well as the high dependency of major income-generating sectors on precipitation patterns such as agriculture and fishing in the country, this is regarded as one of the most pressing disasters in Bangladesh. In recent years, there has been an increase in the frequency of droughts, and a significant number of droughts have occurred in the country since its independence in 1971 [1]. In particular, the northwest region of the country is prone to drought since the region usually receives very less precipitation than the rest of the country [1, 2]. The region regularly experiences meteorological drought, which is associated with the precipitation pattern and other climatic factors [1, 3]. It has been observed that natural droughts have impacted about 53% of people and around 47% of the nation in the past [4]. The effect of droughts may be minimized through drought preparation and mitigation strategies in which drought indices act as one of the vital factors and effective risk control techniques [5–7].

Drought indices are the simplest way to monitor drought conditions, as they can provide a numerical depiction of the beginning and end of a drought event along with its severity [8]. During the last several decades, numerous statistical approaches have been employed to quantify meteorological, agricultural, and hydrological droughts [9, 10]. Among them, the standardized precipitation index (SPI) is the most widely adopted index all over the world [11] to monitor precipitation-based meteorological droughts because it requires only precipitation records and can analyze droughts across different time periods [12, 13]. However, there are some questionable issues with how SPI is calculated. Several factors, including the length of the data series, the time scales, the probability distribution function used to fit the data, and the parameter estimation approach could be responsible for the induced uncertainty in drought assessment using SPI [13–17]. It has been understood that the mostly used Gamma distribution function should not be adopted extensively since other distribution functions often perform better than the standard Gamma distribution function. For example, [18] demonstrates that the Pearson Type-III distribution is the optimal fit for American precipitation data, whereas [19–21] found that the weibull, log-normal, and generalized-normal distributions are the most appropriate fits for European, Guadiana (Portugal), and Brazilian precipitation data, respectively [22]. Showed that the Gumbel distribution is best suited for maximum monthly precipitation in India, whereas [23] found the log-normal distribution to be the best-fitting distribution function for pre-monsoon precipitation in Bangladesh.

Variability in parameter estimation methods also induces uncertainty as it would yield different SPI values, which would eventually affect the drought estimation. Parameter stability and validity have been the focus of research in several published works [24]. Discovered that there is a non-linear relationship between the record length and the stability of parameter estimation. Moreover, it has been shown that the duration and diversity of precipitation data used to fit the probability distribution contribute to the uncertainty, and that the larger the variability of precipitation, the greater the uncertainty [25]. The World Meteorological Organization (WMO, 2012) suggests a minimum time series of 50 or 60 years for this purpose; nevertheless, a 30-year time series is regarded as sufficient for most purposes.

Even though all of the aforementioned research has greatly enhanced the uncertainty analysis in drought assessments, the majority of it focused on the implications of sources of uncertainty on the drought index itself and the failure to take into account their effects on the occurrences and characteristics of droughts. Although [13] considered the effects of uncertainty, their analysis was limited to a 12-month time scale and 55-year data period. To the best knowledge of the authors, there has not been any previous research conducted in Bangladesh that takes into consideration the consequences of uncertainty in drought estimation. Therefore, the objective of the current research is to evaluate the implications of uncertainty in drought assessment, particularly concerning the SPI values and the characteristics of drought.

2 Materials and Methods

2.1 *Study Area and Data Used*

The northwest region of Bangladesh is particularly drought prone and receives less rainfall compared to the other parts of the country. Thus, the current study is mainly focused on the two important stations, namely Bogra and Ishurdi, in this region, which are shown in Fig. 1. The area includes very pronounced seasonal shifts and has high temperatures, moderate rainfall, and typically considerable humidity. Precipitation amounts vary both geographically and seasonally in this region. In the study area, the highest mean annual rainfall is recorded at about 1743 mm in Bogra and 1545 mm at Ishurdi.

Since SPI-based drought estimation requires only the precipitation data, the precipitation data for the Bogra and Ishurdi stations are collected from the Bangladesh Meteorological Department (BMD) for a period of 45 years, from 1975 to 2019. The details of the two aforementioned stations are given in Table 1.

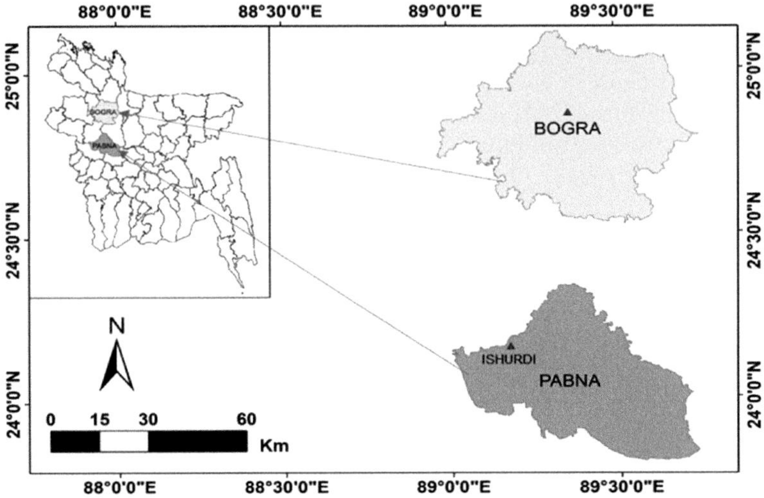

Fig. 1 The study area with the precipitation stations

Table 1 Details of the BMD climatic stations in the study area

Station	Area covered (km^2)	Latitude (Deg.)	Longitude (Deg.)	Data used	Data period
Bogra	2898.68	24.88	89.39	Precipitation	1975–2019
Ishurdi	2376.13	24.12	89.04	Precipitation	1975–2019

2.2 *Standardized Precipitation Index (SPI) Calculation*

The SPI [26] is a widely used indicator to assess meteorological drought at different time scales (e.g., 3-, 6-, 12-, 18-, and 24-months). The procedure for calculating SPI is outlined in the following:

(a) Given a certain time scale a, the cumulative precipitation at the n*th* month during different years is calculated by:

$$x_a^n(t) = \sum_{i=j-a+1}^{j} x(i), \; j = 12(t-1) + n \tag{1}$$

where x is the monthly precipitation record of N years, t is the annual index (from 1 to N), and m is a given month (January, February,..., December)

(b) Then the cumulative precipitation series are fitted by a probability density function where Gamma and Log-normal distributions are taken into consideration for the current study, which are represented by Eqs. (2) and (3), respectively.

Table 2 Drought classification according to SPI values

SPI Value	Drought category	Probability (%)
−0.99 to 0.00	Mild drought	34.1
−1.49 to -1.00	Moderate drought	9.2
−1.99 to -1.50	Severe drought	4.4
≤ 2.00	Extreme drought	2.3

$$f\left(x_a^n|\mu,\sigma\right) = \frac{1}{\sigma^{\mu}\Gamma(\mu)}\left(x_a^n\right)^{\mu-1}e^{-\frac{x_a^n}{\sigma}}, x_a^n > 0 \tag{2}$$

$$f\left(x_a^n|\mu,\sigma\right) = \frac{1}{x_a^n\sigma\sqrt{2\pi}}e^{\left(-\frac{1}{2}\left(\frac{\ln x_a^n-\mu}{\sigma}\right)^2\right)}, x_a^n > 0 \tag{3}$$

where, Γ is the Gamma function and μandσ are the shape and scale parameters, respectively.

(c) The cumulative probability of a certain precipitation event is then determined for a specified time scale and month. The SPI value is obtained by using Eq. (4) to convert the cumulative probability distribution (CDF) into a standard normal distribution with a mean of zero and a variance of one.

$$x_k = \varphi^{-1}(k) \tag{4}$$

where the quantile of the cumulative probability k is denoted by x_k, and the inverse function of the CDF for a normal distribution φ is denoted by φ^{-1}. The SPI-based drought classification is presented in Table 2.

2.3 *Assessment of Drought Characteristics*

Drought can be described in a variety of ways, including its drought event number, and severity, intensity, peak, duration, frequency, etc., which are recommended by various researchers to use for drought assessments [13, 27–29] and thus selected for this study. According to [30–32], a drought event is defined as a period when SPI is continuously below 0, with the lowest SPI value less than -1.0. Once drought events are identified, the characteristics of the drought can be calculated using Run's theory [33]. Run's theory demonstrated that the total sum of all the SPI readings below a threshold level represents the severity (Se), which is shown in Eq. (5). Duration is the length of time during which the SPI value is continuously below the threshold level. The intensity of a drought event (DI_e) is the mean value of SPI below the threshold level, which may be obtained by using Eq. (6). The worse the drought, the higher the DI_e value. The peak is the minimum value of SPI below the threshold level.

$$S_e = \left| \sum_{i=1}^{l} SPI_i \right| \tag{5}$$

$$DI_e = \frac{S_e}{l} \tag{6}$$

where e is regarded as a drought event, i is a month, SPI_i is the SPI value in month i, l, S_e *and* DI_e are the duration, severity, and intensity of a drought event e, respectively. The frequency of drought can be obtained from Eq. (7) given in the following.

$$F_d = \frac{n_d}{N_s} \times 100\% \tag{7}$$

where n_d is the number of drought events, N_s is the number of years in the study period, and s represents a station.

3 Results and Discussion

The SPI values obtained by using two different types of distributions, such as gamma and lognormal distributions, at different time scales for the 45 year period at Bogra and Ishurdi stations produce the same type of results. Drought estimation in Bogra station by SPI-6 (i.e., SPI values for 6-month time scales) and SPI-24 (i.e., SPI values for 24-month time scales) is shown in Fig. 2. As can be seen from the figure, the SPI values found using these distributions are almost identical to each other for a particular type of time scale, except at some extreme ends. Furthermore, it is seen that the droughts are more frequent at shorter time scales and get less frequent as the time scale gets longer. It is also evident that the lognormal distribution is primarily responsible for yielding the higher values of SPI at the extreme ends.

The differences derived from $SPI_{Lognormal}$ minus SPI_{Gamma} for the selected SPI time scales were compared to analyze the irregularity of the SPI values for both Bogra and Ishurdi stations, which are shown in Fig. 3. It is seen from the figure that

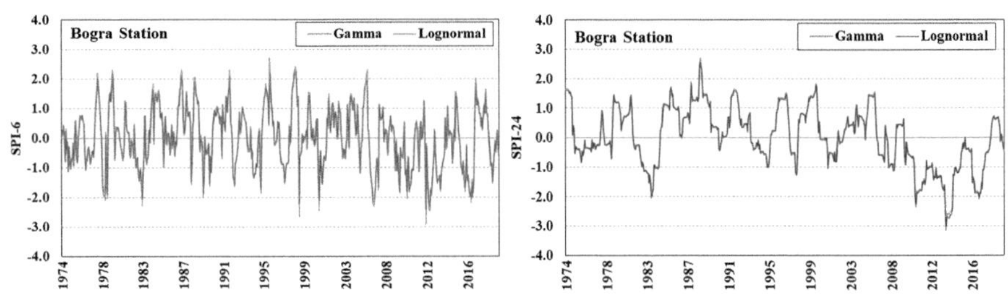

Fig. 2 SPI-based drought calculation at Bogra station using Gamma and Lognormal distributions

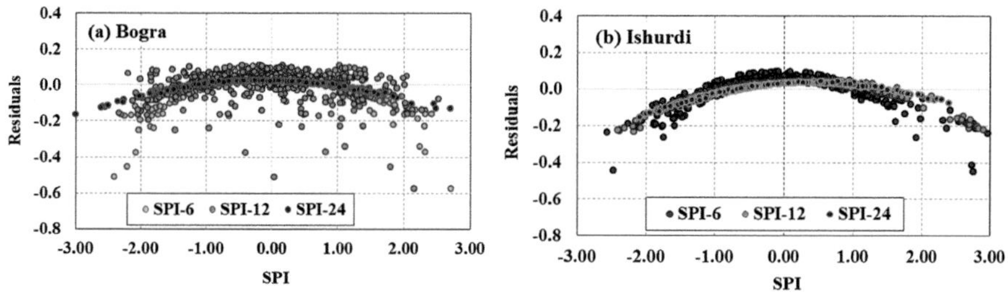

Fig. 3 Residuals of SPI calculations for different time scales for Bogra and Ishurdi stations

the disturbances are more obvious for the extremes and less evident for the normal and moderate classifications of dry and wet periods. Furthermore, it is observed that this induced uncertainty tends to reduce as the time scale gets larger. On the other hand, the parameters of the distributions (i.e., shape and scale) are greatly affected by the type of estimation used along with the length of the data and time scales considered to compute them. It is found that the parameters vary significantly when the distribution is changed from the gamma to the lognormal distribution, as well as when the time scales and the data period change.

The shape (μ) and scale (σ) parameters as given by Eqs. (2) and (3) can be varied based on different time scales and data ranges, which results in different values of probability distributions (i.e., gamma and lognormal distributions in the current study) and hence computations of different SPI values. This ultimately influences the drought assessment using SPI. The variation of shape and scale parameters for different time scales at Bogra station is presented in Fig. 4. It is observed from the figure that due to the change in distribution, the shape parameter varies from 74 to 80%, while it varies almost 88% when the time scale changes. Furthermore, it varies from 4 to 29% when the data period is changed. Similarly, the scale parameters change by almost 97%, 60%, and 19% maximum as the distribution, time scale, and data period change, respectively.

The effect of uncertainty has been observed on drought characteristics. The drought characteristics obtained for both Ishurdi and Bogra stations are shown in

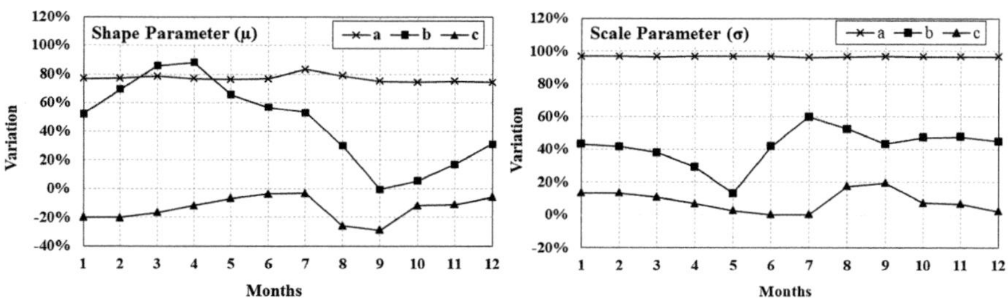

Note: a. changes in probability distribution used at 12-month time scale, b. changes in time scales used in between 6-month and 12-month, c. changes in data records from 35-year to 45-year ranges

Fig. 4 Variation of shape and scale parameters for different time scales at Bogra station

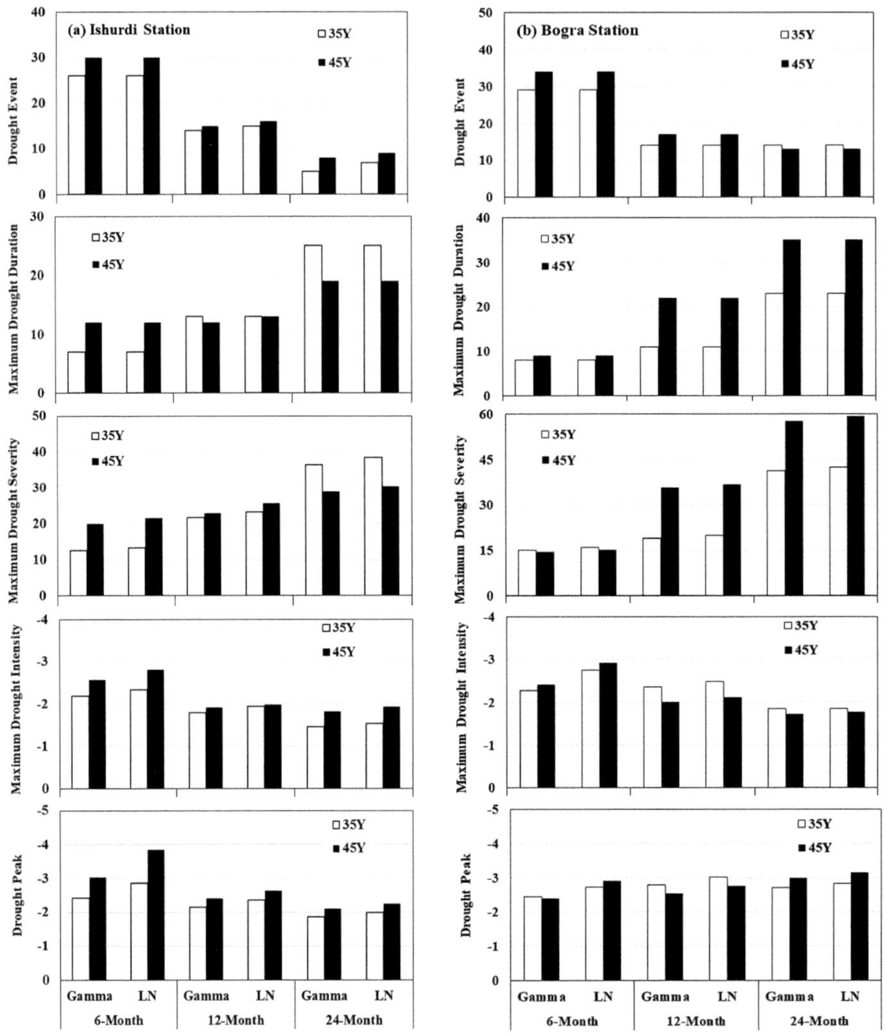

Fig. 5 SPI based drought characteristics at Ishurdi and Bogra stations

Fig. 5. It is evident from the figure that the lognormal distribution yields a slightly greater number of drought events, duration, severity, intensity, and drought peaks in comparison to the gamma distribution for a specific time scale and length of data period. In addition to this, it is seen that the drought characteristics tend to decrease as the time scale increases for a particular length of data period. Furthermore, most of the drought characteristics are found to be higher when considering the 45 year data period. The result also shows that the uncertainty has less effect on drought frequency. However, almost the same frequency of drought is encountered using both gamma and lognormal distributions. Though the highest frequency of drought

Table 3 Most severe droughts obtained in the study area

Time period	Station	Start–End	Most severe drought	Duration (months)	Intensity	Distribution used
35 Y	Bogra	July 1979—April 1984	42.38	23	−1.84	Lognormal
	Ishurdi	April 1994—September 1996	38.39	25	−1.54	Lognormal
45 Y	Bogra	April 2012—March 2015	59.12	35	−1.69	Lognormal
	Ishurdi	September 2009-July 2011	30.50	19	−1.61	Lognormal

obtained is for the mild category, whereas it becomes less frequent for the moderate, severe, and extreme categories, and this tends to remain more or less constant as the time scales and data period change. It is also evident from the results that the most severe type of drought is found at the 24-month time scale for both stations.

Table 3 presents the severity of drought at 35- and 45-year periods for a 24-month time scale and their corresponding intensity, duration, and the distribution used to result in that severity for both stations.

4 Conclusions

The current study focuses on the identification of the sources of uncertainty in SPI-based drought assessment and analyzes their effects on drought characteristics. The study is demonstrated through two selected climate stations, namely Bogra and Ishurdi, located in the northwest region of Bangladesh. Gamma and log-normal distributions are adopted for SPI calculation for 6-, 12-, and 24-month time scales. Parameters of the distribution functions are obtained by the maximum likelihood method, and the characteristics of the drought are determined using Run's theory. It is evident from the obtained results that the estimation process of SPI includes uncertainty that cannot be eliminated. These uncertainties are attributable to SPI's probability distribution functions, parameter estimation techniques, time scales, and length of the data period. The current study incorporates the aforementioned uncertainties into the SPI calculation process and evaluates their influence on SPI values and the corresponding drought characteristics. This study also uses the lognormal distribution in addition to the standard gamma distribution to fulfill its goals. It is evident from the results that the lognormal distribution behaves similarly to the gamma distribution at larger time scales, but exhibits more sensitivity at the extreme ends. This demonstrates that the log-normal distribution can be adopted as a viable alternative to the standard gamma distribution for drought assessment. The results also show that uncertainty greatly affects extreme droughts. The greater the SPI value, the greater the possibility of uncertainty. As extremes are a matter of concern nowadays and uncertainties tend to decrease at extremes at greater time scales and longer data periods, it is recommended to take this into account for the SPI-based drought assessment.

References

1. Habiba U, Shaw R, Takeuchi Y (2011) Drought risk reduction through a socio-economic, institutional and physical approach in the northwestern region of Bangladesh. Environ Hazards 10:121–138. https://doi.org/10.1080/17477891.2011.582311
2. Shahid S, Behrawan H (2008) Drought risk assessment in the western part of Bangladesh. Nat Hazards 46:391–413. https://doi.org/10.1007/s11069-007-9191-5
3. Shahid S, Chen X, Hazarika MK (2005) Assessment aridity of Bangladesh using geographic information system. GIS Development 9(12):40–43
4. Dey N, Alam MS, Sajjan AK, Bhuiyan MA, Ghose L, Ibaraki Y, Karim F (2012) Assessing environmental and health impact of drought in the Northwest Bangladesh. J Environ Sci Nat Resour 4(2):89–97. https://doi.org/10.3329/jesnr.v4i2.10141
5. Wilhite DA (2003) Moving toward drought risk management: The need for a global strategy. Natl Drought Mitig Cent Univ Neb, Lincoln, USA
6. Wilhite DA, Sivakumar MVK, Pulwarty R (2014) Managing drought risk in a changing climate: The role of national drought policy. Weather Clim Extrem 3:4–13. https://doi.org/10.1016/j.wace.2014.01.002
7. Das S, Das J, Umamahesh NV (2021) Nonstationary modeling of meteorological droughts: application to a region in India. J Hydrol Eng 26:05020048. https://doi.org/10.1061/(ASCE)HE.19435584.0002039
8. Tabari H, Nikbakht J, Hosseinzadeh TP (2013) Hydrological drought assessment in northwestern iran based on Streamflow Drought Index (SDI). Water Resour Manage 27:137–151. https://doi.org/10.1007/s11269-012-0173-3
9. Wells N, Goddard S, Hayes MJ (2004) A Self-Calibrating palmer drought severity index. J Clim 17:2335–2351. https://doi.org/10.1175/1520-0442(2004)017%3c2335:ASPDSI%3e2.0.CO;2
10. Jain SK, Keshri R, Goswami A, Sarkar A (2010) Application of meteorological and vegetation indices for evaluation of drought impact: a case study for Rajasthan, India. Nat Hazards 54:643–656. https://doi.org/10.1007/s11069-009-9493-x
11. Kumar MN, Murthy CS, Sesha Sai MVR, Roy PS (2009) On the use of standardized precipitation index (SPI) for drought intensity assessment. Meteorol Appl 16:381–389. https://doi.org/10.1002/met.136
12. Zhang Y, Li W, Chen Q, Pu X, Xiang L (2017) Multi-models for SPI drought forecasting in the north of Haihe River Basin, China. Stoch Env Res Risk Assess 31:2471–2481. https://doi.org/10.1007/s00477-017-1437-5
13. Zhang Y, Li Z (2020) Uncertainty analysis of standardized precipitation index due to the effects of probability distributions and parameter errors. Front Earth Sci 8:76. https://doi.org/10.3389/feart.2020.00076
14. Wu H, Hayes MJ, Wilhite DA, Svoboda MD (2005) The effect of the length of record on the standardized precipitation index calculation. Int J Climatol 25:505–520. https://doi.org/10.1002/joc.1142
15. Vergni L, di Lena B, Todisco F, Mannocchi F (2017) Uncertainty in drought monitoring by the Standardized Precipitation Index: the case study of the Abruzzo region (central Italy). Theoret Appl Climatol 128:13–26. https://doi.org/10.1007/s00704-015-1685-6

16. Stagge JH, Tallaksen LM, Gudmundsson L, van Loon AF, Stahl K (2015) Candidate distributions for climatological drought indices (SPI and SPEI). Int J Climatol 35:4027–4040. https://doi.org/10.1002/joc.4267
17. Beyaztas U, Bickici Arikan B, Beyaztas BH, Kahya E (2018) Construction of prediction intervals for palmer drought severity index using bootstrap. J Hydrol 559:461–470. https://doi.org/10.1016/j.jhydrol.2018.02.021
18. Guttman NB (1999) Accepting the standardized precipitation index: A calculation algorithm. JAWRA J Am Water Resour Assoc 35:311–322. https://doi.org/10.1111/j.1752-1688.1999.tb03592.x
19. Sienz F, Bothe O, Fraedrich K (2012) Monitoring and quantifying future climate projections of dryness and wetness extremes: SPI bias. Hydrol Earth Syst Sci 16:2143–2157. https://doi.org/10.5194/hess-16-2143-2012
20. Angelidis P, Maris F, Kotsovinos N, Hrissanthou V (2012) Computation of drought index SPI with alternative distribution functions. Water Resour Manage 26:2453–2473. https://doi.org/10.1007/s11269-012-0026-0
21. Blain GC, Meschiatti MC (2015) Inadequacy of the gamma distribution to calculate the standardized precipitation index. Rev Bras Eng Agrícola E Ambient 19:1129–1135. https://doi.org/10.1590/1807-1929/agriambi.v19n12p1129-1135
22. Bhakar SR, Iqbal M, Devanda M, Chhajed N, Bansal AK (2008) Probability analysis of rainfall at Kota. Indian J Agric Res 42:201–206
23. Mandal S, Choudhury BU (2015) Estimation and prediction of maximum daily rainfall at Sagar Island using best fit probability models. Theoret Appl Climatol 121:87–97. https://doi.org/10.1007/s00704-014-1212-1
24. Carbone GJ, Lu J, Brunetti M (2018) Estimating uncertainty associated with the standardized precipitation index. Int J Climatol 38:e607–e616. https://doi.org/10.1002/joc.5393
25. Ghasemnezhad F, Bazrafshan O, Fazeli M, Parvinnia M, Singh V (2021) Uncertainty analysis of hydrological drought due to record length. Time Scale, Probab Distrib Funct Using Mt-Carlo Simul Method. https://doi.org/10.21203/rs.3.rs-131396/v1
26. McKee TB, Doesken NJ, Kleist J (1993) The relationship of drought frequency and duration to time scales. In: Proc. of the Ninth Conference on Applied Climatology, American Meteorological Society, 17–22
27. He Z, Liang H, Yang C, Huang F, Zeng X (2018) Temporal–spatial evolution of the hydrologic drought characteristics of the karst drainage basins in South China. Int J Appl Earth Obs Geoinf 64:22–30. https://doi.org/10.1016/j.jag.2017.08.010
28. Sheffield J, Wood EF (2008) Global trends and variability in soil moisture and drought characteristics, 1950–2000, from Observation-Driven simulations of the terrestrial hydrologic cycle. J Clim 21:432–458. https://doi.org/10.1175/2007JCLI1822.1
29. Loukas A, Vasiliades L (2004) Probabilistic analysis of drought spatiotemporal characteristics inThessaly region, Greece. Nat Hazard 4:719–731. https://doi.org/10.5194/nhess-4-719-2004
30. Mesbahzadeh T, Mirakbari M, Mohseni Saravi M, Soleimani Sardoo F, Miglietta MM (2020) Meteorological drought analysis using copula theory and drought indicators under climate change scenarios (RCP). Meteorol Appl 27. https://doi.org/10.1002/met.1856
31. Nedealcov M, Răileanu V, Sîrbu R, Cojocari R (2015) The use of standardized indicators (SPI and SPEI) in predicting droughts over the republic of moldova territory. Present Environ Sustain Dev 9:149–158. https://doi.org/10.1515/pesd-2015-0032
32. Lee S-H, Yoo S-H, Choi J-Y, Bae S (2017) Assessment of the impact of climate change on drought characteristics in the hwanghae plain, North Korea using time series SPI and SPEI: 1981–2100. Water (Basel) 9:579. https://doi.org/10.3390/w9080579
33. Yevjevich VM (1969) An objective approach to definitions and investigations of continental hydrologic droughts: Vujica Yevjevich: Fort Collins, Colorado State University, 1967, p 19. (Hydrology Paper No. 23), J Hydrol 7:353. https://doi.org/10.1016/0022-1694(69)90110-3

Spatial Distribution of Heavy Metals and Identifying Factors that Influence Soil Properties at Rajbandh Landfill Site in Khulna

Ashif Adnan Khan Aoyon and Islam M. Rafizul

Abstract Almost all the landfills in Bangladesh are not controlled/ engineered/ sanitary landfills. Low-lying open spaces, unclaimed land, riverbanks, and roadside areas are all used as landfill sites in and around the city. With the rapidly growing population pollution is also increasing day by day and as a result landfill sites are increasing and heavy metals contaminate the earth\'s crust badly, which leads to the risk to both human and animal health. This study focused on the intensity of the heavy metals such as Fe, Mn, Cr, Cu, Pb, Zn, Ni, Cd, As, Hg, Co, Na, K, Ca, Al, Ti, Sb, Sc, Sr, V, Ba Ba, in soil by spatial distribution analysis and the factors that influence their migration. ArcGIS was used to perform geostatistical analyses such as Inverse Distance Weighting (IDW). Heavy metals in soil are shown graphically using ArcGIS software. The spatial distribution of heavy metals was performed by Inverse Distance Weighting (IDW) in Rajbandh and the surrounding area. The results showed that the intensity of the heavy metals was very high in the middle portion of the landfill for both dry and rainy seasons. The intensity also decreases as the distance increases. Factors identification that influences the heavy metal lateral migration could help to find the sources of soil pollution. To determine the variables that affect soil heavy metals, a categorical regression was performed in this study. Before utilizing a categorical regression to examine the degree to which various factors influenced the soil heavy metal levels in Khulna, environmental factors were first linked to data on soil heavy metals collected during both dry and wet seasons. The findings indicated that pH, land use type, and soil type were the key contributing factors, indicating that these three variables were significant sources of soil heavy metals in Khulna. Cd in soil during the dry period was significantly influenced by the soil parent material, soil type, and pH. However, aside from pH, soil types and soil parent material had a significant impact on Pb throughout the dry season.The types of soil, soil parent material, and pH all play a significant role in Cd during the

A. A. K. Aoyon (✉) · I. M. Rafizul
Department of Civil Engineering, Khulna University of Engineering and Technology, Khulna, Bangladesh
e-mail: aoyonce2k16@gmail.com

I. M. Rafizul
e-mail: imrafizul@ce.kuet.ac.bd

S. Arthur et al. (eds.), *Proceedings of the 6th International Conference on Advances in Civil Engineering*, Lecture Notes in Civil Engineering 368,
https://doi.org/10.1007/978-981-99-3826-1_11

rainy season. Neither of these characteristics, though, have a significant value for Pb. The primary variables influencing soil heavy metals were found in this study, and their relative importance was graded. A viable approach for determining the variables that affect soil heavy metal concentrations is categorical regression, which may also be used to investigate the genetics of local soil heavy metal contamination.

Keywords Landfill · Heavy metals · Categorical regression · Khulna

1 Introduction

Heavy metals in soil can be absorbed by plant roots, such as rice and vegetables, and then enter the food chain, posing a concern to humans. Identifying the major influences on soil heavy metals could serve as a starting point for investigating the causes of soil heavy metals. The presence of high levels of heavy metals in soil is caused by a variety of sources, the majority of which are natural and human. The soil parent material is one of the natural factors. Statistics and geostatistical studies, the isotope tracer technique, and the geographical detector method are the most common methods for detecting factors that influence soil heavy metals. Spatial interpolation is widely used when data are collected at distinct locations (c.g. soil profiles) for producing continuous information. According to [2], geostatistics has been applied in the case of spatial interpolation for more than 20 years. Research conducted by [7] also verified this statement. In the present investigation of heavy metals in the waste disposal site, deterministic and geostatistical interpolation techniques were used. Correlation analysis and principal component analysis are two statistical analysis methodologies. Both categorical and numerical factors have an impact on soil heavy metals. Nominal, ordinal, and numerical variables can be scaled sequentially applying categorical regression (CATREG). The location of domicile and maternal age, for eg, could be affirmed as substantial and self-reliant factors of cord blood lead, while the location of domicile was the only major correlate of maternal blood lead [4]. The above research used CATREG to assess the contribution of the regression model to the variability of the response variables [5]. Employed CATREG to examine if sleep disruptions influence depression and found a clear association between depression scores and the IS global insomnia score. For such grounds, CATREG was suggested as a possible way to identify variables that affect soil toxic metals. This is the first occasion the CATREG strategy has been used for soil toxic substances. After applying environmental variables to soil heavy metal data, the degree of effect of various elements on soil heavy metal levels was assessed. We find that this methodology is particularly suitable for studying both numerical and categorical variables. We looked at how well it may be applied to figuring out what influences soil heavy metal levels. Even though this methodology is no more often utilized, it can be turned to when establishing when soil contamination with heavy metals first started.

2 Research Methodology

The study's methodology includes extracting the spatial distribution of heavy metals & identifying factors that cause heavy metal migration from the Batiaghata subdistrict area, which includes the Rajbandh landfill. A flow diagram is used to depict the systematic development of this study's methodology in the figure. The following section describes a brief overview of the methodology adopted in this study. Detail description of the workflow of the method is described in Fig. 1

2.1 Study Area

For municipal solid waste (MSW), the disposal method is separated into two sections: open dumping and sanitary landfill. A sanitary landfill is among the most significant and secure locations for the disposal of MSW. Eight kilometers outside the city center, at Rajbandh Khulna, on the north side of the Khulna-Satkhira road, is a pilot-scale sanitary landfill. Bangladesh's Khulna is its third-oldest metropolis. It's in the Khulna Division. Its 4394.45 km^2 is surrounded by the districts of Jessore and Narail to the north, the Bay of Bengal to the south, Bagerliat District to the east, and Satkhira District to the west. The geographical coordinates of Khulna are 22.350N and 89.300E. Rupsa, Arpangachhia, Shibsa, Pasur, and the Koyra surround the city. A significant amount of garbage is being generated as urban expansion creeps towards the North and West adjacent zones. KCC and KCPA have respective areas of 45.65 and 69.50 square kilometers. The city of Khulna's expanding population tends to generate more municipal solid waste (MSW) and leachate, a type of liquid waste. These MSW have a high concentration of heavy metals that are exposed to the environment. The environment and human health may be significantly impacted by this (Fig. 2).

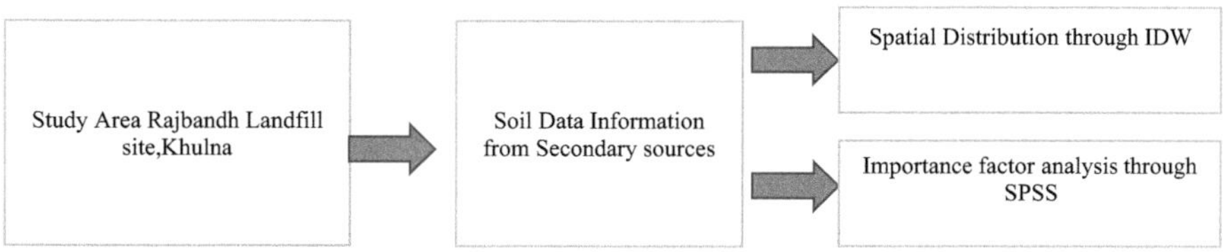

Fig. 1 Workflow of the methodology

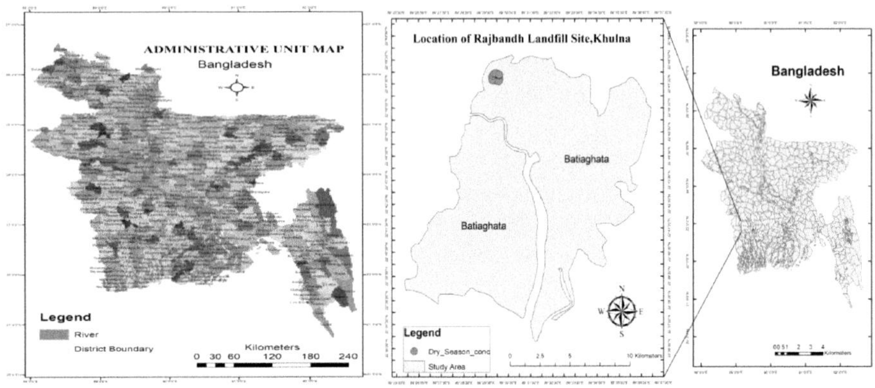

Fig. 2 Map showing selected location of the study area at Rajbandh, Khulna, Bangladesh
Location:
Latitude: 22°47′43.17”.
Longitude: 89°29′58.35”.

2.2 Data Collection

The Rajbandh landfill has been in operation since 1984. Because the major goal of this study is to document the Rajbandh landfill's environmental factors that are responsible for the heavy metal migration. The data were analyzed with ArcGIS and SPSS 22.0 software to get a better output.

2.3 Catreg

Certain survey elements are often classified when it comes to the factors that influence soil toxic metals. Conventional regression analysis might not be suitable in this circumstance for data gathering. When variables are completely categorized or both categorized and quantitative, CATREG, a—anti multiple linear regression approach, may be used. Therefore, CATREG may represent a favored substitute modeling framework. In the simple linear regression model, we look for a linear combination of Xb that interacts with z the most in predicting dependent variables z from m regressors in X. In the Gifi system, “optimal Scaling” maximizes the correlation between (Z) and $\sum_{j=1}^{m}(b_j \Phi_j(X_j)$ spanning realistic nonlinear functions, stated in Eq. 1.

$$||\mathrm{X} * \mathrm{b} - \mathrm{Z}*||^2 \text{where } ||\mathrm{X} * \mathrm{b} - \mathrm{Z}*|| = \sqrt{(\mathrm{X} * \mathrm{b} - \mathrm{Z})^{\mathrm{T}}(\mathrm{X} * \mathrm{b} - \mathrm{Z})}. \quad (1)$$

A binary indicator matrix Gj with n rows and lj columns, where lj stands for the number of categories, is specified in Eq. 2 by a categorical variable, hj. The following is how components hij define components g_{ir}:

$$\begin{cases} g_{ir(j)=1}h_{ij=r} \\ g_{ir(j)=0}h_{ij\neq r} \end{cases} \quad (2)$$

where the running index (r = 1, 2,…, lj) designates a division number in the variable j. A modified parameter can be expressed as Gjyj if category detection threshold are indicated by the symbol yj, and a sum of predictor variables as

$$\sum_{j=1}^{m} b_j G_j y_j = X * b$$

is the same as the standard linear model. Finally, a CATREG model is equivalent to a linear regression model and can be expressed as given below:

$$\mathrm{Z = X * b} + \varepsilon \quad (3)$$

In Eq. (3), X* stands for the coefficient matrix, Z* for the observational vector, b for the normalized coefficients vector, and for the error function. The optimum scaling procedure converted qualitative characteristics into quantitative approaches. CATREG could evaluate numerical, ordinal, and nominal variables all at once. To accurately capture the traits of the basic categories, categorical variables are defined. Nonlinear transformations make it possible to study variables at several levels to discover the model that fits the data the best [9]. The domains of sociology and medicine have both made extensive use of CATREG [4].

2.4 Idw

The popular deterministic interpolation method IDW (Inverse Distance Weighting) directly implements the principle that things adjacent to one another are more comparable than objects farther away. IDW will forecast values for any unquantified site using the measured values close to the predicted location (ESRI, 2001). According to this theory, the IDW method, also known as a distance reverse value of every point from its neighbors, is proportional to the frequency of connections and commonalities between neighbors. One of the most essential factors in the IDW interpolation method, according [6] is the value of the power parameter. With p = 2, the inverse distance standard weighted interpolation is used as the power default. Since there is no theoretical justification for choosing this number above others, it is important to examine the output and cross-validation result statistics to see how altering p will affect the results (ESRI, 2001). It was suggested that the weights given to interpolating points should be inversely proportional to their separation from the article pointed. Because of this, close points are assigned more importance [1], which increases their

Table 1 Selected Independent variables for both dry and rainy season

Independent variables	Types	Categories
pH	Numeric	< 6.5, 6.5–7, 7–7.5 and > 7.5
Soil types	Nominal	1 = Organic soil,2 = red soil,3 = Paddy soil
Soil parent material	Nominal	1 = Calyley 2 = Paddy fields,3 = Red soil Lands

influence on various distances and vice versa. Since values smaller than one are closer to a simple average guess, the estimation with the closest sampled points is only regarded as an integer value for the parameter [6].

2.5 Environmental Factor

Both natural and man-made causes have an impact on soil heavy metals. Three widespread impacting factors were chosen for this investigation, as indicated in Table 1. Styles of soil, types of land usage, and pH were among the natural influences.

(1) Soil Heavy metals in soil are mostly derived from parent material. Each soil's composition affects the topsoil's heavy metal content differently [9]. There are typically three different types of soil parent materials (1 = Calyley Lands, 2 = Paddy Fields, and 3 = Red Soil Lands) based on the properties of the soil in Rajbandh.
(2) Every sort of soil has unique characteristics, like oxide, organic matter, and clay mineral composition. Varied heavy metals have various degrees of soil adsorption, which has an impact on how they migrate in the environment at the top [3] In Rajbandh, there are three varieties of soil (Mostly Organic soil, red soil, and paddy soil).
(3) The pH level has an impact on the characteristics of heavy metals. Five categories of pH values were used: 6.5, 6.5–7, 7–7.5, and > 7.5.

2.6 Data Sources and Processing

Dry topsoil (10 g) and deionized water combinations were used to calculate the pH of the soil (100 mL). The solutions then shaken for 30 min at room temperature, following which they were allowed to stand for 2 h. A pH meter was then used to test the pH levels. The research area's soil heavy metal data were gathered from secondary sources. Throughout Rajbandh, 60 soil samples were taken in both the dry and wet seasons. Metal concentrations for Fe, Mn, Cr, Cu, Pb, Zn, Ni, Cd, As, Hg, Co, Na, K, Ca, Al, Ti, Sb, Sc, Sr, V, and Ba were measured. The contributing variables were found using CATREG and the SPSS 22.0 program. Utilizing SPSS 22.0, a factor interaction analysis was carried out using multi-factor variance analysis.

Using ArcGIS 10.3.1, the data were analyzed for hotspots (Moran's I), geographical distribution, and mapping.

3 Results and Discussion

3.1 Spatial Distribution of Heavy Metals

Figure 3 depicts the graphical depiction of Cadmium (Cd) in the Dry season using IDW from the Rajbandh vicinity. According to the figure, the intensity was divided into four separate classes and set out by various green, bright green, saffron, and red hues. The magnitude of the concentrations of Cd in the soil sample ranges from 5.90 to 7.02 mg/kg at the center point (about 0 m), and from 2.55 to 3.67 mg/kg at the furthest point (roughly 190 m).

Figure 4 displays a graphical representation of Cadmium (Cd) during the rainy season together with IDW from Rajbandh's neighborhood. According to the figure, the concentration was divided into four separate classes and set out by various hues of green, light green, saffron, and red. The magnitude of the Cd content for the soil sample ranges from 1.20 to 1.87 mg/kg at the farthest location (about 220 m), and from 3.22 to 3.89 mg/kg at the center (about 0 m).

Figure 5 displays a graphical depiction of Lead (Pb) in the Dry Season with IDW from the Rajbandh neighborhood. According to the figure, the concentration was divided into four separate classes and set out by various green, bright green, saffron, and red hues. The magnitude of the Pb content for soil samples ranges from 73.22 to 90.53 mg/kg at the center position (about 0 m) and from 21.29 to 38.60 mg/kg at the furthest location (roughly 190 m).

Figure 6 displays a graphical representation of Lead (Pb) during the rainy season together with IDW from Rajbandh's neighborhood. According to the figure, the concentration was divided into four separate classes and set out by various green, bright green, saffron, and red hues. The magnitude of the Pb content for the soil sample spans from the closest position (about 0 m) to the furthest point (almost 220 m) from 10.88 to 14.23 mg/kg.

3.2 Identification of the Main Factors Affecting the Spatial Distribution of Soil Heavy Metals

The R^2 values of the fitting models for the two heavy metals ranged from 0.762 to 0.997, and all of the evaluating models passed the F test ($p < 0.05$) and were statically important. The tolerance of each variable was high enough to guarantee that now the multi-colinearity issue was not present. The regression model and significance of several contributing factors varied for the same heavy metal [2]. In the dry season,

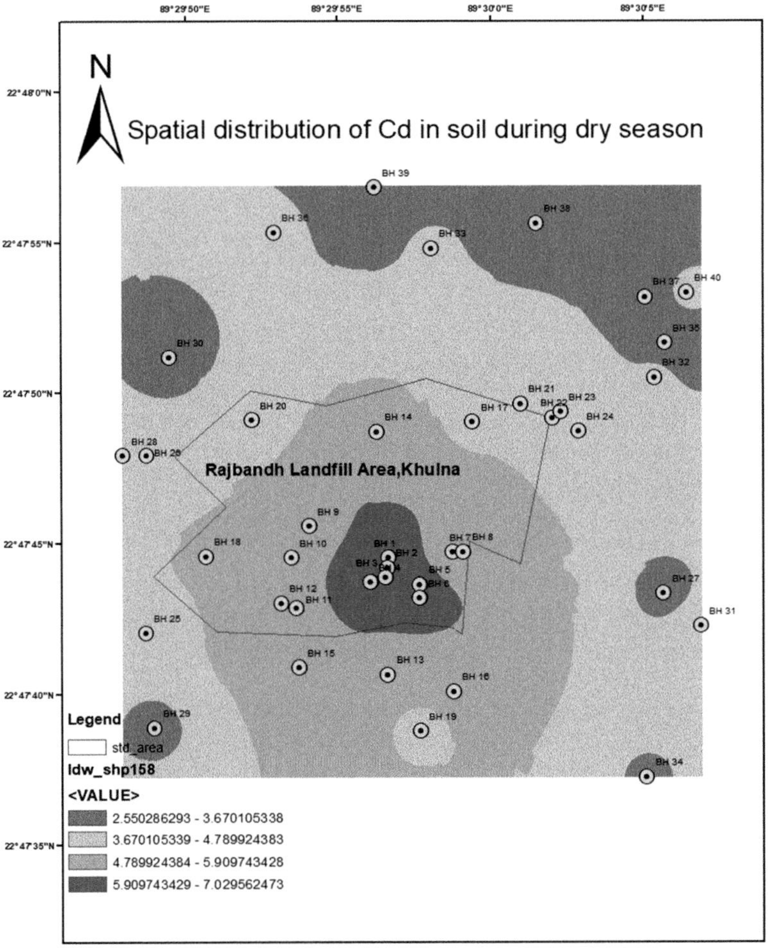

Fig. 3 Spatial distribution of Cd during the dry season

the soil parent material, soil type, and pH all had a big impact on the soil Cd. On the other hand, except for pH, the soil types and soil parent material also showed a lot of promise. The kinds of soil, soil parent material, and pH have a substantial impact on cd during the wet season, however, none of these criteria have an impact on pb. Pratt's measure of relative relevance, which was based on the significance test, helped evaluate the predicted contributions to the regression Table 3. Predictors that are essential to the regression have large individual significance values relative to certain other important values (Pratt, 1987). The numbers represent how important these factors are concerning one another. For instance, the soil type, pH, and soil parent material had a substantial impact on the soil Cd level during the dry season. The same factors that had a substantial impact on the other heavy metals were found

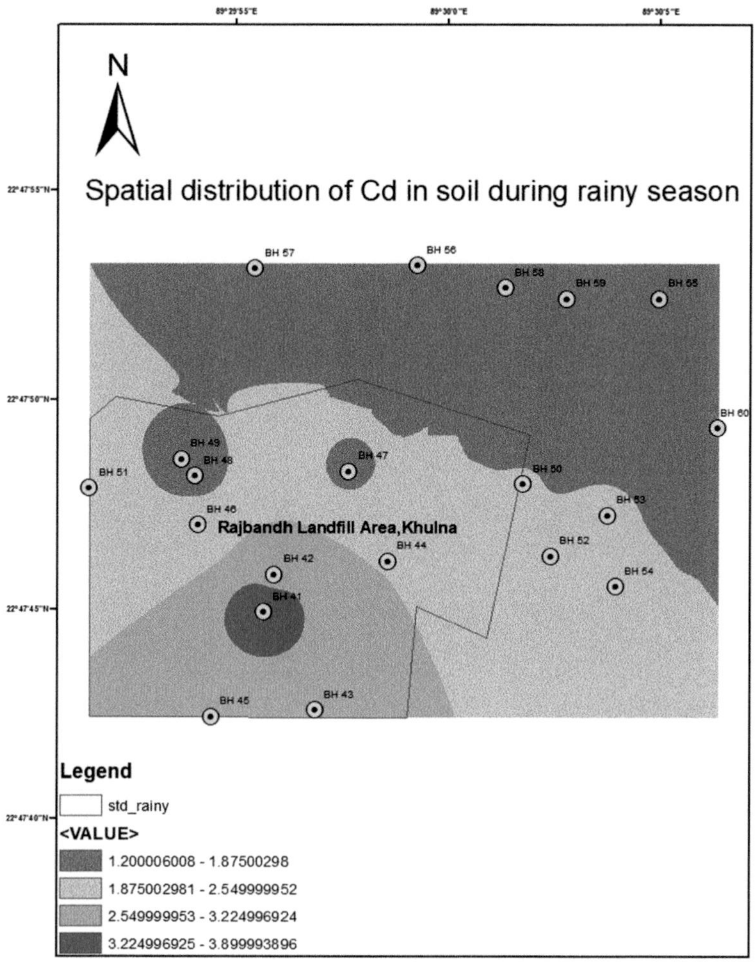

Fig. 4 Spatial distribution of Cd during the rainy season

in Tables 4 and 5. In Rajbandh, the primary influences on soil heavy metals were generally the soil parent material, soil type, and pH (Table 2).

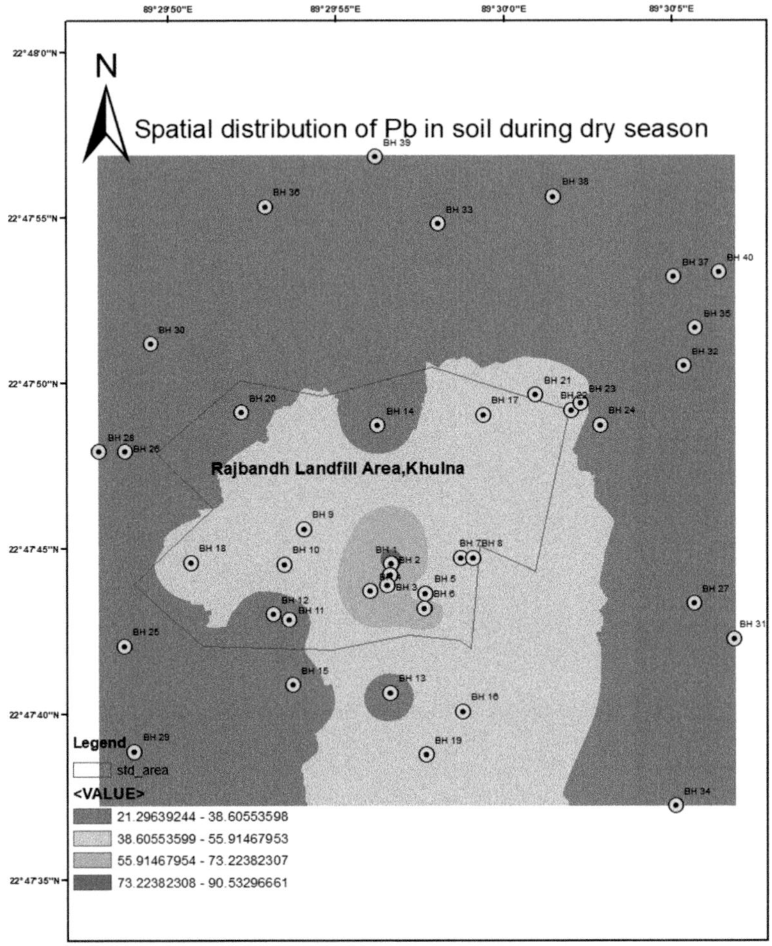

Fig. 5 Spatial distribution of Pb during the dry season

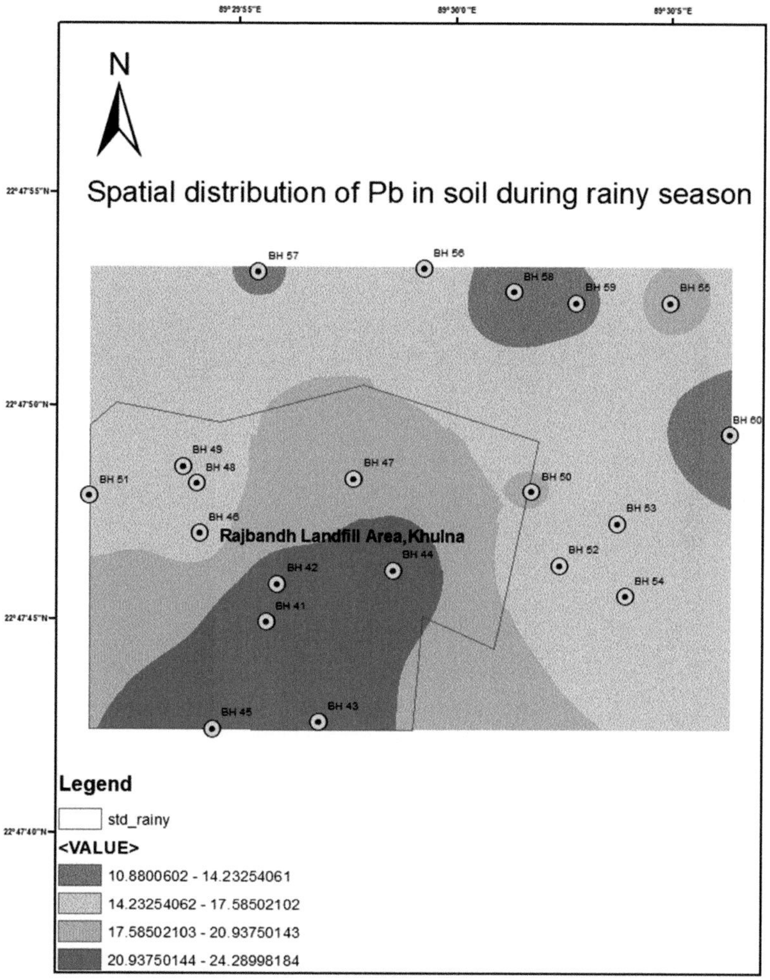

Fig. 6 Spatial distribution of Pb during the rainy season

Table 2 CATREG coefficients and significance test of different factors(Dry Season)

Influence factor	Pb		Cd	
	Beta	Sig.	Beta	Sig.
Soil types	0.344	0.168	0.631	0.004
Soil parent material	0.429	0.129	0.563	0.033
pH	0.751	0.442	0.225	0.000

Table 3 CATREG coefficients and significance test of different factors (Rainy Season)

	Beta	Sig	Beta	Sig
Soil types	0.019	0.983	0.983	0.022
Soil parent material	0.019	0.955	0.983	0.019
pH	0.059	0.889	0.951	0.050

Table 4 Importance of different factors in the Dry season

Influence Factors	Pb	Cd
Soil Types	0.323	0.676
Soil parent material	0.415	0.585
ph	0.195	0.806

Table 5 Importance of different factors in the Rainy season

Influence Factors	Pb	Cd
Soil types	0.015	0.984
Soil parent material	0.015	0.985
ph	0.047	0.952

4 Conclusion

Despite the significant consequences of heavy metal contamination in the soil of the chosen waste disposal site, no adequate procedures are made to regulate or minimize heavy metal element expansion in and around the disposal site's soil. As a result, a thorough investigation and monitoring of the nature and extent of such heavy metal elements in and around the disposal site are required. The intensity of heavy metals in soil is quite high in the middle of the landfill area throughout both the dry and rainy seasons, and it is relatively low at the farthest point of the landfill. The soil parent material, pH, and soil types were the main factors that influenced the spatial distribution of soil heavy metals in the Rajbandh Landfill Site. The soil parent material, soil type, and pH had a significant influence on Cd in soil for the dry season. On the other hand, excepting pH, soil types, and soil parent material had great significance for Pb in the dry season. During the rainy season in the case of Cd, the types of soil, soil parent material, and pH have a great significance on it. However, for Pb, none of these parameters has a significant value.

References

1. Bannari, Abdou, et al. (1995) A review of vegetation indices. Remote sensing reviews. 13 1–2 95–1

2. Burges TM, Webster R (1980) Optimal Interpolation and arithmetic mapping of soil properties. J Soil Sci, 31:315331
3. Dube A, Zbytniewski R, Kowalkowski T, Cukrowska E, Buszewski B (2001) Adsorption and migration of heavy metals in soil. Pol J Environ Stud. 10(1): 1–10
4. Gundacker C, Fröhlich S, Graf-Rohrmeister K, Eibenberger B, Jessenig V, Gicic D, Prinz S, Wittmann KJ, Zeisler H, Vallant B, Pollak A, Husslein P (2010) Perinatal lead and mercury exposure in Austria. Sci Total Environ. 408(23): 5744–5749
5. Ikeda H, Kayashima K, Sasaki T, Kashima S, Koyama F (2017) The relationship between sleep disturbances and depression in daytime workers: A cross-sectional structured interview survey. Ind Health, 55(5): 455–459
6. Isaake EH, Srivastava RM (1989) An introduction to applied geostatistics Oxford University Press, New York
7. Mccormick K, Salcedo J (2017) SPSS statistics for data analysis and visualization. Indianapolis, Indiana: John Wiley & Sons, Inc
8. Wang XS, Qin Y, Sang SX (2005) Accumulation and sources of heavy metals in urban topsoils: A case study from the city of Xuzhou, China. Environ Geol, 48(1): 101–107
9. Zhang, Ye (2011) Introduction to Geostatistics—Course Notes. Dept. of Geology & Geophysics, University of Wyoming

Use of Water Treatment Plant Sludge (WTPS) as a Cementitious Material in Concrete

M. H. Masum, I. J. Ruva, A. Tahsin, and J. Ferdous

Abstract Water treatment plants are generating increasing amounts of sludge, which is creating a significant environmental problem globally. Disposing of WTPS is now a significant environmental issue in developing countries like Bangladesh. The major objective of this study was to investigate the characteristics of WTP sludge, such as physical, chemical, and morphological, and evaluate its impact on concrete performance and mechanical properties. The SEM–EDX testing method is used to determine the physical, elemental, and morphological properties of the WTP sludge. The results revealed that the particles of the sludge possessed an irregular shape with a predominantly rough texture on their porous surface. The results of the EDX experiment indicated the presence of seven distinct elements within the WTP sludge. These elements were found to be present in the following descending order of percentage mass: Oxygen (41.95%) > carbon (40.05%) > silicon (7.80%) > luminium (5.80%) > iron (2.89%) > potassium (1.08%) > magnesium (0.43%). The ACI 211.1 code was used in this study for the design of a concrete mix with strength of 3000 psi after 28 days. During the experiment, cement was replaced with WTP sludge (WTPS) at varying percentages ranging from 5 to 40%. The results indicated that as theamount of WTPS replacement increased, the compressive strength of the concrete gradually decreased. However, the maximum workability was achieved at a 15% replacement rate.

M. H. Masum (✉)
Center for River, Harbor and Landslide Research (CRHLSR), Chittagong University of Engineering & Technology (CUET), Chattogram-4349, Bangladesh
e-mail: mehedi.ce.cuet@gmail.com

I. J. Ruva · A. Tahsin · J. Ferdous
Department of Civil Engineering, Port City International University, Chaatogram-4225, Bangladesh
e-mail: isratjahanruva19@gmail.com

A. Tahsin
e-mail: atanika96@gmail.com

J. Ferdous
e-mail: jannatsonia12345@gmail.com

S. Arthur et al. (eds.), *Proceedings of the 6th International Conference on Advances in Civil Engineering*, Lecture Notes in Civil Engineering 368,
https://doi.org/10.1007/978-981-99-3826-1_12

Keywords WTPS · SEM–EDX · Workability · Compressive strength · Lightweight concrete etc.

1 Introduction

Growing populations in urban areas have increased drinking water consumption, which has caused the corresponding water treatment plants to produce more sludge, which is subsequently deposited nearby and makes the land unproductive. Moreover, the treatment of sludge is not only costly but also has potential hazards to the environment as well as to public health [1]. Researchers at the national and international levels are looking for alternatives to conventional building materials like sand, cement, bricks, aggregates, ceramics, and tiles in order to reduce the environmental impact of these compounds and to fulfill the waste disposal requirements set by the relevant concerned authorities [2]. Different types of sludge (i.e., industrial sludge, waste water treatment sludge, sewage sludge etc.) have been found as potential substitutes for the fundamental components of concrete in the quest for lightweight building materials [3–6]. The amount of $Ca(OH)_2$ in cement paste drops when modified Water Treatment Plant Sludge is added, while SiO_2 and Al_2O_3 react chemically with $Ca(OH)_2$ to create C-A-S–H gel which demonstrates the pozzolanic activity properties of the WTPS [7]. Similar pozzolanic activity characteristics for sewage sludge were documented when a certain proportion of sewage sludge was substituted with cement in mortar [8]. Additionally, the sludges have been tested as prospective substitutes for construction materials by several researchers worldwide in concrete [1, 2, 6, 9–11], hollow concrete block [12], mortar [7, 8], and exterior tiles [13]. The effects of using air-dried sludges as a partial replacement for sand in concrete for 7, 14, and 28 days of compressive strength were explored by [6]. The study revealed that the dry density of the concrete decreased as the sludge content increased and indicated that a maximum of 7.5% of sand content could be substituted with dry wastewater sludge. [6]. The compressive strength values ranged from 3.40 to 2.90 MPa when using 10% to 40% of water treatment sludge as a cement substitute in the production of exterior tiles, which met the requirement of achieving a minimum breaking compressive strength of 2.8 MPa [1, 13]. Found that up to 25% of concrete fine aggregate can be replaced with dry sewage sludge by incorporating 0.5% fibers, resulting in compressive and tensile strengths of 30.7 MPa and 3.58 MPa, respectively.

In comparison to the control group of the mortar, the compressive strength with 20% and 30% modified water treatment sludge decreased by 10.54% and 16.20%, respectively, and the shrinkage value of the mortar after 175 days of drying was reduced by 30.9% with 10% modified water treatment sludge [7]. While, the compressive strength of 10% sewage sludge ash mortar or concrete achieved the designed 28-day compressive strength when cured for longer than 28 days [8, 12]. Recycled the sludge generated from a water treatment facility to manufacture hollow concrete blocks. The study discovered that although water treatment sludge combinations of

10% and 20% may be utilized to create hollow load-bearing concrete blocks, mixtures of other proportions can generate hollow non-load-bearing concrete blocks.

Economically, the combinations of 10% and 20% sludge used in concrete mix can save costs significantly. The water treatment plants in Chattogram, Bangladesh, produce a significant quantity of Water Treatment Sludge (WTPS). The WTPS dumps on the soil after recycling in the water treatment plants. The WTPS dumped across the area is contaminating the soil and creating an unfavorable environment. Quite often, the WTPS has clogged the soil media's subsurface pores, limiting the amount of water that can flow into the subsurface through the pores. An investigation of alternative applications of water treatment plant sludge has been spurred by space constraints for existing dumping sites in urban centers and growing associated environmental hazards such as groundwater contamination from leachate, air pollution, and land pollution. In this study, water treatment sludge was included into concrete as one of the fundamental elements in an effort to make effective and useful use of the material in order to lessen the hazardous impact on the environment and prevent landfills. The major objective of this study is to elevate the value of the sludge generated by water-treatment plants by developing a long-term alternative for cement in concrete produced from water treatment sludge.

2 Methods and Materials

2.1 Material Collection

Water Treatment Plant Sludge (WTPS) was collected from Sheikh Rasel Water Treatment Plant (SRWTP) in Chattogram, managed by Chattogram Water and Sewerage Authority (CWASA), and receives its raw water from the Halda River.

The preparation of modified WTPS from the raw WTPS is shown in Fig. 1a. The raw sludge was dried in an oven at 105°C for 24 h and the dewatered WTPS was then placed in a furnace, heater at 500°C for 4 h. The first step of drying eliminates moisture, while the subsequent high-temperature treatment decomposes organic matter and eliminates harmful microorganisms including most of the pathogens, leading to a material that is more secure and sterile.Then, heat-treated WTPS was grinded into smaller size and the grinded WTPS was sieved by mechanical shaker. The materials passed through #200 sieve was only taken as modified WTPS that was used as replacement of cement.

2.2 SEM and EDX Test

Scanning Electron Microscopy (SEM) analysis was performed to obtain microstructure of WTP sludge concrete, and hence to visualize the size, shape and texture

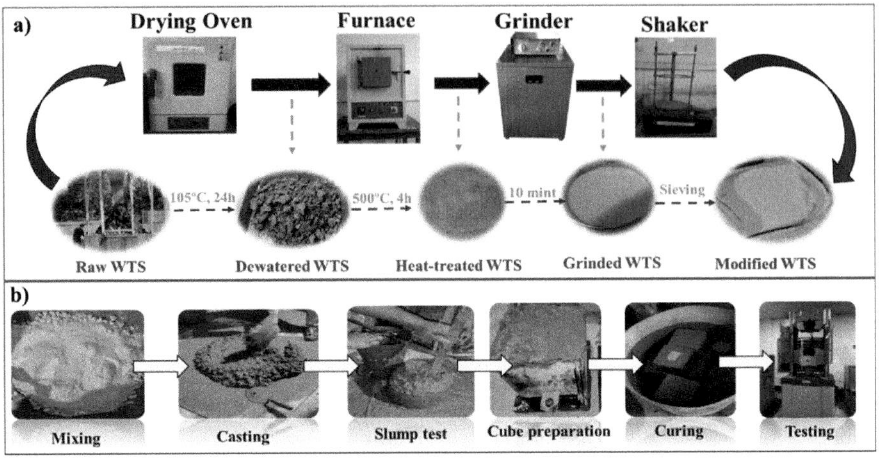

Fig. 1 **a** Schematic diagram for the preparation of modified WTPS from raw WTPS, **b** Schematic diagram for step by step procedure for concrete casting and testing

of the particle. Energy Dispersive X-ray (EDX) test was along with SEM test also carried out to determine the elemental composition of WTPS. Both SEM and EDX tests were carried out at Bangladesh Council of Scientific and Industrial Research (BCSIR), Dhaka by an electronic microscope (JEOL JSM-7610F) at an acceleration voltage of 15 keV with a counting rate of 1096 cps. SEM test was conducted for the modified WTPS for magnification factor ranging from 50–30,000 times.

2.3 *Properties Tests, Material Proportioning, Mixing, Casting and Testing*

The physical property tests for Fine aggregate (FA), Course Aggregate (CA) and Cement were conducted at Port City International University laboratory. The results for physical properties of CA and FA were shown in Table 1. The specific gravity of WTP sludge powder and cement were tested according to C 77–40 (ASTM 1998) and found as 1.98 and 3.2 respectively.

Schematic diagram from the concrete mixing to the compressive strength testing including mixing, casting, curing etc. is shown in Fig. 1b. Concrete of 20 Mpacompressive strength was designed following ACI 211.1 method. Physical properties of Fine aggregate and Course Aggregate used is shown in Table 1.The concrete mix proportions werewater: 206 kg/m^3, cement: 468 kg/m^3, fine aggregate: 493 kg/m^3, coarse aggregate: 829 kg/m^3 (Cement: Fine Aggregate: Coarse Aggregate = 1: 1.1: 1.8; w/c = 0.45). The WTP sludge powder was weighed according to mix design and used in mix to partially replace cement. There were five replacement level utilized. 0%, 5%, 10%, 15%, 25% & 40% replacement of cement with WTP sludge powder

Table 1 Physical properties of fine aggregate and course aggregate used for concrete casting

Parameters	Course Aggregate (CA)	Fine Aggregate (FA)
Specific gravity	2.11	2.12
Moisture content	0.41%	1.60%
Absorption capacity	0.29%	0.79%
Surface moisture	0.12%	0.81%
Dry rodded unit weight	1590 kg/m^3	1353 kg/m^3

have been selected and prepared manually in the laboratory. To compare the test results control specimen (0% replacement level) were also cast. Before mixing, the aggregates were soaked in water to ensure SSD condition. In a dry sheet, the materials were thoroughly mixed. Water was carefully added to the mix according to the water/binder ratio. The workability test (slump test), was performed immediately after mixing and the results were recorded. The test specimens (mold size: 150 mm x 150 mm) were immersed in clean fresh water for 3, 14 and 28 days. The specimens were later subjected to a series of tests after a specified curing period. The results of the experiment were later nalysed through the use of statistical tests (principal component analysis and correlation analysis).The correlation analysis is done with a significance level of 0.05 (95% confidence interval), indicating that the likelihood of observing a correlation as strong as the one found by chance alone is less than 5%.

3 Results and Discussion

3.1 Investigation of Physical, Elemental and Morphological Properties

The results of SEM and EDX test are presented in Figs. 2 and 3. Figure 2 shows, the SEM image of the WTP sludge. The sizes of the particles of WTPS are varying from < 1 μm to 37.8 μm whereas the sizes of the particles of OPC cement vary from 1 μm to 100 μm [14]. This result indicates that the particles size is smaller than cement particles which helps to improve the durability and workability of the replaced concrete. The overall morphological shapes of the WTP sludge were mostly irregular (Fig. 2a), some particles are semi-circular (Fig. 2b) with the diameter varying from 2.81 μm to 4.81 μm, and few fibers (Fig. 2c) are also seen.It is also seen that some sharp edgeshaped particles with pores are visible at higher magnification (30,000x) (Fig. 2d). The edge texture for the particles is found rough for the magnification of 2000 × and smooth for 30,000 × magnification, respectively. For 2000 × cross sectional properties were found porous and that of 30000x, it was found hollow. The cross section of the hole seen in Fig. 2(f) vary from 0.0015 μm^2 to 0.012 μm^2.

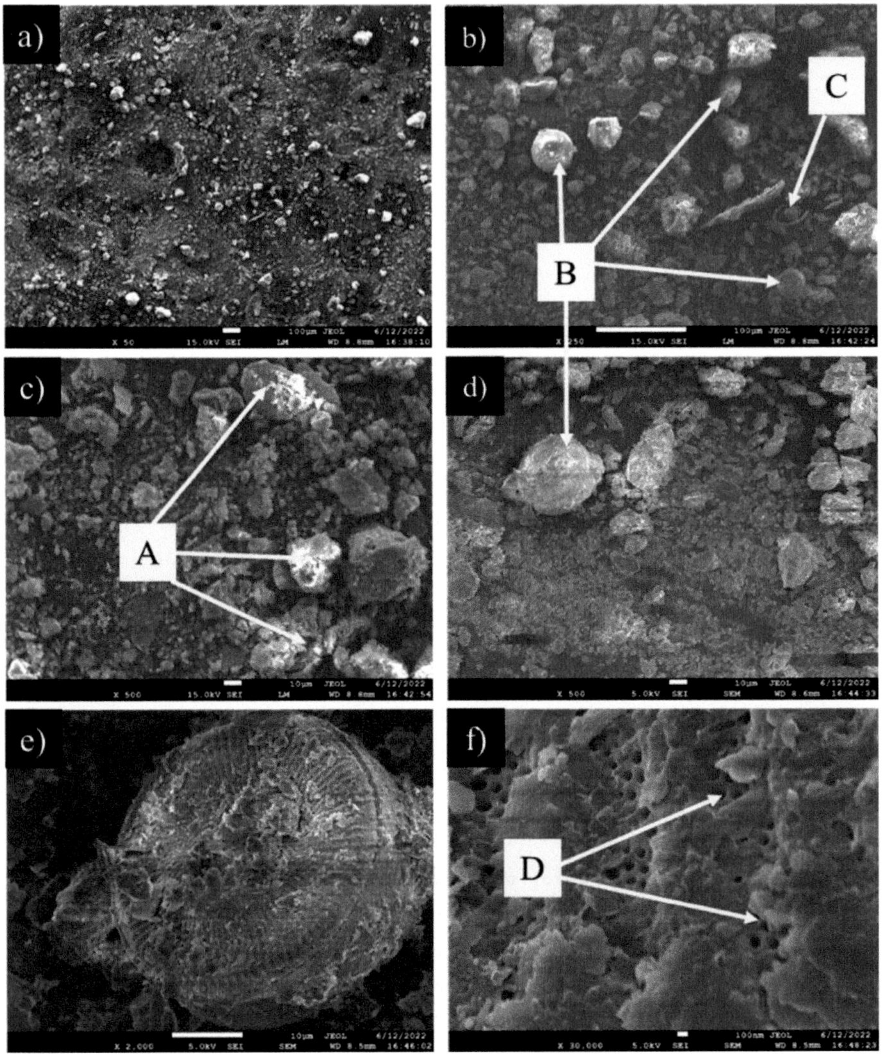

Fig. 2 SEM image of collected WTP sludge sample with magnification of **a** 50 times, **b** 250 times, **c** 500 times, **d** 500 times, **e** 2,000 times and **f** 30,000 times

Based on EDX testing shown in Fig. 3, it can be concluded that WTPS sludge has higherOxygen (O) content (Table 2). The higher Oxygen content in WTPS indicates that there might have interlocking oxides (SiO_2, CaO, Al_2O_3, Fe_2O_3, MgO etc.) which would help to make bonds among the constituents of the concrete materials. The content of carbon is as similar as oxygen is. Furthermore, a notable amount (7.80%) of Silicon (Si) is also present which is mostly responsible for the gain of strength of concrete. While, Murugaboopathy (2020) [14] discovered that Ordinary Portland

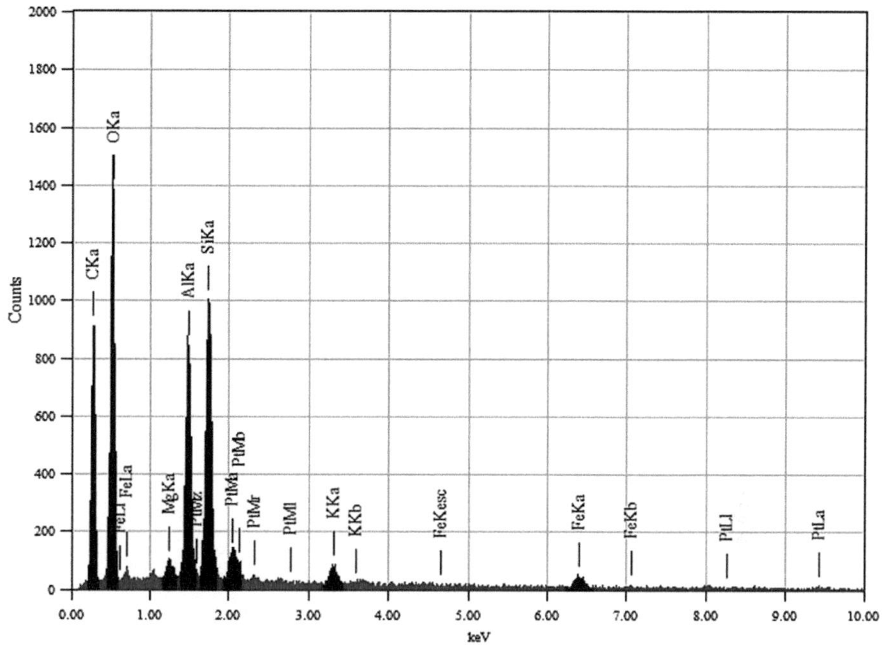

Fig. 3 Representative EDX spectrum and relevant elements of the collected WTP sludge sample corresponding to area shown in Fig. 1a, with magnification of 50 times

Cement (OPC) contains 8.45% Silicon (Si) content. The abundance of silicon in WTPS may lead to an increase in the compressive strength of concrete since the C_2S and C_3S compounds that are responsible for strengthening the concrete are influenced by silicon. [14, 16]. The aluminum (Al) content found in WTPS is comparable to that in OPC cement, which may contribute to the faster hardening of concrete as the C3A compounds present in both materials accelerate the initial hardening of concrete [17]. EDX analysis of the paint sludge revealed that the elements present in the sludge can be ordered by mass as follows: O (41.95%) > C (40.05%) > Si (7.80%) > Al (5.80%) > Fe (2.89%) > K (1.08%) > Mg (0.43%). Alternatively, when ordered by atomic percentage, the elements are ranked as follows: C (50.94%) > O (40.06%) > Si (4.24%) > Al (3.28%) > Fe (0.79%) > K (0.42%) > Mg (0.27%). The EDX analysis further suggests that many chemicals are absorbed on the surface of the sludge and contribute to the stability of the synthesized sludge.

Table 2 Result of the elements analysis by EDX spectrum for the collected WTP sludge Sample

Sl. No	Name of the elements	WTPS [This study]		OPC Cement [14]		OPC Cement [15]
		Mass (%)	Atom (%)	Mass (%)	Atom (%)	Atom (%)
1	Carbon (C)	40.05	50.94	–	–	–
2	Oxygen (O)	41.95	40.06	38.96	61.57	70.08
3	Magnasium (Mg)	0.43	0.27	–	–	1.21
4	Aluminium (Al)	5.80	3.28	4.76	4.46	3.37
5	Silicon (Si)	7.80	4.24	8.45	7.60	8.35
6	Potassium (K)	1.08	0.42	0.87	0.56	–
7	Iron (Fe)	2.89	0.79	4.49	2.04	1.72
8	Sodium (Na)	–	–	0.55	0.61	0.38
9	Sulphur (S)	–	–	1.81	1.43	0.45
10	Calcium (Ca)	–	–	31.66	19.97	14.45
11	Antimony (Sb)	–	–	8.46	1.76	–
Total		100	100	100	100	100

3.2 *Investigation of Workability, Weight and Compressive Strength of Replaced Concrete*

The workability of concrete is classified based on the slump value obtained from slump cone test and the variation of slump value of concrete samples for different percentages of WTP sludge replacement is shown in Fig. 4a shows. It is observed that, concrete having zero presence of sludge has a slump value of 130 mm representing average consistency. On the other hand, with the increase in percentage of sludge from 5–15%, slump gradually increases from 120 to 140 mm. The phenomenon of increasing slump value can be explained by the finer particle size of the WTPS as compared with OPC cement. Moreover, fineness of WTPS and OPC cement are 213 m^2/kg and 338 m^2/kg respectively. The finer size and fineness of WTPS may responsible for the higher slump value, thus increasing the workability of the replaced cement concrete [18, 19]. However, further increase in percentage from 15–40%, slump value gradually decreases. Although 40% replaced concrete shows the slump value greater than the control concrete block. Higher percentage of sludge could be responsible for higher absorption of water from the concrete which eventually results in lower workability. Moreover, the SEM/EDX results show an irregular, porous morphology of the WTP sludge which has caused a negative effect on the workability of concrete [9]. However, the rate of increase in slump is higher than the rate of decrease. As compared to the control concrete having slump value of 130 mm (Fig. 4a), a sludge replacement of around 15% maybe considered optimum in terms of achieving greater consistency. Besides, the varying percentage of cement replacement with sludge can be suitable for concrete works demanding average to higher consistency of mix.

Figure 4b represents the variation of weight of concrete with the increase in replacement of cement with sludge. For the concrete at initial stage, the lowest weight value is obtained to be 1976 kg/m^3 for 40% replacement, whereas for the concrete at 28 days, the lowest weight value is obtained to be around 2131 kg/m^3 for the same replacement percentage. Although the weight increases from initial to final stage of the curing time, but there exists a gradual decline in weight is observed with the increase in sludge replacement up to 40%. The specific gravity of WTPS is almost 1.66 times lower than the OPC cement and the pores in the WTPS particles as seen in SEM images, altogether may causes having the lower weight of the concrete [20]. Utilizing sludge as a substitute for cement in concrete may produce lightweight concrete, as a higher percentage of sludge used as partial replacement for cement can lead to a decrease in the weight of the resulting concrete. Figure 3c shows the

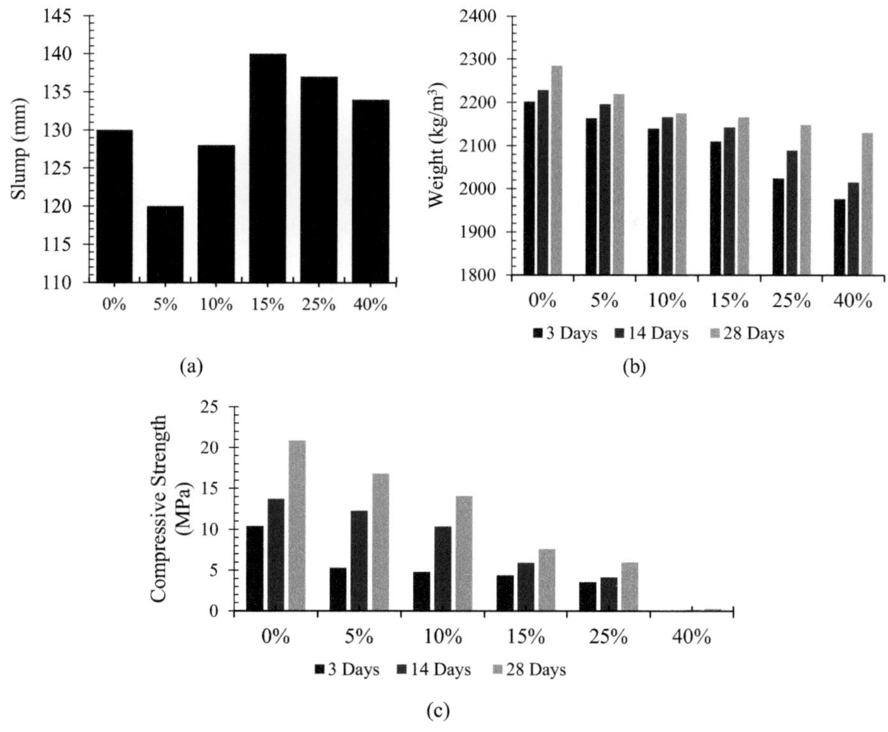

Fig. 4 **a** Slump value, **b** Weight of the cube specimen (150 mm × 150 mm), and **c** Compressive strength at different durations (3,14,28 days) for the partially replaced (WTP sludge is replaced with cement by 0–40%) concrete

compressive strength for the partially replaced concrete at various durations (3, 14, and 28 days) and shows a downward trend in strength development as the proportion of sludge increases. The control concrete is shown to have developed well from the beginning to the final stages and met the targeted strength of 3000 psi for the concrete. However, the value of compressive strength appears to be declining as the sludge percentage increases from 5 to 40%, with the lowest value of compressive strength at 40% sludge replacement. The rate of compressive strength decline is seen to be uniform for samples with 5–10% sludge; however, the rate dramatically decreases when the proportion of sludge increases over 10%. WTP is found to be finer than OPC cement, with fineness values of 332 m^2/kg for WTPS and 211 m^2/kg for OPC cement, respectively. This causes the heat of hydration to be greater. The greater cross-sectional area of the WTPS particles may be the reason for the increased hydration, which results in a decrease in initial strength and an increase in ultimate final strength [21]. In addition, not every WTPS particle has a rough surface; some have a smooth surface, and a small number of particles are crystallized. This may cause the compressive strength to gradually decrease with an increase in WTPS concentration in concrete [14, 21]. In addition, less final strength may result from higher hydration heat [21].

3.3 Correlation and PCA Investigation

Figure 5 depicts the correlation plot between the WTPS replaced concrete parameters. The amount of WTPS replacement is strongly ($p < 0.05$) associated with the concrete's compressive strength, slump value, and weight. The negative connection between the concrete's compressive strength and weight implies that both metrics decrease as the replacement quantity of WTPS increases. However, as can be seen in Fig. 5, there is a considerable positive association between the concrete slump value and the amount of WTPS replacement. This substantial positive connection ($p < 0.05$) shows that the slump value will rise in tandem with the replacement of WTPS in concrete.

4 Conclusion

In this study, the effects of using modified air-dried WTP sludge as a partial replacement for cement (passing through #200 sieve) in concrete were investigated. The study's main results indicate that incorporating modified air-dried WTP sludge into concrete as a partial replacement for cement can improve the workability of the concrete and reduce its weight. The main findings of the study are given as follows:

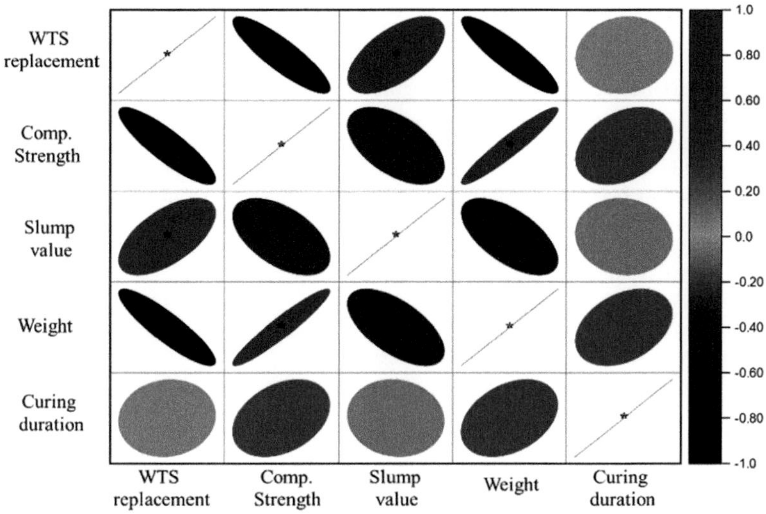

Fig. 5 Correlation matrix of amount WTPS replacement and different concrete properties *$* p < 0.05$ level.*

1. The study found that substituting up to 15% of cement with modified air-dried WTP sludge in concrete can increase workability and reduce the weight of the concrete by 6–10%, making it a promising alternative for developing lightweight concrete.
2. There is a positive correlation between the slump value of the concrete and the WTPS replacement rate, and the compressive strength decreases as the amount of WTP sludge increases. However, using WTP sludge at a lower proportion is preferable for structural applications.
3. The usage of WTP sludge in concrete has the potential to benefit both the environment and the construction industry by utilizing a waste product as a key component in concrete, and can also be an effective method for producing lightweight concrete. The physical, chemical, and morphological characteristics of the WTP sludge were also studied and identified using SEM–EDX testing.

However, it was shown that using WTP sludge in place of cement at a lower proportion was preferable for structural applications. It may be inferred that the usage of WTP sludge will be advantageous for both the environment and the construction industry by using one of its waste products as a key component of concrete.

References

1. Sivaramakrishnan S, Pondeepak P, Sivaguru C, Rajesh J (2020) Performance of concrete with partial replacement of sand by dry sewage sludge. Ilkogr Online—Elem Educ Online 19(4):6761–6767. https://doi.org/10.17051/ilkonline.2020.04.765079
2. Breesem KM, Faris FG, Abdel-Magid IM (2014) Reuse of alum sludge in construction materials and concrete works: A general overview. Infrastruct Univ Kuala Lumpur Res J 2(1):20–30
3. Sogancioglu M, Yel E, Yilmaz-Keskin US (2013) Utilization of andesite processing wastewater treatment sludge as admixture in concrete mix. Constr Build Mater 46(2013):150–155. https://doi.org/10.1016/j.conbuildmat.2013.04.035
4. Lynn CJ, Dhir RK, Ghataora GS, West RP (2015) Sewage sludge ash characteristics and potential for use in concrete. Constr Build Mater 98(2015):767–779. https://doi.org/10.1016/j.conbuildmat.2015.08.122
5. Asadollahfardi G, Delnavaz M, Rashnoiee V, Ghonabadi N (2016) Use of treated domestic wastewater before chlorination to produce and cure concrete. Constr Build Mater 105(2016):253–261. https://doi.org/10.1016/j.conbuildmat.2015.12.039
6. Mojapelo, KS, Kupolati WK, Ndambuki JM, Sadiku ER, Ibrahim ID (2021) Utilization of wastewater sludge for lightweight concrete and the use of wastewater as curing medium, Case Stud Constr Mater, 15(1–11) p e00667. https://doi.org/10.1016/j.cscm.2021.e00667
7. hai He Z et al. (2021) Recycling hazardous water treatment sludge in cement-based construction materials: Mechanical properties, drying shrinkage, and nano-scale characteristics, J Clean Prod, 290(2021) p 125832, https://doi.org/10.1016/j.jclepro.2021.125832
8. Chang FC, Lin JD, Tsai CC, Wang KS (2010) Study on cement mortar and concrete made with sewage sludge ash. Water Sci Technol 62(7):1689–1693. https://doi.org/10.2166/wst.2010.459
9. Vasudevan G (2019) Performance of alum sludge as partial replacement for cement adding superplasticizer. IOP Conf Ser Mater Sci Eng 652(1):1–7. https://doi.org/10.1088/1757-899X/652/1/012056
10. Martins JV, Garcia DCS, Aguilar MTP, Santos WJ (2021) Influence' study of heterogeneity on concrete sludge waste used as supplementary cementitious material, Constr Build Mater, 303(2021) p 124519, https://doi.org/10.1016/j.conbuildmat.2021.124519
11. Srinivasan K, Vazhviniyan R, Palpandi K (2016) Replacement of fine aggregate using sludge in concrete, Int Res J Eng Technol, 03(05), pp 1989–1993, 2016, Available: www.irjet.net
12. Kaosol T (2010) Reuse water treatment sludge for hollow concrete block manufacture. Energy Res. J 1(2):131–134. https://doi.org/10.3844/erjsp.2010.131.134
13. Alqam M, Jamrah A, Daghlas H (2011) Utilization of cement incorporated with water treatment sludge. Jordan J Civ Eng 5(2):268–277
14. Murugaboopathy R (2020) SEM and EDX analysis of waste paper sludge ash and cement, Publ Int J Trend Sci Res Dev, 4(5), pp 763–765, Available: https://www.ijtsrd.com/papers/ijtsrd32963.pdf%0Ahttps://www.ijtsrd.com/engineering/civil-engineering/32963/sem-and-edx-analysis-of-waste-paper-sludge-ash-and-cement/r-murugaboopathy
15. Subagdja A, Sofyan A, Rusmanto A (2020) Compressive strength and permeability of concrete by using GGBFS against seawater Compressive strength and permeability of concrete by using GGBFS against seawater. IOP Conf Ser Mater Sci Eng 830(022045):1–6. https://doi.org/10.1088/1757-899X/830/2/022045
16. Meddah MS, Zitouni S, Belâabes S (2010) Effect of content and particle size distribution of coarse aggregate on the compressive strength of concrete. Constr Build Mater 24(4):505–512. https://doi.org/10.1016/j.conbuildmat.2009.10.009
17. Budnikov PP, Erschler EY (1996) Studies of the processes of cement hardening in the course of Low-Pressure steam curing of concrete, Highw Res Board Spec Rep, 90(36), pp 431–446, Available: https://trid.trb.org/view/101495
18. Mardani-Aghabaglou A, Son AE, Felekoglu B, Ramyar K (2017) Effect of cement fineness on properties of cementitious materials containing high range water reducing admixture. J Green Build 12(1):142–167. https://doi.org/10.3992/1552-6100.12.1.142

19. Ehikhuenmen SO, Igba UT, Balogun OO, Oyebisi SO (2019) The influence of cement fineness on the structural characteristics of normal concrete. IOP Conf Ser Mater Sci Eng 640(1):1–8. https://doi.org/10.1088/1757-899X/640/1/012043
20. Abas M, Salem A, Pandey RK (2015) Influence of specific gravity on weight of proportions of concrete, Int J Eng Res Technol, 4(2), pp 1–3, 2015, Available: www.ijert.org
21. Kadri EH, Duval R (2009) Hydration heat kinetics of concrete with silica fume. Constr Build Mater 23(11):3388–3392. https://doi.org/10.1016/j.conbuildmat.2009.06.008

A MCDM Based Approach to Prioritizing National Highways for Road Safety Improvements

Md. Rifat Hossain Bhuiyan, Md. Asif Raihan, and Moinul Hossain

Abstract Selection of the most vulnerable alternative for implementation of road safety projects considering financial and technical availability often put the road authorities in dilemma. This paper proposes a two-step approach based on multi-criteria decision-making methods to overcome the difficulties in location prioritization. The objective of this study was to rank the national highways based on their vulnerability in terms of road safety and identify the location that requires the most attention. The study area covered Cumilla-Feni section of N1 national highway, Gazipur-Elenga section of N4 national highway, Natore-Nawabganj section of N8 national highway, and Barisal-Madaripur section of N8 national highway. These four alternatives were evaluated under five criteria- average annual daily traffic (AADT), crash per thousand vehicles, percentage of corridor without median, heavy vehicles percentage in the corridor, and percentage of non-motorized vehicles in the corridor. The required data for analysis were collected from some secondary sources along with drive-through video footage from a probe vehicle. To evaluate the relative importance of each criterion, criteria weights were calculated using the Analytic Hierarchy Process (AHP). AHP uses pair-wise comparison based on subjective judgment to establish relative importance and the consistency ratio provides a measure of the consistency of the judgment. Here, the consistency ratio was found 0.02 which does not exceed CR = 0.1, indicative of consistent judgment. The criteria weights and the criteria values were combined using the Technique for Order Preference

Md. Rifat Hossain Bhuiyan (✉)
Department of Civil and Environmental Engineering, Islamic University of Technology, Gazipur, Bangladesh
e-mail: rifathossain@iut-dhaka.edu

Md. Asif Raihan
Accident Research Institute, Bangladesh University of Engineering and Technology, Dhaka, Bangladesh
e-mail: raihan@ari.buet.ac.bd

M. Hossain
Department of Civil and Environmental Engineering, Islamic University of Technology, Dhaka, Bangladesh
e-mail: moinul@iut-dhaka.edu

S. Arthur et al. (eds.), *Proceedings of the 6th International Conference on Advances in Civil Engineering*, Lecture Notes in Civil Engineering 368,
https://doi.org/10.1007/978-981-99-3826-1_13

by Similarity to Ideal Solution (TOPSIS) and ranks of alternatives were achieved. The analysis identified Natore-Nawabganj (N6) as the most vulnerable alternative followed by Gazipur-Elenga (N4), Cumilla-Feni (N1), and Barisal-Madaripur (N8). This study proposes a framework for evaluation of alternatives that is reliable, and data driven. The results indicate that the proposed framework possesses replicability and promises effective decision-making based on scientific approach.

Keywords MCDM · AHP · TOPSIS · Road safety · Ranking · Prioritization

1 Background

An acceptable approach to ensure efficient allocation of limited resources which also imparts maximum benefit to the society often puts the decision makers in disarray. A prioritization framework that is technically sound and easily understandable by the stakeholders can aid decisionmakers in numerous ways. This type of framework is more suitable for safety improvement projects where benefits of project implementation are immeasurable in most cases. Where economic viability of a project can be the most important metric for the selection of a general development project, safety improvement project's feasibility may depend on many more important factors such as- current crash statistics, demography of the road users, existing road safety features, speed differential in the corridor etc.

The application of multi-criteria-decision-making (MCDM) methods in project appraisal or location prioritization is not a very new concept in the transportation research domain. The pragmatic and inclusive nature of the framework has made MCDM methods a top pick for decision makers. Use of MCDM methods such as Analytic Hierarchy Process (AHP), Analytic Network Process (ANP), Technique for Order Preference by Similarity to Ideal Solution (TOPSIS), Best Worst method, Decision Making Trial and Evaluation Laboratory (DEMATEL), Preference Ranking Organization Method for Enrichment of Evaluations (PROMETHEE) are evident in the existing works of literature [1–9]. However, the Analytic Hierarchy Process (AHP) proposed by Saaty [10] is considered the pioneer of all MCDM methods. In AHP, alternatives are evaluated based on several criteria. Later, the wight of each criterion is obtained through a pair-wise comparison matrix developed by experts' consultation. The application of AHP in solving transportation related issues has become very prevalent in recent years [11]. Use of AHP specially in the road transportation mostly encompassed decision and planning related problems [11–14]. From literature, Agarwal et al. [6] proposed an AHP based methodology for ranking hazardous locations. They divided the hazardous condition into different factors and applied the AHP to evaluate the weightage of each factor. Jun et al. [15] assessed the variables involved in virtual road safety audit using AHP. The experts were asked to take survey on AHP based questionnaire based on which the pair-wise comparison matrix was developed. Results indicated that the suggested strategy gave valuable guidance on creating an appropriate experiment for a road safety audit.

The AHP methodology is often presented in conjunction with other approaches to bolster the scientific acceptance of the framework. Fancello et al. [8] aimed to rank road intersections for road safety improvements using Electre III and concordance analysis. For further analysis, the authors used Vikor and TOPSIS and compared the two methods. They found TOPSIS performing the best in determining critical road intersection. Pal et al. [7] considered several techno-economic factors for prioritizing among fifteen state highway sections. They determined the weightage of the factors using AHP and ranked the highway sections using TOPSIS and RIDIT. In this study, AHP has been proposed in conjunction with Technique for Order Preference by Similarity to Ideal Solution (TOPSIS). The AHP was used the evaluate the criterion weights whereas TOPSIS was used to rank the alternatives based on overall score.

2 Study Area and Data Collection

The study area covered Cumilla-Feni section of N1 national highway, Gazipur-Elenga section of N4 national highway, Natore-Nawabganj section of N8 national highway, and Barisal-Madaripur section of N8 national highway. These four alternatives were evaluated under five criteria- average annual daily traffic (AADT), crash per thousand vehicles, percentage of corridor without median, heavy vehicles percentage in the corridor, and percentage of non-motorized vehicles in the corridor. The required data for analysis were collected from some secondary sources along with drive-through video footage from a probe vehicle as shown in Table 1.

3 Methodology

3.1 Analytic Hierarchy Process (AHP)

The AHP methodology adapts the eigenvector method to determine criteria weights [10]. The Perron-Frobenius theory serves as the foundation for this strategy, according to which the greatest eigenvalue may be found given a positive matrix with the weights vector as its associated eigenvector. The eigenvector contains nonnegative entries, and after being normalized, the sum of its parts equals one, resulting in a vector of relative weights [15]. The simple step by step approach within the AHP methodology is described here.

Step 1: Determining the criteria and alternatives for evaluation.

Step 2: Preparing comparison matrix based on experts' opinion.

Table 1 The criteria for evaluation, observed values of criteria and source of data

Criteria and their observed values	Cumilla-Feni (N1)	Gazipur-Elenga (N4)	Natore-Nawabganj (N6)	Barisal-Madaripur (N8)	Data source
Annual average daily traffic	13,995	30,409	21,446	8533	RMMS database
Crash data (2011–2015), Selected section	13	19	7	2	Accident Research Institute (ARI)
Crash per thousand vehicles	0.93	0.62	0.33	0.23	
Length of corridor (KM)	81.2	71.3	61	60	RMMS database
Length of corridor without median (KM)	0	7.41	61	59.7	Satellite imagery and drive through footage
Percentage of corridor without median	0	10.39	100	99.5	
Heavy vehicle AADT	7223	22,223	1798	3271	RMMS database
Percentage of heavy vehicle in the corridor	51.61	73.08	8.38	38.33	
Non-motorized vehicle AADT	197	1111	5027	153	RMMS database
Percentage of non-motorized vehicle in the corridor	1.41	3.65	23.44	1.79	

RMMS Database-road management and maintenance system database by roads and highways department

$$
A = \begin{array}{c} \\ a_1 \\ a_2 \\ \vdots \\ a_n \end{array}
\begin{array}{c} \begin{array}{cccc} c_1 & c_2 & \cdots & c_n \end{array} \\
\begin{vmatrix} \frac{w_1}{w_1} & \frac{w_1}{w_2} & \cdots & \frac{w_1}{w_n} \\ \frac{w_2}{w_1} & \frac{w_2}{w_2} & \cdots & \frac{w_2}{w_n} \\ \vdots & \vdots & \ddots & \vdots \\ \frac{w_n}{w_1} & \frac{w_n}{w_2} & \cdots & \frac{w_n}{w_n} \end{vmatrix} \end{array} \quad (1)
$$

Table 2 Saaty's nine-point scale of preference [10]

Verbal preference	Numerical score
Extremely preferred	9
Very strongly preferred	7
Strongly preferred	5
Moderately preferred	3
Equally preferred	1

Here, $[c_1, c_2, \ldots, c_n]$ indicate the criteria for evaluation and $[a_1, a_2, \ldots, a_n]$ indicate the alternatives that are to be ranked and prioritized. The pair-wise comparison matrix is developed complying with 9-point Saaty scale (Table 2).

Step 3: Developing normalized matrix and calculating priority vectors. As per the Perron-Frobenius theory, the greatest eigenvalue may be found given a positive matrix with the weights vector as its associated eigenvector. In simple words, the sum of rows of the squared comparison matrix are normalized to obtain the priority vectors/ vector of weights.

Step 4: Checking for inconsistency in the subjective judgement. Since the pair-wise comparison matrix is developed by experts' opinion, and the judgement is bound to be kept within the discrete nine-point Saaty scale, inconsistency in the judgement may arise. Hence, Consistency Index (CI) is measured.

$$CI = \frac{\lambda_{max} - n}{n - 1} \tag{2}$$

Here, λ_{max} is the indicator of highest eigenvalue and the number of criteria is represented by n. Consistency Ration (CR) can be calculated further comparing with the Random Index (RI) where if CR < 0.1, the results must be rejected. Random index is the consistency index of a randomly generated pair-wise comparison matrix (Table 3).

$$CR = \frac{CI}{RI} \tag{3}$$

Table 3 Random Index for different number of criteria [16]

n	1	2	3	4	5	6	7	8	9	10
RI	0	0	0.58	0.9	1.12	1.24	1.32	1.41	1.45	1.49

3.2 Technique for Order Preference by Similarity to an Ideal Solution (TOPSIS)

Technique for Order Preference by Similarity to an Ideal Solution (TOPSIS) was initially presented by Chen and Hwang [17] in 1992. The fundamental tenet of this approach is that the optimal compromise alternative should be the furthest away from the negative-ideal solution and the closest to the ideal solution [8]. Simple steps followed in TOPSIS to complete alternative ranking are shown here [18].

Step 1: Formation of a decision matrix.

Step 2: Normalized decision matrix development. If the normalized value is r_{ij}.

$$r_{ij} = \frac{x_{ij}}{\sqrt{\sum_{i=1}^{n} x_{ij}^2}}$$

Here, x_{ij} indicates the objective function value i for an alternative.

Step 3: Developing a weighted normalized decision matrix.

$$v_{ij} = w_{ij} r_{ij}$$

Here, ith objective function's weight is w_i

Step 4: Generation positive ideal and negative ideal solution. Here, $A^* = \{(\max v_{ij} | j \in J), (\min v_{ij} | j \in J\prime)\}$ and $A^- = \{(\min v_{ij} | j \in J), (\max v_{ij} | j \in J\prime)\}$

Here, $J = 1, 2, 3,\ldots\ldots.$,n; J indicates benefit criteria and $J\prime = 1, 2, 3,\ldots\ldots.$,n; $J\prime$ indicates disbenefit criteria.

Step 5: Determining the separation of each alternative from positive and negative ideal solution.

Here, $D_i^* = \sqrt{\sum_{j=1}^{n} (v_{ij} - v_j^*)^2}$ is the separation from positive ideal solution and, $D_i^- = \sqrt{\sum_{j=1}^{n} (v_{ij} - v_j^-)^2}$ is the separation from negative ideal solution.

Step 6: Finding the ideal closeness.

$$C_j^* = \frac{D_j^-}{(D_j^* + D_j^-)}$$

C_j^* ranges between 1 and 0 and it the measure of ranking alternatives. The larger C_j^* indicates better performance.

4 Analysis and Result

This study aimed to identify the most vulnerable highway intersection from road safety perspective base on five assessment criteria: Average Annual Daily Traffic (AADT), Crash per thousand vehicle, percentage of corridor without median, percentage of heavy vehicle in the corridor and percentage of non-motorized vehicle in the corridor (Table 4).

The criteria selection was done through extensive literature review and suggestion from the stakeholder organizations- World Bank (WB), Dhaka Transport Coordination Authority (DTCA), Roads and Highways Department (RHD), and Accident Research Institute (ARI). Due to the absence of authentic safety critical data, AADT was considered as a measure of traffic exposure in the corridor. Crash data provided by the ARI for selected highway chainage was converted into crash per thousand vehicles for gaining consistency among the alternatives. Absence of median, percentage of heavy and non-motorized vehicles were included into the analysis because of the safety threats posed by speed variation in the multi-modal carriageway and chance of head on collision. Accident reports showed that 77% of the highway accidents had involvement of heavy vehicles such as bus or truck and 50% of the accidents were associated with pedestrians, cyclists, motorcyclists or NMVs [19].

4.1 *Determination of Criteria Weights*

In this study, AHP was used for determining the weights of each criterion. A pair-wise comparison matrix was developed by consulting with the stakeholders. Prior briefing sessions were arranged to ensure that the stakeholders are familiar with AHP framework and can deliver consistent judgement. Lastly, the matrix was finalized by the authors after consulting with the transportation safety experts with know-how in ranking, prioritization, and road safety (Table 5).

The comparison scale proposed by Saaty [10] was used to develop the comparison matrix. From the matrix we can see that, AADT is three times less preferred than crash per thousand vehicles. In a similar process, through consultation and surveys, comparison scores were assigned to other pairs. Once the pair-wise comparison

Table 4 Criteria names and criteria codes

Criteria name	Criteria code
Annual average daily traffic	AADT
Crash per thousand vehicles	CPTV
Percentage of corridor without median	% CWM
Percentage of heavy vehicle in the corridor	% HV
Percentage of non-motorized vehicle in the corridor	% NMV

Table 5 The pair-wise comparison matrix for AHP

	AADT	CPTV	% CWM	% HV	% NMV
AADT	1	1/3	1/2	1/2	1/2
CPTV	3	1	2	3	3
% CWM	2	1/2	1	2	2
% HV	2	1/3	1/2	1	1
% NMV	2	1/3	1/2	1	1

matrix was formed, priority vectors/eigen vectors were calculated by simple matrix operations through trial and error (Tables 6 and 7).

The trial-and-error approach to determine priority vector was stopped after second trial as no major difference among the priority vectors were observed. The AHP methodology identified crash per thousand vehicles (CPTV) as the most important criteria (weightage = 40%) followed by percentage of corridor without median (weightage = 24%), percentage of heavy vehicle in the corridor (weightage = 14%),

Table 6 Fist trial of priority vector determination

Trial-1

	AADT	CPTV	% CWM	% HV	% NMV	Sum of row of squared matrix	Priority vector
AADT	4.99	1.24	2.16	3.49	3.49	15.37	0.09
CPTV	22	4.97	8.5	14.5	14.5	64.47	0.4
% CWM	13.5	2.98	5	8.5	8.5	38.48	0.24
% HV	7.99	1.9	3.16	4.99	4.99	23.03	0.14
% NMV	7.99	0.66	3.16	4.99	4.99	21.79	0.13
Total						163.14	1

Table 7 Second trial of priority vector determination

Trial-2

	AADT	CPTV	% CWM	% HV	% NMV	Sum of Row of Squared Matrix	Priority vector
AADT	137.11	27.722	54.175	88.585	88.585	396.1773	0.1
CPTV	565.58	114.431	223.905	365.805	365.805	1635.526	0.4
% CWM	336.255	68.211	133.21	217.655	217.655	972.986	0.24
% HV	204.07	41.542	80.745	132.095	132.095	590.5473	0.14
% NMV	176.79	5.273	70.205	114.115	114.115	480.4983	0.12
Total						4075.7349	1

Table 8 Consistency ratio determination

	AADT	CPTV	% CWM	% HV	% NMV	Row avg.
AADT	1	0.33	0.5	0.5	0.5	0.566
CPTV	3	1	2	3	3	2.4
% CWM	2	0.5	1	2	2	1.5
% HV	2	0.33	0.5	1	1	0.966
% NMV	2	0.33	0.5	1	1	0.966
Row avg.		Priority values		Priority value/row avg. (X)		Avg. (X) λ_{max}
0.566		3.074		5.431		5.094
2.4		12.894		5.373		
1.5		7.696		5.131		
0.966		4.606		4.768		
0.966		4.606		4.768		
CI	0.0235	RI	1.12	CR	0.02	

percentage of non-motorized vehicles in the corridor (weightage = 12%) and lastly Average Annual Daily Traffic (weightage = 10%).

4.2 *Consistency of Judgement*

Consistency Ratio is the measure of consistency of the judgement in AHP. The developed matrix is compared with the consistency index of a randomly developed one to determine the consistency ratio. If CR < 0.1, the developed comparison matrix cannot be claimed any better than a randomly developed matrix (Table 8).

From analysis, the consistency ratio was found CR = 0.02 < 0.1, indicative of a consistent judgement by the stakeholders.

4.3 *Ranking of the Alternatives*

The weight of the criteria obtained from AHP analysis were applied in the TOPSIS framework to rank the highway sections (Table 9).

The TOPSIS analysis identified Natore-Nawabganj section of N6 national highway as the most vulnerable highway section among the four alternatives. This highway corridor had zero percent road median coverage, remarkably high NVM volume with a moderate crash record. With a score very close to the N6 national highway, Gazipur-Elenga section of N4 national highway was identified as the second

Table 9 Determination of the rank of alternatives by TOPSIS

Alternative/Criteria	AADT X1	CPTV X2	% CWM X3	% HV X4	% NMV X5
Cumilla-Feni (N1)	13995	0.93	0	51.61	1.41
Gazipur-Elenga (N4)	30409	0.62	10.39	73.08	3.65
Natore-Nawabganj (N6)	21446	0.33	100	8.38	23.44
Barisal-Madaripur (N8)	8533	0.23	99.5	38.33	1.79
Normalized Decision Matrix	NDX1	NDX2	NDX3	NDX4	NDX5
Criteria Weights From AHP	0.1	0.4	0.24	0.14	0.12
Cumilla-Feni (N1)	0.3442	0.7829	0	0.5283	0.0592
Gazipur-Elenga (N4)	0.7479	0.5219	0.0735	0.7481	0.1532
Natore-Nawabganj (N6)	0.5274	0.2778	0.707	0.0858	0.9836
Barisal-Madaripur (N8)	0.2099	0.1936	0.7034	0.3924	0.0751
A*	0.0748	0.3132	0.1697	0.1047	0.118
A-	0.021	0.0774	0	0.012	0.0071
Relative Closeness	D*	D-	C	Rank	
Cumilla-Feni (N1)	0.2088	0.9335	0.8172	3	
Gazipur-Elenga (N4)	0.2095	1.1378	0.8445	2	
Natore-Nawabganj (N6)	0.2234	1.3249	0.8557	1	
Barisal-Madaripur (N8)	0.27	0.8326	0.7551	4	

most vulnerable location with very high traffic volume in the corridor, relatively high accident record and extremely high heavy vehicle movement.

5 Conclusions and Policy Implications

The goal of this study was to develop a data driven location prioritization framework for road safety improvement projects that can guide policymakers in the decision-making process. The proposed framework adapted a two-step approach based on multi-criteria decision-making methods to overcome the difficulties in location prioritization. In the initial stage, decisions were made by the major stakeholders to evaluate the candidate highway sections based on five important criteria. The AHP methodology was followed to determine the weightage of each criterion through developing pair-wise comparison matrix. Once the criteria weightage was achieved, state of the art ranking method TOPSIS was employed to rank the alternatives. Results showed that the Natore-Nawabganj section of N6 national highway was the most vulnerable location as this highway corridor had zero percent road median coverage, very high NVM volume with a moderate crash record. Furthermore, Gazipur-Elenga section of N4 national highway was identified as the second most vulnerable location with very high traffic volume in the corridor, relatively high accident record and extremely high heavy vehicle movement.

The proposed framework, if adapted by the transportation professionals and the decision-makers, can reduce significant delay in determining suitable candidate for project implementation. Besides, the common issues faced at the policy level such as lack of safety critical data, incorporating stakeholders input into the decision-making process, overcoming delay in the decision-making process etc. can be easily marginalized by this framework. Overall, the proposed methodology provides decision-makers with a tool that is data driven, intuitive and adjustable to user's demand.

Acknowledgements The authors would like to acknowledge World Bank (WB) and Roads and Highways Department (RHD) for their assistance in conducting this study.

References

1. Morfoulaki M, Papathanasiou J (2021) Use of promethee mcda method for ranking alternative measures of sustainable urban mobility planning. Mathematics 9. https://doi.org/10.3390/math9060602
2. Broniewicz E, Ogrodnik K (2021) A comparative evaluation of multi-criteria analysis methods for sustainable transport. Energies 14. https://doi.org/10.3390/en14165100
3. Bhuiyan MRH, Raihan MA, Hossain M (2022) Prioritizing locations for safety improvement: an integrated disutility-based approach. In: International conference on transportation and development 2022. American Society of Civil Engineers, Reston, VA, pp 287–300. https://doi.org/10.1061/9780784484333.026
4. Ivanović I, Grujičić D, Macura D, Jović J, Bojović N (2013) One approach for road transport project selection. Transp Policy 25:22–29. https://doi.org/10.1016/j.tranpol.2012.10.001
5. Sara J, Stikkelman RM, Herder PM (2015) Assessing relative importance and mutual influence of barriers for CCS deployment of the ROAD project using AHP and DEMATEL methods. Int J Greenh Gas Control 41:336–357. https://doi.org/10.1016/j.ijggc.2015.07.008
6. Agarwal PK, Patil PK, Mehar R (2013) A methodology for ranking road safety hazardous locations using analytical hierarchy process. Procedia—Soc Behav Sci 104:1030–1037. https://doi.org/10.1016/j.sbspro.2013.11.198
7. Pal S, Maitra B, Sarkar JR (2016) An approach for prioritization of state highways and its application. Transp Dev Econ 2:1–10. https://doi.org/10.1007/s40890-016-0017-6
8. Fancello G, Carta M, Fadda P (2019) Road intersections ranking for road safety improvement: comparative analysis of multi-criteria decision making methods. Transp Policy 80:188–196. https://doi.org/10.1016/j.tranpol.2018.04.007
9. Gan A, Alluri P, Raihan MA, Liu K, Saha D, Jung R (2017) Automated system to prioritize highway improvement locations and to analyze project alternatives. Transp Res Rec 2654:65–75. https://doi.org/10.3141/2654-08
10. Saaty RW (1987) The analytic hierarchy process—what it is and how it is used. Math Model 9:161–176. https://doi.org/10.1016/0270-0255(87)90473-8
11. Barić D, Pilko H, Strujić J (2016) An analytic hierarchy process model to evaluate road section design. Transport 31:312–321. https://doi.org/10.3846/16484142.2016.1157830
12. Holguín-Veras J (1995) Comparative assessment of AHP and MAV in highway planning: case study. J Transp Eng 121:191–200. https://doi.org/10.1061/(ASCE)0733-947X(1995)121:2(191)
13. Klungboonkrong P, Taylor MAP (1999) An integrated planning tool for evaluating road environmental impacts. Comput Civ Infrastruct Eng 14:335–345. https://doi.org/10.1111/0885-9507.00152
14. Wang WC, Der YuW, Yang IT, Lin CC, Lee MT, Cheng YY (2013) Applying the AHP to support the best-value contractor selection-lessons learned from two case studies in Taiwan. J Civ Eng Manag 19:24–36. https://doi.org/10.3846/13923730.2012.734851

15. Jun Y, Go J, Yeom C (2022) Experimental variables assessment for virtual road safety audit using analytic hierarchy process. J Transp Saf Secur 14:1002–1021. https://doi.org/10.1080/19439962.2021.1883169
16. Saaty TL (1982) Priority Setting in Complex Problems. IEEE Trans Eng Manag EM-30:140–155. https://doi.org/10.1109/tem.1983.6448606
17. Chen S-J, Hwang C-L (1992) Fuzzy multiple attribute decision making methods. In: Chen S-J, Hwang C-L (eds). Springer, Berlin, Heidelberg, pp 289–486
18. Wangchen Bhutia P (2012) Appication of AHP and TOPSIS method for supplier selection problem. IOSR J Eng 02:43–50. https://doi.org/10.9790/3021-021034350
19. Pervaz S, Al A, Ashek N (2021) overview of the highway crashes in Bangladesh. In: 5th international conference on civil engineering for sustainable development (ICCESD 2020), Bangladesh

Predicting Future Land Use and Land Cover Changes and Their Effects on Land Surface Temperature in Chattogram City of Bangladesh

T. Das, M. I. Islam, and D. R. Raja

Abstract Heatwave is one of the most dangerous "natural hazards" with the greatest impact on people and other living beings that rarely receive adequate attention worldwide. Climate change & intense landcover conversion result in extreme heatwaves in urban areas. Chattogram, which is located in the southeast of Bangladesh, is a prominent coastal city and key economic hub. Chattogram City Corporation (CCC) consists of 41 wards with a large population, and most of the heavy, medium, and light industries of Bangladesh are situated in Chattogram. Due to this commercial and industrial development, and rapid urbanization, the landcover of the CCC area has changed expeditiously which is responsible for the urban heatwaves hazard. The study intends to comprehend the future heat wave scenario. For a better understanding of heatwaves, it has been analyzed the temporal changing and relationship between the Land Surface Temperature (LST) and the Land Use & Land Cover (LULC) from 2000 to 2020. Based on this data, the Artificial Neural Network (ANN), Markov chain model (MC), and Cellular Automata (CA) models have been applied to the prediction of future LST and LULC in different two years, 2030 and 2040. The temporal analysis result shows that the LST of the CCC area has risen with the decrease of vegetated areas and water bodies. The model simulation result suggests that the total buildup area of the CCC will be increased around 45% and 60% in 2030 and 2040 respectively, and 13 and 43% of the CCC area will suffer temperatures higher than 36° C in 2030 and 2040, respectively which are considered to be more vulnerable to the people. Policymakers will find this research useful in interpreting the effects of LULC change on LST and in recommending management frameworks. It will also be favorable for analyzing future vulnerability in the CCC area due to heat stress.

T. Das (✉) · D. R. Raja
Department of Urban and Regional Planning, Chittagong University of Engineering and Technology (CUET), Chattogram, Bangladesh
e-mail: trisa.das.td@gmail.com

D. R. Raja
e-mail: debasishroy@cuet.ac.bd

M. I. Islam
Center for Environmental and Geographic Information Services (CEGIS), Dhaka, Bangladesh

S. Arthur et al. (eds.), *Proceedings of the 6th International Conference on Advances in Civil Engineering*, Lecture Notes in Civil Engineering 368,
https://doi.org/10.1007/978-981-99-3826-1_14

Keywords Land Surface Temperature (LST) · Land Use and Land Cover (LULC) · Heatwave · Markov Chain Model (MC) · Artificial Neural Network (ANN)

1 Introduction

World temperature is getting warmer with the rhythm of time. Incidents of extreme heat are becoming more often and more intense on a global scale, and this trend is predicted to continue into the twenty-first century [1, 2]. Currently, 50% (or 3.5 billion) people live in urbanized regions around the world, and by 2030, that number is anticipated to rise to 60% (4.9 billion). This large populations with their significant social inequality will make them more vulnerable to climate change on a social level [3]. The frequency of heat waves and their harmful impacts would increase due to the rise in global temperatures. The effects of heat waves is reliant on the severity and duration of the individual events along with environmental and socio-demographic factors [4]. As we all know heat waves can cause numerous heat-related illnesses and deaths which can be a major problem for developing countries like Bangladesh. Land surface temperature (LST) is a major contributing factor attributed to an increase in urban heat and microclimatic warming [5]. The conversion of natural land cover into artificial materials, causes an increase in sensible heat flow as well as air temperature rise [6]. Between 2000 and 2016, at least 136,835 people in Europe died of heat-related health issues, which represents more than 87% of all disaster-related deaths in that area [7].

Bangladesh is one of the foremost countries susceptible to the unpleasant effects of global warming. In 2003 around 62 people died due to heat waves across Bangladesh and the death rates rise by approximately 20% during heat waves [8]. Heat wave hazard mitigation is becoming a significant policy concern for Bangladesh's policy-makers, since it is a hazard to the country's ecosystem and habitats. Chattogram is the country's biggest port and the primary location for the development of heavy, medium, and light industries. Additionally, Chattogram is the site of Bangladesh's only oil refinery and steel factory. According to research, Chattogram city has observed significant increases in surface urban heat island intensity at night by 1.9 °C during the last 20 years [9].

This study has tried to understand the future heat wave scenario in Chattogram City for mitigating future heat related disaster. The present study aims to analyze the temporal change of land use & land cover (LULC) and land surface temperature (LST) during the year 2000–2020 in Chattogram City Corporation area. This research also aims to predict the future heat wave scenario in 2030 and 2040 in Chattogram City Corporation Area, Bangladesh applying the Markov chain model, Cellular Automata model, and Artificial Neural Network (ANN). Also, this study will help to investigate future heat wave vulnerability of this area.

2 Materials and Methods

The methodology of the study includes outline of the study area, data collection and analysis procedure have been mentioned during this section.

2.1 *Study Area Profile*

Chattogram is the 'Commercial Capital' and the second-biggest metropolitan city declared by the government of Bangladesh. It is considering the industrial, commercial and institutional potentials of the country. Chattogram is also known as a port city of Bangladesh and the third busiest international seaport in South Asia [10]. Most of the heavy, medium, and light industries of Bangladesh are situated in Chattogram. As well as Chattogram is also the location of Bangladesh's only steel mill and oil refinery. Chattogram City Corporation (CCC) has a total area of 160.99 km^2 and 41 wards with a large population. The amount of vegetation and water bodies has been continuously decreasing because of the rapid urbanization, commercialization, and industrial development. And this the main reason for increasing land surface temperature as the urban and industrial areas cannot absorb heat. So, increasing land surface temperature is a threatening issue for this city for now and in the nearest future (Fig. 1).

2.2 *Data and Methodology*

In this research, three Multi-spectral Landsat satellite data were obtained the United States Geological Survey (USGS) website. Landsat Thermal Infrared Sensor (TIRS), Thematic Mapper (TM), and Operational Land Imager (OLI) data were collected from 2000 to 2020 in 10-year intervals with 0% cloud coverage to analyze the land use and land cover (LULC) change and land surface temperature (LST) of the study area.

The acquired satellite images were categorized into five broad land cover types (built-up 01, built-up 02, water body, vegetation & bare soil) for the year of 2000, 2010, and 2020 on the basis of the Maximum Likelihood Supervised Classification (MLSC) technique. After the classification, an accuracy assessment has done for each LULC image for the validation purpose. For the ground truth verification using Google Earth, a total number of 400 sample pixels have been generated based on random sampling [11]. The land surface temperature (LST) has been obtained from the thermal bands of Landsat-5 TM (band 6) and Landsat-8 OLI (band 10). All the LST extraction was done in ArcGIS 10.7 with the help of model builder and raster calculator tools.

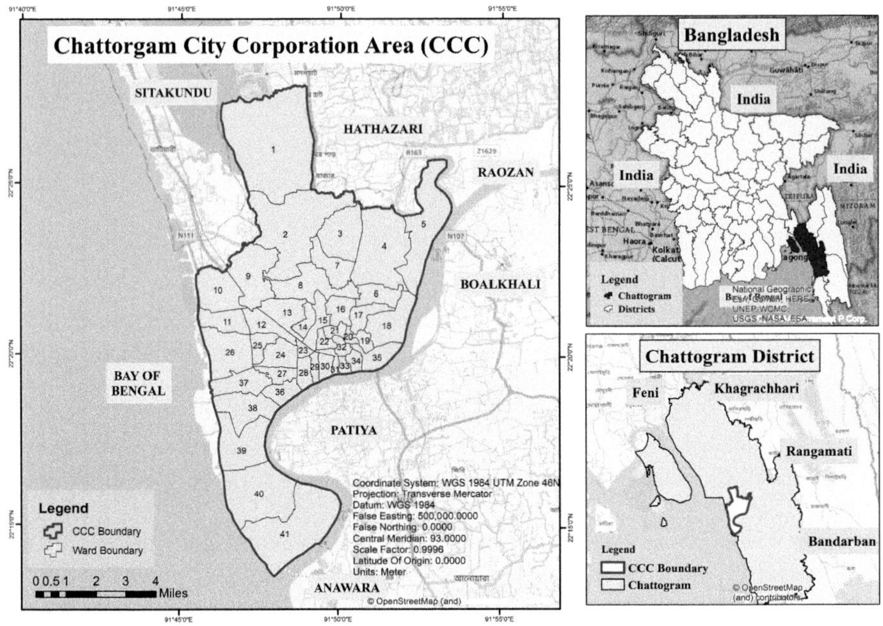

Fig. 1 Location of the study area

For predicting future LULC changes, a combination of the multilayer perceptron (MLP) model, which is an artificial neural network model [12] and the Markov Chain (MLP-MC) model have been used. The Land Change Modeler (LCM) tool has been used to run the model in IDRISI Selva V.17 software. There are two variables in the MLP model which are the land cover images are input as dependent variables and the variables controlling the land cover changes are input as independent variables [12]. For this study, 6 independent variables have been used that control the land cover change such as DEM, slope, aspect, distance to water bodies, major roads, buildup area. Distance to major roads, buildup areas, and water bodies have a great impact on future urban growth or development. An accuracy assessment has performed using the existing dataset for ensuring the model's acceptance. For the acceptance of the model, at first, the LULC of 2020 has been simulated. A validation has been done with the help of predicted LULC 2020 and existing LULC 2020. After validation with an acceptable level, the LULC 2030 and LULC 2040 have been simulated.

In this study, the prediction of LST for the years 2030 & 2040, an Artificial Neural Network (ANN) model has been applied with the assistance of MOLUSCE plugin in QGIS software. As input parameters, the study has been used LULC images, NDVI, NDWI, NDBI, NDBaI, SAVI, BUAI, and MNDWI. For the model's acceptance, a validation process had completed with the simulated LST of 2020 and the existing LST of 2020. After validation with an acceptable level, the LST 2030 and LST 2040 have been simulated as the same way of 2020 prediction (Fig. 2).

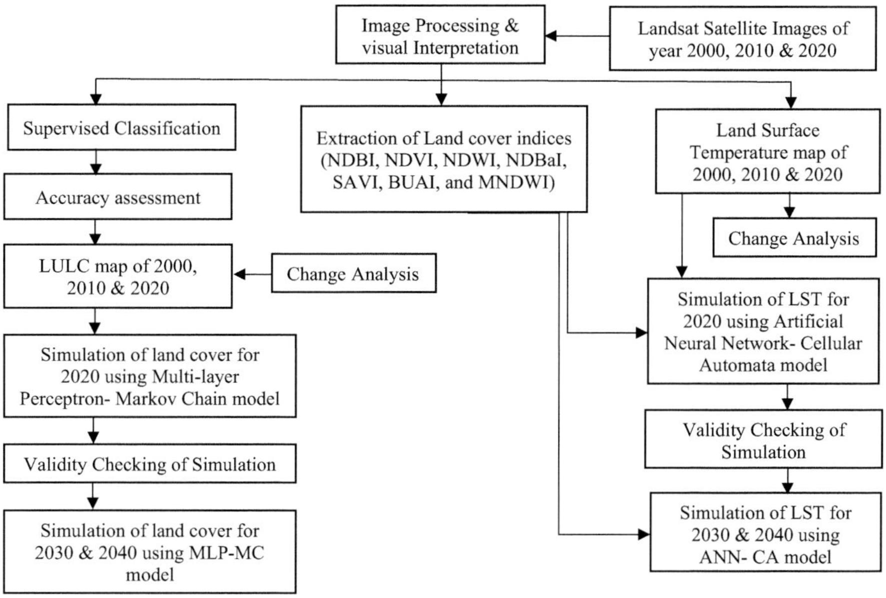

Fig. 2 Methodological flowchart of the overall study

3 Results and Discussions

According to the methodology, the spatiotemporal distribution of LST and LULC pattern as well as simulations of the future distribution LST and LULC were calculated for the area of study.

3.1 Temporal LULC and LST Change Analysis of CCC Area

Maximum Likelihood Supervised Classification method was applied to evaluate the LULC of year 2000–2020. The overall accuracies of the classified images (2000, 2010, and 2020) are, respectively 86.48%, 90.69%, and 94.83%, with Kappa coefficients of 0.86, 0.91, and 0.95. Figure 3 (1a–1c) represent the land classification of CCC area for three different years of 2000, 2010, and 2020. From the Fig. 4 it had found that between 2000 and 2020, the total area covered by the vegetation has reduced gradually. In year 2000, the total percentage of vegetation cover was around 57.37% which decreased into 39.01% in 2020. The buildup 01 areas which are mostly residential, commercial, roads, rail areas have increased 28.07% to 37.48% between 2000 to 2020. Most of the vegetation and waterbody area turned into buildup 01, buildup 02, and bare soil in the year 2020. The expansion of the urban area is the consequence of unplanned population expansion, industrialization and the trend for rural residents to move to the cities.

Landsat Thermal bands were used to determine the Land Surface Temperature for the years 2000 to 2020 using different formulas. The spatial distribution of LST in the CCC area present in Figs. 3(2a–2c). For better visual interpretation the LST range of each is divided into six uniform classes respectively <22, 22–24, 24–26, 26–28, 28–30, and >30 °C. The maximum LST of the year 2000 is 30.84 °C but

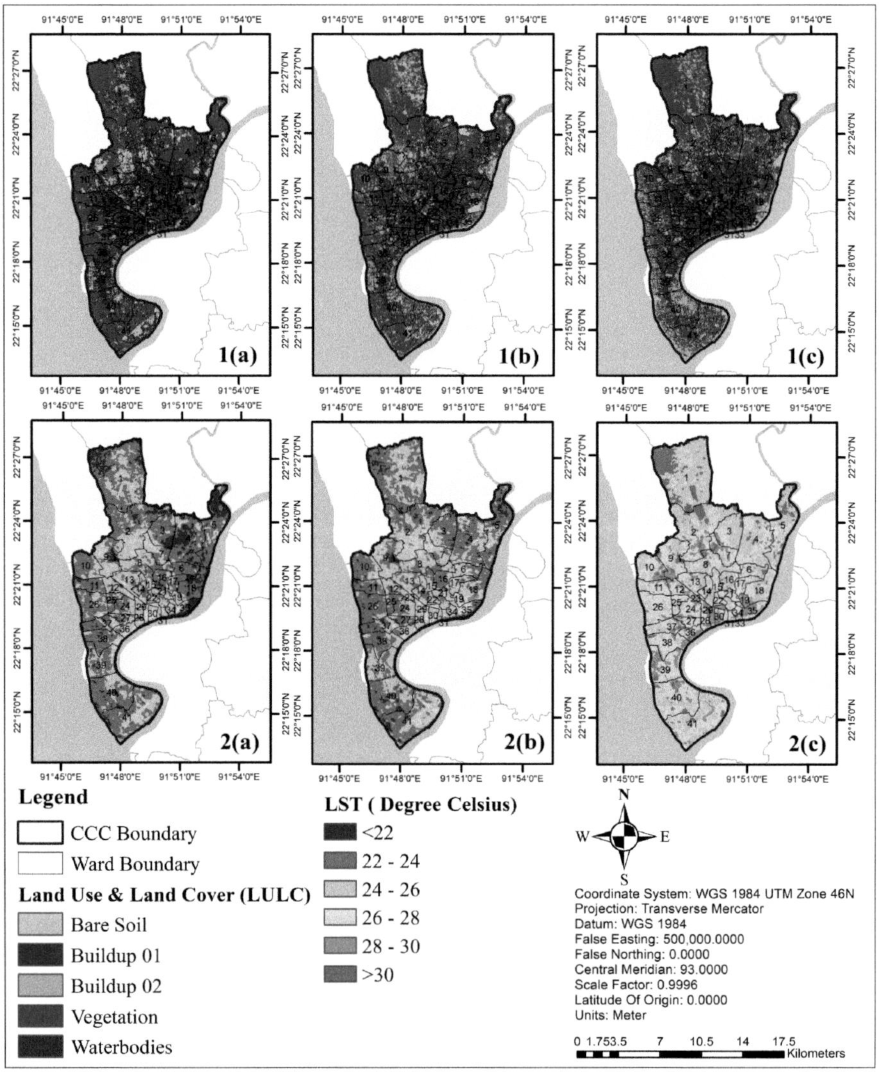

Fig. 3 LULC maps for years **1(a)** 2000, **1(b)** 2010 and **1(c)** 2020. Variation of LST for years **2(a)** 2000, **2(b)** 2010 and **2(c)** 2020

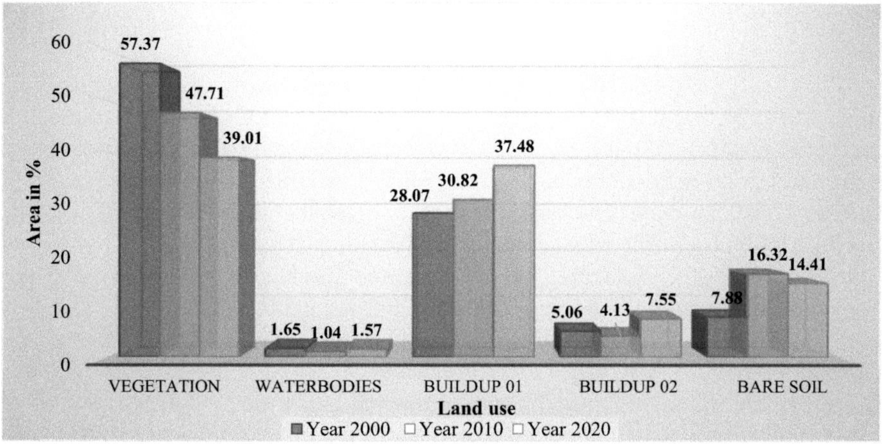

Fig. 4 Percentage area of different land cover

there are fewer areas with this temperature. The LST of the maximum area in the year 2010 is more than 24 °C. In the year 2020, the value of maximum LST increased to 35.95 °C and the maximum area's LST is more than 26 °C. The maximum LST in 2020 is very close to the mild heat wave according to Bangladesh Meteorological Department (BMD). According to Bangladesh Meteorological Department (BMD) 'heat wave' means when maximum day temperature attains 36 °C or more. From 2010 to 2020 the maximum LST increased around 5.11 °C between 10 years. The rapid rate of urbanization can be a major reason behind this increased LST in 2020. We have also found in the LULC map of the year 2020 that 45% of the total area of CCC is buildup area. These buildup areas generate more heat than other land uses as the conversion of the natural land cover into man-made materials like concrete and asphalt, increasing the surface temperature.

3.2 Simulation of LULC and LST of the Year 2030 and 2040

For the simulation of LULC of the year 2030 and 2040, the Multi-layer Perceptron-Markov Chain (MLP-MC) model were used. The Land Change Modeler (LCM) tool had been used to run this model in IDRISI Selva V.17 software. For the validation of the model, the LULC of the year 2020 were simulated first. The MLP model has two variables: independent and dependent. The LULC images used as the dependent variable and 6 independent variables had been used that control the land cover change such as DEM, aspect, slope, distance to water bodies, major roads, buildup area. All the input helped to run the model. After the prediction of LULC 2020, a validation process had done for the acceptance of the model with the help of exiting the LULC 2020 map. The overall kappa coefficient was 0.82 which is an almost perfect agreement [13].

After the LULC prediction of 2020, it can be said that the model is suitable for the prediction of 2030 and 2040. So, the LULC of 2010 and LULC of 2020 have input into the model as dependent variable and DEM, aspect, slope, distance to water bodies, major roads, and buildup area as the independent variable. The predicted LULC of 2030 and 2040 has been represented in Figs. 5. From the visual interpretation of the map in 2040, most of the area converted into buildup 01.

In 2030 44% of the total area will be converted into buildup 01 which are mostly residential, commercial, road, rail, and other infrastructures. And in 2040 it will be increased by 49% of the total CCC area (Table 1). In 2020 the 37% of the total area was buildup 01. The bare soil, vegetation, water bodies have decreased day by day. This rapid rate of buildup growth is the prime reason for the LST increase.

For the LST simulation of the year 2030 and 2040, the Artificial Neural Network-Cellular Automata (ANN-CA) model applied in QGIS using the MOLUSCE plugin.

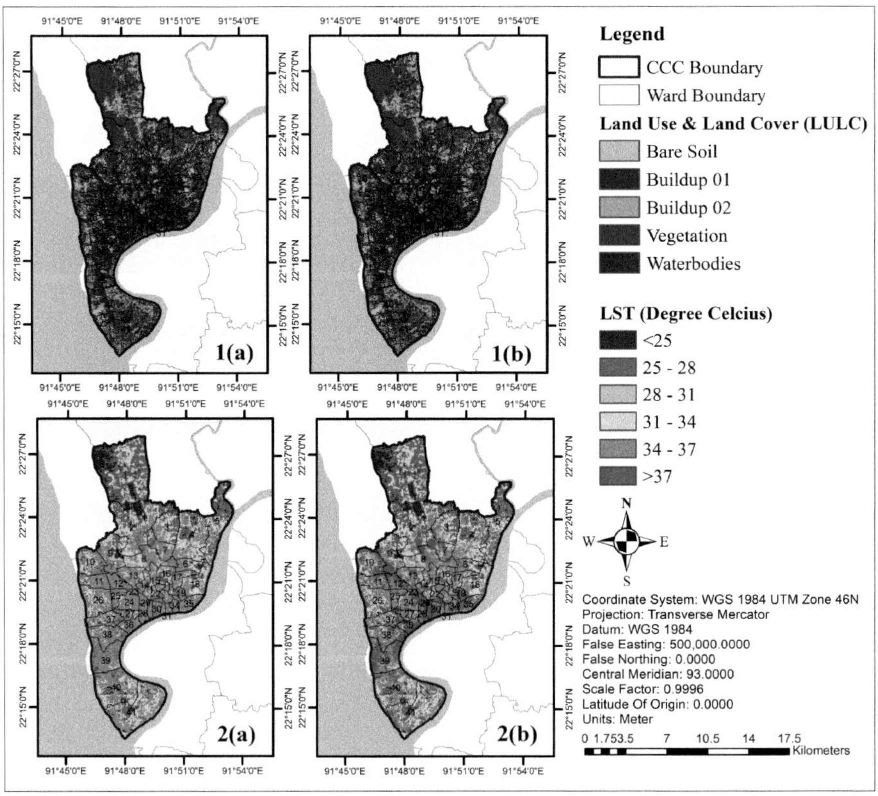

Fig. 5 Predicted LULC maps for years **1(a)** 2030 and **1(b)** 2040. Predicted of LST for years **2(a)** 2030 and **2(b)** 2040

Table 1 Predicted area of LULC 2030 and 2040

Land use	Area in %	
	Year 2030	Year 2040
Vegetation	35.03	33.28
Waterbodies	1.91	1.88
Buildup 01	44.99	49.87
Buildup 02	10.30	10.56
Bare soil	7.76	4.41

Same as the LULC prediction, the LST of 2020 predicted at first for the validation of the model. For the LST simulation of the year 2020, the LST map of 2000 used as the initial data and the LST map of 2010 used as the final data. Some spatial variable which influences the prediction process such as LLC images, NDVI, NDWI, NDBI, NDBaI, SAVI, BUAI, and MNDWI of the year 2000 and 2010 were used as the spatial variable. The kappa value of the model was 0.764 and the correctness with the existing LST of 2020 was 70.91%.

For 2030 and 2040 simulation the LST of 2010 input as initial data and LST 2020 input as final data. LULC images, NDVI, NDWI, NDBI, NDBaI, SAVI, BUAI, and MNDWI of the year 2000 and 2010 input as the spatial variable. Then the ANN model was run with the highest iteration of 1000 and 0.001 momentum for the model's better performance. The current validation kappa value of the model was 0.7037 which is a substantial agreement. After running the ANN model, the result input into the CA simulation for the final prediction output.

From the simulation of LST of year 2030 and 2040, the maximum temperature of the CCC area will be 39.61 °C in 2030 which is moderate heat wave according to Bangladesh Meteorological Department (BMD). And in 2040 the maximum LST will be 41.326 °C which is Severe heat wave. In 2030 most of the area's LST will be more than 34 °C and in 2040 it will be increased in 37 °C. In 2030 and 2040, around 13 and 43% of the CCC area will have temperatures higher than 35 °C.

4 Conclusion

With the increasing heat wave the normal lifestyle of people will be devastated in the future. From the analysis it can be seen that in 2030 and 2040 the temperature will be higher than 36 °C in the maximum area which is not suitable for our living condition. As well as from the analysis it can be realized that in 2040 the maximum area of the CCC will be converted into buildup area. Such continuous growth of buildup area along with LST will create several medical, economic, environmental problems for the city. So, from this study the most vulnerable area can be identified and proper mitigation strategies can be taken. From this study the similar study of mitigation and resilience capacity can be done by the researchers. It can be helpful to reduce the adverse effect of heat wave which is useful for inhabitants of these places.

Acknowledgements (if any) This research is the output of the author's undergraduate thesis. The authors gratefully acknowledge the Department of Urban and Regional Planning, Chittagong University of Engineering and Technology for facilitating all the requirements of this research.

References

1. Luber G, McGeehin M (2008) Climate change and extreme heat events. Am J Prev Med 35(5):429–435. https://doi.org/10.1016/j.amepre.2008.08.021
2. Weber S, Sadoff N, Zell E, de Sherbinin A (2015) Policy-relevant indicators for mapping the vulnerability of urban populations to extreme heat events: a case study of Philadelphia. Appl Geogr 63:231–243. https://doi.org/10.1016/j.apgeog.2015.07.006
3. Aubrecht C, Özceylan D (2013) Identification of heat risk patterns in the U.S. National Capital Region by integrating heat stress and related vulnerability. Environ Int 56:65–77. https://doi.org/10.1016/j.envint.2013.03.005
4. Macnee RGD, Tokai A (2016) Heat wave vulnerability and exposure mapping for Osaka City, Japan. Environ Syst Decis 36(4):368–376. https://doi.org/10.1007/s10669-016-9607-4
5. Maduako I (2015) Simulation and prediction of land surface temperature (LST) dynamics within Ikom City in Nigeria using artificial neural network (ANN). J Remote Sens GIS 5(1). https://doi.org/10.4172/2469-4134.1000158
6. Fabrizi R, Bonafoni S, Biondi R (2010) Satellite and ground-based sensors for the Urban Heat Island analysis in the city of Rome. Remote Sens 2(5):1400–1415. https://doi.org/10.3390/rs2051400
7. Nissan H, Burkart K, de Perez EC, Van Aalst M, Mason S (2017) Defining and predicting heat waves in Bangladesh. J Appl Meteorol Climatol 56(10):2653–2670. https://doi.org/10.1175/JAMC-D-17-0035.1
8. Rajib MA, Mortuza MR, Selmi S, Ankur AK, Rahman MM (2011) Increase of heat index over Bangladesh: impact of climate change. Int J Civ Environ Eng 5(10):434–437
9. Dewan A, Kiselev G, Botje D, Mahmud GI, Bhuian MH, Hassan QK (2021) Surface urban heat island intensity in five major cities of Bangladesh: patterns, drivers and trends. Sustain Cities Soc 71(April):102926. https://doi.org/10.1016/j.scs.2021.102926
10. Sarwar IM, Billa M, Paul A (2016) Urban land use change analysis using RS and GIS in Sulakbahar ward in Chittagong city, Bangladesh. Int J Geomat Geosci 7(1):1–10
11. Madow WG (1968) Elementary sampling theory. Technometrics 10(3):621–622. https://doi.org/10.1080/00401706.1968.10490610
12. Pereira e Silva L, Xavier APC, da Silva RM, Santos CAG (2020) Modeling land cover change based on an artificial neural network for a semiarid river basin in northeastern Brazil. Glob Ecol Conserv 21. https://doi.org/10.1016/j.gecco.2019.e00811
13. Das N, Mondal P, Sutradhar S, Ghosh R (2021) Assessment of variation of land use/land cover and its impact on land surface temperature of Asansol subdivision. Egypt J Remote Sens Sp Sci 24(1):131–149. https://doi.org/10.1016/j.ejrs.2020.05.001

Assessment of Air Quality and Noise Level in Chattogram City, Bangladesh

M. J. Alam, M. A. Aziz, A. Haque, and M. H. Masum

Abstract This study evaluated air quality and noise parameters in the city of Chattogram, Bangladesh. Public health studies are associated with both air pollution and noise pollution in terms of general health consequences due to the severity of pollution, including increased blood pressure, heart disease and other respiratory illnesses. In the urban areas, air pollution and noise pollution from various causes such as transportation, industry and construction activities are gradually and surprisingly increasing. The purpose of this study was to analyze air pollution and noise pollution in 15 different locations classified into industrial, commercial, and institutional areas. The evaluation was carried out for 6 weeks in the dry season. Noise levels, suspended particulate matter ($PM_{2.5}$, PM_{10}), formaldehyde (HCHO), carbon monoxide (CO), carbon dioxide (CO_2), total volatile organic carbon (TVOC), temperature, light, wind velocity, and relative humidity were measured. Measurements were collected in 15 specific locations within Chattogram city for both daytime and nighttime periods during the dry season (November–December 2021). These locations were chosen to represent three distinct types of land use. Average concentrations of $PM_{2.5}$, PM_{10}, CO, CO_2, HCHO and TVOC in the air of Chattogram city are 99, 115 $\mu g/m^3$, 46, 479, 16 and 1.47 mg/m^3, respectively. On the other hand, the noise levels in different locations shows significant variation (55–87 dB). Suspended particulate matter and gaseous pollutants (TVOC, HCHO, CO, CO_2) were found in greater concentration during the evening period than the day period respectively. Meteorological parameters (Temperature, Relative Humidity) were found hostile during the day period. The

M. J. Alam (✉) · M. A. Aziz · A. Haque
Department of Civil Engineering, Port City International University, Chittagong 4225, Bangladesh
e-mail: jahedm150@gmail.com

M. A. Aziz
e-mail: yasinmdabdulaziz@gmail.com

A. Haque
e-mail: akramul109183@gmail.com

M. H. Masum
Center for River, Harbor and Landslide Research (CRHLSR), Chittagong University of Engineering and Technology (CUET), Chattogram 4349, Bangladesh
e-mail: mehedi.ce@cuet.ac.bd

S. Arthur et al. (eds.), *Proceedings of the 6th International Conference on Advances in Civil Engineering*, Lecture Notes in Civil Engineering 368,
https://doi.org/10.1007/978-981-99-3826-1_15

equivalent noise level (Leq) was found to exceed the permissible limit. The maximum average concentration of gaseous pollutants exceeded acceptable limits, which poses environmental and health concerns for local residents. The air quality in those areas was poor and should be monitored on a regular basis. Therefore, regulatory agencies need to enforce environmental standards.

Keywords Air quality · Air pollution · Noise pollution · Public health

1 Introduction

Air pollution can be defined as the presence of contaminants such as sulfur dioxide (SO_2), particulate matter (PM), nitrogen oxides (NOx), carbon monoxide (CO), volatile organic compounds (VOCs), and ozone (O_3) in the air we breathe to a level that could have some adverse effects on the environment and human health [1]. From different sources, this pollutant emits and adversely affects human health and also leads to permanent chronic diseases [2]. Air pollution seriously threatens human health as it is responsible for low birth weight, high mortality, sudden infant death, cardiac arrest, and brain stroke [3]. About 80% of people in urban areas are exposed to air pollution exceeding air quality standards set by the World Health Organization (WHO), and 98% of middle-income countries fail to meet WHO guidelines [4]. Due to potential high levels of exposure to air pollution, road noise, and high traffic congestion, different modes of transportation may be related with health and safety issues [5]. Estimated over 4.2 million (outdoor) air pollution impacts cause deaths worldwide every year [6]. According to the World Health Organization (WHO), noise pollution is the third most hazardous type of pollution after air and water pollution [7]. Increased levels of air pollution are directly associated with high death rates and decreased life expectancy [8]. Residents who reside less than 300 m from a major road are more likely to be exposed to greater concentrations of pollutants such as particulate matter, carbon monoxide, and nitrogen oxide [9]. Particularly at peak traffic times, higher particle number concentration values were seen [3]. Numerous studies carried out by researchers worldwide have established a correlation between air pollutants and meteorological parameters, indicating that the concentration of particulate and gaseous pollutants in the air is significantly influenced by meteorological conditions [10–13].

A significant amount of literature contends that the health and wellbeing of people are adversely affected by traffic noise [14]. A substantial amount of literature suggests that noise from traffic has a negative impact on one's health and well-being [14]. Few studies have found that the level of noise pollution in Chattogram city is between 53–97 dB [15] and 65–85 dB [16]. A total of 85 bicycle journeys were examined, in a study of Montreal where 70.5 dB (A) was the average exposure level after roughly collecting 25 h of data [17]. Furthermore, noise can affect human health in many ways, including, for example, sleep and mental disorders, irritable reactions, hearing loss, and reduced productivity [7]. Therefore, it is essential to monitor air and noise levels in urban environments to assess the environment, leading to effective planning for the respective areas of concern. Chattogram (formerly Chittagong) (latitude 22.22 N, longitude 91.47 E) is the largest port in Bangladesh, and especially the central city area covering about 10 km^2, is heavily trafficked with persistent traffic jams most of the day [6]. Bangladesh's largest cities have over 5 million inhabitants and the population is growing all the time [7]. It is a densely populated city facing high air pollution problems [18]. Residents of Chittagong City are facing high levels of noise pollution after air and water pollution. Noise pollution is becoming a major concern in the city, as the industry is located near residential areas and traffic is heavy [7].

A study on air and noise pollution in Chattogram city, Bangladesh, is crucial due to its potential health, environmental, and economic impacts. High levels of air and noise pollution can cause serious health problems, affect the environment and wildlife, and lead to a decline in revenue for local businesses. Therefore, understanding the extent of air and noise pollution in Chattogram city is essential for assessing and mitigating its impact. Continuous monitoring of pollution levels is also critical to ensure the ongoing effectiveness of pollution reduction efforts and the continued protection of public health and the environment. It allows for the identification of any potential new sources of pollution, measurement of changes in pollution levels over time, and immediate action to address any sudden increase in pollution. The movement and accumulation of air pollutants are significantly impacted by meteorological parameters, which have a significant influence on the formation and dispersion of these pollutants in the atmosphere. Therefore, a study is planned to assess air quality and noise levels with the surrounding meteorological parameters in the selected locations in Chattogram City to determine pollution levels. The research will help city residents become more aware of the impact of pollution levels and will encourage policymakers to take the necessary steps to control pollution to ensure better health.

2 Materials and Methodology

2.1 Study Area and Sampling Points Selection

Chittagong (as seen in Fig. 1) is a port city of Bangladesh (latitude and longitude of the city are 22.22 N, and 91.47 E, respectively) and the second largest city in southeastern Bangladesh [5]. The Chittagong metropolitan area covers an area of about 5,282.92 km^2, with an urban area of 2,054.90 km^2 [5]. The average population density is 19,800 km^2 [5]. To identify the environmental pollution level of urban roads in different land uses in the Chattogram city area, fifteen (15) different locations (as seen in Fig. 1) from three land use (commercial, residential, and industrial) were selected for data collection. The selection of these locations is based on a random and systematic sampling method, which helps to ensure that the data collected is unbiased and representative of the larger population. The data collected from these 15 locations can provide valuable insights into the overall pollution levels in the

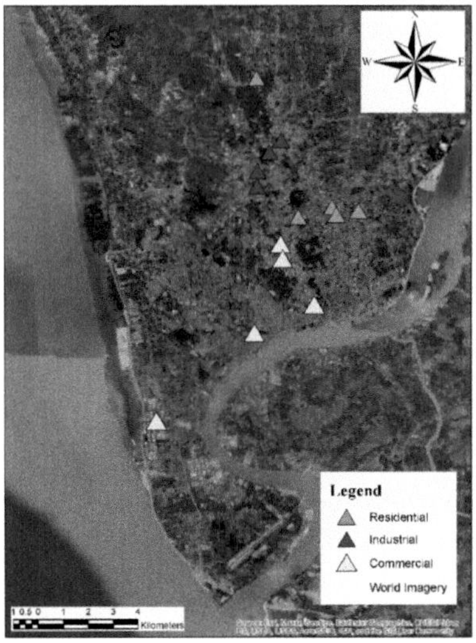

Land use types	Name of the location
Commercial	GEC
	NEW Market
	Agrabad
	WASA mor
	CEPZ
Residential	Asian Housing
	Arakan Housing
	Shugandha
	New Chandgaon
	Nasirabad Housing
Industrial	Oxygen
	Shershah
	BSRM Circle
	Bayzid
	Ctg Polytechnic

Fig. 1 Map showing the geographical location of the study area and sampling points (left) and details of sampling locations (right)

different land uses of Chattogram city and can help to inform policymakers and city planners on how to address and mitigate pollution in these areas.

2.2 *Data Collection and Analysis*

Noise Level, Particulate Matter ($PM_{2.5}$ and PM_{10}), Gaseous air pollutants (CO, CO_2, TVOC, and HCHO), meteorological parameters (Temperature, Humidity, light and wind speed) data were collected at two different times, from 9.00 am to 3.00 pm considered as daytime and from 3.00 pm to 9.00 pm considered as evening time. The weather conditions were dry in all sampling locations. The data was collected during the winter season (October 2021–November 2021). The technical specifications of the uses devices for the collection of the mentioned data and standard limits for different considered parameters are shown in Table 1. All the equipments were calibrated before taking them to the field for data collection. Variation of air pollutants and meteorological parameters were analyzed for both day and night periods for three different land use as mentioned earlier. Finally, the pollution levels were compared with standard values.

Table 1 Technical specifications of the devices and standard limit for different parameters

Parameters	Name of the instrument	Range	Accuracy	Standard limits
$PM_{2.5}$	Air quality monitor (Temtop-M2000)	0–999 $\mu g/m^3$	$\pm 10\ \mu g/m^3$	65 μgm^{-3} [a]
PM_{10}		0–999 $\mu g/m^3$	$\pm 15\ \mu g/m^3$	150 μgm^{-3} [a]
CO_2		0–5000 ppm	±50 ppm	1200 ppm[b]
TVOC	Air quality detector (PC-S3)	0.001–9.999 mgm^{-3}	±0.001 mgm^{-3}	0.25–0.5 mgm^{-3} [c]
HCHO		0.001–3.888 mgm^{-3}	±0.001 mgm^{-3}	0.95 mgm^{-3}
CO		0–9999 mgm^{-3}	±10 ppm	10 ± 50 ppm[a]
Noise	Sound level meter (AR814)	35–130 dBA	±1 dB	50–75 dB[d]
Temperature	Air quality monitor (Temtop-M2000)	0–100 °C	±0.1 °C	n/a
Humidity		0–100%	±0.1%	n/a
Wind speed	UNI-T Wind Speed Meter (UT363BT)	0–30 m/s	±1 m/s	n/a
Light intensity	Data logging light intensity meter TA8133	0–10,000 lx	±1 lx	n/a

Note [a] CASE-DoE [19]
[b] Goh et al. [20]
[c] WHO [21]
[d] BECR [22]

3 Results and Discussion

3.1 Variation of Air Pollutants in Different Land Use

The analysis of air quality such as particulate Matter ($PM_{2.5}$, PM_{10}), CO, CO_2, TVOC, and HCHO lead to the identification of pollution levels in Chattogram City. Here all the air quality parameters were compared during the day and evening period (Fig. 2) for three different types of areas such as commercial, industrial and residential. Figure 2a–b illustrates the variation of $PM_{2.5}$ and PM_{10}. The maximum value of $PM_{2.5}$ and PM_{10} was found 148 and 164 $\mu g/m^3$ during the evening time in the New Chandgaon residential area and Shershah industrial area. The minimum value of $PM_{2.5}$ and PM_{10} was found 57 and 65 $\mu g/m^3$ both in CPEZ commercial area during the evening time. The average value of $PM_{2.5}$ and PM_{10} was found 83 and 110 $\mu g/m^3$, respectively. Residential areas were having more $PM_{2.5}$ and PM_{10} than others. The values of $PM_{2.5}$ and PM_{10} crossed the Bangladesh standards which were 65 and 150 $\mu g/m^3$, respectively. In Fig. 2c, the maximum and minimum value of CO was found 194 and 7 ppm in BSRM industrial area and Asian Housing Society residential area during the day and evening period, respectively. So the average value of CO is 53 ppm. Compared with the day, the value obtained during the evening was more satisfactory; where the majority of values during days exceeded the Bangladesh standard of 40 ppm. The industrial areas during the day were facing high CO pollution than commercial and residential. The major sources of CO are smokes from motor engines, burning coal, gas, oil and wood, industrial activities [4].

In Fig. 2d, the maximum and minimum value of CO_2 was found 138 and 58 ppm in Oxygen industrial area and Nasirabad Housing residential area during the day and evening period, respectively. So the average value of CO_2 is 74 ppm. The residential areas were facing more CO_2 emissions than others which were so alarming for the residence dweller. The CO_2 emission had increased because of stationary and mobile combustion. In Fig. 2e, the maximum value of TVOC was found 0.97 ppm in the

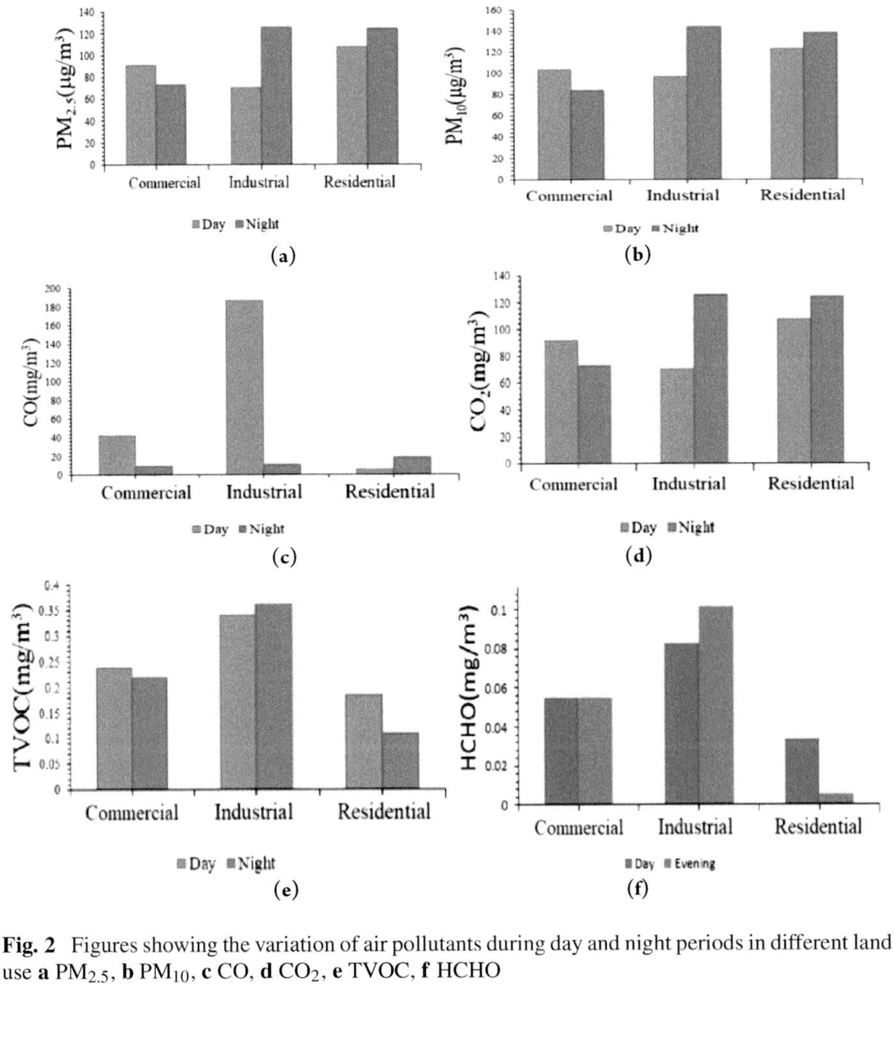

Fig. 2 Figures showing the variation of air pollutants during day and night periods in different land use **a** $PM_{2.5}$, **b** PM_{10}, **c** CO, **d** CO_2, **e** TVOC, **f** HCHO

Arakan housing Society residential area during the evening period. The minimum value was found 0.09 ppm in the Bayzid Industrial area during the evening period. The industrial areas were facing a high concentration level of TVOC compared to others. In Fig. 2f, the maximum and minimum value of HCHO was found 0.258 and 0.003 ppm in Ctg Polytechnic industrial area and New Chandgaon R/A area both during the day period. With TVOC, the industrial areas were facing a high concentration level of HCHO compared to others. Selected air quality parameters varied significantly ($p < 0.05$) for different locations. Compared with commercial and residential areas, the industrial areas exceeded the maximum air quality parameters. Maximum values of air quality exceeded the standard because of the increased number of vehicles in the roadside, increased number of industries, burning fuels, domestic activities, dust re-suspension, and machinery operation. Arif Hossen et al. [18] showed the environmental condition was unhealthy for Chattogram city as the yearly average value of AQI was above 100.

3.2 *Variation of Noise Level in Different Land Use*

The mean noise level (Leq) varied between 55 and 85 dB. The mean noise levels of commercial, industrial and residential areas were around 80, 80 and 70 dB (Fig. 3), which exceed the Bangladesh standard 70, 75 and 55 dB, respectively. Because of the increased amount of population, traffics, construction work, the ambient noise has increased more, which created an alarming issue for the city dwellers. Abdul Aziz et al. [7] showed the average noise level in Chattogram City was 72.3 dB (A) and all the parameters of noise level exceeded the standard [5].

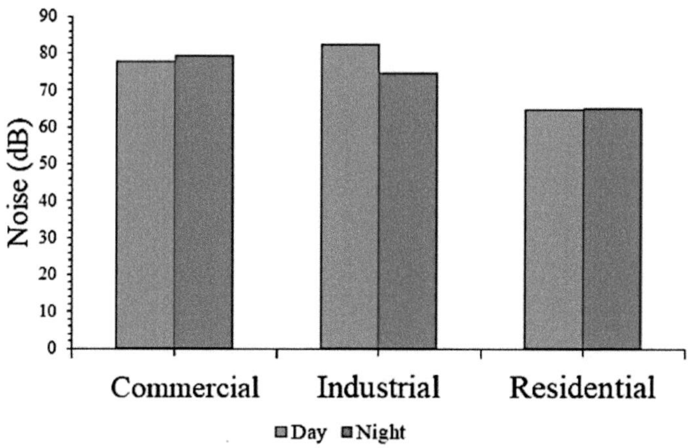

Fig. 3 Figures showing the variation of noise level (L_{eq}) during day and night periods in different land use

3.3 Variation of Meteorological Parameters in Different Land Use

Figure 4 provides some insight into how relative humidity and temperature levels can vary across different types of areas and at different times of the day. The study found that the average relative humidity levels (see Fig. 4a) varied from 48 to 82% across the different areas and there was a difference in humidity levels between day and evening periods, with evenings generally having higher humidity levels. Additionally, industrial areas were found to have higher humidity levels compared to other types of areas during the night. Regarding temperature levels, the study also found that (see Fig. 4b) industrial areas had the highest average temperatures, ranging from 25.2 to 33.8 °C.

The industrial areas had higher temperatures than other areas during the day period because of having lots of industries. In Fig. 4c, light intensity found a maximum value of 9500 lx in industrial Areas. Industrial areas were facing higher light intensity than the other two land use areas. This is likely due to the presence of artificial lighting used by industries, such as street lamps, light poles, and outdoor lighting for security

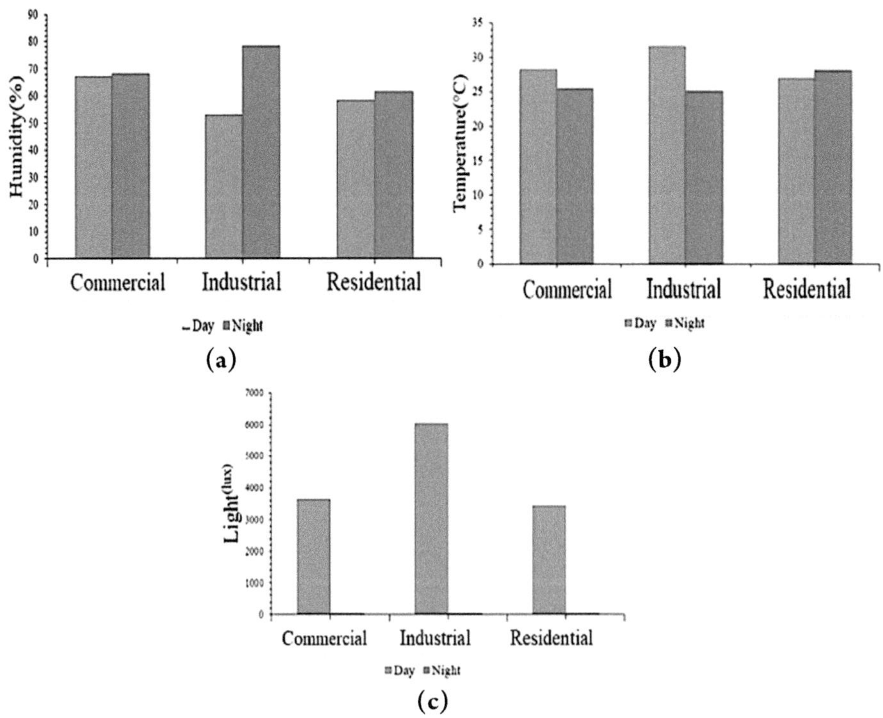

Fig. 4 Figures showing the variation of meteorological parameters during day and night periods in different land use **a** Humidity, **b** Temperature, **c** Light intensity

purposes. Higher light intensity in industrial areas can also be attributed to the larger number of buildings and structures compared to other areas, which can reflect and refract light in various ways. Monitoring light intensity is important for ensuring safety and security in industrial areas, as it allows for better visibility and reduces the risk of accidents and crime. The average wind speed of all the locations were measured 13.6 km/h. Understanding wind patterns and speeds can also aid in the development of urban planning strategies, such as determining the best location for industries and reducing the impact of pollutants on nearby residential areas.

The relationship between air pollution levels and meteorological conditions is complex and can vary depending on the specific pollutant and meteorological parameter being considered. However, in general, certain meteorological conditions can influence the concentration and dispersion of air pollutants in the atmosphere [12]. In the context of the study described above, the higher humidity levels found in industrial areas during the night may have contributed to the higher levels of particulate matter and other pollutants observed in those areas [11, 13]. Similarly, the higher temperatures observed in industrial areas may have contributed to increased formation of certain pollutants [10]. Wind speed and direction may have also played a role in the dispersion of pollutants across different areas [12].

4 Conclusion

The air quality parameters, meteorological parameters and noise level analysis of Chattogram City provided information about the concentrations of the pollutants, weather conditions and noise limit mostly studied worldwide, such as particulate Matter ($PM_{2.5}$, PM_{10}), CO, CO_2, TVOC, HCHO, temperature, humidity, light, wind and noise. Chittogram city is experiencing several problems because of air pollution [6]. Overall concentrations of pollution levels in the environment showed a negative impact for the living standards, which exceeded the permissible limits and therefore it will create serious environmental and health problems in the respective areas. Particulate matter $PM_{2.5}$ and PM_{10} and gaseous pollutants CO, CO_2, TVOC, and HCHO were found in greater concentrations during the evening period and the day period, respectively. Climatic parameters temperature and relative humidity were found to be hostile during the day period. The industrial areas are in highly risk zone comparing with commercial and residential areas. This study was conducted during dry season as the rainfall amount is low in dry season and concentrations of particulates were higher at that time, so the air quality and climatic parameters should also be carried out during wet season to see the variations. The t noise level was found to exceed the permissible limit. To reduce the air pollution vehicular emission have to be reduced, more electronic and hybrid vehicles should be introduced on the road, industries should be located far away from the city areas and solid waste management should be efficient. To reduce noise pollution improving air quality is therefore of great concern. Therefore, regulatory agencies need to enforce environmental standards.

References

1. Tezel MN, Sari D, Ozkurt N, Keskin SS (2019) Combined NOx and noise pollution from road traffic in Trabzon, Turkey. Sci Total Environ 696(134044):1–11. https://doi.org/10.1016/j.scitotenv.2019.134044
2. Zhu Y, Hinds WC, Kim S, Sioutas C (2013) Concentration and size distribution of ultrafine particles near a major highway PAPER concentration and size distribution of ultrafine particles near a major highway. J Air Waste Manage Assoc 52(9):1032–1042. https://doi.org/10.1080/10473289.2002.10470842
3. Shi JP, Khan AA, Harrison RM (1999) Measurements of ultrafine particle concentration and size distribution in the urban atmosphere. Sci Total Environ 235:51–64
4. WHO (2016) Air pollution levels rising in many of the world's poorest cities. Health Bull 2016. https://www.who.int/news/item/12-05-2016-air-pollution-levels-rising-in-many-of-the-world-s-poorest-cities. Accessed 5 Aug 2022
5. Apparicio P, Carrier M, Gelb J, Séguin AM, Kingham S (2016) Cyclists' exposure to air pollution and road traffic noise in central city neighbourhoods of Montreal. J Transp Geogr 57. https://doi.org/10.1016/j.jtrangeo.2016.09.014
6. Rehnuma M, Riad AA, Shakur RI (2020) Assessment of air quality parameters at different locations of Tangail Sadar Upazila, Tangail. J Environ Sci Nat Resour 13(1&2):118–122. https://doi.org/10.3329/jesnr.v13i1-2.60699
7. Md. Abdul Aziz, Karim A, Masum MMH, Newaz KK (2021) Evaluation of noise environment and noise mapping with ArcGIS in Chittagong City, Bangladesh. J Korean Soc Environ Eng 43(4):230–240. https://doi.org/10.4491/KSEE.2021.43.4.230
8. Zhai M, Wolff H (2021) Air pollution and urban road transport: evidence from the world's largest low-emission zone in London. Environ Econ Policy Stud 23(4):721–748. https://doi.org/10.1007/s10018-021-00307-9
9. Zhu Y, Hinds WC, Kim S, Sioutas C (2002) Concentration and size distribution of ultrafine particles near a major highway. J Air Waste Manag Assoc 52(9). https://doi.org/10.1080/10473289.2002.10470842
10. Islam N, Toha TR, Islam MM, Ahmed T (2023) Spatio-temporal variation of meteorological influence on PM2.5 and PM10 over major urban cities of Bangladesh. Aerosol Air Qual Res 23(1):1–20. https://doi.org/10.4209/aaqr.220082
11. Kirešová S, Guzan M (2022) Determining the correlation between particulate matter PM10 and meteorological factors. Eng 3(3):343–363. https://doi.org/10.3390/eng3030025
12. Johnson AC (2022) Correlation study of meteorological parameters and criteria air pollutants in Jiangsu Province, China. Pollution 8(1):341–354. https://doi.org/10.22059/POLL.2021.321137.1048
13. Oji S, Adamu H (2020) Correlation between air pollutants concentration and meteorological factors on seasonal air quality variation. J Air Pollut Heal 5(Winter):11–32. https://doi.org/10.18502/japh.v5i1.2856
14. Boogaard H, Borgman F, Kamminga J, Hoek G (2009) Exposure to ultrafine and fine particles and noise during cycling and driving in 11 Dutch cities. Atmos Environ 43(27):4234–4242. https://doi.org/10.1016/j.atmosenv.2009.05.035
15. Aziz MA, Karim A, Masum MMH, Newaz KK (2021) Evaluation of noise environment and noise mapping with ArcGIS in Chittagong City, Bangladesh. J Korean Soc Environ Eng 43(4):230–240. https://doi.org/10.4491/ksee.2021.43.4.230
16. Masum MH, Pal SK, Akhie AA, Ruva IJ, Akter N, Nath S (2021) Spatiotemporal monitoring and assessment of noise pollution in an urban setting. Environ Challenges 5(July):100218. https://doi.org/10.1016/j.envc.2021.100218
17. Apparicio P, Carrier M, Gelb J, Séguin A, Kingham S (2016) Cyclists ' exposure to air pollution and road traffic noise in central city neighbourhoods of Montreal. JTRG 57:63–69. https://doi.org/10.1016/j.jtrangeo.2016.09.014

18. Arif Hossen M, Hoque A (2018) Variation of ambient air quality scenario in Chittagong city: a case study of air pollution. J Civil Constr Environ Eng 3(1):10–16. https://doi.org/10.11648/j.jccee.20180301.13
19. CASE-DoE (2018) Ambient air quality in Bangladesh. Dhaka. http://case.doe.gov.bd/
20. Goh CC, Kamarudin LM, Shukri S, Abdullah NS, Zakaria A (2018) Monitoring of carbon dioxide (CO_2) accumulation in vehicle cabin. In: 3rd international conference on electronic design, ICED 2016, 2016, June 2018, pp 427–432. https://doi.org/10.1109/ICED.2016.7804682
21. WHO (1997) World Health Organization, assessment of exposure to indoor air pollutants. Regional Office for Europe, Copenhagen. https://www.euro.who.int/
22. BECR (1997) The environment conservation rules. Bangladesh Department of Environment Ministry Environment Forest Government of People's Republic Bangladesh, pp 179–227

Scenario of Medical Waste Management in Bangladesh During Covid-19

K. N. Mim, Md. Z. Alam, and N. Nowshin

Abstract Corona virus outbreak has substantially affected the conventional waste management method. As the number of new cases rises, so does the use of medical supplies such as disposable gloves and masks, syringes, and personal protective equipment resulting into a larger bio medical waste production. The study conducted an online survey in Bangladesh with adults aged 18 or older to gather information about the proper disposal of BMW generated at households by COVID-19 affected patients. The research employed a quantitative approach, including Principal Component Analysis (PCA) to identify the factors influencing volume of biomedical waste produced in home quarantine. Patients with mild symptoms under home care across the entire country generated infectious wastes of 2.2 and 0.76 kg/day in urban and rural areas respectively. Among these 63.35% of patients placed their waste in a plastic bag with two layers and labels. Furthermore, only 0.8% of these patients retained infectious wastes for 72 h which is the standard waste disposal method according to WHO guidelines during the pandemic. Among various factors that are affecting the generation of wastes during the quarantine period, 5 principal components are identified by Principal Component Analysis (PCA). It makes the process simple to explore the data set to determine the main variables. The infectious waste that was not handled separately but merged with general waste creates a possible secondary source of contamination. The study shows the lack of the current waste management system practiced in developing countries increases the spread of COVID-19 through infected wastes which may be useful for policymakers in developing nations.

Keywords COVID-19 · Medical waste · Management · Principal component analysis · Home quarantine · BMW (Biomedical waste)

K. N. Mim · Md. Z. Alam (✉) · N. Nowshin
Department of Civil Engineering, Rajshahi University of Engineering and Technology, Rajshahi, Bangladesh

S. Arthur et al. (eds.), *Proceedings of the 6th International Conference on Advances in Civil Engineering*, Lecture Notes in Civil Engineering 368,
https://doi.org/10.1007/978-981-99-3826-1_16

1 Introduction

Every civilized society must have the foundational element of human health care. The wastes produced by medical procedures can be dangerous, poisonous and even fatal due to the high risk of disease transmission [1]. When the first case of severe acute respiratory syndrome (SARS) disease, named COVID-19 was discovered in the city of Wuhan, China it took no time in spreading all over the globe [2]. Consequently, the medical waste dramatically raised from 40 tons/day to about 240 tons/day [3]. Bangladesh being a third world country it was already struggling with its insufficient medical and waste management infrastructure trying to serve its massive population [4]. It has been stated that, the average waste generation rate for all the medical care facilities evaluated is 0.5 kg/patient/day [5]. The first COVID 19 case in Bangladesh was reported on March 08, 2020 [6]. Along the increase in new cases, it increases the usage of medical supplies such as disposable gloves and masks, respirators, syringes, PPE, collection swabs, singular use equipment for various lab tests etc. resulting in bulk production of biomedical waste than before [7]. Viable virus was detected up to 72 h after application to different surfaces, although infectivity decay was also observed [8]. Since most of the biomedical waste consists of plastic and other material over which the virus can survive for days, it becomes imperative to manage infected waste to prevent transmission of corona virus [9]. Indirect hands contact with surfaces contaminated by infectious droplets and then eventually touching the mouth, nose or eyes seems to be another route of an indirect contact transmission [10]. In developing countries like Bangladesh, the waste collectors are not properly equipped or even dressed correctly to protect themselves from being infected while doing their job [11]. They are at high risk and might be source of community transmission. Inappropriate collection of infected waste holds risk of contaminating general municipal wastes which might result into spreading of the virus [12]. Massive volumes of medical waste from COVID-19 are continuously produced and dumped into the environment and poor waste management procedures, particularly in developing nations can raise the danger of contamination [13, 14]. When the wastes are not disposed properly following the methods suggested by WHO there is possibility of communication of the virus [15]. It is considered that wastes from quarantine homes with infected patients would fall within the category of biomedical waste [16]. It raises the possibility that, if not adequately managed, waste from COVID-19 quarantined patients treated in homes or other settings could serve as a source of virus contamination [10]. The research pursued quantitative approach to find out the amount of waste generated by the patients under home care. As many people has endured home quarantine, it becomes essential to determine the variables fundamentally influencing the BMW generation during pandemic using practical alternative method such as Principal Component Analysis. The PCA approach offers a large multivariate data set to be reduced into a smaller dimension while retaining most of the variability [17]. The study could be beneficial to the policy makers of developing countries like Bangladesh or similar socio-economic and cultural systems to manage the contaminated bio medical waste during crisis period.

2 Methodology

2.1 Sample Collection

The survey was conducted to obtain information about proper disposal of BMW generated at households by COVID-19 affected patients. When the study's topic has been conceptualized and the goals for the study area have been chosen, the process of gathering data begun. It involved design of a questionnaire using a literature review [18]. Online survey using a questionnaire were conducted with adults in Bangladesh who were 18 years of age or older [19]. It was prepared using Google Forms to conduct the survey across the entire country in order to reach persons who were infected with the Corona virus and were in their homes under quarantine. It was disseminated by email and popular social media sites among Bangladeshis, including Facebook, LinkedIn, What's App, and Messenger. The respondents were drawn from a variety of social groups, such as academics and government officials, practitioners in the field of development, physicians, engineers and technologists, students and youth leaders, business people and industry officials and people in the banking and finance industries [20]. It was optional to respond to the survey's questionnaire. A variety of statistical approaches, including PCA were used to analyze the datasets. There were 503 respondents in the sample. Through this online survey, data from 517 respondents were originally gathered using a straightforward random sample procedure of which 14 were eliminated due to incompleteness [21]. To ensure that the sample size reflected the widest range of scenario possibilities, pertinent individuals were chosen for targeted sampling based on their Bangladeshi citizenship, age of 18 wlor older, current activities, occupation, social and economic obligations, and engagement [22]. The list of questions prepared can be categorized into four groups as showed in Table 1.

2.2 Data Analysis

To comprehend the traits of respondents, descriptive statistics (frequency, percentage, and T-test) were used [23]. To investigate the relationship between items, principal component analysis (PCA) was used. In large sample datasets, PCA is a data reduction tool that illustrates the potential and degree of confidence of each parameter [24]. Kaiser-Maier Olkin (KMO) and Bartlett's sphericity tests were used to validate the need for this analysis prior to conducting the PCA analysis. The KMO > 0.5 results (in this work, the KMO value was 0.725) and the significance of Bartlett's sphericity test at p0.01 validated the fitting of the datasets for the PCA [25]. According to Kaiser's approach, which only took into account components with eigenvalues >1.0, a number of factors were considered [26].

Table 1 Category and description of the survey questionnaire

Number of parts	Sector	Description
Part 1	Personal information	Email address
		Age
		Place of work and role
		Division of participant
		Number of infected family members
		Whether he was a city dweller or not
		Organization that is responsible for collecting their waste
Part 2	Awarness	Whether they took the vaccine or not
		Whether the patient used a separate Washroom from other family members
		Frequency of temperature check
		Frequency of oxygen saturation level check
		Frequency of face mask disposal
Part 3	Waste generation	Whether they used facemasks or not, and the type of mask used
		List of all sorts of waste generated during the patient's quarantine period
		Amount of waste generation (kg/patient/day)
Part 4	Waste management system	Waste retention period
		Whether they retained waste in a double-layered labeled plastic bag or not
		Waste collection frequency
		Waste disposal location by individual

3 Results and Discussions

Global public environmental health is threatened by biomedical waste, particularly in lower-middle-income nations like Bangladesh. The majority of healthcare facilities lack suitable and efficient systematic methods for disposing of medical waste. The municipal waste is simply blended with the medical waste in the roadside collection bins and part of it is buried without following proper procedure or burned in the open [27]. Healthcare facilities can treat hazardous-infectious medical waste on-site or off-site [6]. During the pre-pandemic period, sanitary landfills, incinerators, and autoclaves were frequently utilized to treat medical waste [28]. The pandemic has changed the dynamics of waste generation. It included the probable criticality in the waste collection and treatment stage management [29]. It has been indicated that the pandemic has caused the decline of recycling and waste diversion efforts [30]. The most preferred measure for BMW treatment is incineration as it ensures complete destruction of the pathogen [31].

3.1 Descriptive Outcome of Waste Generation, Management, and Disposal

The age of participants varied between (18–67) years old with a mean of 37.9 (SD $\pm$ 13.63). People living in the cities generated 2.2 kg of waste/day on average. The waste was disposed at their doorstep which was then collected by city corporation workers separately or it was disposed of in the local waste bin along with other non-infectious waste. People from rural areas tend to produce less waste (0.76 kg/day) but the disposal management was not satisfactory. Among them, 57.04% of rural participants chose to dispose of it mixed with other household wastes. The remaining 28.17 and 14.79% of the rural people discarded the waste in the river and disposed into an excavated hole respectively. 63.35% of the total infected people managed to put their waste in a double-layer of labeled plastic bag as recommended by WHO guidelines while the remaining 36.65% did not follow the instruction which makes it difficult for the waste collector to distinct the infectious waste. It was essential to retain wastes for 72 h so that the virus decayed during that period. The retention period has been shown in Fig. 1. The standard is to retain the waste for 72 h prior to disposal which is followed by only 0.80% of people. In Bangladesh, most people retain their waste for 24 h. Around 49.60% family used to retain for 1 day and 3.19% retained it for 12 h only which is not adequate. It could be due to lack of knowledge or the interval of waste collection service provided into their area. 27.49% of them didn't retain their waste as they were eager to get rid of the infectious waste but did not care about the spread of the virus into the environment.

3.2 Relation Among Waste Generation, Waste Management, Awareness, and Demographic Characteristics

The scree plot (Fig. 2) demonstrated the retention of a total of five components (determined by components with eigenvalues >1). Table 1 showed the results of the items that performed well on each component. According to studies, the loading scores were divided into three categories: weak (0.50–0.30), moderate (0.75–0.51), and strong (> 0.75) scores [7].

Through its moderate positive and negative loading, the PC1 (First) explained 14.841% of the variance, representing the variables of waste generation including age, likely to be vaccinated, the practice of putting garbage in a labeled double-layered plastic bag, and amount of waste generated.

After that, the PC2 (Second) loaded relatively positively and explained 14.352% of the overall variation. It portrayed whether he resided in a city or not, the duration of waste retention, and the organization responsible to collect the waste.

Moderately positively loaded, the PC3 (Third) explained 9.372% of the variance with the frequency of mask disposal and oxygen saturation of blood examination. Additionally, negatively loaded with types of masks used.

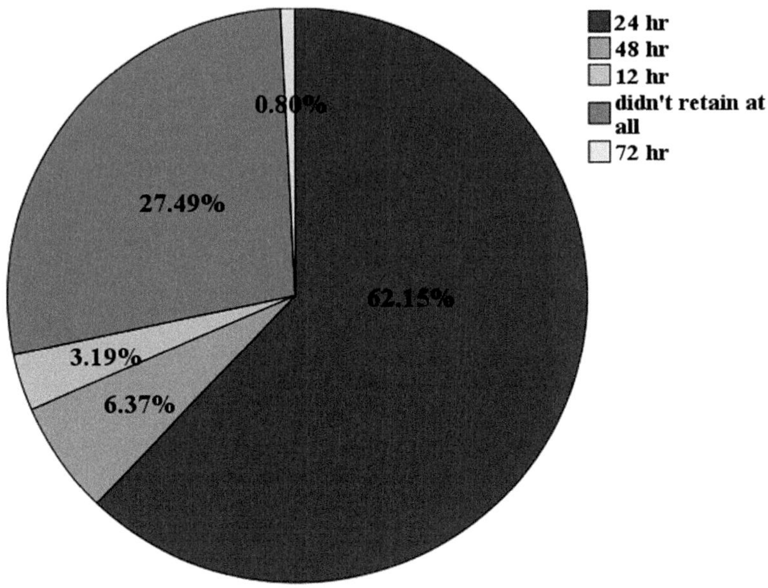

Fig. 1 Retention period before the waste disposal

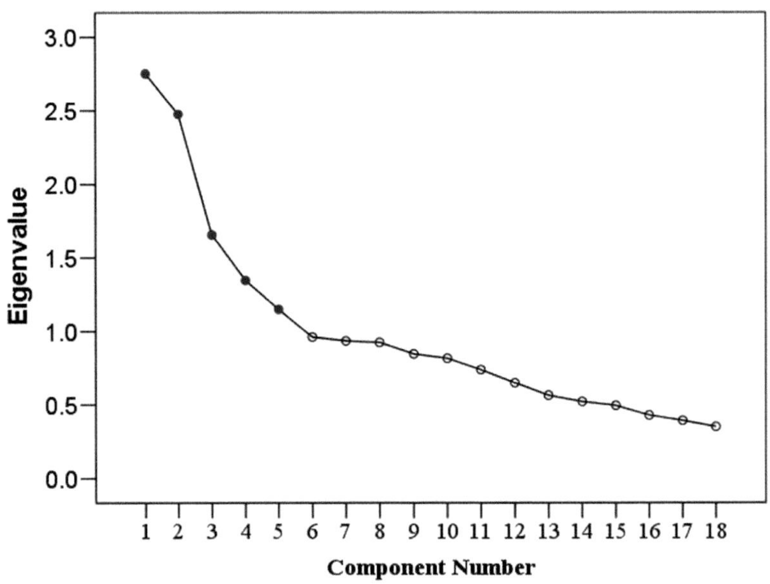

Fig. 2 The scree plot

PC4 (Fourth) was slightly positive including hygiene of the waste collector and frequency of temperature checks which accounted for 6.811% of the variation. At the same time, it was negatively loaded with the division the participant lived in.

While PC5 (Fifth) demonstrated moderate positive loadings. It explained 5.676% of the total variations with using a separate toilet and the number of family members who had been infected by Corona. A significant correlation between all of the PCs (Table 2) has been revealed from the t-Test (Table 3).

Corona virus has already limited the options for reusing and recycling [30]. Tons of surplus waste are now being produced due to bio medical waste associated with corona-affected people staying in home quarantine. It resulted in 1.52–4.4 times increase in waste generation. When these wastes are not managed properly, it might give rise to a secondary source of contamination. BMW disposed of randomly as well as discarded in rivers results in polluting the river water which should be checked immediately. People living in rural areas have no minimum waste disposal facility available when they were suffering from Corona virus. Open burning has been a preferred choice by waste management authorities to get rid of infectious wastes which are forbidden due to the emissions of chemicals or hazardous pollutants.

Table 2 Retained items after principal component analysis

Items	PC1	PC2	PC3	PC4	PC5
Age	−0.77	0.306	0.049	−0.035	0.066
Dose of vaccine	−0.697	0.091	−0.039	0.009	−0.058
Amount of waste (kg/day)	0.686	−0.115	−0.155	0.163	0.035
Double-layer labeled plastic bag	0.501	0.206	0.105	−0.007	0.192
Disposal location	0.429	0.199	−0.104	−0.162	0.003
City dweller or not	0.077	0.81	0.09	−0.248	0.119
Retention period prior to disposal	0.044	0.642	0.305	−0.304	−0.164
Waste collection organization	−0.149	0.641	−0.176	0.179	0.101
Collection frequency of organization	−0.362	0.407	−0.284	0.224	0.061
Frequency of mask disposal	−0.029	0.005	0.726	−0.031	−0.032
Type of mask used	0.232	0.049	−0.675	0.224	−0.186
Frequency of oxygen saturation check	0.408	0.036	0.522	0.228	−0.011
Hygiene of waste collector	−0.199	−0.1	0.16	0.646	0.084
Frequency of temperature check	0.126	−0.094	−0.107	0.54	−0.076
Participant's location	−0.154	0.24	0.386	−0.525	0.001
Type of waste generated	0.013	−0.158	0.05	−0.476	0.206
Use of separate washroom	0.26	0.176	0.228	−0.109	0.731
No. of infected family members	−0.031	−0.012	−0.102	−0.075	0.876
Eigenvalues	2.671	2.583	1.687	1.226	1.022
Variance %	14.841	14.352	9.372	6.811	5.676
Cumulative %	14.841	29.193	38.565	45.376	51.052

Table 3 Test of association between each component and the demographic characteristic using a T-test

Items	T-test					
	T	Df	Sig. (2-tailed)	Mean difference	95% Confidence Interval of the Difference	
					Lower	Upper
PC1	1.994	17	0.006	0.21672	−0.0126	0.4461
PC2	3.104	17	0.006	0.29717	0.0952	0.4991
PC3	4.539	17	0.000	0.32861	0.1759	0.4814
PC4	2.494	17	0.023	0.22128	0.0341	0.4085
PC5	1.630	17	0.002	0.16044	−0.0473	0.3682

The rural waste management authority (Pouroshova) has to be more spontaneous to manage the waste actively. 91.7% of the corona-affected people generated infected face masks. Additionally, 7.2% had PPE disposed of as bio-medical waste. 60.6% of them chose to use a disposable mask even when they were living in isolation. The used mask was thrown once (28%) or twice (22.2%) each day to ensure less buildup of the infectious virus around victims. The large amount of waste has to be disposed of safely without causing health hazards to the waste collector and adverse environmental effects. Controlled landfill and deep burial may be adopted in areas where incinerator has surpassed their capacity. The five defined principal components for the sample in this study show more straightforward implications than the eighteen characteristic clusters. However, it was not possible to reach every class of people, especially those who are not accustomed to the use of the internet as the survey was performed online. As a result, elderly persons and people who do not have the privilege of using the internet could not become a part of the study. The inclusion of them might reflect additional concerns in BMW management.

4 Conclusion

The management of waste has a significant impact on the environment and public health. The unexpected increase in medical waste caused by the COVID-19 outbreak has overloaded many waste management systems, illustrating the significance of efficient waste disposal. Biomedical waste, often known as BMW, is one kind of waste that needs specific treatment since it has the potential to transmit illness and damage the environment. According to the study, city dwellers generated 2.2 kg of waste on average every day, compared to 0.76 kg per patient each day in rural areas. Retaining the waste for 72 h is an effective method to let the virus decompose and minimize the risk of contamination Only 0.8% of patients retained COVID-19-generated trash for the suggested duration to lower the possibility of exposure. Lastly, using techniques like Principal Component Analysis can improve the characterization of data sets and provide a more comprehensive understanding of waste production patterns. 5 key factors that affected the production of waste among numerous factors throughout the quarantine period were found by PCA. It is essential to educate the general people on BMW disposal best practices in order to improve waste management. Public education campaigns can successfully convey the detrimental repercussions of improper disposal through media including television, newspapers, and striking posters. The COVID-19-generated wastes may be treated at the source and disposed of, or it may be transferred off site to be discarded. After incineration, the residue must be removed and safely cremated to prevent air and soil pollution. Additionally, the waste materials can be disposed of using a burial procedure. In such circumstances, the site should be chosen such that it is remote from the neighborhood and won't be used for other activities in the ensuing years. Waste management facilities need to be able to quickly raise their capacity to handle the sudden rise in waste when recycling is restricted.

References

1. Babanyara YY (2015) Poor Medical Waste Management (MWM) practices and its risks to human health and the environment: a literature review. Int J Environ Chem Ecol Geol Geophys Eng 11(7):1–8
2. Sharma HB, Vanapalli KR, Cheela VS, Ranjan VP, Jaglan AK, Dubey B, Goel S, Bhattacharya J (2020) Challenges, opportunities, and innovations for effective solid waste management during and post COVID-19 pandemic. Resour Conserv Recycl 162:105052. https://doi.org/10.1016/j.resconrec.2020.105052
3. Hantoko D, Li X, Pariatamby A, Yoshikawa K, Horttanainen M, Yan M (2021) Challenges and practices on waste management and disposal during COVID-19 pandemic. J Environ Manag 286:112140. https://doi.org/10.1016/j.jenvman.2021.112140
4. Rana MMP (2011) Urbanization and sustainability: challenges and strategies for sustainable urban development in Bangladesh. Environ Dev Sustain 13:237–256. https://doi.org/10.1007/s10668-010-9258-4
5. Hassan MM, Ahmed SA, Rahman KA, Biswas TK (2008) Pattern of medical waste management: existing scenario in Dhaka City, Bangladesh. BMC Pub Health 8:36. https://doi.org/10.1186/1471-2458-8-36
6. Taghipour H, Mohammadyarei T, Asghari Jafarabadi M, Asl Hashemi A (2014) On-site or off-site treatment of medical waste: a challenge. J Environ Health Sci Eng 12(1):68. https://doi.org/10.1186/2052-336X-12-68
7. Bodrud-Doza M, Shammi M, Bahlman L, Islam ARMT, Rahman MM (2020) Psychosocial and socio-economic crisis in Bangladesh due to COVID-19 pandemic: a perception-based assessment. Front Pub Health 8. https://doi.org/10.3389/fpubh.2020.00341
8. di Maria F, Beccaloni E, Bonadonna L, Cini C, Confalonieri E, la Rosa G, Milana M R, Testai E, Scaini F (2020) Minimization of spreading of SARS-CoV-2 via household waste produced by subjects affected by COVID-19 or in quarantine. Sci Total Environ 743. https://doi.org/10.1016/j.scitotenv.2020.140803
9. Ferronato N, Torretta V (2019) Waste mismanagement in developing countries: a review of global issues. Int J Environ Res Pub Health 16:1060. https://doi.org/10.3390/ijerph16061060

10. Das AK, Islam MN, Billah MM, Sarker A (2021) COVID-19 pandemic and healthcare solid waste management strategy—a mini-review. Sci Total Environ 778:146220. https://doi.org/10.1016/j.scitotenv.2021.146220
11. Azuma K, Yanagi U, Kagi N, Kim H, Ogata M, Hayashi M (2020) Environmental factors involved in SARS-CoV-2 transmission: effect and role of indoor environmental quality in the strategy for COVID-19 infection control. Environ Health Prev Med 25(1):1–16
12. Kaza S, Yao L, Bhada-Tata P, Van Woerden F (2018) What a waste 2.0: a global snapshot of solid waste management to 2050. World Bank Publications
13. Hu L, Deng W-J, Ying G-G, Hong H (2021) Environmental perspective of COVID-19: atmospheric and wastewater environment in relation to pandemic. Ecotoxicol Environ Saf 219:112297. https://doi.org/10.1016/j.ecoenv.2021.112297
14. Corburn J, Vlahov D, Mberu B, Riley L, Caiaffa WT, Rashid SF, Ko A, Patel S, Jukur S, Martínez-Herrera E, Jayasinghe S, Agarwal S, Nguendo-Yongsi B, Weru J, Ouma S, Edmundo K, Oni T, Ayad H (2021) Correction to: slum health: arresting COVID-19 and improving well-being in urban informal settlements. J Urban Health 98:309–310. https://doi.org/10.1007/s11524-020-00491-1
15. Rowan NJ, Laffey JG (2021) Unlocking the surge in demand for personal and protective equipment (PPE) and improvised face coverings arising from coronavirus disease (COVID-19) pandemic—implications for efficacy, re-use and sustainable waste management. Sci Total Environ 752:142259. https://doi.org/10.1016/j.scitotenv.2020.142259
16. Nghiem LD, Morgan B, Donner E, Short MD (2020) The COVID-19 pandemic: considerations for the waste and wastewater services sector. Case Stud Chem Environ Eng 1:100006. https://doi.org/10.1016/j.cscee.2020.100006
17. Jolliffe I (2005) Principal component analysis. In: Encyclopedia of statistics in behavioral science. Wiley, Chichester, UK
18. Zahanggir Alam Md, Alinur Rahman Md, Abdullah Al Firoz Md (2013) Water supply and sanitation facilities in urban slums: a case study of Rajshahi city corporation slums. Am J Civil Eng Architect 1:1–6. https://doi.org/10.12691/ajcea-1-1-1
19. Shammi M, Bodrud-Doza Md, Towfiqul Islam AR Md, Rahman Md M (2020) COVID-19 pandemic, socioeconomic crisis and human stress in resource-limited settings: a case from Bangladesh. Heliyon 6:e04063. https://doi.org/10.1016/j.heliyon.2020.e04063
20. Shammi M, Bodrud-Doza Md, Islam ARMT, Rahman M Md (2021) Strategic assessment of COVID-19 pandemic in Bangladesh: comparative lockdown scenario analysis, public perception, and management for sustainability. Environ Dev Sustain 23:6148–6191. https://doi.org/10.1007/s10668-020-00867-y
21. Keeble C, Law GR, Barber S, Baxter PD (2015) Choosing a method to reduce selection bias: a tool for researchers. Open J Epidemiol 05:155–162. https://doi.org/10.4236/ojepi.2015.53020
22. Rahman MM, Chisty MA, Sakib MS, Quader MA, Shobuj IA, Alam MA, Halim MdA, Rahman F (2021) Status and perception toward the COVID-19 vaccine: a cross-sectional online survey among adult population of Bangladesh. Health Sci Rep 4(4):e451. https://doi.org/10.1002/hsr2.451
23. D'Amico EJ, Neilands TB, Zambarano R (2001) Power analysis for multivariate and repeated measures designs: a flexible approach using the SPSS MANOVA procedure. Behav Res Methods Instrum Comput 33:479–484. https://doi.org/10.3758/BF03195405
24. Islam SMD-U, Safiq MB, Bodrud-Doza Md, Mamun MA (2020) Perception and attitudes toward PPE-related waste disposal amid COVID-19 in Bangladesh: an exploratory study. Front Pub Health 8. https://doi.org/10.3389/fpubh.2020.592345
25. Islam ARMT, al Mamun A, Rahman M Md, Zahid A (2020) Simultaneous comparison of modified-integrated water quality and entropy weighted indices: implication for safe drinking water in the coastal region of Bangladesh. Ecol Indicat 113:106229. https://doi.org/10.1016/j.ecolind.2020.106229
26. Constantin C (2014) Principal component analysis-a powerful tool in computing marketing information. Bull Transilvania Univ Brasov. Econ Sci. Ser V 7(2):25

27. Hamajima N (2014) Evaluation of knowledge, practices, and possible barriers among health-care providers regarding medical waste management in Dhaka, Bangladesh. Med Sci Monitor 20:2590–2597. https://doi.org/10.12659/MSM.890904
28. Windfeld ES, Brooks MS-L (2015) Medical waste management—a review. J Environ Manage 163:98–108. https://doi.org/10.1016/j.jenvman.2015.08.013
29. Peng J, Wu X, Wang R, Li C, Zhang Q, Wei D (2020) Medical waste management practice during the 2019–2020 novel coronavirus pandemic: experience in a general hospital. Am J Infect Control 48:918–921. https://doi.org/10.1016/j.ajic.2020.05.035
30. Chen C, Chen J, Fang R, Ye F, Yang Z, Wang Z, Shi F, Tan W (2021) What medical waste management system may cope with COVID-19 pandemic: lessons from Wuhan. Resour Conserv Recyc 170:105600. https://doi.org/10.1016/j.resconrec.2021.105600
31. Thind PS, Sareen A, Singh DD, Singh S, John S (2021) Compromising situation of India's bio-medical waste incineration units during pandemic outbreak of COVID-19: associated environmental-health impacts and mitigation measures. Environ Pollut 276:116621. https://doi.org/10.1016/j.envpol.2021.116621

Factors Influencing Pedestrian Crashes in Dhaka City: A Multiple Correspondence Analysis Approach

S. A. Iqra, A. S. Huq, and S. H. Iqra

Abstract Pedestrian road safety is a major concern in Dhaka, where pedestrians account for 65% of fatal crashes. To alleviate this situation, pedestrian safety aspects must be identified. Using a multiple correspondence analysis (MCA) approach, this study examines 631 pedestrian accidents in the Dhaka metropolitan area from 2017 to 2020. In this study, twenty categorical variables were used, including pedestrian age, location and action at the time of the accident, driver intoxication, road geometry, and lighting condition. In the first two dimensions, pedestrian location and action at the time of the crash, injury severity, lighting condition, and time of crash are found to be the most discriminant. The eight distinct clusters were later identified in this study. By analyzing these clusters, research showed that while crossing the street, pedestrians aged forty to sixty are associated in fatal crashes. Pedestrian crashes are also more common at night when visibility is low. Male pedestrians are more likely to be involved in accidents on uncontrolled roadways, while female pedestrians are more vulnerable at cross and tee junctions. Elderly drivers are more likely to collide with a pedestrian while making a right turn in an unusual road geometry. Furthermore, it has been discovered that overtaking behavior in car drivers results in severe pedestrian accidents. Some key countermeasures suggested in this study to assuage pedestrian safety include the installation of street lights for nighttime visibility, ramps for senior pedestrians, guardrail, retro reflective marking for elderly drivers, adequate sight distance at intersections, speed calming devices, and so on.

S. A. Iqra (✉)
Graduate student, Department of Civil, Environmental and Construction Engineering, University of Central Florida, Orlando, FL, USA
e-mail: sa920932@ucf.edu

A. S. Huq
Assistant professor, Accident Research Institute, Bangladesh University of Engineering and Technology, Dhaka, Bangladesh
e-mail: ashuq@ari.buet.ac.bd

S. H. Iqra
Graduate student, Department of Civil Engineering, Bangladesh University of Engineering and Technology, Dhaka, Bangladesh
e-mail: 1704012@ce.buet.ac.bd

S. Arthur et al. (eds.), *Proceedings of the 6th International Conference on Advances in Civil Engineering*, Lecture Notes in Civil Engineering 368,
https://doi.org/10.1007/978-981-99-3826-1_17

Finally, it is expected that the outcomes of this study will assist the policymakers to understand the factors that contribute to pedestrian crash severity and fatality, as well as aid in the implementation of necessary measures to reduce unexpected and avoidable road losses.

Keywords Pedestrian crash · Road safety · Multiple correspondence analysis · Dhaka City

1 Introduction

The vast majority of pedestrian accidents (approximately 80%) take place in urban settings, whereas the remaining 20% take place in rural areas [1]. In the Dhaka metropolitan region, 65% of all fatal traffic accidents include pedestrians [1, 2]. Previous research also shows that, in Dhaka city, 67% of the pedestrian collisions and 66% of the pedestrian casualties occur in the intersections [3]. Many statistical models, including partial proportional odds model, logit, decision tree, time series, binomial, perpetual cycle model and multinomial models, are used to analyze pedestrian road traffic crashes [6–10]. Findings of such studies suggests several parameters such as bad weather, dawn/dusk, night (without street lights), off-peak, no road divider, straight and flat roads, heavier vehicles, increased exposure, small lanes, mountainous terrains, obstructed multifunctional complicated footpaths, inadequate crossing facilities, and no median barriers; increase the likelihood of collision. However, other factors, such as; group crossings behavior, effective traffic control system, adequate lighting reduces the risk of pedestrian accidents [6–10]. But, these statistical models identify the association of single independent variable with the dependent variable. As a result, they fail to determine how these variables can affect a pedestrian accident while they act in a group, which is the case in reality.

Multiple Correspondence Analysis (MCA) is an outstanding method for displaying the relative association of categorical variables in any dataset [11]. MCA is used to identify systematic relationships between variables and variable categories with no initial assumptions, whereas traditional hypothesis testing is used to verify an initial hypothesis regarding relationships between variables [13–15]. Without any prior hypothesis, the MCA approach can identify complex patterns and associations between categorical variables. The relationship between the variables is further simplified by MCA's graphical overview, making interpretation simple [12]. In contrast to bivariate analysis, MCA can examine multiple data types and dimensions at once. Since MCA is multivariate, it can identify associations between various nominal variables [11].

2 Methodology

In MCA, it is unnecessary to discriminate between dependent and independent variables. It necessitates the development of a matrix based on pairwise cross-tabulation of each variable. For example, the dimension of this study's final dataset is 631 × 20. MCA can be described for a table of qualitative or categorical variables with dimension 631 × 20, by picking an individual record (in row), i (i = 1 to 631), where 20 categorical variables (represented by 20 columns) have varying category sizes. Based on these 20 variables, MCA may create the spatial distribution of points by multiple dimensions. Let X be the number of variables (i.e., columns) and I is the number of observations (i.e., rows). This will create a matrix of I' times X'. If P_x represents the number of categories for variable x, the total number of categories for all variables is, $P = \sum^{X}_{x=1} P_x$. It will spawn another matrix I' times P'. Each variable will have various columns in this matrix to showcase all of its potential categorical values.

$$W_K = B_K/(BX);\ \sum_{K \in K_q} W_k = \frac{1}{X} \text{ and } \sum_{k \in K} W_K = 1$$

$$\left(A^k A^{k'}\right)^2 = \frac{2B_{KK'} + B_K + B_{k'}}{B_k B_{k'}/B}$$

The cloud formed by MCA technique is considered as a weighted combination of K points. Category k is represented by a point denoted by A^K with weight of B_k. For each of the variables, the sum of the weights of category points is B. In this way, for the whole set K the sum is BX. The relative weight W_K for point A^K is $W_K = B_K/(BX)$. The sum of the relative weights of category points is 1/X, which makes the sum of the whole set as 1.

3 Results and Discussions

In this research, pedestrian crash data of Dhaka metropolitan area in the span of 2017 to 2020 was analyzed with MCA. The data was collected from Accident Research Institute (ARI). ARI maintains crash data in an excel format that includes four unique information criteria: crash, driver, pedestrian, and passenger information. Crash data provides information about the spatial and temporal features of a crash, such as time, day, season, location, junction type, traffic control system, and so on, as well as the severity of the crash.

Driver information includes both vehicle data (vehicle loading, maneuvering, vehicle type, and so on) and driver personal information (age, gender, alcohol status, seat belt status etc.). Both pedestrian and passenger data include age, gender, injury level, and so forth. However, passenger information is not necessary for this investigation. Only pedestrian crashes that happened at DMP locations in Bangladesh are the subject of this study. Using the variable “Collision type,” the crash data that fit for the study were retrieved from the original dataset.

When the “Collision type” variable’s value is 9, a crash involving a hit pedestrian is indicated. To extract pedestrian crashes, all collision reports with the code “Collision type = 9” were filtered. To build a spreadsheet that only contained pedestrian crashes, all records of crashes involving pedestrians that were coded in this way were filtered and extracted. Using the above-mentioned extraction approach, 631 observations were obtained from a total of 1030 DMP crashes between 2017 and 2020. The missing data are included in the variable’s maximum frequency category. Thus, in this study, 631 observations were analyzed.

The pedestrian’s demographic features include: gender (male or female); age, which is divided into four groups (less than 20 years; 20–40 years; 40–60 years; and above 60 years). Other variables included were the pedestrian’s behavioral covariate at the moment of the crash (referred to as pedestrian action), which is classified into four categories: no action, crossing the road, walking along the roadside, and other action. The “other action” category comprises walking along the road and playing on the road. Because the frequency of these two groups is so low, they have been combined into one group. There are three categories of pedestrian location at the time of an accident (roadside, road center and other location).

Table 1 shows that 93.34% of pedestrians were in the center of the road. When the collision happened, 89.86% of pedestrians were crossing the street. Additionally, it has been noted that 74.33% of all pedestrian accidents resulted in fatalities. It was discovered that 92.71% of incidents occurred on good surface quality. In 55.47% of accidents, the bus was the single most prevalent vehicle. Furthermore, going ahead vehicle maneuvers were responsible for 70.21% of all pedestrian-related accidents. Among the seven types of traffic control systems, the police-controlled systems have had the most accidents. Furthermore, 96.67% of the accidents occurred in favorable weather conditions.

The eigen values measure, which ranges from 0 to 1, reflects how much of the categorical information is explained for by each dimension. The greater the eigen value the higher the variance. The eigenvalues in the MCA method are arranged in a descending manner, with the first two dimensions conveying the most relevant information due to a higher degree of variance. Table 2 displays the eigenvalues and percentages of variance for the first ten dimensions. The first dimension in this study has an eigenvalue of 0.109, and the second dimension has an eigenvalue of 0.093. The first principal axis accounted for 4.74% of the principal inertia, whereas the second contributed for 4.05%.

Together, these two dimensions account for approximately 8.78% of data variability, while the remaining dimensions have descending variability. Table 3 depicts the coordinates of the first five dimensions for the top 10 categories. The value of R^2 can vary between 0 and 1, where 0 indicates no relationship between the variable and MCA dimension, and 1 indicates a strong relationship.

Table 1 Descriptive statistics for pedestrian crashes

Category	Frequency	Percentage	Category	Frequency	Percentage
Junction type			*Vehicle maneuver*		
Cross	80	12.68	Unknown	94	14.90
Tee	50	7.92	*Vehicle loading*		
Roundabout	27	4.28	Legal	626	99.21
Staggered	15	2.38	Illegal	5	0.79
Not a junction	321	50.87	*Driver age*		
Unknown	138	21.87	Young driver (<30)	102	16.16
Traffic control system			Middle aged driver (30–50)	518	82.09
No control	141	22.35	Elderly driver (>50)	11	1.74
Centerline	50	7.92	*Alcohol*		
Police + traffic light	12	1.90	Suspected	64	10.14
Police controlled	348	55.15	Not suspected	567	89.86
Pedestrian crossing	22	3.49	*Seat belt*		
Traffic light	8	1.27	Worn	35	5.55
Unknown	50	7.92	Not worn	596	94.45
Movement			*Pedestrian sex*		
One way	297	47.07	Female	149	23.61
Two way	334	52.93	Male	482	76.39
Weather			*Pedestrian age*		
Bad weather	21	3.33	Less than 20	69	10.94
Good weather	610	96.67	20–40	231	36.61
Lighting condition			40–60	257	40.73
Daylight	313	49.60	More than 60	74	11.73
Other light	318	50.40	*Injury severity*		
Road geometry			Fatal	469	74.33
Straight	584	92.55	Grievous	146	23.14
Not straight	47	7.45	Simple	16	2.54
Surface quality			*Pedestrian location*		
Good	585	92.71	Road center	589	93.34
Rough	46	7.29	Road side	29	4.60
Vehicle type			Other location	13	2.06
Bus	350	55.47	*Pedestrian action*		
Truck	83	13.15	Crossing road	567	89.86
Car	65	10.30	Walking roadside	24	3.80
Motor cycle	43	6.81	No action	30	4.75

(continued)

Table 1 (continued)

Category	Frequency	Percentage	Category	Frequency	Percentage
NMV	15	2.38	Unknown	10	1.58
Others	75	11.89			
Vehicle maneuver					
Going ahead	443	70.21			
Left turn	25	3.96			
Right turn	14	2.22			
U turn	8	1.27			

Table 2 Inertia values for top ten dimensions

Dimensions	Eigenvalue	Percentage of variance	Cumulative percentage of variance
Dimension 1	0.1091	4.7431	4.7431
Dimension 2	0.0931	4.0492	8.7922
Dimension 3	0.0828	3.6007	12.3930
Dimension 4	0.0792	3.4421	15.8351
Dimension 5	0.0762	3.3125	19.1477
Dimension 6	0.0732	3.1837	22.3313
Dimension 7	0.0698	3.0342	25.3656
Dimension 8	0.0666	2.8966	28.2621
Dimension 9	0.0648	2.8193	31.0815
Dimension 10	0.0644	2.7998	33.8813

Figure 1 is an illustration of the associations among study variables as computed by MCA. Several variables have clustered at the origin; this means that their contributions (the eigen values) are smaller than those variables that reside away from the origin. The most discriminant variables for dimension 1 are: pedestrian action, pedestrian location, injury severity; regarding dimension 2 the most discriminant variables are: time, lighting condition, vehicle type. By observing the relative closeness of the variables, it is found that the traffic control system, vehicle maneuver, seat belt usage, pedestrian age are closer in the two-dimensional space. Based on the positioning of the variable clusters, different clouds were formed using the most discriminating variables and less distinct variables as shown in Fig. 2. The clouds formed with discriminating variables are shown as ellipses and less distinct variables are discussed in Figs. 3, 4. These clouds form the category groups that most likely to contribute to the crash.

First cloud forms association between six significant factors. The association suggests that pedestrians in roundabout junctions having both traffic police and traffic light as a traffic control system are more likely to be involved in accidents. Trucks are found to be involved in the accidents of these systems according to this cloud.

Table 3 Significance of key variables on the first plane

Dim1	R^2	P value	Dim2	R^2	P value
Pedestrian action	0.3325	<0.001	Time	0.3524	<0.001
Pedestrian location	0.2795	<0.001	Lighting condition	0.3303	<0.001
Injury severity	0.1941	<0.001	Vehicle type	0.2413	<0.001
Driver's age	0.1909	<0.001	Pedestrian location	0.1667	<0.001
Lighting condition	0.1685	<0.001	Pedestrian action	0.1615	<0.001
Time	0.1628	<0.001	Junction type	0.1713	<0.001
Pedestrian age	0.1303	<0.001	Seat belt	0.0983	<0.001
Vehicle type	0.1383	<0.001	Driver's age	0.0549	<0.001
Vehicle maneuver	0.1362	<0.001	Traffic control system	0.0657	<0.001
Divider	0.1055	<0.001	Vehicle maneuver	0.0574	<0.001
Junction type	0.1152	<0.001	Pedestrian age	0.0406	<0.001
Traffic control system	0.1059	<0.001	Driver's alcohol status	0.0289	<0.001
Movement	0.0339	<0.001	Pedestrian sex	0.0231	<0.001
Surface quality	0.0303	<0.001	Divider	0.0229	<0.001
Seat belt	0.0204	<0.001	Injury severity	0.0252	<0.001
Road geometry	0.0192	<0.001	Vehicle loading	0.0177	<0.001
Vehicle loading	0.0101	<0.001			

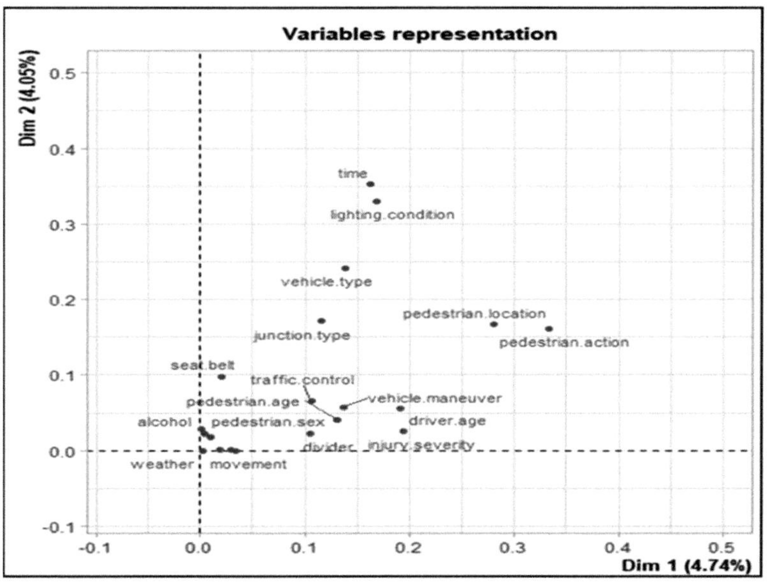

Fig. 1 MCA plot for the variables

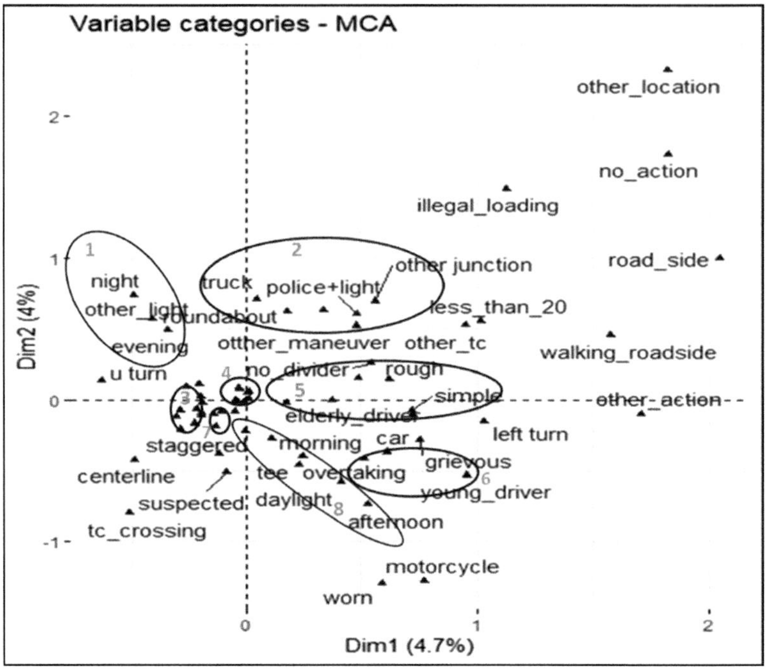

Fig. 2 Principal MCA plot for the variable categories

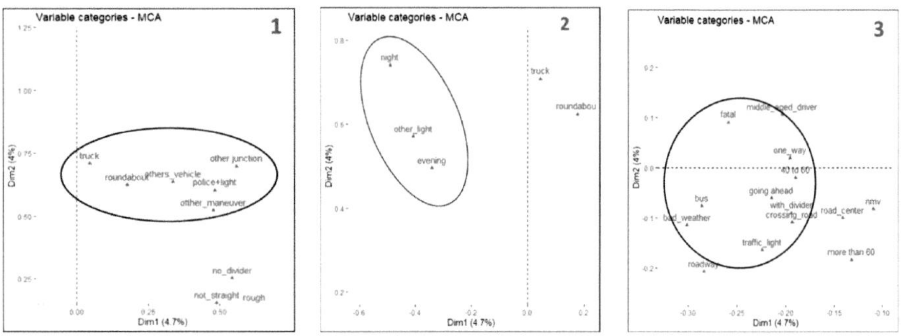

Fig. 3 **a** Cloud 1 of MCA factor map; **b** Cloud 2 of MCA factor map **c** Cloud 3 of MCA factor map

Second cloud explains that pedestrian crashes may occur during evening and night when the lighting condition of the road is nighttime lighting system (lit/unlit). Many earlier research also provides similar findings. Prevention measures such as targeted traffic enforcement and incorporating median separators into roadways, and street light installation can help reduce pedestrian accidents. FHWA insists to remove trees and other obstruction to visibility. Third cloud indicates ten prevalent factors that

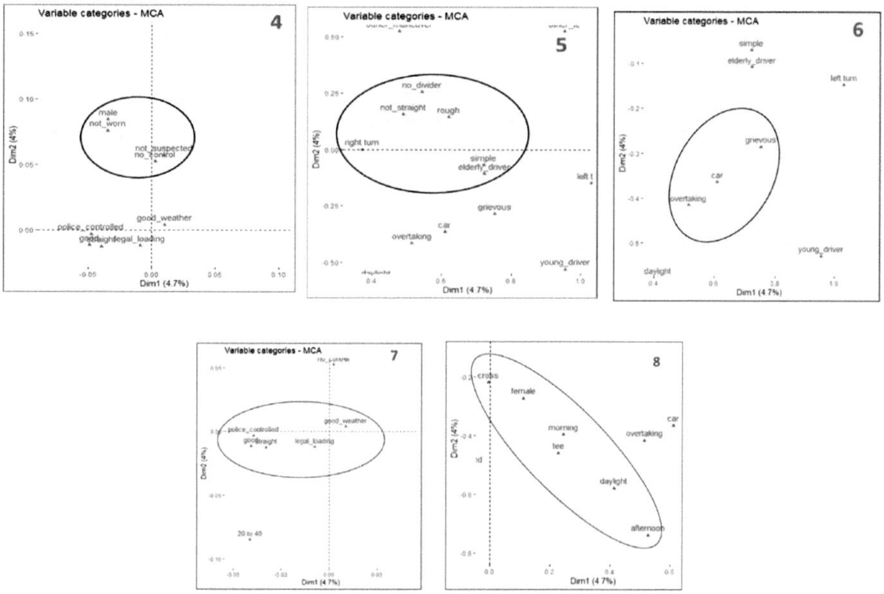

Fig. 4 **a** Cloud 4 of MCA factor map; **b** Cloud 5 of MCA factor map; **c** Cloud 6 of MCA factor map; **d** Cloud 7 of MCA factor map; **e** Cloud 8 of MCA factor map

are involved with pedestrian accidents. The correlation dictates that pedestrian aging between 40 and 60 while crossing the road are more likely to suffer from fatal crashes in a divided road during bad weather condition. Middle aged bus drivers while going ahead in a road are more likely to hit a pedestrian in this scenario. FHWA suggests provision of sidewalk with adequate width and visibility along both side of the road for safe pedestrian movement. Also, continuous and gradually sloped ramps should be available for physically impaired senior pedestrians. Proper visual guidance such as retro reflective sign and marking is necessary for safe pedestrian movement in bad weather condition. Signal timing should be established considering senior pedestrians. Road crossing facilities such as zebra crossing needs to be clearly visible by all ages of pedestrians.

Fourth cloud displays the association among four factors (pedestrian gender = male, seat belt/helmet = not worn by the driver, alcohol status of driver = not suspected, traffic control system = no control). Pedestrian crashes are common among male pedestrians in junctions with no traffic control system. Drivers who had worn seat belt/helmet in a sober condition are more likely to be associated in those accidents. Presence of effective traffic control system is proved to be conducive in reducing crash risk by many studies. Also, from the descriptive analysis it is perceived that male pedestrians are more vulnerable to crashes. From the socioeconomic scenario of Bangladesh male trip generation is more likely compared to female. As a result, they suffer more from RTA. Awareness among road users should be increased. Fifth cloud presents correlation among six categories (divider = no divider, road geometry = not straight, surface quality = rough, vehicle maneuver = right turn, injury severity = simple, driver age = elderly driver). This connection among these categories indicates that elderly drivers (>50 years) during making a right turn in a undivided rough surfaced road are involved in simple injury accidents. Enough sight distance must be maintained to ensure safe passage of pedestrian in the junction. Clearly visible signs and markings is needed along right turn in unusual road geometry. Sixth cloud portrays interrelation among three factors (vehicle type = car, vehicle maneuver = overtaking, injury severity = grievous).

This cloud indicates that overtaking behavior of car drivers results in grievous pedestrian accidents. Speed limitations needs to be maintained. Speed calming devices such as speed ramps, rumble strips, channelization needs to be installed in pedestrian crowded area. Overtaking and lane changing behavior is to be prohibited strictly. Seventh cloud deals with five factors (traffic control system = police controlled, weather condition = good, surface quality = good, road geometry = straight, vehicle loading = legal) that have strong interconnection among each other. Pedestrian accidents are more likely to occur in police controlled, straight roadway system in a good weather condition. In Dhaka city most of the junctions are police controlled. So, accident density in those junctions are higher than other type of junctions. According to the findings of this cloud pedestrian accidents are more likely to occur in good weather condition at a straight roadway. This makes sense because at a straight road in a good meteorological condition drivers are intended to drive in a higher speed resulting in a crash. Cloud eight deals with factors associated with female pedestrian accidents. It is displayed by these association during morning and afternoon in cross and tee junctions, female pedestrians are more likely to be associated with accidents. To ensure female pedestrian friendly movement in these junction signal timing can be increased considering presence of kids with female pedestrian. Guardrail, over bridge, curb, discernible delineation should be maintained.

4 Conclusion

This study analyzes total 631 pedestrian accidents of Dhaka metropolitan area in the period of 2017–2020 with multiple correspondence analysis (MCA) approach. Among fifty-eight categories; pedestrian location and action at the moment of accident, injury severity, lighting condition and time of accident are found to be most significant in the first two dimensions. Variable that are clustered around the origin confirms that alcohol status, road geometry and vehicle loading are less discriminant in explaining pedestrian accidents. Later, in this research the eight distinct clusters were identified using FactoMineR package in R software. By analyzing these clusters, research showed that pedestrians aged forty to sixty are more likely to be involved in fatal accidents while crossing the street. Pedestrian accidents are also more common at night when visibility is low. Moreover, male pedestrians are more likely to be involved in accidents on uncontrolled roadways, while female pedestrians are more vulnerable at cross and tee junctions. In addition, pedestrian accidents are more likely to occur in a straight roadway system in a good weather condition. Also, elderly drivers are more likely to collide with a pedestrian while making a right turn in an unusual road geometry. To alleviate pedestrian safety many countermeasures are suggested in this study considering the above-mentioned findings. Some key countermeasures are installing street lights for nighttime visibility, ramps for senior pedestrian, guardrails for lateral separation of pedestrian from mainstream traffic, retro reflective markings for ensuring clear visibility of elderly drivers, ensuring adequate sight distance at junctions, and speed calming devices etc. The findings of this research are indeed to provide policymakers with insights into the influencing factors of pedestrian crash severity and fatality, as well as assist in the execution of necessary measures to decrease unexpected and avoidable losses on the roads.

References

1. Hoque MM, Hossain S, Islam S, Rahman MA (2022) Safe system for sustainable pedestrian safety in Bangladesh. https://trid.trb.org/view/1286878
2. Debnath M et al (2021) An investigation of urban pedestrian behaviour in Bangladesh using the Perceptual Cycle Model. Saf Sci 138, None
3. Barua U, Tay R, Hoque M, Mamun M (2007) Analysis of pedestrian safety on five major arterial roads in Dhaka. Can Transp Res Forum, Bangladesh
4. Pour-Rouholamin M, Zhou H (2016) Investigating the risk factors associated with pedestrian injury severity in Illinois. J Saf Res 57:9–17
5. Tulu GS, Washington S, Haque MM, King MJ (2015) Investigation of pedestrian crashes on two-way two-lane rural roads in Ethiopia. Accid Anal Prev 78:118–126
6. Velloso MS, Jacques MAP (2012) On-the-spot study of pedestrian crashes on Brazilian Federal District rural highways crossing urban areas. Transp Res Part F Traffic Psychol Behav 15(5):588–599
7. Liu Y-C, Tung Y-C (2014) Risk analysis of pedestrians' road-crossing decisions: effects of age, time gap, time of day, and vehicle speed. Saf Sci 63:77–82
8. Moradi A, Soori H, Kavousi A, Eshghabadi F, Jamshidi E (2016) Spatial factors affecting the frequency of pedestrian traffic crashes: a systematic review. Arch Trauma Res 5(4):e30796
9. Zafri NM, Prithul AA, Baral I, Rahman M (2020) Exploring the factors influencing pedestrian-vehicle crash severity in Dhaka, Bangladesh. Int J Inj Contr Saf Promot 27(3):300–307
10. Sarkar S, Tay R, Hunt JD (2011) Logistic regression model of risk of fatality in vehicle-pedestrian crashes on national highways in Bangladesh. Transp Res Record: J Transp Res Board 2264(1):128–137. https://doi.org/10.3141/2264-15
11. Das S, Sun X (2014). Exploring clusters of contributing factors for single-vehicle fatal crashes through multiple correspondence analysis. In: Transportation research board 93rd annual meeting
12. Das S, Avelar R, Dixon K, Sun X (2018) Investigation on the wrong way driving crash patterns using multiple correspondence analysis. Accid Anal Prev 111:43–55
13. Jalayer M, Pour-Rouholamin M, Zhou H (2022) Multiple correspondence approach to identifying contributing factors regarding wrong-way driving crashes.17-01182, 2017. Accessed: 25 Aug 2022. https://trid.trb.org/view/1437511
14. Baireddy R, Zhou H, Jalayer M (2018) Multiple correspondence analysis of pedestrian crashes in rural Illinois. Transp Res Rec 2672(38):116–127
15. Thibenda M, Wedagama DMP, Dissanayake D (2022) Drivers' attitudes to road safety in the South East Asian cities of Jakarta and Hanoi: socio-economic and demographic characterisation by multiple correspondence analysis. Saf Sci 155:105869. https://doi.org/10.1016/j.ssci.2022.105869

Modeling of Non-stationary Adsorption Behavior of Binary Heavy Metals to Porous Media

F. T. Jahara and M. Kikumoto

Abstract Non-stationary adsorption of various heavy metals, such as nickel and copper, on porous media is modeled both for the single and binary heavy metals. For modeling kinetic adsorption, firstly final equilibrium state using an existing isotherm model incorporating binary adsorption is described. For this, extended Freundlich isotherm showed good agreement with both single and binary systems. Second, the rate form of kinetic equations is applied. The advantage of the rate form is adsorption of heavy metals at an equilibrium state, which needs to be predetermined in the classical pseudo-kinetic equation, is not necessary. The simulated results are compared with the past experimental results on non-stationary adsorption of binary heavy metals, and the ordinary equation of pseudo-second-order kinetic best fits the results.

Keywords Heavy metal · Porous media · Adsorption · Equilibrium · Non-stationary binary adsorption

1 Introduction

Numerous industrial processes heavily rely on the usage of heavy metals, such as the production of steel and iron, electrolysis, mining, metal finishing, electro-osmosis, battery manufacturing, pesticide production, petroleum refining, pigment creation, tanneries, and photograph [1]. However, they are of particular concern because of their toxicity, propensity for bio-accumulation and persistent nature [2]. Heavy metals are toxic and cannot be digested if their concentrations exceed the tolerance limit in a living organism; as a result, the metal bio accumulates in the soft tissues. Furthermore,

F. T. Jahara (✉)
Department of Civil Engineering, Southern University Bangladesh, Chittagong, Bangladesh
e-mail: jaharafatema@gmail.com

M. Kikumoto
Department of Civil Engineering, Yokohama National University, Yokohama, Japan
e-mail: kikumoto-mamoru-f@ynu.ac.jp

S. Arthur et al. (eds.), *Proceedings of the 6th International Conference on Advances in Civil Engineering*, Lecture Notes in Civil Engineering 368,
https://doi.org/10.1007/978-981-99-3826-1_18

when heavy metals come into close touch with people in settings like agriculture, the pharmaceutical industry, or homes, they can permeate into the body through water, air, food, or skin absorption [3]. Therefore, heavy metals must be removed before releasing wastewater into the environment.

Reverse osmosis, coagulation-flocculation, chemical precipitation, ultra-filtration, flotation, electro-dialysis, and ion exchange processes are the traditional methods for treating heavy metal-contaminated water. However, their use is limited because they are typically expensive, unfriendly to the environment, and ineffective when heavy metal concentrations are between one and one hundred milligrams per liter [1]. Meanwhile, adsorption technology has been proven efficient for heavy metals treatment because it is simple, inexpensive, and effective even if there is little metal contamination of the water. Various porous materials such as clays, sands, and ceramics are widely used as an adsorbent, and the adsorption characteristic of heavy metal to such porous media has been studied.

In a practical situation, the waste contains more than single heavy metal simultaneously. The adsorption of various heavy metals must be emphasized. The adsorption equilibrium for single heavy metal has been extended to a multi-component adsorption equilibrium to incorporate the effect of the counter heavy metal. Furthermore, adsorption studies are investigated for multiple metals as well as single metals for non-equilibrium conditions by kinetic modeling. However, all the previous kinetic adsorption studies are concerned with the integrated form of the famous Lagergren pseudo-first-order kinetics and Ho pseudo-second-order kinetics models. The shortcoming of the combined form is final equilibrium adsorption value must be predetermined by the experiment, and this equilibrium value is considered constant for the kinetic study. But in actual cases, the equilibrium value changes for different concentrations at different times. So, the rate form of the kinetic equation is suitable for defining adsorption behavior for multi-component metals as well as single heavy metal. The objective of this study is to conduct a numerical simulation of stationary and non-stationary adsorption (using rate form) of binary heavy metals and to validate the simulation with the past experimental study [4] on the non-stationary adsorption of binary heavy metals.

2 Adsorption Model

There exist two models to predict the nature of the adsorption of heavy metals. One is a stationary model known as the isotherm model, and the other is a non-stationary/kinetic model.

2.1 Stationary/Isotherm Model:

Mono-component adsorption isotherm

This model considers the adsorption of single heavy metal without interacting with other heavy metals in the system. Two prominent equilibrium models are listed below:

(a) **Langmuir Isotherm**

In the adsorbed ions and the monolayer coverage, there is no interaction between on uniform adsorbent surface, which is assumed by the Langmuir model [1]. The Langmuir isotherm model can determine the maximum adsorption. The equation below constitutes the Langmuir model:

$$q_e = \frac{q_m K_L C_e}{1 + K_L C_e} \tag{1}$$

where,
q_e (mg/g) = equilibrium metal ion adsorbed
K_L (L/mg) = Isotherm constant of Langmuir
q_m (mg/g) = maximum capacity of adsorption
C_e (mg/l) = metal ion concentration in equilibrium.

(b) **Freundlich Isotherm**

The Freundlich model assumes a heterogeneous surface with the interaction of different adsorbed ions [5]. The following equation represents the Freundlich model:

$$q_e = K_F C_e^n \tag{2}$$

where,
q_e (mg/g) = adsorbed metal ion in equilibrium.
K_F ($\mathrm{mg^{1-n}L^n g^{-1}}$) = Freundlich constant representing the adsorption capacity of the adsorbent.
C_e (mg/l) = equilibrium metal ion concentration.
n (g/L) = Freundlich constant representing affinity/adsorption intensity (surface heterogeneity)

Multi-component Adsorption Isotherm

A mixture of different heavy metals in wastewater is common rather than single heavy metal. For this reason, the equilibrium isotherm models have been extended from a single component to a multi-component system to see the interaction among heavy metals.

(a) **Non-modified competitive Langmuir isotherm**

This is the primary fundamental Langmuir model for component i in a system of N components where competitive adsorption occurs [6]. The isotherm equation is described as follows:

$$q_{e,i} = \frac{q_{m,i} K_{L,i} C_{e,j}}{1 + \sum_{j=1}^{N} K_{L,j} C_{e,j}} \quad (3)$$

$q_{m,i}$ and $K_{L,}i$ can be estimated from the fitting of the experimental data by the corresponding individual Langmuir isotherm equations.

(b) **Modified competitive Langmuir isotherm**

The interactive effect is incorporated in the individual isotherm equations by an interaction term, ηi [6] as given in Eq. (4).

$$q_{e,i} = \frac{q_{m,i} K_{L,i} \left(C_{e,j}/\eta j\right)}{1 + \sum_{j=1}^{N} K_{L,j} \left(C_{e,j}/\eta j\right)} \quad (4)$$

$q_{m,i}$, and $K_{L,i}$ are determined by using the corresponding individual Langmuir isotherm equations to fit the experimental data, ηi which can be estimated from the competitive adsorption data.

(c) **Extended Langmuir isotherm**

Considering that all of the adsorbate molecules (ions) in the solution compete for the same surface sites and the surface sites are uniform, Yang extended the Langmuir equation [6] for multi-component systems is as

$$q_{e,i} = \frac{q_{\max} K_i C_{e,i}}{1 + \sum_{j=1}^{N} K_j C_{e,j}} \quad (5)$$

The optimal fitting of Eq. (5) to the experimental data for the equilibrium adsorption of components in the multi-component system yields the values of $q_{\max}$ and K_i.

(d) **Extended Freundlich isotherm**

The extended Freundlich equation [7] provided below can also be used to model the equilibrium adsorption from binary mixtures:

$$q_{e,i} = \frac{K_{F,i} x C_{e,i}^{\eta_i + x_i}}{C_{e,i}^{x,i} + y_i C_{e,j}^{z_i}} \quad (6)$$

$K_{F,i}$ and η_i are determined using the corresponding individual Freundlich isotherm equation for component i and other three parameters (x_i, y_i, z_i) representing the multi-component Freundlich isotherm constant in the solution that contains i and j components [6].

2.2 Non-stationary/Kinetic Adsorption Model

Kinetic study offers useful insight into the mechanism of adsorption and subsequent investigation of the governing mechanism of the sorption process as either mass transfer or chemical reaction [8].

Pseudo-first-order kinetics

The adsorption rate for first-order adsorption proposed by Lagergren [9] is shown below:

$$\frac{\mathrm{d}q_t}{\mathrm{dt}} = K_1 * (q_e - q_t) \tag{7}$$

where, k_1 (min^{-1}) is the pseudo-first-order adsorption rate constant

q_t (mg g^{-1} of dry weight) is the adsorbed amount (metal ion) on the sorbent surface at time t (min).

q_e (mg g^{-1} of dry weight) is the equilibrium adsorbed amount (metal ion).

Integrating Eq. (7) with respect to boundary conditions: q = 0 at t = 0 and q = q at t = t, Eq. (7) becomes

$$\ln(q_e - q_t) = \ln q_e - K_1 * t \tag{8}$$

Most of the previous studies [10] have considered the integrated form of this equation where q_e is assumed constant. However, this integrated form of this equation cannot predict the real scenario since equilibrium adsorption is constantly changing for different initial concentration of heavy metal. Furthermore, final adsorption in equilibrium must be determined by batch test before conducting the simulation. Thus, the rate form of the kinetics equation is used in our simulation, which can be applied directly to the actual field.

Pseudo-second-order kinetics

The following equation is known as the pseudo-second-order kinetic equation proposed by Ho [11]:

$$\frac{\mathrm{d}q_t}{\mathrm{dt}} = K_2(q_e - q_t)^2 \tag{9}$$

where K_2 (g mg^{-1} min^{-1}) is the adsorption rate constant of pseudo-second-order.

Again, most of the researchers focused on the integrated form of this equation [12]:

$$\frac{t}{q_t} = \frac{1}{K_2 * q_e^2} + \frac{t}{q_e} \tag{10}$$

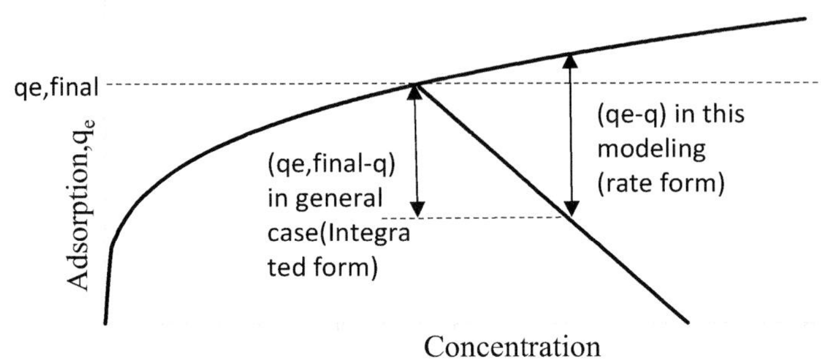

Fig. 1 Difference between the general integrated form of the kinetic model and the present modeling

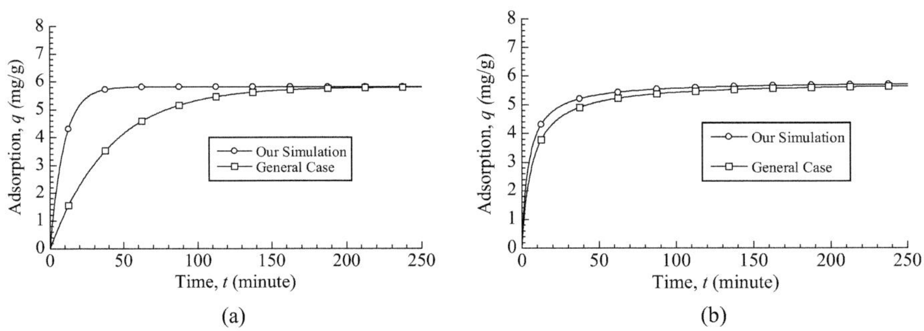

Fig. 2 Difference between rate form (our simulation) and integrated form (general case) using **a** Pseudo first-order kinetic; **b** Pseudo second order kinetic

where q_e was constant. The difference between integrated form and rate form is indicated in Fig. 1.

The rate form is applied in the simulation regarding q_e as a variable in this study. Thus, adsorption is slightly higher in the kinetic curve of the simulation compared to the general case (Fig. 2).

3 Results and Discussions

For the simulation of adsorption, peat is chosen as an adsorbent to adsorb two heavy metals nickel and copper. The simulation is validated using the experiments conducted by Ho and Forster [4]. The experiment was mainly focused on the kinetic behavior of Ni and Cu in the single and binary systems. For the single solute system, 1.2 gm peat was used as an adsorbent in 300 ml metal ion solutions, and the initial concentration of Ni and Cu was kept at 10, 25, 50, and 100 mg/l. For the bi-solute system, 3.2 gm of peat was mixed with a larger volume (800 ml) of metal ion solution containing an equal concentration of two heavy metals. The peat suspension was shaken for 3 h, and adsorption was measured periodically within 3 h.

3.1 Stationary Adsorption

For the Simulation of equilibrium adsorption, Ni and Cu concentrations are varied in the range of 0–100 mg/l in a single system. For the binary system, for every initial concentration of 1, 10, and 100 mg/l, the counter heavy metals are varied from 0 to 100 mg/l.

The simulation is conducted using the modeling technique of Microsoft Excel 2016. The parameters are fitted by minimizing the sum of square error (SSE).

$$SSE = \sum_{i=1}^{N} \left(q_{e(\exp,i)} - q_{e(cal,i)}\right)^2 \tag{11}$$

$q_{e(exp,i)}$ represent the equilibrium adsorption value from the reference paper [4]and $q_{e(cal,i)}$ is calculated value by the simulation. The simulation is performed using all the isotherm models described above. Among them, the extended Freundlich model shows better agreement with the experiment for Ni and Cu for both single and binary systems (Figs. 3 and 4). The isotherm parameters for Ni and Cu in binary and single systems are listed in Table 1.

Both Ni and Cu exhibit antagonistic behavior in the binary system. The adsorption of Ni and Cu decreases as the concentration of counter heavy metal in the binary system increases. It is apparent from Fig. 5 that nickel adsorption is decreased much more for the effect of copper than the reduction of Copper adsorption in the binary

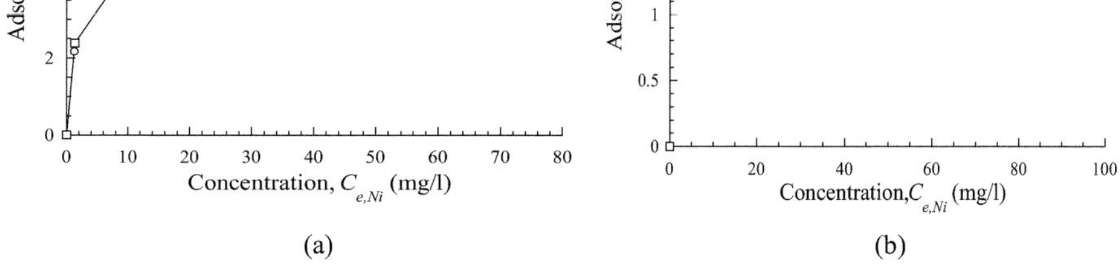

Fig. 3 Stationary adsorption of Ni: **a** Single **b** binary

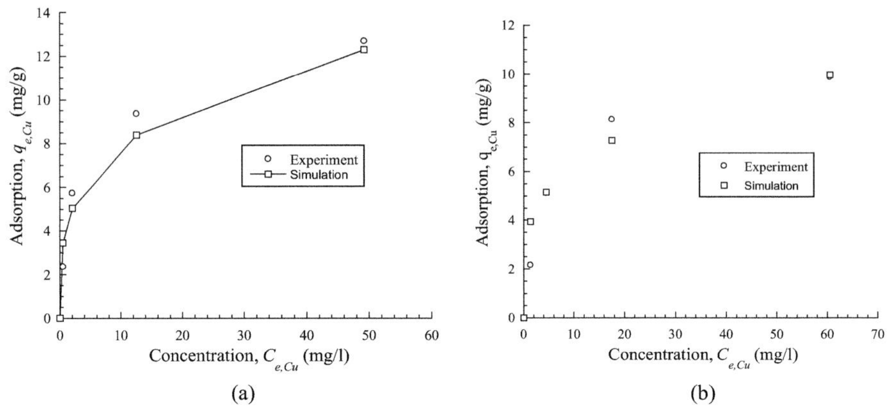

Fig. 4 Stationary adsorption of Cu: **a** Single **b** binary

Table 1 Extended Freundlich isotherm parameters

Adsorbate	K_F	n	x	y	z
Ni	2.17	0.3	0.0085	0.65	0.3
Cu	4.13	0.28	0.01	0.13	0.2

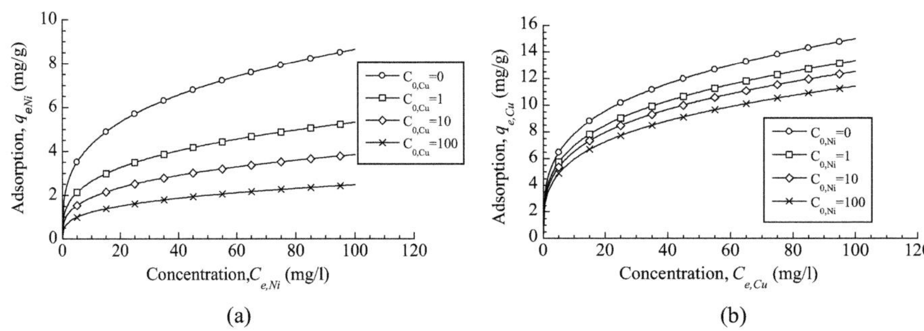

Fig. 5 Binary stationary adsorption: **a** Ni **b** Cu

system. A similar effect has been found in the study by Vijayaraghavan [1] that Cu has a more competitive effect on Ni adsorption than in the opposite situation.

3.2 *Non-stationary Adsorption*

Kinetic simulations are carried out by the rate form of kinetic equations using an initial concentration of 50 mg/l. For the binary system, the ratios of Ni and Cu concentrations are kept to 1:1 using an initial concentration of 50 mg/l. The simulation is verified by digitizing the kinetic data (Fig. 1) of the reference paper. The pseudo-second-order kinetic equation shows better agreement with the experiment than the pseudo-first-order kinetic model. In past literature also, the pseudo-second-order model provides the best correlation with the experiment for the entire period [13–15], whereas for the initial period pseudo-first-order model is applicable only [16].

In kinetic modeling, adsorption increases with time elapse and finally reaches equilibrium. At the initial stage, adsorption is found to be rapid and then becomes stagnant and slow with the rise in time. The abundance of adsorption sites may be responsible for the high initial uptake rate [17]. Adsorption becomes slow when the sites are gradually filled up and slowly approaching equilibrium. In a single system, the equilibrium is reached in 6.5 and 10 h for Ni and Cu. Ni takes more time (approximately 25 h) for the binary system to attain equilibrium than Cu (10 h). The antagonism in the mixture is substantiated in the kinetic model also. The kinetic rate constant (k_2) in the second order of Ni decreased from 0.025 to 0.012 in a binary system, and Ni adsorption was reduced by more than 50% for the inhibitory effect of Cu. The kinetic adsorption of Ni and Cu for the single and binary systems is presented in Figs. 6 and 7.

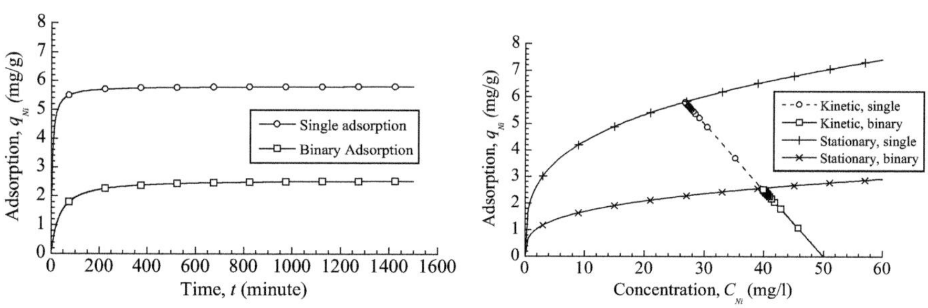

Fig. 6 Kinetic adsorption of Ni (Single and Binary)

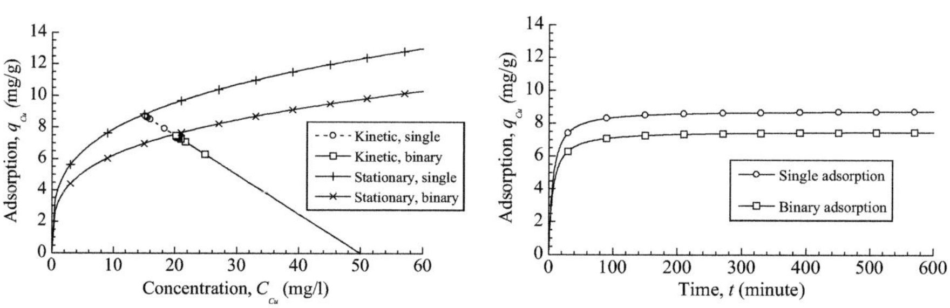

Fig. 7 Kinetic adsorption of Cu (Single and Binary)

4 Conclusion

The present study is focused on the modeling of the non-stationary adsorption of Ni and Cu metal ions in binary systems. The model considers the constant changing of equilibrium adsorption by using the rate form of kinetic equations. Besides, stationary adsorption is simulated for binary heavy metals. The experiment verifies both stationary and non-stationary models. The extended Freundlich model shows a good fit with an experiment in a stationary case. On the other hand, for non-stationary adsorption, the pseudo-second-order equation is best fitted with the experiment compared to the first-order kinetic equation. In future, this model will be further checked by the own experiments with various porous media as an adsorbent.

References

1. Vijayaraghavan K, Ashokkumar T, Rangabhashiyam S, Arockiaraj J (2016) Mono- and multi-component biosorption of lead(II), cadmium(II), copper (II) and nickel(II) ions onto coco-peat biomass. Sep Sci Technol 51:2725–2733. https://doi.org/10.1080/01496395.2016.1212889
2. Sud D, Kaur M, Mahajan G (2008) Agricultural waste material as potential adsorbent for sequestering heavy metal ions from aqueous solutions—a review. Bioresour Technol 99:6017–6027. https://doi.org/10.1016/j.biortech.2007.11.064
3. Taha AA, Ahmed AM, Heiba SMA, HF, (2016) Multi-component adsorption of Pb(II), Cd(II), and Ni(II) onto Egyptian Na-activated bentonite; equilibrium, kinetics, thermodynamics, and application for seawater desalination. J Environ Chem Eng 4:1166–1180. https://doi.org/10.1016/j.jece.2016.01.025
4. Ho YS, Forster CF, Wase DAJ (1996) Kinetic studies of competitive heavy metal adsorption by sphagnum moss peat. Environ Technol 17:71–77. https://doi.org/10.1080/09593331708616362
5. Gharin Nashtifan S, Maghsoudi A, Azadmehr A (2017) Comparative and competitive adsorptive removal of Ni2+ and Cu2+ from aqueous solution using iron oxide-vermiculite composite. Appl Clay Sci 140:38–49. https://doi.org/10.1016/j.clay.2016.12.020
6. Srivastava VC, Mishra IM, Mall ID (2006) Equilibrium modelling of single and binary adsorption of cadmium and nickel onto bagasse fly ash. Chem Eng J 117:79–91. https://doi.org/10.1016/j.cej.2005.11.021
7. Fritz S-U, W, (1974) Simultaneous adsorption equilibria of organic solutes in dilute aqueous solutions on activated carbon. Chem Eng Sci 29:1279–1282. https://doi.org/10.1016/0009-2509(74)80128-4
8. Montazer-Rahmati MM, Abdolali A, Rabbani P, Keshtkar AR (2011) Kinetics and equilibrium studies on biosorption of cadmium, lead, and nickel ions from aqueous solutions by intact and chemically modified brown algae. J Hazard Mater 185:401–407. https://doi.org/10.1016/j.jhazmat.2010.09.047
9. Alyüz B, Veli S (2007) Adsorption of copper and zinc from aqueous solutions by using natural clay. J Hazard Mater 149:226–233. https://doi.org/10.1016/j.jhazmat.2007.04.109
10. Feng N, Zhu Y, Guo X, Liang S, Liu J (2011) Biosorption of heavy metals from aqueous solutions by chemically modified orange peel. J Hazard Mater 185:49–54. https://doi.org/10.1016/j.jhazmat.2010.08.114
11. Ho YS, McKay G (1999) Pseudo-second order model for sorption processes. Process Biochem 34:451–465. https://doi.org/10.1016/S0032-9592(98)00112-5

12. Caliskan N, Sogut EG, Kul AR, Alkan S, Alacabey İ (2011) Adsorption of Zinc(II) on diatomite and manganese-oxide-modified diatomite: A kinetic and equilibrium study. J Hazard Mater 193:27–36. https://doi.org/10.1016/j.jhazmat.2011.06.058
13. Bulgariu L, Bulgariu D (2012) Bioresource technology equilibrium and kinetics studies of heavy metal ions biosorption on green algae waste biomass. Bioresour Technol 103:489–493. https://doi.org/10.1016/j.biortech.2011.10.016
14. Reddad Z, Andres Y, Gerente C, Cloirec PLE (2002) Adsorption of several metal ions onto a low-cost biosorbent: kinetic and equilibrium studies 36:2067–2073
15. Babazadeh H, Manshouri M, Nazemi AH (2011) Isotherm and kinetic studies on adsorption of Pb Zn and Cu by Kaolinite 9:243–255
16. Taha AA, Ahmed AM, Shreadah MA, Fathy H (2016) Journal of environmental chemical engineering multi-component adsorption of Pb (II), Cd (II), and Ni (II) onto Egyptian Na-activated bentonite; equilibrium, kinetics, thermodynamics, and application for seawater desalination. Biochem Pharmacol 4:1166–1180. https://doi.org/10.1016/j.jece.2016.01.025
17. Sen S, Bhattacharyya KG (2008) Influence of acid activation on adsorption of Ni (II) and Cu (II) on kaolinite and montmorillonite : kinetic and thermodynamic study. 136:1–13. https://doi.org/10.1016/j.cej.2007.03.005

Morphological Analysis of Bishkhali River Using Multi-temporal Satellite Images and Historical Cross-Sectional Data

I. Sarkar and M. A. Rahman

Abstract Bangladesh lies at the downstream of GBM basin which carries a large number of sediments through its river network which changes the morphological characteristics of rivers. Bishkhali River also experiences morphological changes throughout time. Bishkhali River is located at the south-central region of Bangladesh and discharges directly into the Bay of Bengal. Being a tidal river, its morphological characteristics are governed not only by the changes in flow from upstream but also by the impact of sea-level and tidal characteristics of the Bay of Bengal. Bishkhali River is mainly known for its bank erosion, bar formation, and shifting of thalwegs. This study is aimed at planform analysis and channel bar analysis for the period between 1995 to 2020. This study also includes thalweg analysis for the time period from 2002 to 2020 (based on the availability of cross-sectional data). For the estimation of erosion-accretion along both right-bank, left-bank, and channel bar analysis, LANDSAT images have been collected from USGS (United States Geological Survey) and these two analyses have been conducted by using GIS. Also, for thalweg analysis, historical cross-sectional data has been collected from BWDB (Bangladesh Water Development Board) and this analysis has been performed by using Microsoft Excel and GIS. It has been found from the planform analysis that total erosion and accretion for the right bank from 1995 to 2020 are 1073.68 and 664.61 ha, respectively. Also, for the left bank, total erosion and accretion from 1995 to 2020 were 1062.50 and 831.82 ha. From channel bar analysis, maximum areas for both bars situated near Ramna Launch Ghat and Kalikabari have been recorded in 2020, and these are 105.15 and 74.66 ha, respectively. From thalweg analysis, at (Rajapur Upazila, Daskinpara, Kalikabari, Mar bazar, Patharghata Upazila) stations, thalwegs are shifting towards the left bank, at (Aurabunia Bazar, Kathalia Launch Ghat, Karuna Khal, Patharghata Upazila) stations, thalwegs are shifting towards the

I. Sarkar (✉) · M. A. Rahman
Department of Water Resources Engineering, Bangladesh University of Engineering and Technology, Dhaka, Bangladesh
e-mail: indrorahul7777@gmail.com

M. A. Rahman
e-mail: mataur@wre.buet.ac.bd

S. Arthur et al. (eds.), *Proceedings of the 6th International Conference on Advances in Civil Engineering*, Lecture Notes in Civil Engineering 368,
https://doi.org/10.1007/978-981-99-3826-1_19

center of the river width; and at (AL-Madina Jama Mosque, Kakchira, Rupdhan High School) stations, thalwegs are shifting towards the right bank. To achieve sustainable coastal zone development, policymakers and administrators may use the study's findings to establish integrated coastal zone management (ICZM), limit the effects of coastal hazards, and develop effective mitigation techniques.

Keywords Bishkhali River · Landsat images · Erosion · Accretion · Thalweg

1 Introduction

1.1 Background of the Study

Bangladesh is a country of rivers. It is situated in the delta of the three greatest rivers in the world, the Ganges, the Brahmaputra, and the Meghna [1]. Channels, floodplains, bars, and vegetation all contribute to rivers self-organizing overall planform and pattern [2]. River course modification and bank erosion-accretion are connected to flow amount, sediment volume and type conveyed by the river, climate change, and the local hydrological circumstance [3]. For instance, SIDR (a very severe cyclonic storm in the Bay of Bengal in 2007) caused considerable morphological alterations when it made ashore at the Bishkhali outlet [4]. Also, the Lower Meghna River which is part of the GBM system downstream and well connected to Bishkhali River, is quite active and goes through significant structural changes [5]. River planform modifications may have a variety of negative effects on the environment and society [6]. People migrate as a result of the displacement caused by the erosion of existing rural, agricultural, riparian and urban areas, as well as the degradation of hydraulic structures and transportation infrastructure [7]. However, bank accretion creates fresh alluvial land that may be used for farming and raising livestock [5]. Land loss, meander formation, short- and long-term channel adjustments, sedimentation issues downstream, and sediment dynamics of the river basin are all significantly impacted by riverbank erosion [8]. The Bishkhali River runs through the south-central region of Bangladesh, and the aim of this article is to pinpoint key bank erosion spots and areas where new lands may arise. Due to their ability to effectively combine a variety of factors, geospatial tools and techniques (GIS and RS) are very helpful in identifying river shape as well as dynamics at different spatial–temporal scales [9]. Nihal et al. (2015) studied for investigating morphological changes for the Bishkhali River, it was discovered that both bank lines are gradually migrating, resulting in a change in planform [10]. Hassan and Mahmud-ul-Islam (2016) found that human habitation, woodland, seasonal crops, and agricultural characteristics had decreased while river coverage had greatly increased in their study of riverbank erosion and bar formation in the Sirajganj District [11]. Dewan et al. (2016) monitored the Ganges-Padma river system's channel alterations, and it became evident that both of the river's banks had lost a significant amount of land [12].

1.2 Study Area

In the south-central part of Bangladesh, the Bishkhali River is one of the significant rivers. The river is 96 km long in total. The river's average width from its source to the first 30 km is roughly 1 km, while the last km is roughly 2 km [4]. It comes from Sugandha River, which flows through the Gabkhan Dhansiri Union neighborhood of the Jhalokati Sadar Upazila in Bangladesh's Jhalokati district. The river then flows through the Upazilas of Rajapur, Kathalia, and Betagi until it enters the Bay of Bengal at Patharghata union in the Patharghata Upazila of the Barguna district. The river reach length between Jhalokati Sadar and Patharghata Upazila was considered for this study. Using Arc-GIS, a map of the research area has been produced, shown in Fig. 1.

1.3 Objective of the Study

I. To perform planform analysis by estimating erosion-accretion of riverbank from 1995 to 2020 at 5 years interval.
II. To perform channel bar analysis from 1995 to 2020 at 5 years interval.
III. To perform thalweg analysis by using available cross-sectional data from some specific years.

2 Materials and Methods

The USGS (United States Geological Survey) obtained georeferenced LANDSAT satellite pictures for the dry season of January to March, which we utilized for the study. During this time, there is a good chance of taking crisp pictures, and the bank lines are also easily apparent. Using Arc-GIS, the polylines from the Landsat photos were meticulously digitized in the Geographic Coordinate System GCS WGS 1984, a map is used to indicate the digitization of planform which includes the left bank and right bank in 1995 and 2020, and also the digitization of two channel bars in 2020, as shown in Fig. 2. Then, for both banks, the areas of erosion and accretion have been calculated every five years between 1995 and 2020, and the area of channel bars has also been assessed every five years between 1995 and 2020. The thalweg movement analysis component of this study also contains bathymetry data of the Bishkhali River, which is displayed in Table 1, which was gathered from various stations between 1995 and 2020. For thalweg analysis, bathymetry data was analyzed and a thalweg shifting map was generated by Arc-GIS. The overall process is shown as a flowchart in Fig. 3.

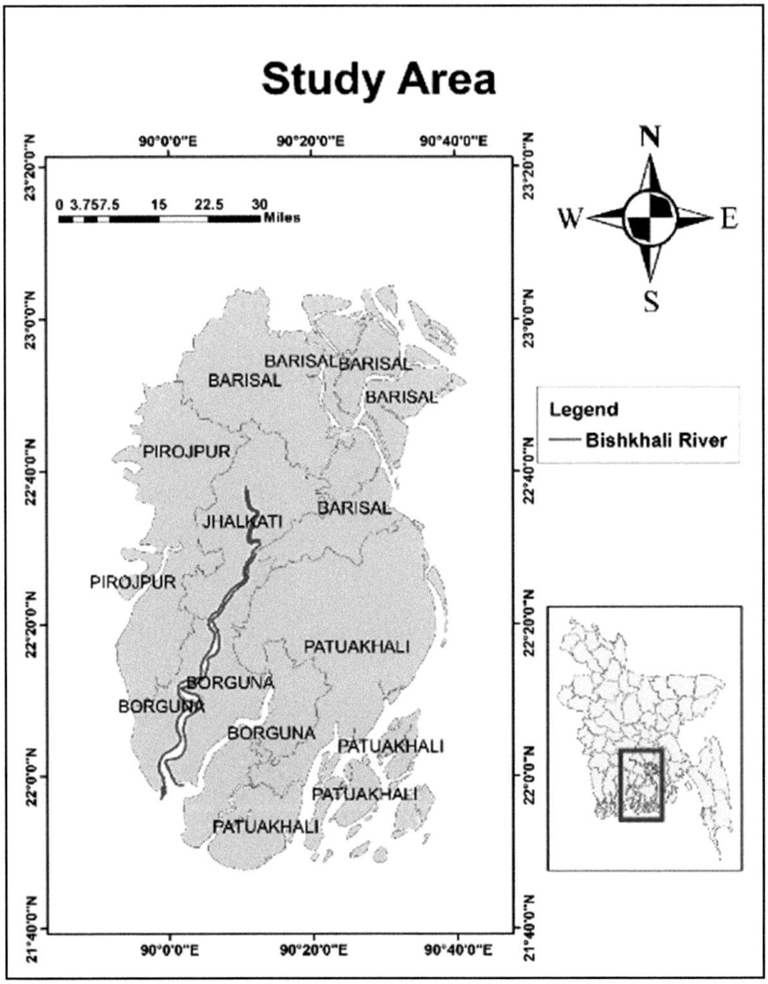

Fig. 1 Map of the study area showing Bishkhali River

3 Results and Discussions

3.1 Planform Analysis

For the right bank, the dominance of erosion decreases over time. Erosion suddenly dropped from 2000 to 2005, which is 94.82 ha as compared to the erosion of 397.69 ha from 1995 to 2000, shown in Fig. 4. After 2010, accretion dominates over erosion, although the maximum amount of accretion is 241.28 ha throughout 2000–2005. But as a whole, erosion dominates over accretion for the right bank, as displayed in Fig. 6. For the left bank, maximum accretion is 396.98 ha from 2000 to 2005, whereas

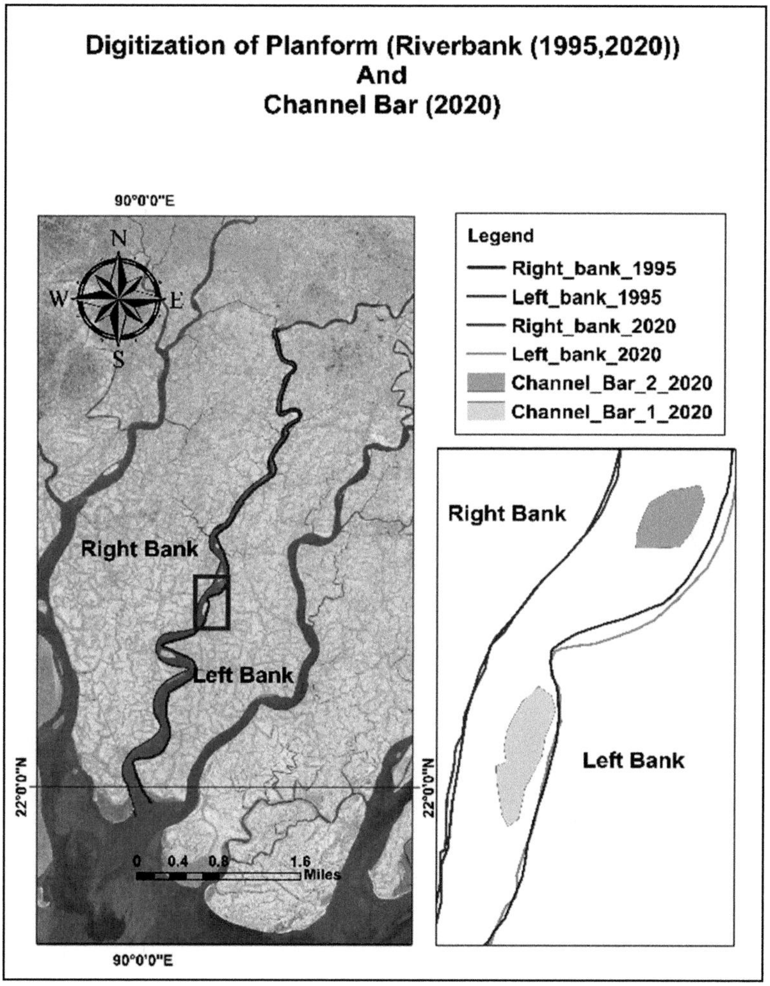

Fig. 2 Digitization of planform (Riverbank (1995, 2020)) and Channel Bar (2020)

Table 1 Summary of all required data

Data type	Location	Data source	Station ID	Period
Satellite images	Bishkhali River	USGS (30 m Resolution)	–	(1995–2020)
Bathymetry data	Bishkhali River	BWDB	RMBIS 5 to RMBIS 16	19,951–2020)

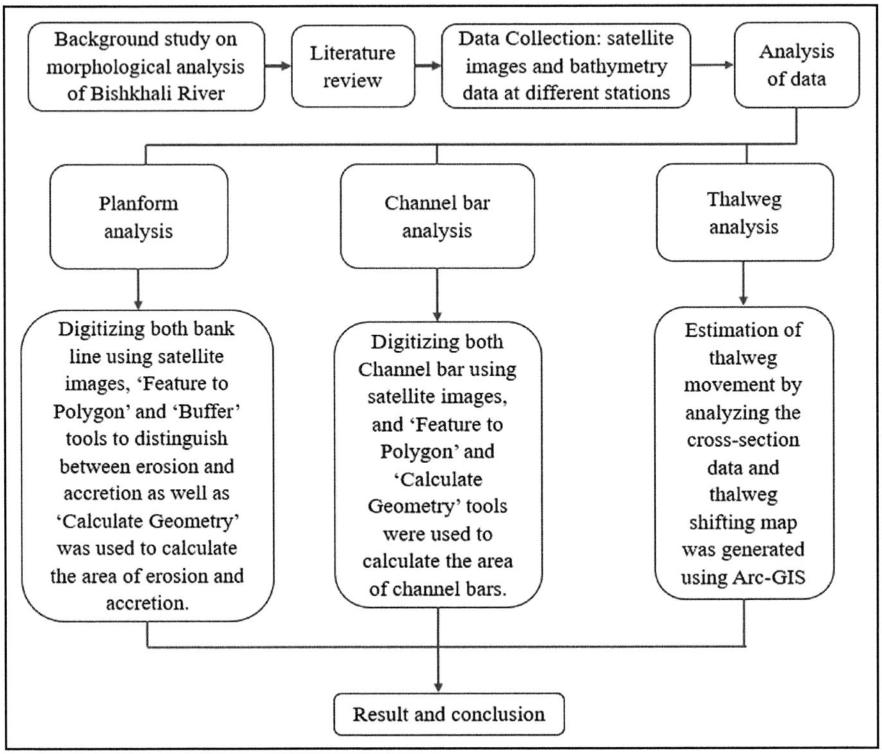

Fig. 3 Steps and processes of the study

maximum erosion is 341.04 ha from 1995 to 2000. In recent times, accretion has been gradually increasing and is currently carrying approximately the same amount as erosion, displayed in Fig. 5. However, the total amount of erosion dominates over accretion for the left bank from 1995 to 2020, as shown in Fig. 7.

3.2 Channel Bar Analysis

This bar chart shows the areas of channel bar 1 (near Ramna Launch Ghat, Bamna Upazila, Bargua District, Bishkhali River) and channel bar 2 (near Kalikabari, Bamna Upazila, Bargua District, Bishkhali River) at a five-year interval, as shown in Fig. 8. From this chart, the area of channel bar 1 is increasing in 2005, 2015, and 2020 and decreasing in 2000 and 2010. The area of channel bar 2 is increasing in 2000, 2005, 2015, and 2020 and decreasing in 2010. Maximum areas for both channel 1 and channel 2 bars have been recorded in 2020 and these are 105.15 and 74.66 ha, respectively.

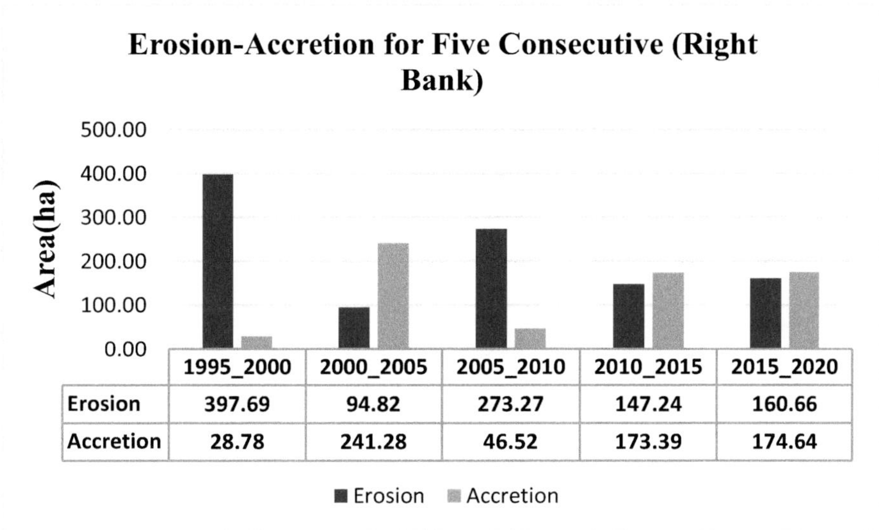

Fig. 4 Erosion-accretion for 5 years (right bank)

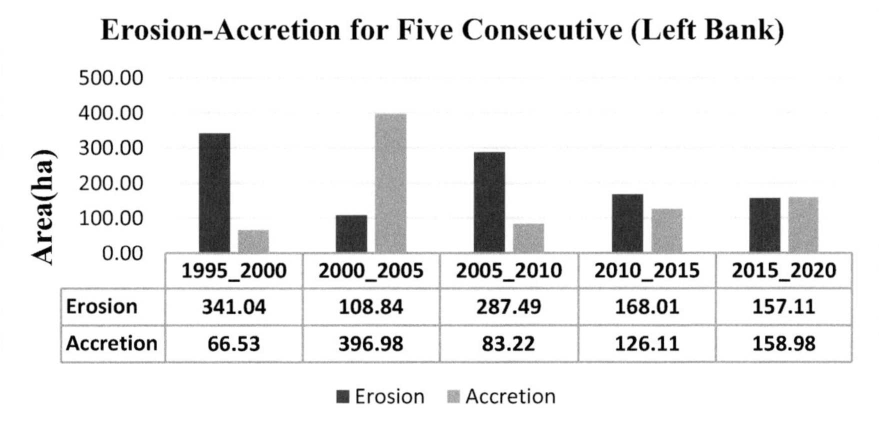

Fig. 5 Erosion-accretion for 5 years (left bank)

3.3 Thalweg Analysis

Thalwegs are actually the extreme deep points of any individual cross-section. From cross-sectional change analysis, it is observed that how thalwegs are moving and what are the patterns which are actually followed by these. For sections like Rajapur Upazila (RMBIS5), Daskinpara (RMBIS7), Kalikabari (RMBIS10), Mar bazar (RMBIS12), and Patharghata Upazila (RMBIS15)), thalwegs have a tendency to shift

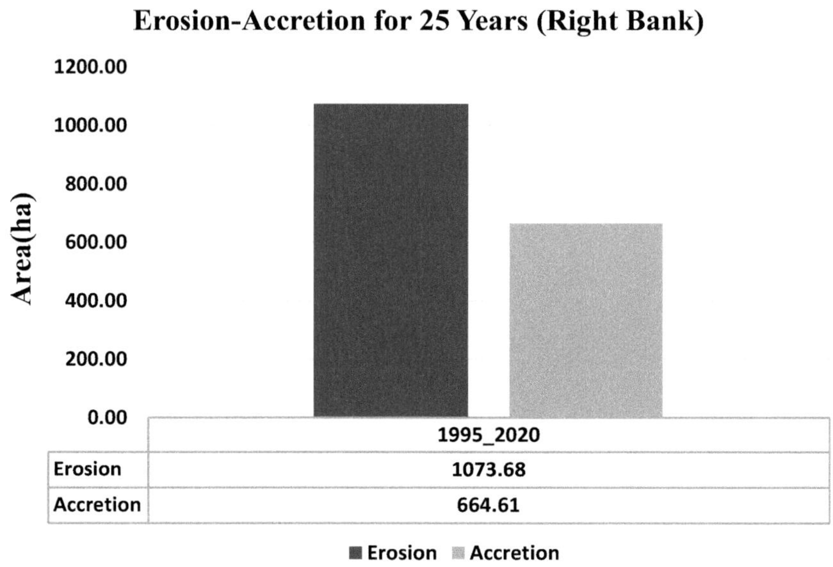

Fig. 6 Erosion-accretion for 25 years (right bank)

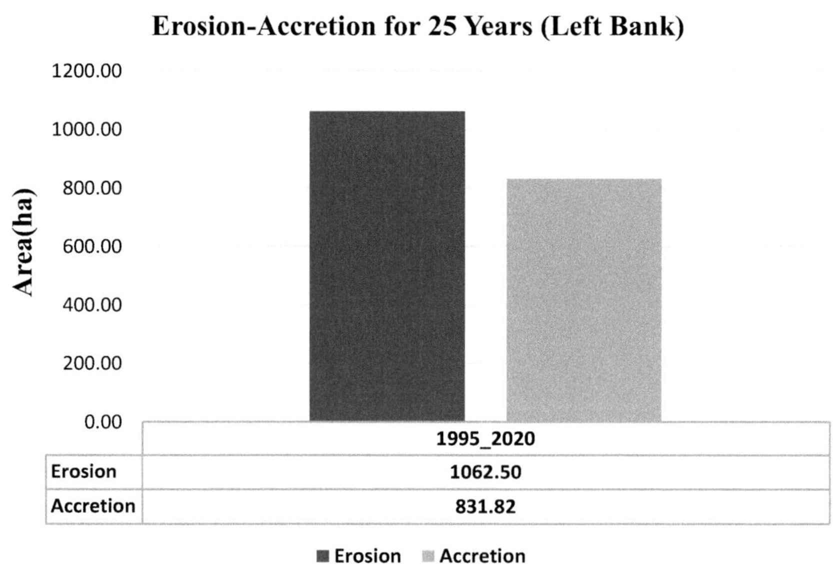

Fig. 7 Erosion-accretion for 25 years (left bank)

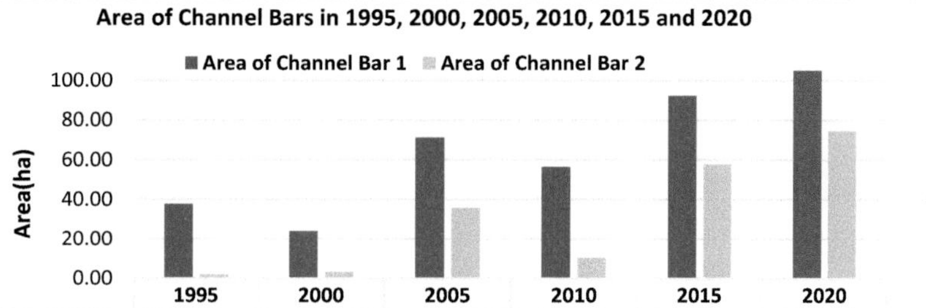

	1995	2000	2005	2010	2015	2020
Area of Channel Bar 1	37.74	24.10	71.17	56.46	92.44	105.15
Area of Channel Bar 2	1.96	3.33	35.68	10.47	57.83	74.66

Fig. 8 Area of channel bars in year 1995, 2000, 2005, 2010, 2015 and 2020

towards the left bank. For this reason, erosion of the left bank in that particular area can be observed at the same time accretion rates at that location have increased throughout the years. The thalwegs for sections (Aurabunia Bazar (RMBIS6), Kathalia Launch Ghat (RMBIS8), Karuna Khal (RMBIS9), and Patharghata Upazila (RMBIS16)) are mainly located at the center portion of the channel. It is also observed that at the same location, the rates of accretion have increased in both the right bank and the left bank. For sections (AL-Madina Jama Mosque (RMBIS11), Kakchira (RMBIS13), and Rupdhan High School (RMBIS14)), thalwegs have a tendency to shift towards the right bank, Fig. 9.

According to the findings, both the right bank and the left bank are seeing a considerable pattern development in morphological changes. Numerous significant structures that are located along the banks are constantly in danger due to river bank erosion. The ability to examine and evaluate patterns of riverbank erosion and accretion has been made possible by this work. This study also serves as an illustration of how GIS tools can be used to identify structural influences on riverbank erosion-accretion rates. Thalweg shifting occasionally provides us with further hints concerning morphological alterations. Thalweg shifting analysis has also been used in this work to better understand the regional morphological alterations.

Satellite pictures were used to manually digitize the shorelines for this study because high-resolution Landsat photographs were unavailable. A better outcome can be obtained by applying automated bank line extraction algorithms and higher resolution Landsat photos, as they can assure more precise delineation of the bank lines.

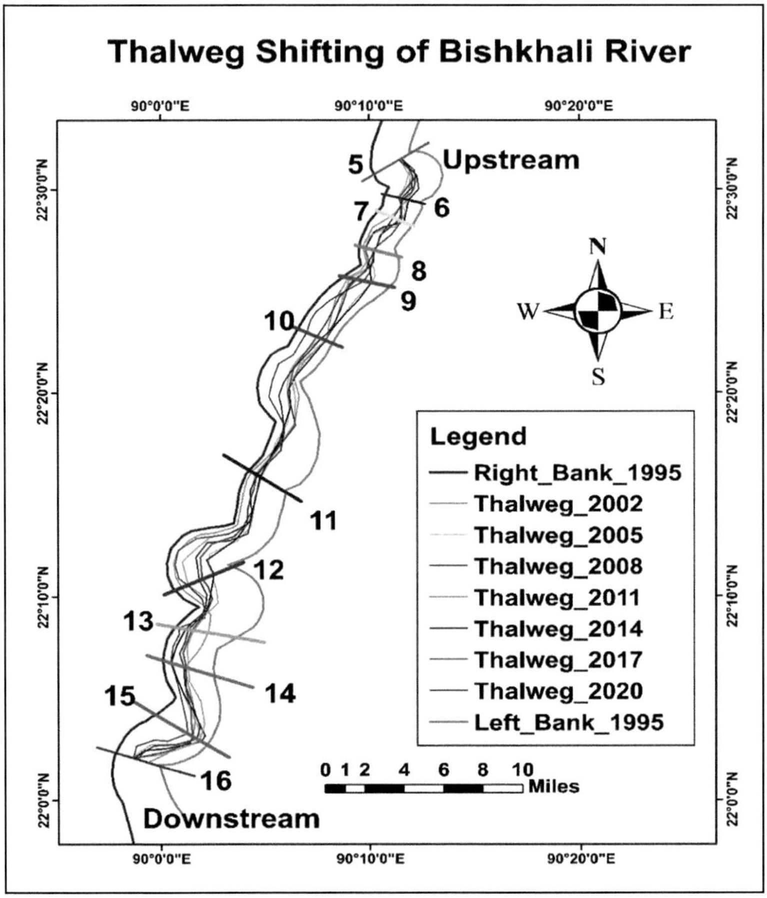

Fig. 9 Thalweg movement of Bishkhali River

4 Conclusion

In morphological analysis, it includes planform analysis, channel bar analysis and thalweg movement analysis.

I. From planform analysis, it can be seen that erosion predominates over accretion for both the right bank and left bank, with the difference between erosion and accretion for both the left bank and right bank being, respectively, 230.68 and 409.07 ha from 1995 to 2020.
II. From channel bar analysis, it is observed that channel bars have an increasing trend and the greatest area of both channel bars which are situated near Ramna Launch Ghat and near Kalikabari have been recorded in 2020 and these are 105.15 and 74.66 ha respectively. The formation of channel bars reflects the

dominance of accretion over erosion. Both new channel bars and existing channel bars are growing larger.

III. From thalweg analysis, it can be noticed that thalweg shifts rapidly and randomly in 2002, 2008 and 2020 but in 2002 the shifting of thalweg was more rapid and more random. After observing the results of total erosion and accretion changing rate, Bishkhali River is considered to be a moderately unstable river in terms of river morphology.

The study will be put into practice to counteract the consequences of morphological changes, it can be inferred from the conclusion. By putting more vegetation along riverbanks and building bank protection structures, local authorities can take preventative measures to deal with riverbank erosion. The results of this investigation show that riverbanks are seriously eroding and channel bars are beginning to appear. Consequently, it is advisable to relocate people and new construction from at-risk areas to secure channel bar zones.

References

1. Nath B, Naznin SN, Alak P (2013) Trends analysis of river bank erosion at Chandpur, Bangladesh: a remote sensing and GIS approach. Int J Geomat Geosci 3:454
2. Kleinhans MG (2010) Sorting out river channel patterns. Prog Phys Geogr Earth Environ 34:287–326. https://doi.org/10.1177/0309133310365300
3. Nones M (2021) Remote sensing and GIS techniques to monitor morphological changes along the middle-lower Vistula river, Poland. Int J River Basin Manag 19:345–357. https://doi.org/10.1080/15715124.2020.1742137
4. Elahi MW, Mohiuddin S, Nihal F, Karim R, Haque A, Rahman MM Changes of morphological characteristics of Bishkhali Estuary due to cyclone SIDR
5. Mahmud M, Mia AJ, Islam M, Peas MH, Farazi AH, Akhter SH (2020) Assessing bank dynamics of the Lower Meghna River in Bangladesh: an integrated GIS-DSAS approach. Arab J Geosci 13:1–19
6. Li L, Lu X, Chen Z (2007) River channel change during the last 50 years in the middle Yangtze River, the Jianli reach. Geomorphology 85:185–196. https://doi.org/10.1016/j.geomorph.2006.03.035
7. Lawler DM (1991) A new technique for the automatic monitoring of erosion and deposition rates. Water Resour Res 27:2125–2128
8. Lawler DM, Couperthwaite J, Bull LJ, Harris NM (1997) Bank erosion events and processes in the Upper Severn basin. Hydrol Earth Syst Sci 1:523–534. https://doi.org/10.5194/hess-1-523-1997
9. Batalla RJ, Iroumé A, Hernández M, Llena M, Mazzorana B, Vericat D (2018) Recent geomorphological evolution of a natural river channel in a Mediterranean Chilean basin. Geomorphology 303:322–337. https://doi.org/10.1016/j.geomorph.2017.12.006
10. Nihal F, Sakib M, Elahi MW, Ahmed T, Haque A, Rahman MM, Omar MAT Monitoring of planform changes and computation of erosion/accretion of Bishkhali Estuary using satellite image

11. MdS Hassan S Mahmud-ul-islam 2016 Quantification of River Bank erosion and bar deposition in Chowhali Upazila, Sirajganj District of Bangladesh: a remote sensing study J Geosci Environ Prot 04 50 57 https://doi.org/10.4236/gep.2016.41006
12. Dewan A, Corner R, Saleem A, Rahman MM, Haider MR, Rahman MM, Sarker MH (2017) Assessing channel changes of the Ganges-Padma River system in Bangladesh using Landsat and hydrological data. Geomorphology 276:257–279. https://doi.org/10.1016/j.geomorph.2016.10.017

User Perception and Assessment of Existing Public Toilets in Jessore, Bangladesh

S. A. S. Rafee and M. M. Rahman

Abstract Public toilets are considered an important part of city life. However, access to public toilets is essential in any community. The community's daily activities are severely disrupted if public toilets are unused. This paper deals with the present condition of existing public toilets and user perception of Jessore Paurashava. It was found that except for PT2 and PT5, the rest of the area's public toilets were not clean enough; structural maintenance was not done routinely; flushing equipment was broken and dysfunctional. Only PT2 and PT3 were well maintained, including basin, mirror, and handwash facilities. Four public toilets' physical conditions were moderate, whereas two were poor. The internal consistency of a questionnaire for questionnaire survey was found to be "excellent" (Cronbach's Alpha coefficient = 0.929). Around 61% of public toilet users use public toilets rarely, and most (68%) of the user lives within a 1 km radius of a public toilet. Most of the users agreed that the security of the surrounding environment (55%), service charge (58%), indoor light (61%), water availability (61%), odor (56%), and ventilation (66%) were satisfactory. From an In-Depth Interview (IDI) of caretakers, it was found that caretakers were not paid enough and were dissatisfied with the payment. The caretaker arranged cleaning materials, cleaned the toilet without additional payment, and worked for around 18 h in all public toilets except PT2. There was no arrangement of training regarding cleaning, safety, and emergency.

Keywords Public toilet · Jessore · Sanitation · Questionnaire survey · Maintenance

---

S. A. S. Rafee · M. M. Rahman (✉)
Department of Civil Engineering, Bangladesh Army University of Engineering & Technology (BAUET), Qadirabad, Bangladesh
e-mail: drmahmudur.rits@gmail.com

S. Arthur et al. (eds.), *Proceedings of the 6th International Conference on Advances in Civil Engineering*, Lecture Notes in Civil Engineering 368,
https://doi.org/10.1007/978-981-99-3826-1_20

1 Introduction

Public toilets are one of the most necessary elements of city life, but there are few works about them. Some studies provide adequate justification for public toilets in urban areas as a solution to a sustainable urban environment. Greed [1] mentioned that a public toilet is an essential factor in creating sustainable and accessible towns and cities. The users of a public toilet are generally street workers, homeless people, passing by people, etc. [2–4]. Research has shown that a public toilet can play an important role in public health and cleanliness as people may urinate on walls and sidewalls in the absence of a public toilet [5], for disease prevention by properly operating and managing it [6, 7] and for contribution to local economies [8]. Moreia [2] reported that the use of public toilets depends on their provision, accessibility, and hygiene aspect and suggested that the local authority consider workers and homeless people for planning public toilets.

It was reported that the existing public toilets in large cities in Bangladesh were mainly constructed by the local authority and leased out to private operations. However, the public toilets were oppressed by negligence and mismanagement [9]. In Dhaka city, male users were used to urinating in open places, which led to diarrheal. Public toilets become home to bacteria due to non-maintenance for a long time [10]. Hasan et al. [11] reported that rickshaw puller was the main group of roadside urinating due to insufficient public sanitation facilities. Hasan and Rahman [12] investigated the problems faced by users of public toilets in Dhaka, such as unsanitary public toilet conditions, lack of social norms, accessibility, inadequate proportions, active and inactive public toilets, lack of security, and lack of universal accessibility. Another study was conducted by the Disaster Management Watch [13] in Kushtia Paurashava to assess the current status of public toilet facilities provided by Kushtia Paurashava and measured current and future public toilet needs and potential sites that require public toilets.

Jessore Paurashava is the first paurashava of Bangladesh [14] and one of the central business hubs and is a transit point on the border between Bangladesh and India, where Benapole's land port is located. Benapole's land port can only be accessed via Jessore Paurashava. Also, Jessore Paurashava is a class "A" Paurashava, [15] and authorities should ensure that people in this community enjoy the essential services the community must provide. Public toilets are one of the basic facilities offered by the Paurashava authorities and are directly linked to SDG Goal 6 [16]. However, research on the situation of public toilets and public awareness of the use of public toilets has not yet been conducted. For this reason, this study focused on the access to the existing public toilet facilities provided by Jessore Paurashava and community perceptions by the users of public toilets.

2 Methodology

2.1 Study Area

The study area that has been selected is the Jessore Paurashava Fig. 1 which is a first-class Paurashava with an area of 14.72 km^2. Geographically, it lies from 22° 49' north latitude to 23° 23' north latitude and from 88° 50' east longitude to 89°34' east longitude. The number of populations in this Paurashava is 2,86,163. This Paurashava has nine public toilets, of which seven are functional, and two are out of service. In Table 1, location and toilet ID are shown.

Fig. 1 Study area map (Jessore Paurashava) (Extracted from Google Earth)

Table 1 List of Public Toilets at Jessore Paurashava

Toilet ID	Location	Coordinate
PT1	Palbari Mor Bus Stand	N23.18, E89.19
PT2	Paura Park	N23.16, E89.20
PT3	Monihar Bus Stand	N23.16, E89.22
PT4	General Hospital	N23.16, E89.21
PT5	New Market Bus Stand	N23.17, E89.21
PT6	Sadar Thana	N23.16, E89.21
PT7	Notun Bazar	N23.14, E89.23
PT8	Chachra Bus Stand	N23.14, E89.19
PT9	Palbari Bus Stand	N23.18, E89.19

2.2 *Data Collection Method and Tools*

Before collecting data, it is essential to know what data is required and what data to ignore. The aspects to be covered by the study also need to find. To assess the current state of the public toilet establishment, a list of indicators [13] was prepared, as shown in Table 2.

The primary data was collected by using three tools which were physical observation, a questionnaire survey, and a depth interview, as shown in Table 3. In physical observation, six public toilets (PT1, PT2, PT3, PT4, PT5, and PT6) were selected to observe the surrounding area and entrance of the public toilet, the common area of the public toilet, urinal/cubicle information, and maintenance of the Public Toilet. A checklist and scoring criteria were used to record the physical condition of existing public toilets which was adopted from a similar study by Disaster Management Watch [13]. According to the scoring criteria, below 30 is poor, from 30 to 45 is moderate and 45 to 60 is good. Eight (8) key performance characteristics were chosen as the basis for the scoring system since they are crucial from the user's perspective for a working or ideal public toilet. The eight key performance characteristics were the functionality of toilet equipment, internal cleanliness of the toilet, internal lighting and ventilation, availability of washing facilities, communication, and signage, surrounding

Table 2 Indicators and sub-indicators of the study

Indicator	Sub-indicator
Accessibility	Facility at entry
	Physical inclusiveness
	Facilities inside the toilet
	Basic Design consideration
Operation and regular maintenance	Day-to-day operation
	Cleanliness
	Health, Safety. Security of the users, Caretakers and cleaner

Table 3 Overview of the primary data collection process

Tools	Participants	Number of participants
Physical observation		6 public toilets
Questionnaire survey	Existing public toilet user	196 users
In-depth interview	Public toilet caretaker	6 Caretaker/existing and functioning public toilet

area, safety, security and privacy, and maintenance. A variety of components were mentioned under these eight key performance indicators, and the corresponding data was taken from a physical observation checklist. Each key performance indicator was given an initial score based on the components highlighted, and the final score for the indicators was calculated using the component's overall score. A questionnaire survey for existing users was carried out to know the present condition and useability of public toilet facilities that included opinions about the surrounding environment, traceability, lighting, ventilation, cleanliness, functionality, and facility of public toilets. The questionnaire consisted of a total of 29 questions. For the questionnaire survey, 196 users were interviewed and selected using the Bill Godden formula for the infinite population [17]. The caretaker of six public toilets (PT1, PT2, PT3, PT4, PT5, and PT6) was interviewed in an In-depth interview (IDI) (consisting of 35 questions) to collect detailed information on the current operation and maintenance practices and the barriers encountered, maintenance costs, frequency of use, security, frequency of maintenance, inspection, maintenance funding, source of funding, and other challenges to maintaining public restrooms.

2.3 *Data Analysis*

Cronbach's Alpha analyzed the questionnaire survey for existing users for reliability tests [18] using IBM SPSS (Statistical Package for the Social Sciences) to measure the internal consistency of a questionnaire, demonstrated mainly by the 5-point Likert scale questions [19]. There were 13 Likert scale questions selected from the 29 questions of the questionnaire survey regarding the surrounding and internal environment, functionality, useability, and security of public toilets (excluding demographic questions). Microsoft Office Excel 2019 was also used for data processing and analysis.

3 Results and Discussions

3.1 Physical Observation

From the physical observation, it was seen that the surrounding environment was not clean enough at PT1 Fig. 2a. Structural maintenance was not done routinely. Except for PT2 and PT5, the rest of the area's public toilets were not clean enough. There was a lack of maintenance inside of cubicles Fig. 2b. The flushing equipment was found to be broken and dysfunctional. Except for PT2 and PT3 at Jessore Paura Park, all other public toilets are not maintained properly by the authority. There is a separate entrance for males and females only in PT2 Fig. 2c and PT6. No dedicated sitting arrangement for the caretaker was found in any public toilet, as shown in Fig. 2d. Only PT2 and PT 3 were found to be well maintained, which include basin, mirror, and handwash facilities Fig. 2e. In PT2, PT6 urinals were placed in separate cubicles, and in PT1, PT3, PT4, and PT5, urinals were placed in an open place near the hand wash facility. Besides the overall observation, detailed surveillance of the public toilets has been done on a common ground based on the scoring criteria documented in Disaster Management Watch [13]. A score of 30 or less is considered bad, a score of 30–45 is considered intermediate, and a score of 45–60 is considered good. Rafee and Uddin's [20] dissertation has comprehensive individual profiles that were created using the data gathered from the public restroom locations. Table 4 provides a list of the state of several public toilets, with the engineer's assessment of each one's general physical condition based on observations made on-site. Four of the public toilets were judged to be in a moderate physical condition which scores were found in between 31 to 41, while two were in poor condition which scores were 23 and 27 (Table 4), which suggests that the conditions of the public toilets at Jessore Paurashava require improvement.

3.2 Questionnaire Survey

The Cronbach's Alpha coefficient of 13 Likert scale questions was found to be 0.929, which was "excellent" according to Cronbach's Alpha reliability test, which indicates that the internal consistency of a questionnaire for questionnaire survey was excellent.

Demographic data refers to statistical socioeconomic information, including employment, education, income, marriage, birth and death rates, and more [21]. From Fig. 3a, it was found that the maximum number of respondents were from the 30–39 age group (35%), which was similar to the Islam study [22] at Jessore Sadar Upazila. Most of the respondents (88%) were male Fig. 3b similar to Sharif's [23] study at Jessore district. Most (46%) users' education level was secondary level than primary level (18%), as shown in Fig. 3c, and most of the respondents were businessmen (44%) than day laborers (37%), as shown in Fig. 3d which was similar

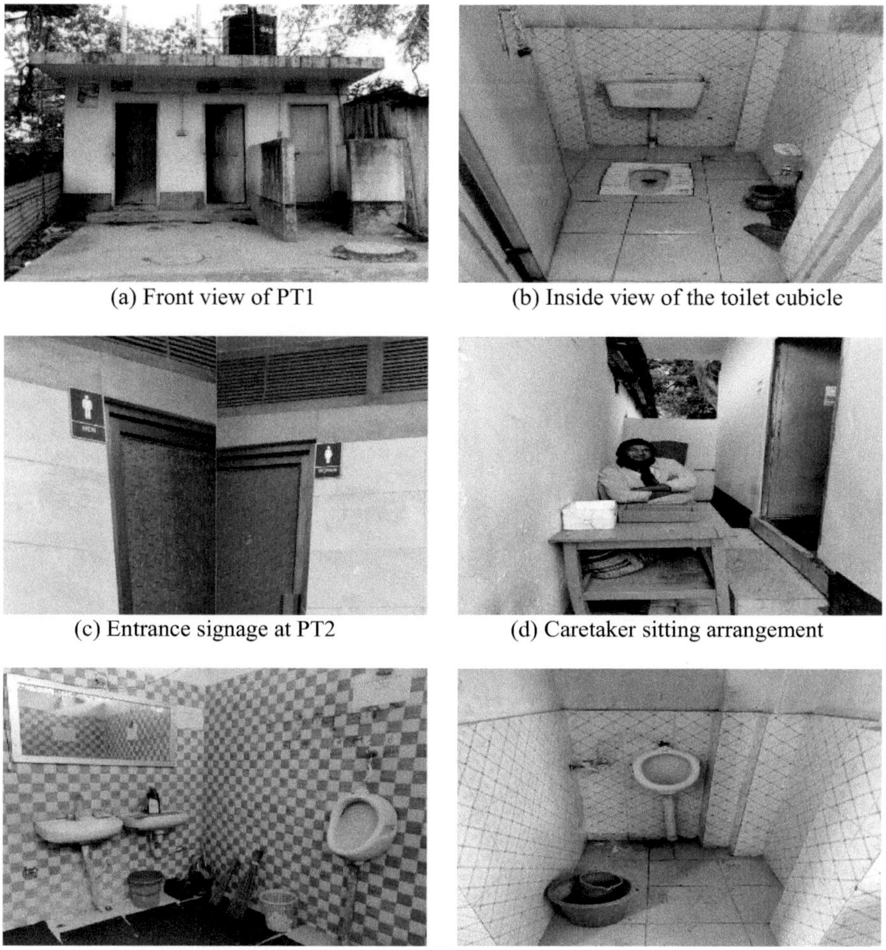

(a) Front view of PT1

(b) Inside view of the toilet cubicle

(c) Entrance signage at PT2

(d) Caretaker sitting arrangement

(e) Hand Wash, Basin and Mirror Facility at PT3

(f) Urinal of PT6

Fig. 2 Physical evidence o/f different existing public toilet

Table 4 Condition of public toilets in Jessore

Toilet ID	Location	Operational status	Overall physical condition	Score (60)
PT1	New Market	Operational	Poor	23
PT2	Paura Park	Operational	Moderate	41
PT3	Monihar	Operational	Moderate	33
PT4	General Hospital	Operational	Moderate	31
PT5	Palbari	Operational	Poor	27
PT6	Sadar Thana	Operational	Moderate	31

to Disaster Management Watch study carried out at Kushtia [13]. From Fig. 4a, it was seen that 61% of the public toilet users use public toilets rarely, 27% use 2/3 times a week, 11% use them daily, and most (68%) of the user lives within a 1 km radius of a public toilet Fig. 4b.

Around 51% of users agreed that the surrounding environment of the public toilet was clean Fig. 5a whereas 49% of users Fig. 5b agreed that the toilets were clean. In addition, it also found that more than 70% of public toilet users were either satisfied or strongly satisfied with the cleanliness of the surroundings and the toilet. Most of the users agreed that the security of the surrounding environment (55%), service charge (58%), indoor light (61%), water availability (61%), odor (56%), and

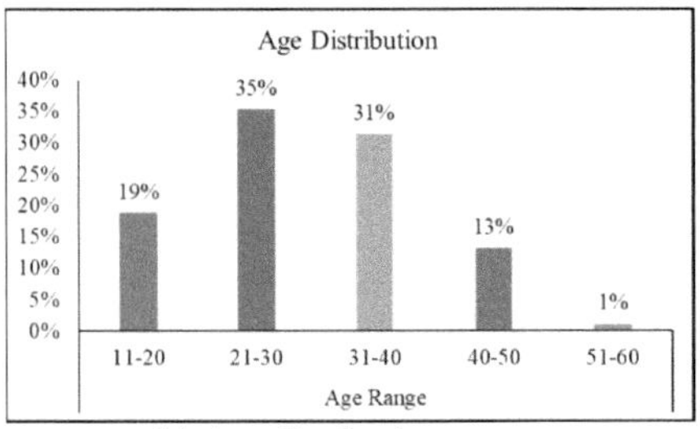

(a) Age distribution

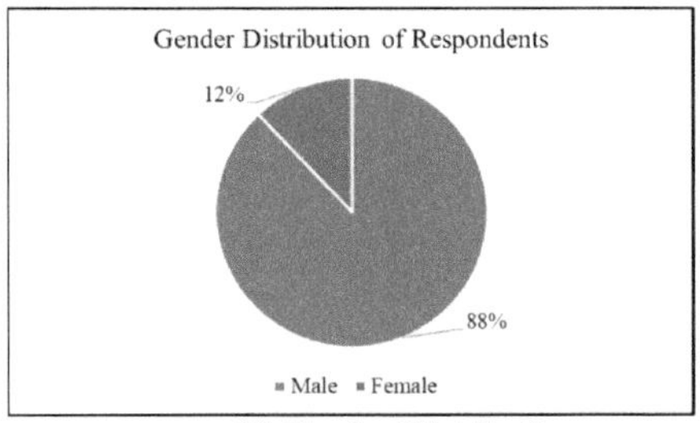

(b) Gender Distribution

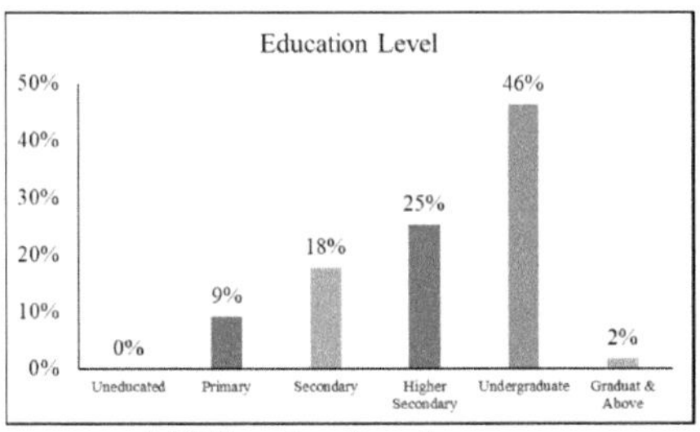

(c) Education Level

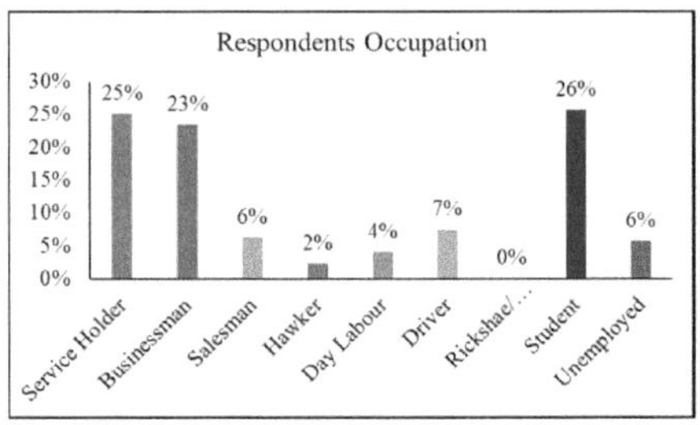

(d) Occupations

Fig. 3 Frequency of use and distance from public toilet

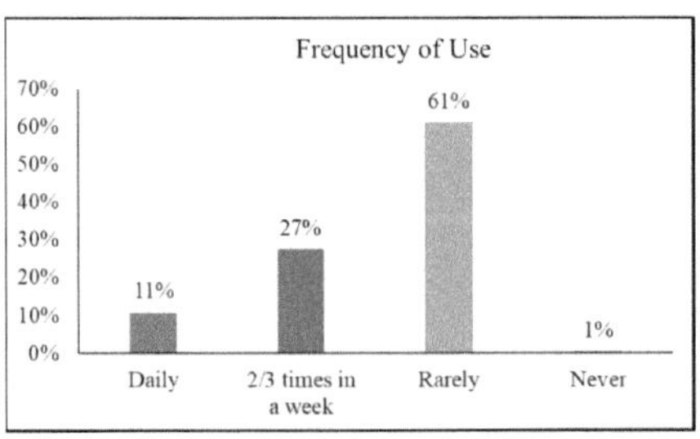

(a) Frequency of use

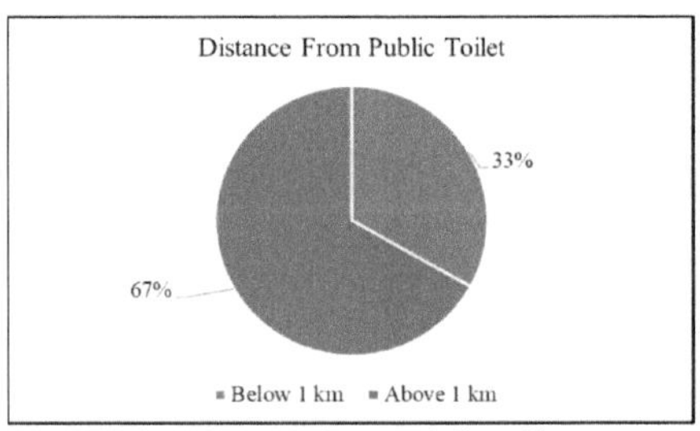

(b) Distance from public toilet

Fig. 4 Frequency of use and distance from public toilet

ventilation (66%) were satisfactory as shown in Fig. 5c–h. The Pearson correlation significance was found to be 0.646, 0.654, and 0.669 respectively for "Cleanliness of surrounding," "Cleanliness of toilet," and "Cleanliness of floor and wall" with respect to "Availability of water" which indicates that the possible variation for condition/ cleanliness to maintain hygiene is moderately dependent on the availability of water.

3.3 In-Depth Interview

Data collected through the in-depth interview with the caretaker was tabulated in [20]. It was found that caretakers were not paid enough and were dissatisfied with the payment. The caretaker arranged cleaning materials in PT1, PT3, PT4, PT5, and PT6, whereas in PT2, cleaning materials were arranged by the paurashava authority. PT1, PT3, PT4, PT5, and PT6 had an operation hour of around 18 h, whereas PT 2 operation hour was 12 h. PT1, PT3, PT4, PT5, and PT6 were cleaned once a day, but PT2 was cleaned three times a week. Only PT2 has a different cleaner provided by the paurashava authority. In the case of PT1, PT3, PT4, PT5, and PT6, they were cleaned by the caretaker, and they do not get paid for cleaning. There was no arrangement of training regarding cleaning, safety, and emergency. Also, there was no fund allowed from paurashava for maintenance except PT2. Any emergency maintenance had to be done by the leaser. The broken apparatus remains unrepaired due lack of funding.

4 Conclusions

The major findings of this paper were the present condition of the existing public toilet and the users' perceptions of the public toilet. From physical observation and IDI of the caretaker, it was found that except for PT2 and PT5, the rest of the area's public toilets were not clean enough; structural maintenance was not done routinely; flushing equipment was found to be broken and dysfunctional. Only PT2 and PT 3 were found to be well maintained, which include basin, mirror, and handwash facilities. Four public toilets' physical conditions were moderate, whereas two public toilets were poor. The caretakers were not paid enough and were dissatisfied with the payment. The caretaker arranged cleaning materials, cleaned the toilet without additional payment, and worked for around 18 h in all public toilets except PT2. There was no arrangement of training regarding cleaning, safety, and emergency for the caretakers. From the questionnaire survey, it was found that the internal consistency of a questionnaire was "excellent" with Cronbach's Alpha coefficient value of 0.929. Around 61% of public toilet users use public toilets rarely, and most (68%) of the user lives within a 1 km radius of a public toilet. Most of the users agreed that the security of the surrounding environment (55%), service charge (58%), indoor light (61%), water availability (61%), odor (56%), and ventilation (66%) were satisfactory.

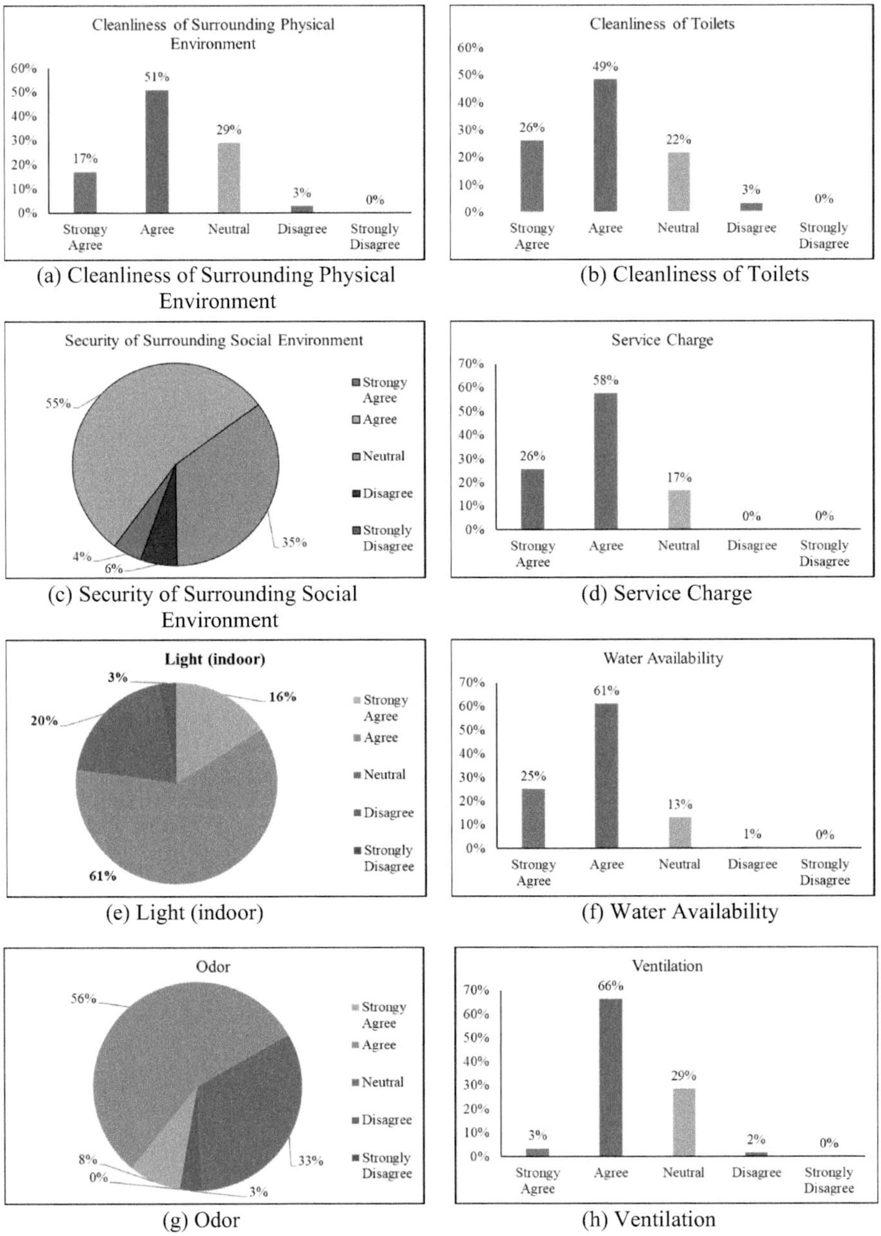

(a) Cleanliness of Surrounding Physical Environment

(b) Cleanliness of Toilets

(c) Security of Surrounding Social Environment

(d) Service Charge

(e) Light (indoor)

(f) Water Availability

(g) Odor

(h) Ventilation

Fig. 5 Operational condition and maintenance of public Toilet

References

1. Greed C (2004) Public toilets: the need for compulsory provision. Proceed ICE—Munic Eng 157(2):77–85. https://doi.org/10.1680/muen.2004.157.2.77
2. Moreira FD, Rezende S, Passos F (2022) Public Toilets from the perspective of users: a case study in a public place. Brazil. J Water Health 20(1):41–53
3. Frye EA, Capone D, Evans DP (2019) Open defecation in the United States: perspectives from streets. Environ Justice 12(5):226–230
4. Drewko A (2007) Resource-oriented public toilets in developing countries: ideas, design, operation and maintenance for Arba Minch, Ethiopia. Doctoral dissertation, Technische Universitat. Hamburgh-Harbug, Germany
5. Stanwell-Smith R (2010) Public toilets down the drain? why privies are a public health concern. Public Health 124(11):613–616
6. Donkor TA, Acheampong C (2015) Assessment of the performance and management of public toilet facilities in Ghana: a case study of Banatama in the Kumasi metropolis. Bachelor dissertation, University of Twente
7. Oloruntoba EO, Folarin TB, Ayede AI (2014) Hygiene and sanitation risk factors of diarrhoeal disease among under-five children in Ibadan. Nigeria. Afr Health Sci 14(4):1001–1011
8. Rotowa O (2020) Assessment of Public Lavatories in Akure, Nigeria. Int J Res Environ Sci (IJRES) 6(1):10–18. http://dx.do18 i.org/https://doi.org/10.20431/2454
9. Basahr MK (2017) Study on developing business models for public toilets in Dhaka and other major cities of Bangladesh. WaterAid
10. Hossain SB, Kabir R, Shahjahan M, Chowdhury A, Arafat SM, Arafat Y Gap between expectation and convenience of open access public toilets in Dhaka: a cross-sectional investigation. Int J Res Methodol Soc Sci 3(1):53–61
11. Hasan MM, Basak K, Ryhan K (2018) Roadside urination and public sanitation facilities in Dhaka, Bangladesh. J Adv Civ Eng Pract Res 7:9–17
12. Hasan SF, Rahman F (2021) Re-thinking the public toilet of Dhaka City: an adaptable prototype design solution to improve the existing situation. Creat Space 9(1):1–14
13. Disaster Management Watch (2018) Report on public toilet demand scan and development of individual profile of public toilets in Kushtia Paurashavas. SNV Netherlands Development Organization
14. Hossain KMD, Hakim MA, Mondal S, Khan MAS, Seddique AA (2014) Water supply and demand gap analysis: a case study on Jessore Pourashava, Bangladesh. J Biosci Agricult Res 2(1):08–4
15. List of Pourashava/City Corporation (Division wise). (n.d.). https://oldweb.lged.gov.bd. Retrieved August 30, 2022, from chrome-extension://efaidnbmnnnibpcajpcglclefindmkaj/ https://oldweb.lged.gov.bd/UploadedDocument/UnitPublication/10/686/List%20of%20Paurashava_City%20Corporation-28May2019.pdf
16. Terms of Reference (ToR) for "Pourashava Sanitation Support Consultancy" (Package No.: BMWSSP/SD-12) under Bangladesh Municipal Water Supply and Sanitation Project (BMWSSP) (2021). Department of Public Health Engineering (DPHE)
17. Godden B (2004) Sample size formulas. J Stat 3:66
18. Cronbach LJ (1951) Coefficient alpha and the internal structure of tests. Psychometrika 16:297–334
19. Robinson J (2014) Likert Scale. In: Michalos AC (ed) Encyclopedia of quality of life and well-being research. Springer, Dordrecht, pp 3620–3621
20. Rafee SAS, Uddin MH (2022) Community perception and assessment of public toilet at Jessore Paurashava among existing public toilet users. Bachelor dissertation, Bangladesh Army University of Engineering & Technology (BAUET), Qadirabad Natore, Bangladesh
21. Hayes A (2021) Demographics. Retrieved from Investopedia. https://www.investopedia.com/terms/d/demographics.asp#:~:text=Demographic%20analysis%20is%20the%20study,and%20death%20rates%2C%20and%20more.

22. Islam MA, Asif A-A, Samad M, Rahman M, Yeasmin S (2014) Socio-economic conditions of the fish farmers in Jessore, Bangladesh. Int J Bus Soc Sci Res 2(2):153–160
23. Sharif B, Islam M, Asif A-A, Vaumik S, Zafar M (2015) Socio economic status of fry collectors at Sundarbans region. Int J Fish Aquat Stud 3(2):89–94

Water Treatment and Scaling in Distribution Pipes—The Impact of Silica and Phosphate on Calcium Carbonate Precipitation

T. Bhattacharjee, S. Tarannum, and M. A. Ali

Abstract Scaling due to the precipitation of $CaCO_3$ in water distribution systems results in flow restrictions and head loss, damage to plumbing fixtures, and clogging membranes/filters in treatment systems. A potential method for controlling $CaCO_3$ precipitation is using scale inhibitors (e.g., PO_4^{3-}, Zn^{2+}, etc.). A recent study found that in NaCl electrolyte solution, both phosphate and zinc inhibited the precipitation of $CaCO_3$; however, neither phosphate nor zinc could reduce the precipitation of $CaCO_3$(s) in natural groundwater. Two possible reasons identified for this were: (a) the presence of silica in groundwater which could enhance the precipitation of calcium, and (b) the possible precipitation of Ca-PO_4 solids in groundwater systems. This study assesses silica's effect on the precipitation of $CaCO_3$, both in the presence and absence of phosphate. This research involved laboratory batch experiments in 0.1 M NaCl solution containing calcium chloride and sodium bicarbonate. Later, similar batch experiments were carried out in synthetic groundwater. In the batch experiments, the mixing time was 30 min, and the concentration of Ca^{2+} was 75 mg/L. The effect of silica and phosphate on the precipitation of $CaCO_3$ was evaluated by varying their concentrations. The precipitation of $CaCO_3$ was slightly enhanced in the presence of silica (without phosphate) in NaCl electrolyte solution. On the contrary, the precipitation of $CaCO_3$ decreased in NaCl electrolyte solution in the presence of phosphate (without silica). In the presence of both phosphate and silica, silica did not significantly affect phosphate's inhibiting ability on precipitation of $CaCO_3$. Experiments carried out in synthetic groundwater also did not find any effect of silica on the precipitation of $CaCO_3$. Further investigations under a broader range of conditions are needed to understand better the chemical processes leading to the different effects of phosphate on the precipitation of calcium carbonate in an electrolyte solution and natural groundwater.

Keywords Silica · Phosphate · Precipitation · Calcium carbonate · Water treatment

T. Bhattacharjee (✉) · S. Tarannum · M. A. Ali
Department of Civil Engineering, Bangladesh University of Engineering and Technology (BUET), Dhaka, Bangladesh
e-mail: trip.bhatt@gmail.com

S. Arthur et al. (eds.), *Proceedings of the 6th International Conference on Advances in Civil Engineering*, Lecture Notes in Civil Engineering 368,
https://doi.org/10.1007/978-981-99-3826-1_21

1 Introduction

Scaling causes several technical problems in water treatment and distribution systems [1]. Scaling in boilers leads to heat exchange failure and a rise in pipe wall temperature, which increases power costs and compromises safety [2]. Most scales consist of carbonates, sulfates, hydroxides, phosphates, and silicates of alkaline earth metals, particularly calcium, and magnesium [3]. Industrial systems frequently experience scale expansion, necessitating scale inhibition, and control [4]. Precipitation of calcium carbonate has been researched for over a century [1]. Scales typically form when minerals grow into hard, dense masses in high-temperature regions [5]. Scale formation in high-pressure membrane systems is strongly influenced by concentration polarization [6].

In Bangladesh, Meherpur Paurashava (municipality) is suffering from high levels of hardness (Ca and Mg) in the groundwater used for the supply of potable water [7]. Citizens of the Paurashava are facing serious problems because of this. The encrustation of calcite is damaging plumbing fixtures, clogging the distribution system, resulting in head loss. The main cause of the problem is the precipitation of $CaCO_3$. A literature review reveals climate change, rising temperatures, and lower use of scaling and corrosion inhibitors contribute to the re-emergence of $CaCO_3$ scaling problems [8]. The possible methods to control scaling include lowering the pH of water, removing dissolved calcium ions, using scale inhibitors, decreasing water temperature, controlling $CaCO_3$ morphology, and controlling the use of surface materials (e.g., pipes). Products with anti-scale qualities are called scale inhibitors. Different types of scale inhibitor (e.g., Mg^{2+}, Ni^{2+}, Co^{2+}, Cu^{2+}, Zn^{2+}, phosphates) have been used for delaying precipitation of $CaCO_3$ [1, 9, 10].

In a recent study [7], it has been found that in NaCl electrolyte solution (in distilled water), precipitation of $CaCO_3$ decreases as the concentration of the inhibitor – phosphate or zinc– increases. The inhibition effect of phosphate appears to diminish to some extent with time. In contrast, for Zn, the inhibition effect seems to become stronger with time (see Fig. 1). In electrolyte solution, the inhibition effect of phosphate is higher than that of Zn for shorter mixing time, but for longer mixing periods, Zn is more effective in reducing precipitation of $CaCO_3$ [7]. However, in this study [7], it was also found that in natural groundwater, neither phosphate nor Zn could reduce the precipitation of $CaCO_3$(s). In fact, precipitation slightly increased. The inhibitory effect of phosphate and Zn (based on experiments in electrolyte solution) does not appear to apply to natural groundwater systems with a range of other dissolved constituents [7]. The possible reasons behind this, as reported by [7], are the presence of dissolved silica (which may enhance the precipitation of calcium carbonate), and the possible precipitation of Ca-PO_4 solids and $ZnCO_3$(s) in natural groundwater.

It was found in some studies that the presence of silica affects the stability of the various polymorphs; silica temporarily stabilizes ACC (Amorphous Calcium Carbonate) particles and transforms them into calcite [11]. The presence of dissolved silica affects the initial appearance of $CaCO_3$ nuclei because it lowers the surface

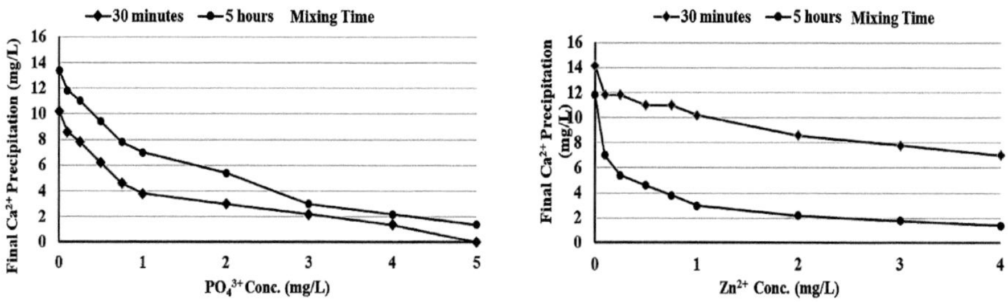

Fig. 1 Concentration of precipitated Ca (calculated) as a function of phosphate and zinc concentration (initial Ca = 75 mg/l) for two different mixing times [7]

free energy necessary for $CaCO_3$ nucleation [12]. So dissolved silica (present in natural groundwater) can potentially counteract the ability of phosphate and zinc to delay the precipitation of calcium carbonate [7]. Therefore, it is important to assess silica's effects on the precipitation of calcium carbonate. The principal objective of this study was to evaluate how silica affects the precipitation of calcium carbonate ($CaCO_3$) in electrolyte (NaCl) solution with or without phosphate (PO_4^{3-}).

2 Methodology

The main objective of this research was to assess the effect of silica on the precipitation of calcium carbonate ($CaCO_3$) with or without the presence of phosphate (PO_4^{3-}), which has the potential to delay the precipitation of calcium carbonate ($CaCO_3$) in electrolyte (NaCl) solution. This objective was achieved by carrying out a series of batch experiments where the effectiveness of (a) silica (SiO_2), (b) phosphate (PO_4^{3-}), and (c) silica and phosphate—on the precipitation of calcium carbonate was assessed in 0.1 M NaCl solution in deionized water. The effect of silica on the precipitation of calcium carbonate ($CaCO_3$) was evaluated under a range of conditions, including with and without pH balancing, using different filter media, varying the concentration of silica (SiO_2), and varying the concentration of phosphate (PO_4^{3-}) with and without silica.

The research was conducted in three phases comprising four scenarios. In the first phase, batch experiments were conducted in 0.1 M NaCl solution in deionized water containing silica (SiO_2) at concentrations of 50 mg/L (with and without adjusting pH), 100 mg/L, 150 mg/L, 200 mg/L, and 500 mg/L. Each sample had an alkalinity of 200 mg/L as $CaCO_3$ (achieved through the addition of $NaHCO_3$ solution) and a Ca^{2+} concentration of 75 mg/L (achieved through the addition of $CaCl_2$ solution). These experiments were used to analyze the influence of silica on the precipitation of $CaCO_3$, with and without pH adjustment.

In the second phase, batch experiments were carried out in 0.1 M NaCl solution in deionized water containing phosphate (PO_4^{3-}) at concentrations of 0.5 mg/L, 2 mg/

L, and 5 mg/L, without silica and with silica at 50 mg/L. As before, each sample had an alkalinity of 200 mg/L as $CaCO_3$ and Ca^{2+} ion of 75 mg/L.

In the third phase, two batch experiments were carried out in 0.1 M NaCl solution in deionized water containing 5 mg/L phosphate (PO_4^{3-}), 75 mg/L Ca^{2+}, and 275 mg/L (as $CaCO_3$) alkalinity. In one of the two experiments, silica was present at 65 mg/l, while in the other, silica was absent. These samples were prepared to mimic the composition of natural groundwater [7]. The second and third phases of experiments were conducted to analyze the influence of silica on the inhibition effect of phosphate ions in the precipitation of $CaCO_3$.

In all cases, the samples were mixed for 30 min on a magnetic mixer. After mixing, the samples were filtered with regular filter paper in scenario-1 and with 0.20 μm nylon membrane filter for all other scenarios for filtering the precipitated $CaCO_3$. The temperature and pH of the samples were measured at the initial, middle (15 min), and end of each experiment. The filtrates were then tested for the determination of dissolved calcium ions. The precipitated amount of calcium ion was calculated by comparing the initial and final dissolved Ca concentrations.

In addition, dissolved phosphate (PO_4^{3-}) ions were measured using Atomic Absorption Spectrometer (AAS) to determine the precipitation of phosphate ions (as Ca-PO_4 solids) in the third phase of the experiment. The mixing process of each batch experiment in the first phase was carried out following a six-step procedure (Fig. 2). A magnetic stirrer was used to mix the chemicals properly.

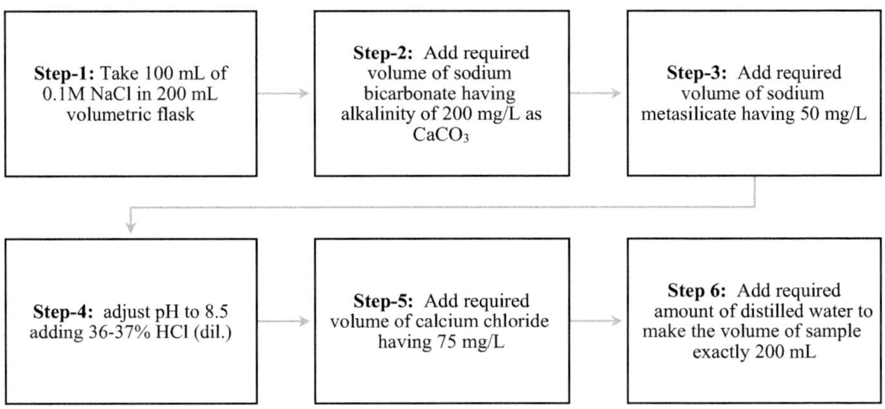

Fig. 2 Sample preparation and mixing process in the first phase of the experiment

3 Results and Discussions

3.1 *Effect of Reagents on pH of Sample Mixtures*

3.1.1 *Effect of Silica (SiO_2) on pH of Mixture*

The variation of pH of the mixture (containing 0.1 M NaCl and $NaHCO_3$) with the change in concentration of silica (SiO_2) can be seen in Fig. 3. With an increase in the concentration of SiO_2, the value of pH increases. In the mixtures with an alkalinity of 200 mg/L as $CaCO_3$, the pH value increased from 9.58 at 50 mg/L SiO_2 to 10.68 for 200 mg/L SiO_2, and to 11.89 for 500 mg/L SiO_2.

This pH variation was due to the addition of sodium metasilicate ($Na_2SiO_3.9H_2O$) as the source of silica. This leads to a rise in the pH value of the mixture. Since pH strongly influences the precipitation of calcium carbonate (with higher pH causing higher precipitation), this change in pH of the mixture with a change in silica concentration needed adjustment before the addition of $CaCl_2$. This pH adjustment ensured that the measured precipitation of $CaCO_3$ was only associated with the varying concentration of silica, and no other factors were involved in the results.

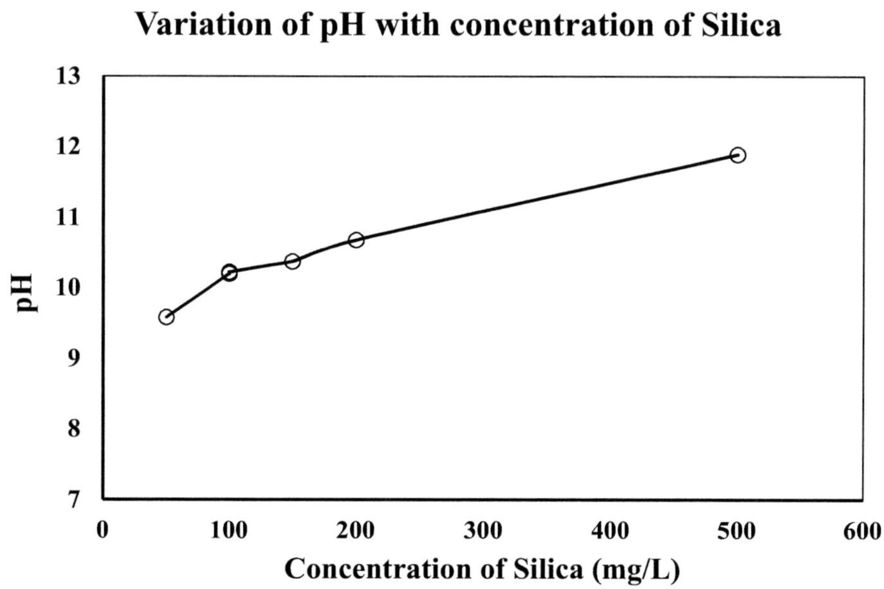

Fig. 3 Variation of pH with the concentration of silica in the mixture

3.1.2 Effect of Different Reagents on pH of Sample Mixtures

Figure 4 shows a summary of the variations in pH of the experimental solutions/mixtures in response to the addition of different reagents. It should be noted that pH has a significant impact on the precipitation of calcium carbonate. Therefore, the pH of the experimental solutions/mixtures was carefully monitored and finally adjusted to the desired value (with the addition of dilute HCl). Temperature also affects the precipitation of calcium carbonate (with higher temperature favoring higher precipitation); therefore, the temperature of the experimental mixtures was also carefully monitored. Figure 4 shows that the pH of the experimental mixtures first increased with the addition of $NaHCO_3$, and further increased with the addition of silica (as its sodium salt). These increases are expected as bicarbonate and silica were added as their respective sodium salt.

From Fig. 4, it can be seen that initially, in 0.1 M NaCl solution, the pH of the solution is in the neutral range. With the addition of sodium bicarbonate ($NaHCO_3$), the pH of the mixture rises to about 8.3. The pH of the mixture further increased with the addition of silica (as its Na salt), which was then balanced to around 8.5 with the addition of dilute HCl. Upon the introduction of $CaCl_2$, the pH of the mixture starts to decrease slightly to become 8.35 ~ 8.4. There was no notable change in the pH of the experimental solution/mixture at any time during the mixing period. From the figure, it can also be seen that in all cases, it was possible to adjust the pH of the samples to the desired pH value. Figure 5 shows no significant variation of temperature in the experimental mixture at any time during the experiment, but with an increase in silica dosage, the mixtures' temperature decreased slightly.

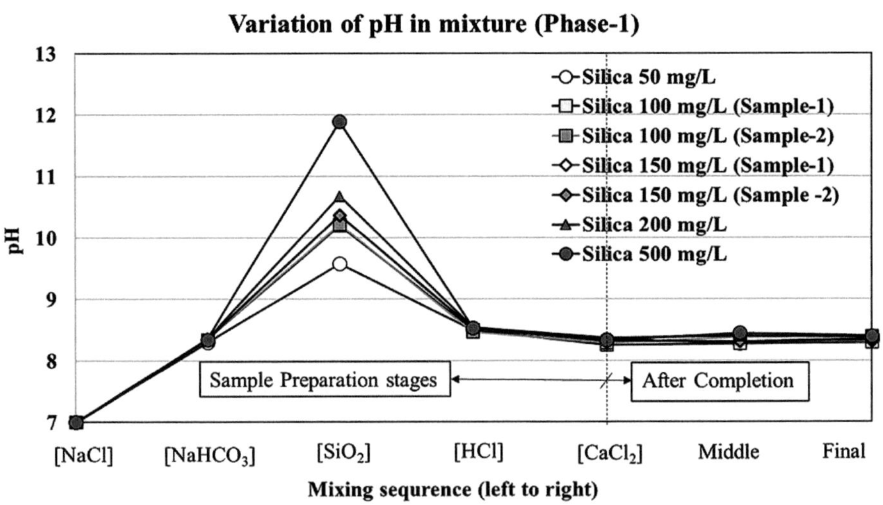

Fig. 4 Summary of the variation of pH in response to the addition of different reagents to the experimental mixtures for different concentrations of silica

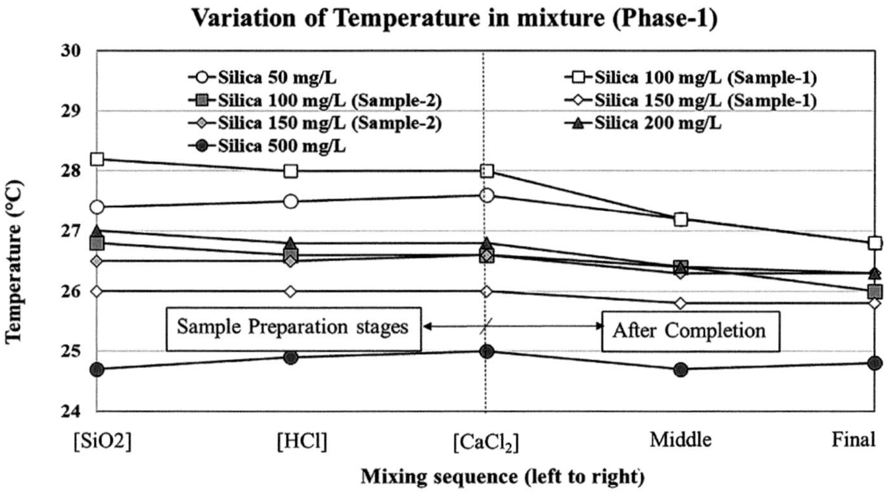

Fig. 5 Summary of variation of temperature in the experimental mixtures with different concentrations of silica

3.2 *Simultaneous Effect of Silica and PO_4^{3-} on Precipitation of $CaCO_3$ in Electrolytic Solution*

3.2.1 *Effect of PO_4^{3-} on the Precipitation of $CaCO_3$*

Table 1 presents the data from batch experiments carried out to assess the effect of PO_4^{3-} concentration on the precipitation of $CaCO_3$ when the other factors are the same. The table shows that the final dissolved Ca^{2+} in samples increases with an increased concentration of PO_4^{3-} inhibitor.

For example, after 30 min of mixing, the concentration of dissolved Ca^{2+} increases from 70 mg/L (at $PO_4^{3-} = 0$ mg/L) to 75.6 mg/L (at $PO_4^{3-} = 2$ mg/L). Again, the

Table 1 Data from the batch experiments carried out to assess the effect of PO_4^{3-} concentrations on the precipitation of $CaCO_3$

Experimental condition:
Initial Ca^{2+} concentration: 75 mg/L; Total Alkalinity: 200 mg/L as $CaCO_3$
Mixing time: 30 min; Without silica

Sample number	$[PO_4^{3-}]$ (mg/L)	Dissolved Ca^{2+} (mg/L)	pH			Temperature		
			Initial	Middle	Final	Initial	Middle	Final
1	0	70	8.33	8.42	8.4	24	23.9	23
2	2	75.6	8.37	8.46	8.51	23.6	23.5	23.4
3	0	72	8.28	8.27	8.21	22.5	22.5	22.4
4	5	74.8	8.3	8.27	8.31	22.5	22	22.2

concentration of dissolved Ca^{2+} increases from 72 mg/L (at $PO_4^{3-} = 0$ mg/L) to 74.8 mg/L (at $PO_4^{3-} = 5$ mg/L), as shown in Table 1.

3.2.2 *Simultaneous Effect of Phosphate and Silica on Precipitation of $CaCO_3$*

Table 2 presents data from batch experiments carried out to assess the precipitation of $CaCO_3$ in the presence of both phosphate and silica. The data shows that in the presence of 2 mg/L phosphate, the concentration of dissolved Ca^{2+} is 75.6 mg/L without silica, and 74.8 mg/L with silica [50 mg/L]. Here, the concentrations of dissolved Ca^{2+} with and without silica are very close.

By comparing the concentrations of dissolved Ca^{2+} in the absence and presence of silica, it is observed that there is no significant effect of silica on the precipitation of calcium carbonate in the presence of phosphate. So, it can be concluded that silica does not significantly affect phosphate's ability to reduce the precipitation of calcium in solution under the experimental conditions employed in this study. Therefore, it appears that silica is probably not responsible for the enhanced precipitation of Ca observed by Rashid (2021) in groundwater systems. However, more carefully controlled experiments under different conditions should be carried out to confirm this observation.

This study also investigated other possibilities for the enhanced precipitation of Ca in groundwater in the presence of phosphate, as observed by Rashid (2021). Batch experiments were carried out in solutions similar to those (in terms of Ca, alkalinity, and silica concentration) used by Rashid (2021) (with groundwater), but this time in NaCl electrolyte solution.

Table 3 presents the data from batch experiments carried out to assess the effect of phosphate and silica on the precipitation of $CaCO_3$ and Ca-PO_4. The objectives of these experiments were: (1) to observe how the results of two similar systems (one in NaCl electrolyte solution and one in groundwater) vary; (2) whether we can

Table 2 Data from the batch experiments carried out to assess the simultaneous effect of PO_4^{3-} and silica on the precipitation of $CaCO_3$

Experimental condition:
Initial Ca^{2+} concentration: 75 mg/L; Total Alkalinity: 200 mg/L as $CaCO_3$
Mixing time: 30 min; With silica

Sample number	$[PO_4^{3-}]$ (mg/L)	$[SiO_2]$ (mg/L)	Dissolved Ca^{2+} (mg/L)	pH			Temperature		
				Initial	Middle	Final	Initial	Middle	Final
1	2	0	75.6	8.37	8.46	8.51	23.6	23.5	23.4
2	2	50	74.8	8.35	8.37	8.42	24	23.9	23.8
3	5	0	74.8	8.3	8.27	8.31	22.5	22.5	22.4
4	5	50	76.4	8.38	8.42	8.34	22.1	22	21.8

Table 3 Data from the batch experiments carried out to assess the simultaneous effect of PO_4^{3-} and silica on the precipitation of $CaCO_3$ and Ca-PO_4

Experimental condition:
Initial Ca^{2+} concentration: 75 mg/L; Total Alkalinity: 275 mg/L as $CaCO_3$
Mixing time: 30 min

Sample number	[PO_4] (mg/L)	[SiO_2] (mg/L)	Dissolved Ca^{2+} (mg/L)	Dissolved PO_4^{3-} (mg/L)	pH			Temperature		
					Initial	Middle	Final	Initial	Middle	Final
1	0	0	73.5	–	8.31	8.32	8.41	21.9	22	22
2	5	0	72	0.68	8.29	8.27	8.28	22	21.9	21.8
3	5	65	74.25	0	8.31	8.23	8.19	21.7	21.7	21.7

detect any precipitation of Ca-PO_4 solids (through measurement of phosphate and Ca concentration) in our experimental system.

From the only data set presented in Table 3, it is difficult to interpret the results as there was no significant precipitation of Ca^{2+} in any of the experiments. These results do not match the previous experiments. The precipitation of phosphate is also negligible for the experiments. From these data, no definite conclusions could be drawn.

4 Conclusion

The primary goal of this study was to assess the effect of silica on the precipitation of calcium carbonate ($CaCO_3$) in electrolyte (NaCl) solution in the presence and absence of phosphate (PO_4^{3-}). Batch experiments were carried out in 0.1 M NaCl solution in deionized water to assess the impact of silica alone or in combination with phosphate (PO_4^{3-}) on the precipitation of $CaCO_3$. The major conclusions from the present study are as follows. In NaCl electrolyte solution (containing dissolved Ca and carbonates), the precipitation of calcium carbonate appears to be slightly enhanced in the presence of silica. In NaCl electrolyte solution (containing dissolved Ca and carbonates), the precipitation of calcium carbonate decreases in the presence of phosphate; this result agrees with the results reported by Rashid (2020). From the experimental setup used for assessing the combined effect of phosphate and silica on the precipitation of calcium, it appears that silica does not significantly impact phosphate's ability to reduce the precipitation of calcium in an electrolyte solution. Therefore, it appears that silica is probably not responsible for the enhanced precipitation of Ca observed by Rashid (2020) in groundwater systems. The single experiment conducted to mimic (in terms of Ca, alkalinity, and silica concentration) one of the experimental systems used by Rashid (2020) did not show any effect of silica on the precipitation of $CaCO_3$ or Ca-PO_4.

References

1. Hamdi R, Tlili M (2019) Investigation of scale inhibitors effect on calcium carbonate nucleation process. Desalin Water Treat 160:14–22
2. Tyusenkov AS, Cherepashkin SE (2014) Scale inhibitor for boiler water systems. Russ J Appl Chem 87:1240–1245
3. Tomson MB (1983) Effect of precipitation inhibitors on calcium carbonate scale formation. J Cryst Growth 62:106–112
4. Saji VS, Meroufel AA, Sorour AA (2020) Corrosion and fouling control in desalination industry
5. Kavitha AL, Vasudevan T, Prabu HG (2011) Evaluation of synthesized antiscalants for cooling water system application. Desalination 268:38–45
6. Antony A, Low JH, Gray S, Childress AE, Le-Clech P, Leslie G (2011) Scale formation and control in high pressure membrane water treatment systems: a review. J Memb Sci 383:1–16
7. Rashid I, Alam T, Ahmed T, Ali MA (2021) Evaluation of the Effectiveness of Zinc as Scale Inhibitor in Delaying Precipitation of CACO3 in Water Treatment and
8. Richards CS (2016) Calcium Carbonate Formation in Water Distribution Systems and Autogenous Repair of Leaks by Inert Particle Clogging, pp 1–62
9. Chaussemier M, Pourmohtasham E, Gelus D, Pécoul N, Perrot H, Lédion J, Cheap-Charpentier H, Horner O (2015) State of art of natural inhibitors of calcium carbonate scaling. Rev Art Desal 356:47–55
10. Sabzi R, Arefinia R (2019) Investigation of zinc as a scale and corrosion inhibitor of carbon steel in artificial seawater. Corros Sci 153:292–300
11. Kellermeier M, Glaab F, Klein R, Melero-García E, Kunz W, García-Ruiz JM (2013) The effect of silica on polymorphic precipitation of calcium carbonate: an on-line energy-dispersive X-ray diffraction (EDXRD) study. Nanoscale 5:7054–7065
12. Lakshtanov LZ, Stipp SLS (2010) Interaction between dissolved silica and calcium carbonate: 1. Spontaneous precipitation of calcium carbonate in the presence of dissolved silica. Geochim Cosmochim Acta 74:2655–2664

Effect of Overweight Vehicles on the Design of PC Girder Bridges

R. S. Tomal and K. M. Amanat

Abstract Overloading has become a major concern for engineers in recent years. The mentality of logistics service providers towards increasing their profit margins and higher transportation and fuel costs in middle-income countries like Bangladesh induce overloading of freight traffic. These result in an increase in the number of accidents due to braking system failure and infrastructure damage such as roads and bridges. Increased frequency of overloads may trigger shorter service life of critical components and perhaps catastrophic bridge failure. This study aims to assess the impact of overweight vehicle on the design of prestressed concrete girder bridges, which comprise the majority of highway bridges. For this purpose, an overweight vehicle has been selected investigating monthly overweight truck status report from a previous study. Bridges of four different spans (30 m, 35 m, 40 m and 45 m) have been modeled and designed in CSiBridge for strength and serviceability combination. To explore the effects of overloading, series of selected overweight vehicles were driven over those bridges. Parameters considered for studying the effect are maximum mid span moment, D/C ratio, maximum reactions at supports for Strength I combination and girder bottom tensile stress at Service III combination. It has been observed that maximum moment has been increased by 20.46% for 30 m bridge, 25.74% for 35 m bridge, 31.59% for 40 m bridge and 37.89% for 45 m bridge. D/C ratio has been increased by 19.76% for 30 m bridge, 35.33% for 35 m, 41.44% for 40 m bridge and 32.41% for 45 m bridge. Substantial changes have been noticed in girder bottom tensile stresses. Stress has been increased by 625.8% for 30 m bridge, 1995% for 35 m bridge, 884.5% for 40 m bridge and 763.8% for 45 m bridge. Maximum support reaction has been increased by 44.33% for 30 m bridge, 37.83% for 35 m bridge, 48.77% for 40 m bridge and 41.63% for 45 m bridge for interior girders and for exterior girders 32.55% for 30 m bridge, 32.48% for 35 m bridge, 37.20% for 40 m bridge and 35.79% for 45 m bridge. This study provides quantitative evidence

R. S. Tomal (✉) · K. M. Amanat
Department of Civil Engineering, Bangladesh University of Engineering and Technology, Dhaka 1000, Bangladesh
e-mail: 0422042418@ce.buet.ac.bd

K. M. Amanat
e-mail: amanat@ce.buet.ac.bd

S. Arthur et al. (eds.), *Proceedings of the 6th International Conference on Advances in Civil Engineering*, Lecture Notes in Civil Engineering 368,
https://doi.org/10.1007/978-981-99-3826-1_22

of damages caused by overweight vehicles to highway bridges and represents the significance of overloading in the design of pc girder bridges.

Keywords Overloading · CSiBridge · Service life · Strength and serviceability combination · Prestressed concrete bridge · Damage · Failure

1 Introduction

1.1 Background and Motivation

Increasing freight movement especially overweight freight is gradually becoming a great concern in Bangladesh. Two main reasons contributing to this increased overweight freight movement are: (1) Major development projects and activities generating movement of freights heavily loaded with construction materials; (2) Recent trends in online trades and shopping generating large numbers of freight heavily loaded with consumers goods. Most of this freight transportation is happening through trucks. The main concern for structural engineers here is that bridges are designed for a certain design traffic load (e.g., HS 20–44, HS 15–44, HL-93 K) which may be crossed by this overweight traffic and cause significant damage. In Bangladesh, although there is axle load rules and axle load control centers, lack of proper implementation of rules and enforcement often permits overweight freight to move through the bridges. Moreover, Government often relaxes this load control rules during special occasions when the highways become congested. This necessitates a requirement of study that can quantitatively show how much damage is occurring due to those overweight freights/trucks.

1.2 Objectives and Scope of This Study

This focus of this study is to determine the extent to which these overweight trucks are crossing different resistance limits provided by AASHTO (American Association of State Highway and Transportation Officials) and to quantify the damage occurring due to heavy truck movements on PC Girder Bridge. If it becomes possible to quantify this damage or if it is possible to represent this mathematically, it will work as evidence and it is expected that this will be helpful in raising consciousness of Government authority, Highway agency and Bridge authority about the severity of this overloading problem. The objectives and scopes of this study can be summarized as:

(1) To study the process of analysis, design, and performance evaluation of prestressed concrete bridges by the methods specified in AASHTO LRFD 2012.
(2) To conduct a parametric study by varying pc girder bridge span and develop a comparison between design truck and overweight truck on different parameters like maximum moment, demand to concrete shear capacity ratio and stresses at service III condition.
(3) Focuses mainly on the superstructure modeling, analysis and design
(4) Substructure elements will not be modeled in detail.

1.3 Findings of Previous Studies

Researchers have investigated the effects of overweight vehicles, but most of these studies were limited in scope. The researchers' observations were only provided during and after the overloading passages. The long-term consequences of overweight cars have not been studied. The following is a list of notable studies on overweight vehicles. Duncan [1] investigated superload effects on bridges in South Africa, emphasizing the significance of precise methodologies for assessing superload effects on bridges in order to use lower margins of strength for regulated superload passages. The effects of overload trucks on deck damage were explored by Kostem [2], who highlighted the necessity of including material nonlinearities in the analysis to depict how forces are redistributed across the structure. According to Kostem [2], The deck slab's flexural cracking as a result of the bridge's overloading is what causes the damage. Ohio's three steel-girder bridges were put to the test by Turer and Aktan [3] to see whether they could handle an 817-kip superload truck with a cooling unit. Cross-frames, which link the girders together and offer stability, were revealed to be the most severely strained parts with a stress range of 10 ksi. Additionally, it was determined that the transverse deck stress produced by transverse bending was significant. On the other hand, the girders' stress level remained at or around 5 ksi. The three bridges were all confirmed to be in good condition. Though this type of damage was not seen, the loss of composite action was described as a frequent damage mechanism. Additionally, the deflections were far lower than the typical limit of the span length divided by 800 [4]. Before, during, and after a superload passage, by Ziehl and Lamana, [5] a Louisiana bridge made of prestressed concrete (PC) had its stresses and deflections examined. The acoustic emission method was also used to evaluate damage. The researchers found no deterioration or alteration in the bridge's response following the passage of the superload.

2 Model Development and Validation

2.1 Model Data

1. Span (c/c distance of bearing), L = 104.5 feet (31.86 m)
2. Roadway width = 24 feet (7.32 m)
3. Total width of the bridge = 31 feet (9.45 m)
4. Numbers of Lane = 2
5. Number of longitudinal girders = 4
6. Future wearing course = 4 inch thick
7. c/c spacing of girders = 7.75 feet (2.36 m)
8. AASHTO Girder type = V
9. Post tensioning steel: 7 wire strand, low relaxation, Grade 270, Nominal diameter = 0.6 inch (15.2 mm)
10. Curb height = 12 inch (3.66 m)
11. Maximum initial concrete stress at transfer: $f_{ci}\prime = 0.8 f_c\prime$
12. Concrete for Deck slab, $f_{c(Deck)}\prime = 5$ ksi (34.5 MPa)
13. Concrete for Girder, $f_{c(Girder)}\prime = 6$ ksi (41.4 MPa)
14. Elastic Modulus for Slab Concrete, $E_{c(Deck)} = w_c^{1.5}\sqrt{f_{c(Deck)}\prime} = 3904$ ksi = 26,920 MPa
15. Elastic Modulus for Slab Concrete, $E_{c(Girder)} = w_c^{1.5}\sqrt{f_{c(Girder)}\prime} = 4276$ ksi = 29,480 MPa
16. Elastic Modulus of Prestressing steel, $E_p = 28{,}500$ ksi (Fig. 1).

2.2 Modeling Steps

See Fig. 2.

2.3 Validation

The model validation will show variation of different parameters from CSiBridge to manual calculation according to Code. This will ensure that the model has been correctly modeled. Validation for section properties, moment, support reaction, stress in tendon after all losses, stress check at transfer, check for final condition (Table 1).

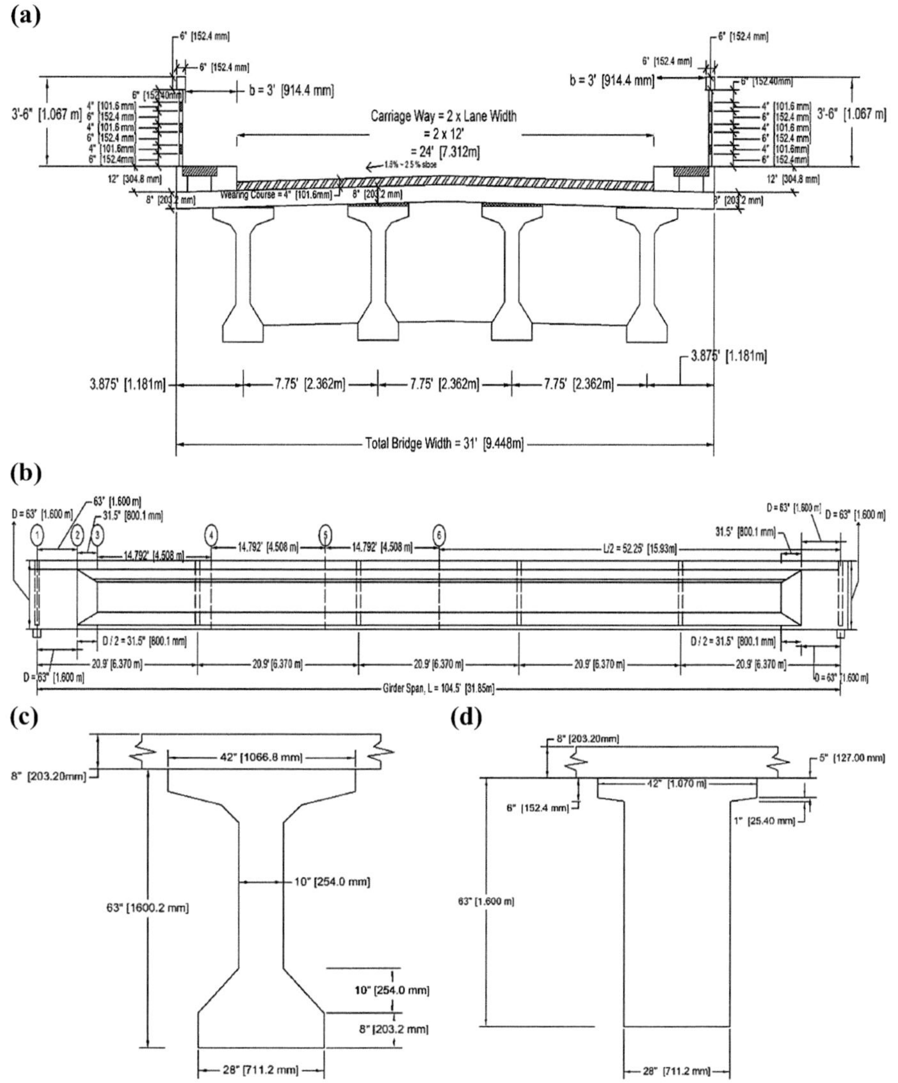

Fig. 1 **a** Deck configuration of model bridge; **b** Longitudinal section of AASHTO Type V girder, **c** Cross section of Mid-section and **d** Cross section of End-section

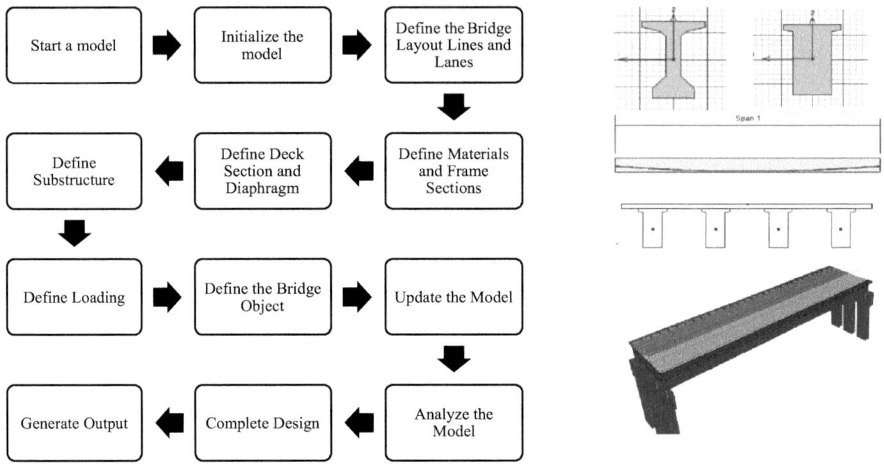

Fig. 2 **a** Steps of implementing the 3D FE model of bridge in CSiBridge; **b** 3D FE model in CSiBridge

	Location	Distance (m)	Manual calculation (KN-m)	CSiBridge (KN-m)	Variation (%)
Check for moment due to girder self-weight	1	0	0	0	0
	2	6.911	1464	1425	2.664
	3	11.42	1974	1935	1.976
	4	15.93	2144	2107	1.726
Check for moment due to Dead load (Girder Self weight + Diaphragms + Deck Self weight)	1	0	0	0	0
	2	6.911	2700	2564	5.000
	3	11.42	3645	3487	4.335
	4	15.93	3955	3795	4.046
Check for moment due to Sidewalk and railing	1	0	0	0	0
	2	6.911	638.8	627.8	1.722
	3	11.42	864.8	859.0	0.671
	4	15.93	940.3	947.3	0.739
Check for moment due to wearing course load	1	0	0	0	0
	2	6.911	454.9	445.3	2.110
	3	11.42	616.4	606.4	1.622
	4	15.93	670.1	660.4	1.448
Check for moment due to pedestrian live load	1	0	0	0	0
	2	6.911	141.5	139.0	1.767
	3	11.42	191.5	190.3	0.627
	4	15.93	208.1	209.8	0.810

(continued)

(continued)

	Location	Distance (m)	Manual calculation (KN-m)	CSiBridge (KN-m)	Variation (%)
Check for stresses in tendon after all losses	1	0	1125	1130	0.4425
	2	6.911	1143	1147	0.3487
	3	11.42	1155	1159	0.3451
	4	15.93	1168	1172	0.3413
Stress check at transfer at bottom fiber of concrete (- means compression)	1	0	−7.786	−7.731	0.7064
	2	6.911	−17.98	−17.85	0.7230
	3	11.42	−20.39	−20.22	0.8337
	4	15.93	−21.36	−21.20	0.7491
Stress check at transfer at top fiber of concrete (- means compression)	1	0	−4.754	−4.678	1.599
	2	6.911	−3.376	−3.367	0.2666
	3	11.42	−1.240	−1.295	4.247
	4	15.93	−0.5000	−0.5259	4.925

From this validation part, it can be summarized that the model has been almost correctly modeled as most of the parametric variations are within 5%. This implies that the model can used further for parametric studies.

3 Parametric Study

3.1 *Overweight Truck Selection and Modeling*

Due to time and resource constraints, it was not possible to conduct an axle load survey. Hence previous survey data has been studied and the most overweight vehicle has been selected. From such a previous survey found in a study by Ullah [6], the most overweight vehicle observed to have a total weight of 33.2 ton which is a two-axle truck having respectively 9.5 ton and 23.7 ton on axle 1 and axle 2. For modeling purpose, it has been assumed that the axle-to-axle distance is 14 feet similar to design truck (Fig. 3).

For parametric study purpose, 4 different length bridges have been selected: 30 m, 35 m, 40 m, and 45 m. They have been designed (Design procedure will be described in the later portion of this chapter). It has been assumed that the total bridge length will be occupied by a series of selected overweight trucks. Truck to truck distance and other dimensions are shown in the figures (Figs. 4, 5 and 6).

In CSiBridge these series of 4 and 5 trucks have been defined as a single vehicle of respectively 8 and 10 axles and having an impact factor of 1.15 moving in both lanes at a speed of 50km/h (13.889 m/s).

Table 1 Validation of model bridge

	Parameters	Manual calculation	CSiBridge	Variation (%)
Section properties for non-composite section	Section modulus of mid-section for bottom fiber	2.729×10^8 mm^3	2.729×10^8 mm^3	0
	Area of mid-section	7.097×10^5 mm^2	7.097×10^5 mm^2	0
	Section modulus of end-section for bottom fiber	3.230×10^8 mm^3	3.230×10^8 mm^3	0
	Area of end-section	1.188×10^6 mm^2	1.188×10^6 mm^2	0
Section properties for composite section	Section modulus of mid-section for bottom fiber	3.780×10^8 mm^3	3.780×10^8 mm^3	0
	Area of mid-section	1.148×10^6 mm^2	1.148×10^6 mm^2	0
	Distance of centroid of mid-section from bottom fiber	1155 mm	1155 mm	0
	Section modulus of end section for bottom fiber	4.813×10^8 mm^3	4.813×10^8 mm^3	0
	Area of end section	1.626×10^6 mm^2	1.626×10^6 mm^2	0
	Distance of centroid of end section from bottom fiber	1065 mm	1065 mm	0

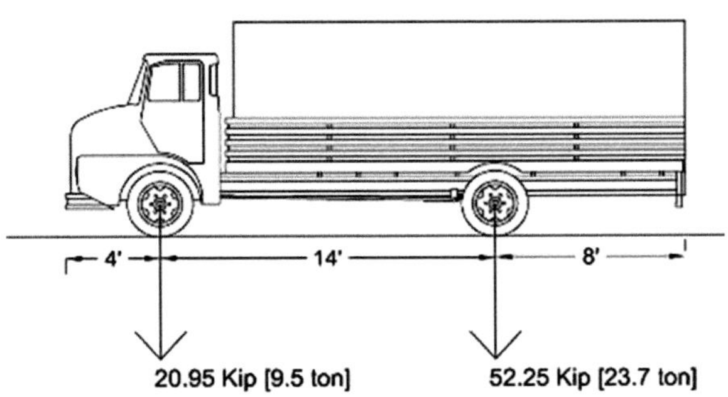

Fig. 3 Two axle overweight truck

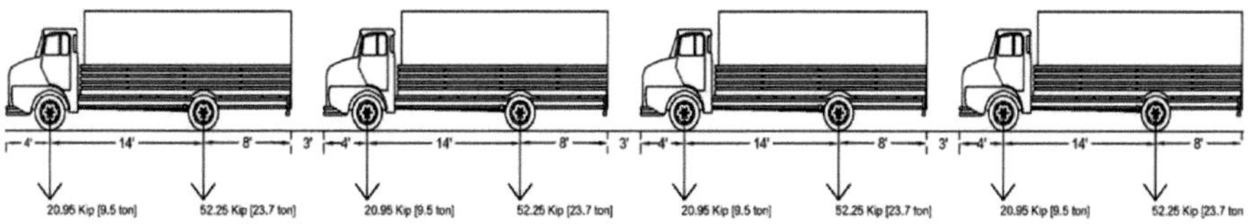

Fig. 4 Series of 4 overweight trucks for 30 and 35 m bridges

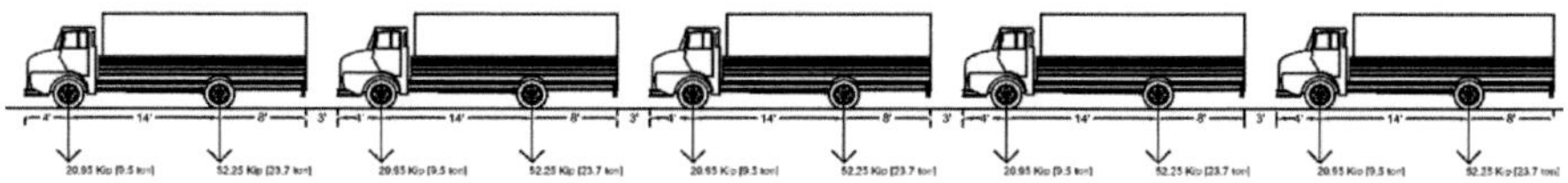

Fig. 5 Series of 5 overweight trucks for 40 m and 45 m bridges

Fig. 6 Vehicle data for vertical loading

4 Results and Discussions

The observations found from this **study** can be summarized as follows (Fig. 7):

1. Flexural moments for strength I combination using a series of overweight trucks have been increased by 20.46% for 30 m bridge, 25.74% for 35 m bridge, 31.59% for 40 m bridge and 37.89% for 45 m bridge from moments produced using Design HL 93 Live load and these moments have also crossed the positive flexural resistance limit.
2. D/C ratios for strength I combination using a series of overweight trucks have been increased by 19.76% for 30 m bridge, 35.33% for 35 m bridge, 41.44% for

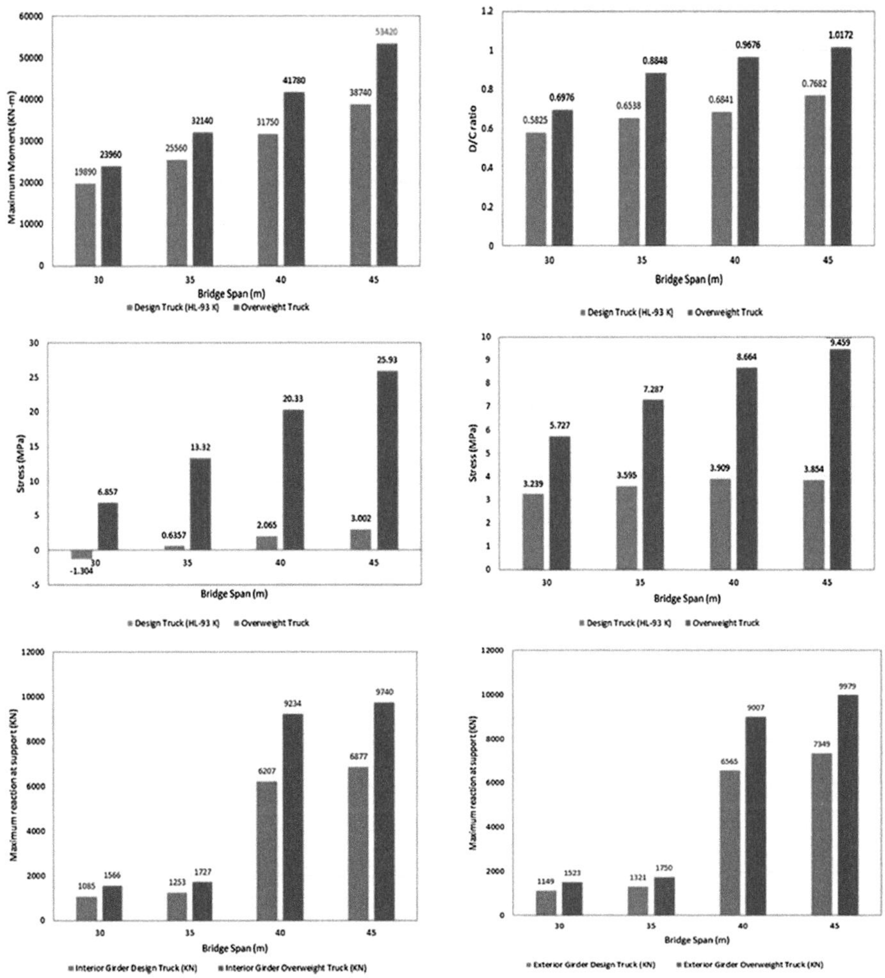

Fig. 7 **a** Maximum moment for strength I combination versus Bridge span; **b** Demand to maximum concrete shear capacity ratio for strength I combination versus Bridge Span; **c** Girder bottom stress for service III combination versus Bridge span; **d** Girder bottom stress for live load versus Bridge span; **e** Maximum reaction at support for strength -I combination for interior girder versus Bridge span; **f** Maximum reaction at support for Strength -I combination for exterior girders versus Bridge span

40 m bridge and 32.41% for 45 m bridge from D/C ratios produced by using design HL-93 Live load. For 30, 35 and 40 m bridges this ratio is within limit but for 45 m bridge it crosses the limit (Figs. 8, 9 and 10).

3. Tensile stresses at girder bottom for Service III combination using a series of overweight trucks, have been increased by 625.8% for 30 m bridge, 1995% for 35 m bridge, 884.5% for 40 m bridge and 763.8% for 45 m bridge from stresses produced by using HL 93 vehicle and also, they have crossed the limit by 2.143

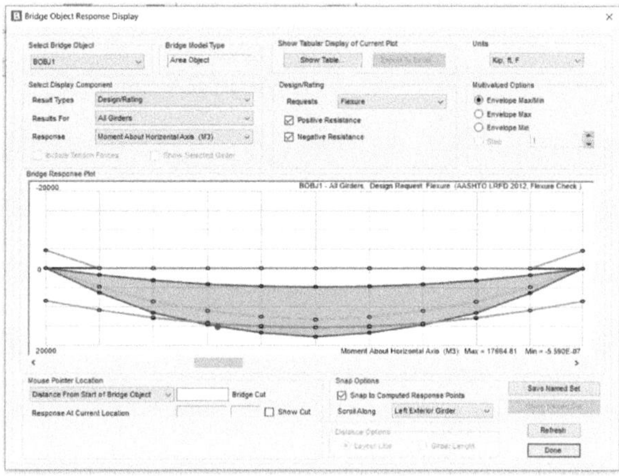

Fig. 8 Flexural moment check due to overweight truck for Strength I combination (including prestress force) for 30 m bridge

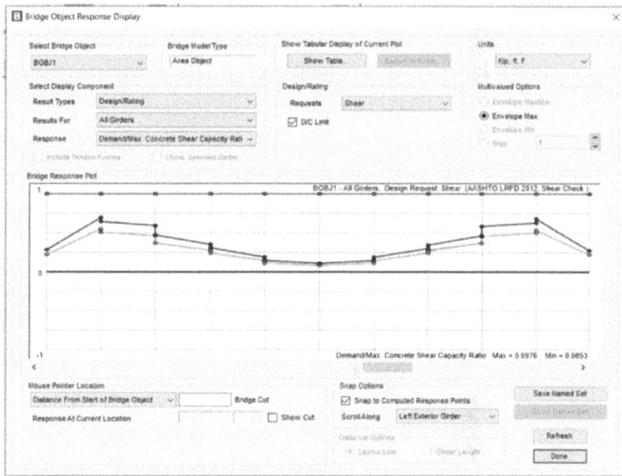

Fig. 9 D/C ratio check due to overweight truck for Strength I combination (including prestress force) for 30 m bridge

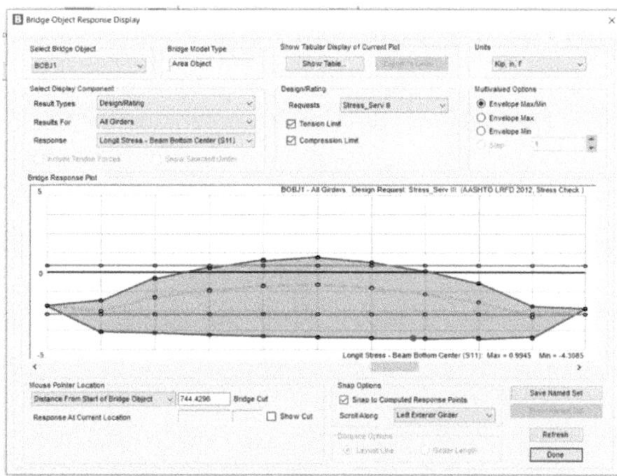

Fig. 10 Stress Check due to overweight truck for service III combination for 30 m bridge

times by 30 m bridge, 4.163 times by 35 m bridge, 6.353 times for 40 m bridge and 8.103 times by 45 m bridge.

4. Maximum reaction at supports for Strength I combination using a series of overweight trucks, have been increased by 44.33% for 30 m bridge, 37.83% for 35 m bridge, 48.77% for 40 m bridge and 41.63% for 45 m bridge for interior girders and for exterior girders 32.55% for 30 m bridge, 32.48% for 35 m bridge, 37.20% for 40 m bridge and 35.79% for 45 m bridge.

5 Conclusion

5.1 *Findings*

The findings of this **study** can be summarized as:

(1) Bridges designed considering single AASHTO Standard HL-93 vehicle in each lane as Live load are not capable of withstanding series of overweight vehicles (both lanes) considered in this study.
(2) All the bridges considered in this study failed in flexure moment check crossing the flexural resistance limit for the series of overweight vehicles considered.
(3) For 30, 35, 40 m bridges overweight vehicles couldn't not cause shear failure, but 45 m bridge failed in D/C limit check due to overweight vehicle. D/C ratio increased with the increase in the bridge span.
(4) All the bridges studied in this **study** failed in Girder bottom tensile stress check. Though these bridges have passed this stress check for design live load, they have failed to meet the same limit when series of overweight moved through them.
(5) Due to series of overweight truck movement, bearing support reaction increased than that is required for design live load. So, bearing pad needs to be redesigned considering overweight vehicle movements.

5.2 *Limitations and Recommendations*

This **study** has a very limited scope to observe all the things about overweight vehicle problems. The fields in which study can be further extended are discussed below:

(1) Due to time and resource constraint axle load survey has been skipped and overweight vehicle data has been collected from previous studies which may not be the real scenario of the recent time. So, a comprehensive survey can be conducted in future to get a more realistic result.
(2) In this study, CSiBridge v23.3.1 has been used for finite element modeling and analysis of bridges. This software has many limitations. It was not possible to check all the stress limits recommended by AASHTO LRFD 2012 code. So, in

future a more accurate and detailed study can be conducted using other software like Midas Bridge, Ansys etc.

(3) It was not possible to replicate the effect of accumulative damages due to live load in highway bridges which is very important in studying the fatigue damage and fatigue life of bridges. So, consideration of this effect in future studies can bring a very interesting and significant result.

From the trend of damages occurring due to a range of overweight vehicles, a **study** can be conducted to predict the reduction in service life of bridge.

References

1. Duncan MAG (1977) Superloads on highway bridges in Southern Africa. J Struct Div 103:2165–2179
2. Kostem CN (1978) Overloading of highway bridges—initiation of deck damage. Transp Res Rec 664:212–220
3. Turer A, Emin Aktan A (1999) Issues in superload crossing of three steel stringer bridges in Toledo, Ohio. Transp Res Rec 1688:87–96
4. Specifications A-LBD, others (2012) American association of state highway and transportation officials. Washington, DC
5. Grimson JL, Commander BC, Ziehl PH (2008) Superload evaluation of the bonnet carré spillway bridge. J Perform Constr Facil 22:253–263
6. Ahad Ullah M, others (2012) Evaluation of pavement design parameters for national highways of Bangladesh

Effect of the Inclination Angle of Hooked Steel Fiber on the Flexural Behavior of Steel Fiber Reinforced Concrete

J. A. Mahmud, A. Islam, and M. S. Hosen

Abstract The inclusion of metallic fibers within concrete is a widespread approach for improving crack resistance, energy-absorbing abilities, and longevity. The geometry of the fibers, which usually have a hooked end, has a significant impact on boosting the fiber-matrix bond as well as the mechanical properties of both the fiber and the concrete. This research looked at the strength and failure behavior of concrete with hooked-end steel fibers. For this, specimens for the flexural test were cast using hooked steel fibers with inclination angles of 120, 135, and 150°, a diameter of 0.75 mm, and multiple hooked lengths (50 and 60 mm). A numerical analysis was conducted to model flexural members with ANSYS 2022. To validate the numerical study, a four-point bending experiment was performed. A better understanding of the effects of hooked length and inclination angle was revealed. As fiber dosages and lengths increase, flexural strength also increases. On the contrary, increasing the hooked angle above 135° had little effect on enhancing flexural strength; as a result, a hooked angle of 135° was regarded as the most suitable option. The optimal steel percentage was revealed to be 2%, the hook length was 60 mm, and the inclination angle was 135° for ultimate strength.

Keywords Hooked end steel fibers · Inclination angle · Fiber-matrix bond · Ultimate strength · Numerical analysis

1 Introduction

Concrete is a semi-brittle material with crimp failure and low tensile and strain endurance. Usually, 8–15% of its compressive strength makes up its tensile strength [1]. By preventing and bridging developed cracks, reinforcement, like steel bars or fibers, is employed to boost tensile strength and ductility [2]. Fiber reinforced

J. A. Mahmud (✉) · A. Islam · M. S. Hosen
Department of Building Engineering & Construction Management, Rajshahi University of Engineering & Technology, Rajshahi 6204, Bangladesh
e-mail: jobaer.ruet2016@gmail.com

S. Arthur et al. (eds.), *Proceedings of the 6th International Conference on Advances in Civil Engineering*, Lecture Notes in Civil Engineering 368,
https://doi.org/10.1007/978-981-99-3826-1_23

concrete without steel rebar reinforcement has been used in low-load bearing parts such as facades, tunnels, raised septic tanks, pavement, and so on [3]. Steel fibers prevent cracks, whereas rebar restricts crack breadth. Scientists at ETH Zurich have developed a lightweight concrete floor solution that does not require steel reinforcement. It is 70% lighter than ordinary concrete flooring [4]. The insertion of steel fibers into concrete results in fiber-reinforced concrete with enhanced energy absorption, deformation before failure, and ductile behavior [5]. Additionally, the steel fibers improve the flexural characteristics of concrete and its tensile strength, which is thought to be its fundamental structural weakness [6, 7]. Utilizing deformed fibers, such as hooked fibers, can enhance the mechanical properties of steel fibers. According to research, the fiber-matrix binding strength of deformed steel fibers is three to seven times greater than that of straight fibers [8, 9]. In the essence of flexural strength and durability, Banthia and Trottier [10] claim that the employment of hooked-end steel fibers outperforms the usage of crumpled and duplex deformed steel fibers in strengthening ordinary and semi-concrete compressive strengths. Theoretically, fibers with hooked ends offer greater mechanical coupling than other types of fiber [11]. Raising the fiber volume can generally enhance mechanical properties. The increase in fiber usage to accommodate loads is to blame for this [12]. Wu et al. [13] demonstrated that raising the fiber content by roughly 3% improved the compressive and flexural characteristics of steel fiber reinforced concrete (SFRC) when combined with a stable superplasticizer (SP) dose despite decreasing flowability. Shi [8] demonstrates that 2% is the optimum fiber content for SFRC and that composites with 2% steel fibers are much more reliable than those with 1 or 3% steel fibers. Lengthier fiber lengths generate a broader interfacial bonding surface and a higher fraction of pullout loading, which leads to improved flexural performance, as demonstrated by Abdallah et al. [14]. Yoo et al. [15] recommended that lengthier steel fibers with significantly greater aspect ratios be utilized to lower responsiveness to the size effect when compared to shorter steel fibers, and that 2% by volume of steel fibers can achieve minor influence on flexural strength unless the fiber distribution properties are constantly governed in spite of the size of the specimen. Murali et al. [16] demonstrated that steel fibers with hooked ends outperform twisted steel fibers in terms of impact energy dispersion capacity at concentrations ranging from 0.5 to 2%. Despite extensive research on SFRC, there has not been much work on the use of varied concrete strengths and angles of fibers of varying aspect ratios, lengths, and contents. The majority of published studies focused on either fiber angles or short fiber lengths. The mechanical characteristics of SFRC with different angles, lengths, and volume percentages of steel fibers are investigated in this work. The primary purpose of this research was to deliver precise information on the mechanical properties of SFRC, taking fiber lengths, angles, and volume percentages into account. The relationships between the numerous important mechanical characteristics of SFRC are established and analyzed in light of previous researchers' outcomes.

Table 1 The composition of OPC

Constituents	SiO_2	Al_2O_3	Fe_2O_3	CaO	MgO	SO_3	Insoluble residue	Free lime	Loss on ignition
Weight (%)	20.76	4.76	3.32	65.12	1.20	2.23	0.36	0.94	0.52

Table 2 Characteristics of coarse and fine aggregate

Parameters	Coarse aggregate	Fine aggregate
Fineness modulus	6.13	2.95
Specific gravity	2.70	2.88
Moisture content	2.03%	1.84%

2 Materials and Methods

2.1 Materials

2.1.1 Cement

A regional cement plant supplied Type I ordinary Portland cement (OPC) per ASTM. The composition of the cement utilized of that analysis is presented in Table 1.

2.1.2 Fine and Coarse Aggregates

For the preparation of the mixtures, fine sand and crushed sand with differing particle sizes were utilized. Crushed sand and fine sand have respective fineness modulus (FM) of 1.46 and 4.61. To give the mixture of fine and crushed sand a FM of 2.56, 10 mm maximum-sized coarse aggregate was added. Table 2 displays the properties of both coarse and fine aggregates.

2.1.3 Fibers

For the purposes of this study, the SFRC was constructed using two varieties of steel fibers with hooked ends. Steel fibers of two different lengths and hooked at different angles were given for this purpose. Specifically, the steel fibers used had a tensile strength of 534 MPa and a Young's modulus of 212 GPa. Other features of the utilized steel fibers are summarized in Table 3. Figure 1 depicts the physical form of the hooked end fibers.

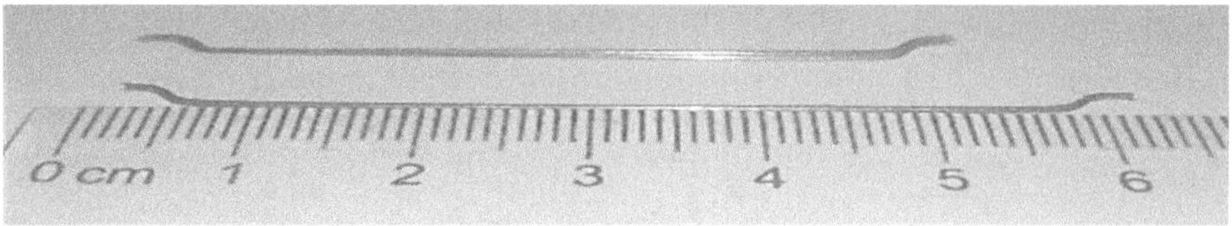

Fig. 1 Visual attributes of the hooked end fiber

Table 3 Characteristics of steel fiber

Length (mm)	Diameter (mm)	Aspect ratio (l/d)
50	0.75	66.67
60	0.75	80.00

2.2 *Mixture Proportions and Specimen Preparation*

The mixture proportion for SFRC that was used in this investigation is shown in Table 4. There were no water-reducing additives applied. All types of concrete were made using a water-to-cement ratio of 0.46. Hooked angles of 120°, 135°, and 150° were applied to steel fibers in five different ratios: 0, 0.5%, 1, 1.5, and 2% by volume of concrete. Twenty-five beams were formed for the flexural strength tests. Three separate rounds of mixing the fibers together produced a uniform SFRC blend. Flexural samples of 100 × 100 × 400 mm were prepared to examine the flexural strength of SFRC. Excess material was removed from the mold's top surface, which was smoothed with a trowel, and plastic was placed over the specimens to keep them moist. When the 24 h were over, the samples were demolded and cured at 22 ºC with 100% relative humidity until the testing age.

Table 4 Mixture Proportion of SFRC

Volume fraction (%)	Cement (kg/m^3)	Fine aggregate (kg/m^3)	Coarse aggregate (kg/m^3)	Hooked end steel fiber (kg/m^3)	Water (kg/m^3)
0	394	784	992	0	185
0.5	394	784	992	11.775	185
1.0	394	784	992	23.55	185
1.5	394	784	992	35.325	185
2.0	394	784	992	47.10	185

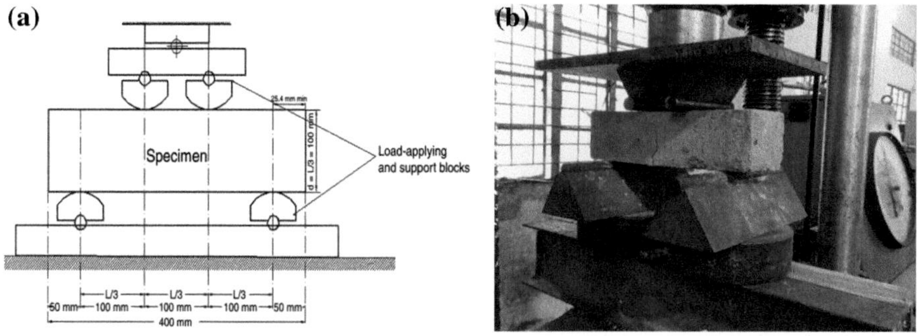

Fig. 2 A and **b** Setting up the four-point flexural test

2.3 Experimental Methods

According to ASTM C1609/C1609M-05, the flexural strengths for each beam specimen were calculated. Four-point bending studies on beam samples with sizes of 100 × 100 × 400 mm and a span length of 300 mm between supports were carried out (Fig. 2). According to ASTM C78-02 requirements, the loading and support system was able to apply third-point loading to the sample without twisting. The experiment was executed on a universal testing machine (UTM) with a 100 KN ultimate load. The rate of change was 0.08 mm per minute due to displacement.

2.4 Numerical Study

The numerical investigation to contrast the experimental results of the beams uses non-linear finite element analysis (FEA) relying on ANSYS 2022. A mathematical tool called FEA was used to approximatively solve boundary value difficulties for partially differential equations. According to published investigations, the SFRC beam modeling utilizing the aforementioned software has been proven to be adequate. As a result, the FEA was performed using full-scale beam models, which were taken into account throughout the experimental investigation.

2.4.1 Modeling of Materials

Concrete

Concrete is prone to both cracking and crushing under tension and compression, respectively. The material was modeled using the Solid 65 module. Each of the eight endpoints on the Solid 65 element has three degrees of freedom, permitting conversions in the x, y, and z nodal dimensions. Tables 5 and 6 display the SolidWorks

Table 5 Parameters for concrete design

Material characteristics	Values
Modulus of elasticity (E)	22,350 MPa
Grade of concrete	M20
Poisson's ratio (ν)	0.22
Shear factor for an open crack (βt)	0.32

Table 6 Properties of 3D hooked-ends steel fiber

Property	Unit	Steel fiber values
Diameter	mm	0.75
Length	mm	50
	mm	60
Aspect ratio	-	66.67
	-	80
Tensile strength	N/mm^2	532
Youngs modulus	KN/mm^2	210

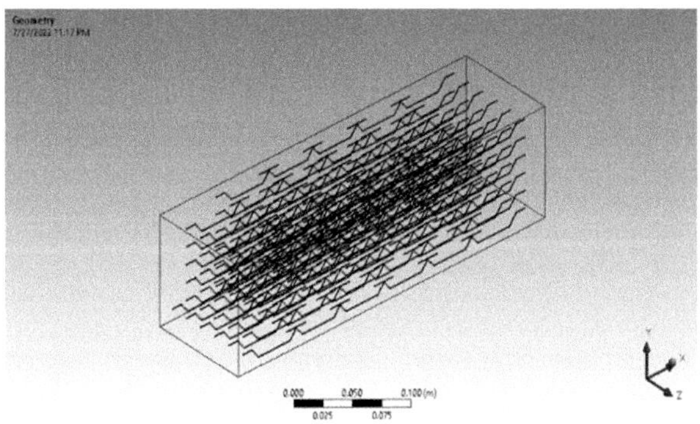

Fig. 3 Fiber design in Solid 65

concrete modeling parameters and properties of hooked end steel fiber that were utilized, respectively.

Steel Fiber

Figure 3 depicts the configuration of hooked steel fibers that were modeled using the Solid 65 program, while Fig. 4 depicts deformation as a result of stress.

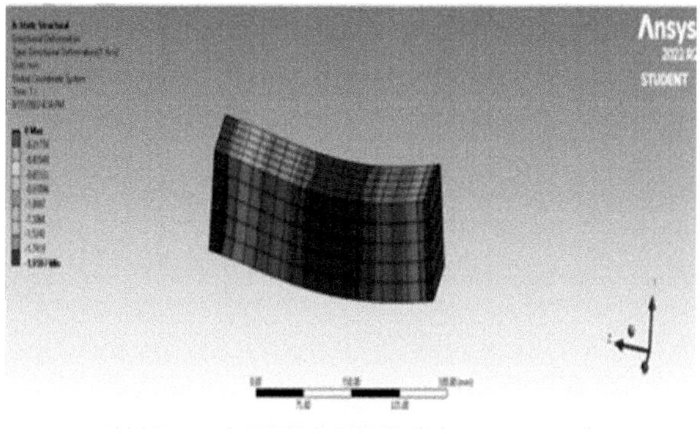

Fig. 4 SFRC beam analysis in Ansys

3 Results and Discussions

3.1 Numerical Findings

3.1.1 First Crack Load

The "first-crack load" of the improved SFRC beam (F0.5L50A20) (where F, L, and A stand for fiber content, length, and angle, respectively) was greater than that of the control beam (unreinforced beam) (Fig. 5) (F0) because of its enhanced stiffness. In unreinforced fibrous beams, an increase in fiber quantity decreased the load required to trigger the initial fracture, making the distinct steel more susceptible to damage. In addition, the load corresponding to the initial failure in fibrous beams was greatest when the fiber concentration ($V_f = 2\%$) was at its maximum. The current investigation revealed that the amount of fiber had a considerable impact on the load at the initial stage of cracking. In this experiment, the quantity of fibers had a clear effect on the load at the first crack phase. Table 7 shows that when comparing unreinforced fibrous beams of the same length and angle, the percentage increase in the first-crack load of the former beam at different V_f is roughly 11.84, 58.68, 118.95, and 172.40%. It was credited to the action of steel fibers during the pre-cracking phase that the preceding rise had a significant effect on increasing the first-crack load of reinforced beams. when V_f exceeds 0.5%, such as when V_f was 1.0, 1.5, and 2%, the first fracture load of reinforced fibrous beams was increased dramatically. This suggests that $V_f = 0.5\%$ was the "threshold limit" for manipulating the first-crack load of fibrous beams.

3.1.2 Ultimate Load

Similar to the first-crack load, the ultimate load of SFRC-strengthened fibrous beams (F0.5L50A20, F1L50A20, F1.5L50A20, F2L50A20, etc.) was increasing relative to control beams. As indicated previously, the largest load found had the highest fiber

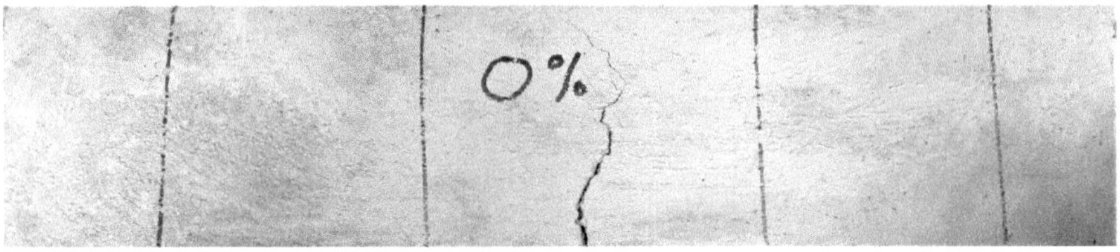

Fig. 5 Unreinforced Beam

Table 7 First crack load and ultimate load obtained from ANSYS 2022 analysis

Mix ID	First crack stage	Percentage increase comparing with F0	Ultimate stage	Percentage increase comparing with F0
	Load, Ffc (MPa)	%	Load, Fu (MPa)	%
F0	3.80	-	5.14	-
F0.5L50A20	4.25	11.84	6.43	25.1
F0.5L50A35	4.85	27.63	6.94	35.02
F0.5L50A50	4.46	17.37	6.49	26.26
F0.5L60A20	4.83	27.10	6.25	21.60
F0.5L60A35	4.97	30.80	6.85	33.27
F0.5L60A50	4.74	24.74	6.56	27.63
F1L50A20	6.03	58.68	8.31	61.67
F1L50A35	6.14	61.58	8.65	68.30
F1L50A50	5.98	57.40	8.27	60.90
F1L60A20	6.12	61.05	8.38	63.04
F1L60A35	6.21	63.42	8.74	70.04
F1L60A50	6.15	61.84	8.68	68.87
F1.5L50A20	8.32	118.95	11.24	118.68
F1.5L50A35	8.58	125.80	11.52	124.12
F1.5L50A50	8.38	120.53	11.45	122.76
F1.5L60A20	8.41	121.32	11.22	118.30
F1.5L60A35	8.66	127.90	11.63	126.26
F1.5L60A50	8.43	121.84	11.52	124.12
F2L50A20	10.35	172.40	13.31	159.00
F2L50A35	10.67	180.80	13.74	167.32
F2L50A50	10.62	179.47	13.62	164.98
F2L60A20	10.55	177.63	13.44	161.50
F2L60A35	10.76	183.16	13.81	168.68
F2L60A50	10.68	181.05	13.76	167.70

content. The first-crack load demonstrates the same behavior as previously described. The highest increase of 25% in the ultimate load of the F0.5L50A20 beam over the F0 beam was attributable to the hardening effect of SFRC bonding. The aforementioned percentage rise appears to be nearly twice as great in the latter stages as it was in the initial stages. It was apparent that strengthening had a superior impact, particularly at the ultimate failure stage, where brittle failure may be to blame. Such achievement in a region prone to earthquakes is unfathomable. Table 7 indicates that for $V_f = 0.5$, 1.0, 1.5, and 2%, the ultimate load of the strengthened fibrous beams (F0.5L50A20, F1L50A20, F1.5L50A20, and F2L50A20) was significantly greater than that of the F0 beam by up to 29, 61.67, 118.68, and 159%, respectively. This demonstrates how steel fibers drastically enhance the flexural capacity of SFRC strength in this work. When a higher V_f (2%) was taken into account, the effect of the above yield was quite strong and could still be seen at a lower V_f (0.5%). This yield was a perfect example of how fibrous beams harden after cracking and being stretched.

3.2 *Investigational Results and Discussions*

3.2.1 Effect of Fiber Dosages, Length of Fiber and Inclination Angle on Flexural Strength

Different composites with various fiber compositions showed a noticeable improvement in flexural strengths. Figures 6 and 7 shows the ultimate load was found to rise along with fiber dosages and lengths, respectively. The flexural strength improved from 5 to 166% for the fiber with a lower aspect ratio of 66.67 when the fiber content was increased from 0.5 to 2%, and from 5 to 180% for the fiber with a higher aspect ratio of 80. That was comparable an increase in flexural strength 3 to 124% increased for fibers with a smaller aspect ratio of 65, but 140% for fibers with a higher aspect ratio of 80 as compared to concrete without fibers [17]. Longer fibers were enhanced the load behavior of concrete, regardless of the fiber volume percentage. When compared to shorter fiber 50 mm, the fiber with a length of 60 mm and an aspect ratio of 80 considerably increased the flexural strength of SFRC. At a fiber length of 60 mm, F2L60A20 had a flexural strength of 12.58 MPa, which was 5% higher than F2L50A20. Although the inclination angle has little impact on the flexural strength of beam, but it is significant for design of a structure. The flexural strength of SFRC may rise by 2 to 8% depending on the angle of the hook (Fig. 8).

Figure 6 shows that the increasing fiber content increases the flexural strength of SFRC beams. The flexural strength of the SFRC beam was 14.46 for 2% fiber volume, whereas it was 4.63 for 0% fiber volume. The addition of steel fibers significantly improves many of the engineering properties of mortar and concrete, notably impact strength and toughness. Flexural strength, fatigue strength, tensile strength, and the ability to resist cracking and spalling were also enhanced.

Figure 7 shows that increasing fiber length increases the flexural strength of SFRC beams. The interlocking area grows as fiber length increases, enhancing the flexural

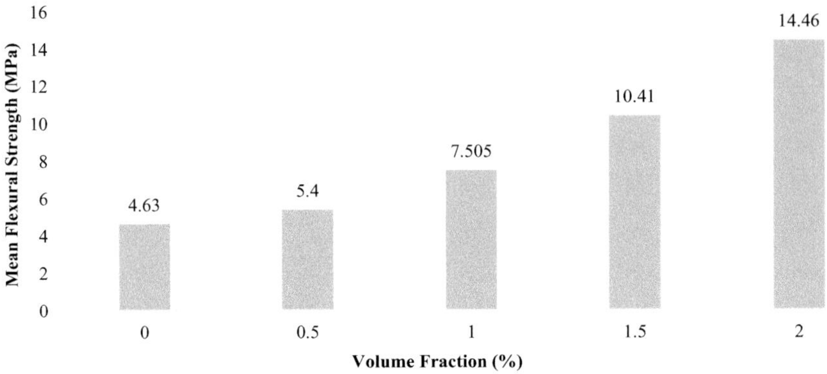

Fig. 6 Mean flexural strength for various fiber quantities

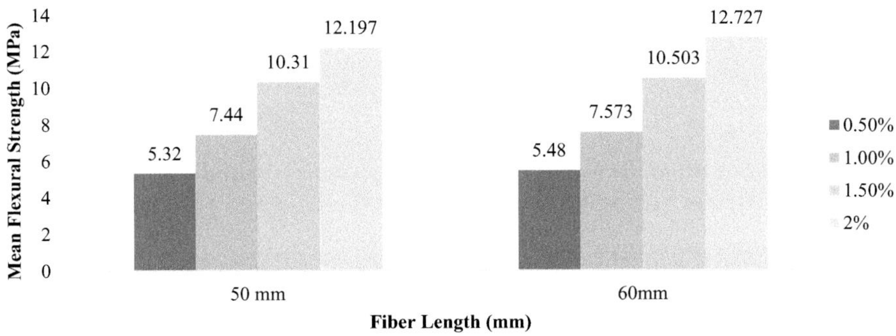

Fig. 7 Variation of flexural strength with fiber length

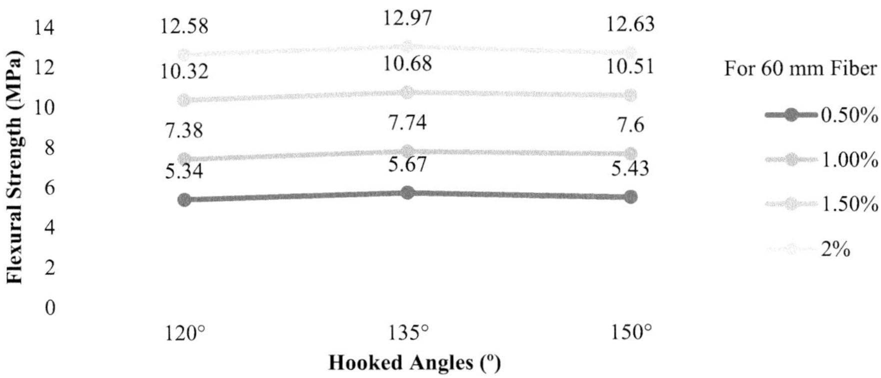

Fig. 8 Variation in flexural strength for hooked angles

strength. Although having a same fiber composition, longer fiber has significantly better strength. Here, a maximum strength of 12.727 MPa was found for 60 mm fiber length at 2% fiber volume, whereas 12.197 MPa for 50 mm fiber.

Figure 8 shows the flexural strength of SFRC beams for different hooked angles. Hooked fibers can improve flexural strength because of an increase in fiber-matrix bond strength caused by mechanical anchoring from the hooked region. The maximum strength for each fiber content was discovered at an angle of 135°. The maximum strength was found for hooked angle 135° (12.97 MPa) that is slightly higher than 150° (12.63 MPa). It is clear from this that the flexural strength relies on both the angle of the hooked fibers and the amount of fiber present.

4 Conclusion

The test results led to the conclusion that fibers were essential to both the mechanical and physical characteristics of beam specimens. The flexural strengths of SFRC increased in proportion to the steel fiber dosages. The addition of 0.5 to 2% hooked steel fibers increased the flexural strength by 5 to 180% when compared to the unreinforced control mixture. In addition, it is important to note that the flexural strength of steel fiber reinforced concrete increases by 1–5% with increasing fiber length from 50 to 60 mm. When the fiber content was increased from 0.5 to 2%, the flexural strength increased from 5 to 166% for the fiber with a lower aspect ratio of 66.67, and from 5 to 180% for the fiber with a higher aspect ratio of 80. Also, increasing the hooked steel fiber inclination angle further to 150 degrees lowers the flexural strength, and the strength of SFRC was obtained in the following order: 135° > 150° > 120°.

References

1. Shah S, Swartz S, Ouyang C (1995) Fracture mechanics of concrete: applications of fracture mechanics to concrete, rock and other quasi-brittle materials
2. Banthia N, Sappakittipakorn M (2007) Toughness enhancement in steel fiber reinforced concrete through fiber hybridization. Cem Concr Res 37:1366–1372
3. Ghods A, Sharbatdar MK (2020) An investigation on the behavior of two fixed end beams cast with HPFRCC composite. Case Stud Constr Mater 13:e00466. https://doi.org/10.1016/j.cscm.2020.e00466
4. Modern Construction Without Steel Reinforcement. https://www.techexplorist.com/modern-construction-without-steel-reinforcement/5368/. Accessed 12 Mar 2023
5. Islam A, Shuvo AK, Chowdhury SA, Sharmin S, Hasan M (2021) A comparative study on the properties of natural, synthetic and steel fibre reinforced concrete. J Civ Eng Constr 10:216–224
6. Lankard DR, Newell JK (1984) Preparation of highly reinforced steel fiber reinforced concrete composites. Spec Publ 81:287–306. https://doi.org/10.14359/6456
7. Lee JH (2017) Influence of concrete strength combined with fiber content in the residual flexural strengths of fiber reinforced concrete. Compos Struct 168:216–225. https://doi.org/10.1016/j.compstruct.2017.01.052

8. Wu Z, Khayat KH, Shi C (2018) How do fiber shape and matrix composition affect fiber pullout behavior and flexural properties of UHPC? Cem Concr Compos 90:193–201
9. Wille K, Naaman AE (2013) Effect of ultra-high-performance concrete on pullout behavior of high-strength brass-coated straight steel fibers. ACI Mater J 110:451–461. https://doi.org/10.14359/51685792
10. Banthia N, Trottier J-F (1995) Concrete reinforced with deformed steel fibres. Part II: toughness characterization. ACI Mater J 92:146–154
11. Boulekbache B, Hamrat M, Chemrouk M, Amziane S (2016) Flexural behaviour of steel fibre-reinforced concrete under cyclic loading. Constr Build Mater 126:253–262
12. Wu Z, Shi C, Khayat KH (2019) Investigation of mechanical properties and shrinkage of ultra-high performance concrete: influence of steel fiber content and shape. Compos Part B Eng 174:107021. https://doi.org/10.1016/j.compositesb.2019.107021
13. Wu Z, Shi C, He W, Wu L (2016) Effects of steel fiber content and shape on mechanical properties of ultra high performance concrete. Constr Build Mater 103:8–14. https://doi.org/10.1016/j.conbuildmat.2015.11.028
14. Abdallah S, Fan M, Rees DWA (2018) Bonding mechanisms and strength of steel fiber-reinforced cementitious composites: overview. J Mater Civ Eng 30:1–15. https://doi.org/10.1061/(asce)mt.1943-5533.0002154
15. Lee S-J, Yoo D-Y, Moon D-Y (2019) Effects of hooked-end steel fiber geometry and volume fraction on the flexural behavior of concrete pedestrian decks. Appl Sci 9:1241
16. Murali G, Venkatesh J, Lokesh N, Nava TR, Karthikeyan K (2018) Comparative experimental and analytical modeling of impact energy dissipation of ultra-high performance fibre reinforced concrete. KSCE J Civ Eng 22:3112–3119
17. Abbass W, Khan MI, Mourad S (2018) Evaluation of mechanical properties of steel fiber reinforced concrete with different strengths of concrete. Constr Build Mater 168:556–569. https://doi.org/10.1016/j.conbuildmat.2018.02.164

Analyzing Historical Temperature Variation in Dhaka, Bangladesh in Climate Change Perspective

A. Rudaiba and R. Mamtaz

Abstract Climate change is a major environmental hazard in the twenty-first century. This long-term weather change threatens all living species. Temperature is the best climatic indicator. Identifying and evaluating historical temperature fluctuations is vital to climate change research, and scientists from many nations are working on it. Dhaka, Bangladesh's capital, is a climate-vulnerable metropolis. In this study, an attempt has been made to analyze temperature variations from 1980 to 2020. Using linear regression and the Mann–Kendall test, the long-term trend of historical temperature change was examined. The findings of both studies were quite comparable, except for March's monthly average maximum temperature. Every month of the year had a noticeable rising trend in the monthly average minimum temperature. Between 2010 and 2019, both the monthly average maximum temperature and the monthly average minimum temperature increased significantly. Over this period, the rate of increase of monthly average maximum temperature for September is 1.3 °C/decade, and the rate of increase of monthly average minimum temperature for December is 1.6 °C/decade. The average maximum temperature during the hottest months—March, April, and May—and the average minimum temperature during the coldest months—December, January, and February—both increased by about 0.15 and 0.45 °C for every decade, respectively. Rising trends of 0.15 °C/decade and 0.28 °C/decade were seen in the average annual maximum and minimum temperatures. The number of hot days per decade with maximum temperature exceeding 30 and 35 °C grew by 7.3 and 30%, respectively, from 1980 to 2020. For the same period, the number of cold days per decade having a minimum temperature of less than 15 and 10 °C dropped by 27.8 and 54.8%, respectively. The study reveals a rising trend in Dhaka's temperature.

Keywords Climate change · Trend analysis · Dhaka · Temperature · Mann–Kendall test

A. Rudaiba (✉) · R. Mamtaz
Department of Civil Engineering, BUET, Dhaka, Bangladesh

S. Arthur et al. (eds.), *Proceedings of the 6th International Conference on Advances in Civil Engineering*, Lecture Notes in Civil Engineering 368,
https://doi.org/10.1007/978-981-99-3826-1_24

1 Introduction

At all geographic and economic dimensions, dealing with climate change and climatic unpredictability is seen as one of the biggest concerns of the next decades [1]. During the industrial era, the atmospheric concentrations of major anthropogenic greenhouse gases like CO_2, N_2O, CH_4, water vapor, and fluorinated gases have grown dramatically [2]. It is anticipated that elevated quantities will stay in the atmosphere for thousands of years. The higher atmospheric concentrations of these gases, particularly CO_2, increase the amount of Infrared (IR) energy absorbed by the atmosphere, resulting in a warming effect at the Earth's surface [3]. Each of the last four decades has been successively warmer than any decade that preceded it since 1850 [4]. The majority of the expected impacts of climate change are negative, according to the Intergovernmental Panel on Climate Change (IPCC). Increased heat waves (and fewer cold spells), stronger storms and cyclones, an increase in weather-related damage, more frequent droughts and flooding, warmer surface temperatures, particularly at higher latitudes, faster disease spread, decreased agricultural productivity in many areas, and a shift toward higher latitudes, and increasing sea levels that might inundate coastal regions are some of these [5].

Bangladesh's capital, Dhaka, is the eighth-largest city in the world [6] and is among the cities in the world that are most at risk from climate change. Due to the high population concentration, climate change's impacts will be particularly severe for the mega-cities of the developing world [7], especially those in Asia. Dhaka tops the risk of climate change in Asia [8]. Knowledge and data on the climatic variance features of an ecosystem are essential for the assessment and planning of environmental research. Consequently, the necessity of comprehending the features of future climatic fluctuations cannot be emphasized. There have been a few studies done on how climate parameters are changing over time in Bangladesh [9–11]. Ahmed and Alam (1999) anticipated an average temperature rise of 1.3 °C in 2030 and 2.6 °C in 2075 for Bangladesh [12]. Basak and Titumir (2013) reported that the average annual maximum temperature changes by approximately 0.0186 °C in a year, whereas the average annual minimum temperature changes by 0.0152 °C in a year [13]. Another study done by Mohiuddin and Bhuiya (2014) showed that Winter (December–February) has the biggest and most substantial temperature variations, with a rising tendency of 13 °C in 100 years [14]. All the above studies are focused on Bangladesh. However, Dhaka is the most populated and polluted city in Bangladesh, and all kinds of urban activities, such as vehicular movement, industries etc. are going on here. Therefore, these anthropogenic activities may have a role in increasing the temperature of this city, which may be higher than that of Bangladesh. Keeping this in mind, this research focused on the changes in Dhaka's temperature by analyzing historical data from 1980 to 2020 and forecasted future changes.

2 Methodology

The study area for this study is the Dhaka Metropolitan Area (Fig. 1) and Dhaka's temperature data was gathered from the Bangladesh Meteorological Department. For the years starting from 1980 to 2020, data on the daily temperatures have been compiled.

A data quality check was carried out prior to index calculation as incorrect outliers might have a significant influence on index calculation and trend. Once patterns had been identified, estimated, and predicted, statistical procedures were used to determine their statistical and physical significance. Using RStudio, simple linear regression was performed to estimate the linear trend (slope), and the non-parametric Mann–Kendall test (MK) was also employed to figure out the significance of the monotonic trend (either linear or non-linear).

Employing Dhaka's monthly maximum and minimum temperatures, the study performed a statistical trend analysis. Using a simple linear regression equation, a trend analysis was performed [17] as shown in Eq. (1). The model is expressed as,

$$y = A + Bx \tag{1}$$

In the equation, y represents the temperature of a certain year, x represents that year, and A represents the intercept of the straight line depicted by the equation. "A" and "B" can be determined from the following equation,

$$A = \frac{(\sum y)(\sum x^2) - (\sum x)(\sum xy)}{n(\sum x^2) - (\sum x)^2} \tag{2}$$

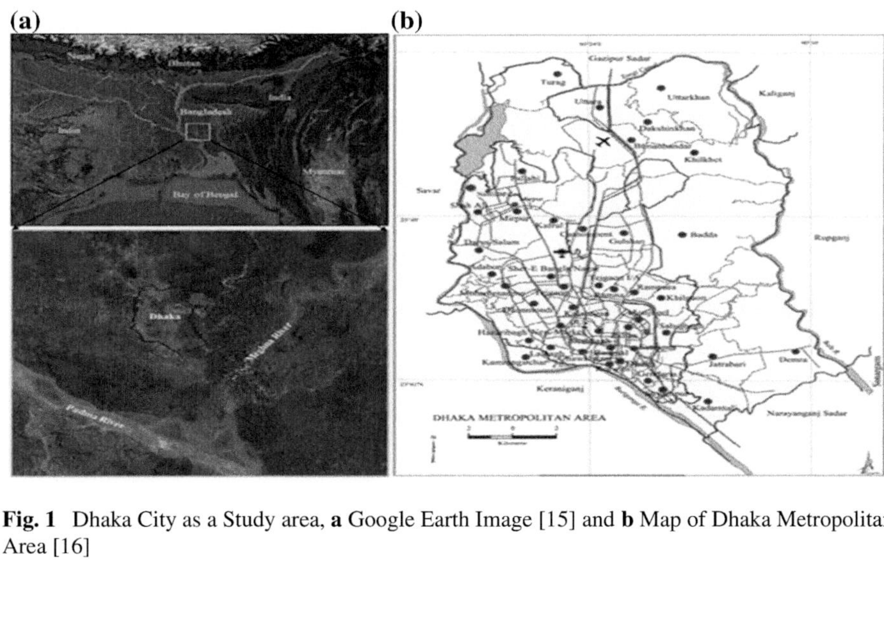

Fig. 1 Dhaka City as a Study area, **a** Google Earth Image [15] and **b** Map of Dhaka Metropolitan Area [16]

$$B = \frac{n(\sum xy) - (\sum x)(\sum y)}{n(\sum x^2) - (\sum x)^2} \tag{3}$$

The Mann–Kendall trend test [18] is used to determine if a group of data values is trending upward or downward over time and whether the trend is statistically significant for either direction. [19]. The magnitude of the change is not evaluated by the MK test. The data values are analyzed as ordered time series. If n data points are presented by $x_1, x_2, x_3 \ldots . x_i$ where x_j represents the data point at time j, then S is defined as follows,

$$S = \sum_{k=1}^{n-1} \sum_{j=k+1}^{n} sign(x_j - x_k) \tag{4}$$

$$Where: sign(x_j - x_k) = \begin{cases} 1 if x_j - x_j > 0 \\ 0 if x_j - x_j = 0 \\ -1 if x_j - x_j < 0 \end{cases} \tag{5}$$

The following formula is used to calculate the normalized test statistic Z:

$$z = \begin{cases} \frac{S-1}{\sqrt{VAS(S)}} if S > 0 \\ 0 if S = 0 \\ \frac{S+1}{\sqrt{VAS(S)}} if S < 0 \end{cases} \tag{6}$$

While negative Z values reflect decreasing trends, positive Z values show increasing trends.

3 Result and Discussion

The monthly temperature variation has been investigated employing time series data. Based on the average maximum and minimum temperature for the previous 41 years (1980–2020), the monthly variation in temperature has been assessed.

Figure 2 shows that the winter months from December to February have low monthly maximum average temperatures. January has the lowest average maximum (24.95 °C) and minimum (13.27 °C) temperatures in a year. The summer season, i.e., the months of March (32.42 °C), April (33.89 °C), and May (33.44 °C), are characterized by higher average monthly temperatures. April has the highest average monthly maximum temperature. Based on this study, Dhaka city's coldest and warmest months are January and April, respectively.

Figures 3 and 4 present the change in average maximum temperature in the summer season (March, April, and May) and the change in average minimum temperature in the winter season (December, January, and February) up to the year

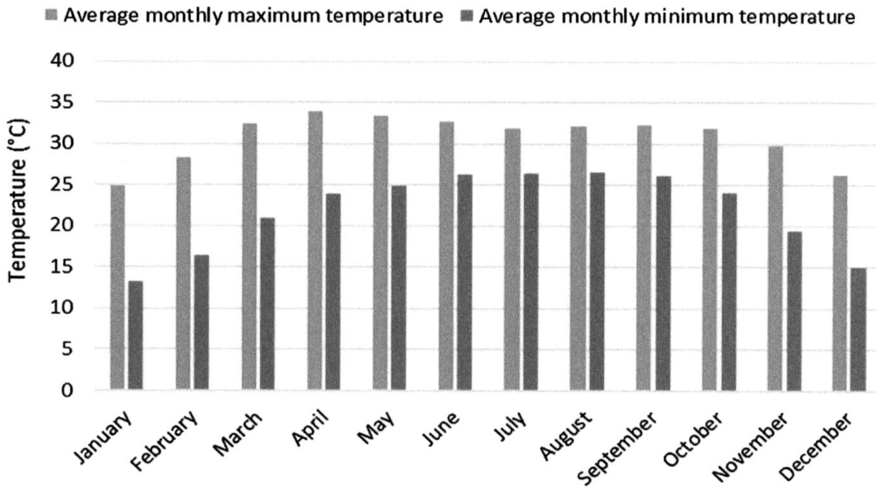

Fig. 2 Average monthly maximum and minimum temperature of Dhaka during the period of 1980–2020

2040, respectively. Both the figures show an increasing trend over time. Both the average summer maximum temperature and the average winter minimum temperature increased by approximately 0.15 and 0.45 °C each decade, respectively. The projected trend line up to the year 2040 shows that the average maximum summer temperature will be 33.9 °C in 2040 (Fig. 3). Furthermore, the minimum average temperature in winter will rise to 16.7 °C in 2040 from 14 °C in 1980 (Fig. 4).

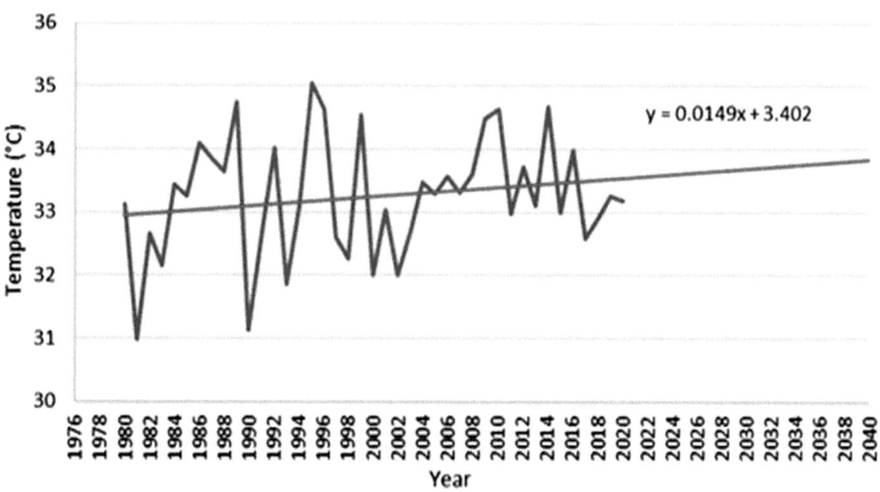

Fig. 3 Trend Analysis of mean maximum summer season temperature

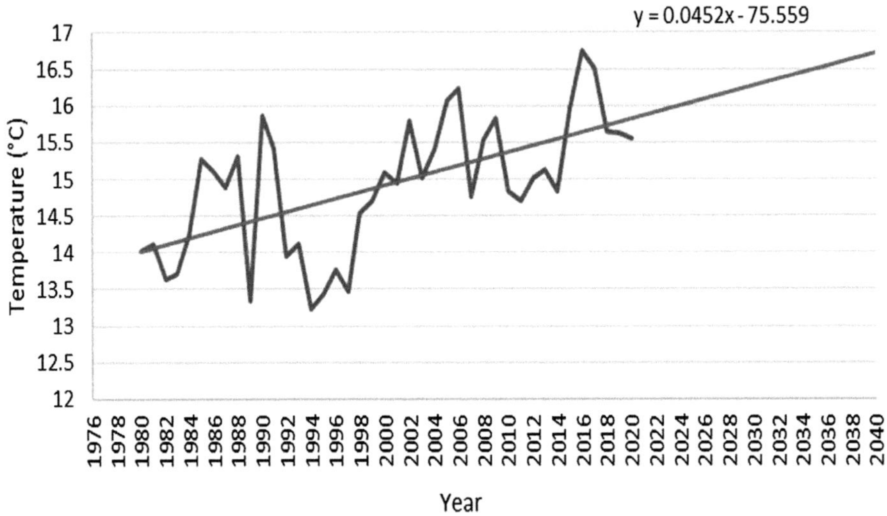

Fig. 4 Trend Analysis of mean minimum Winter season temperature

Although average annual maximum and minimum temperatures fluctuated for Dhaka City throughout the time shown in Fig. 5, the rate of increase for both the maximum and minimum temperatures was 0.15 °C/decade and 0.28 °C/decade, respectively. Both trendlines were projected to continue for the next 20 years, i.e., the year 2040. This trendline predicts that the average annual maximum temperature in 2040 might be 31.42 °C, a rise of 1.13 °C in 60 years. Additionally, the average annual minimum temperature in 2040 is projected to reach 23.02 °C, an increase of 1.63 °C in 60 years.

The Mann–Kendall test and linear regression model findings for the monthly average maximum and minimum temperatures for different months are shown in Table 1. Except for December and January, the monthly average maximum temperature is rising for all other months. Significant increases are seen in May (0.37 °C/decade), July (0.35 °C/decade), and September (0.39 °C/decade). The monthly average minimum temperature is rising throughout the whole year. The study found that December is the month where the average minimum temperature rises the most (rate is 0.53 °C/decade). Compared to the monthly average maximum temperature, the rising trend is more evident for the monthly average minimum temperature.

The results of the simple linear regression model and the Mann–Kendall statistical test are similar. The monthly average maximum temperature only shows a declining trend in January, March, and December. The MK trend test for the remaining nine months indicates an increase in the monthly average maximum temperature. In contrast, for the monthly average minimum temperature, all months exhibit an upward tendency.

Linear regression result shows that September is the month with the highest increase in monthly average maximum temperature, and thus Fig. 6 presents the

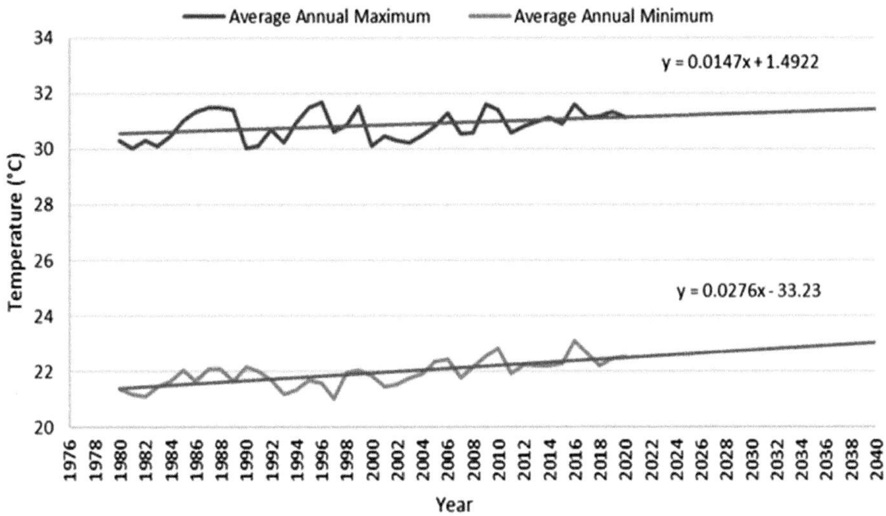

Fig. 5 Average annual maximum and minimum temperature trend of Dhaka during 1980–2040

decade-wise trend of monthly average maximum temperature for September over the period of 1980 to 2020. It has been found that the recent past decade (2010–2019) shows the highest rise in temperature (at a rate of 0.13 °C per year) compared to previous decades.

The average minimum temperature increases the most during December, according to the findings of linear regression. The slope is highest for the 2010–2019 decade. Figure 7 illustrates a decade-wise trend analysis of the monthly minimum temperature for December. The average minimum temperature for December increased at a rate of 0.16 °C per year from 2010 to 2019.

Table 2 displays the overall number of hot and cold days for the four decades examined in this study. It is evident that the number of hot days is increasing from 1980 to 2019, 7.3% for temperatures more than 30 °C and 30% for temperatures more than 35 °C. In contrast, the number of cold days decreased from 1980 to 2019, 27.8% for temperatures less than 15 °C and 54.8% for temperatures less than 10 °C. This indicates that the winter will be warmer and shorter in the coming future, which agrees with the finding of the IPCC [20].

Table 1 Linear Regression and Mann–Kendall test results of monthly average maximum and minimum temperature of different months

Month	Monthly average	Linear regression model result			Mann–Kendall test result			
		Linear regression equation	Slope	Trend	z	S	tau	Trend
January	Maximum	y = −0.023x + 70.95	−0.023	Decreasing	−1.96	−176	−0.21	Decreasing
	Minimum	y = 0.034x −55.48	0.034	Increasing	2.39	214	0.26	Increasing
February	Maximum	y = 0.019x −10.07	0.019	Increasing	0.92	83	0.1	Increasing
	Minimum	y = 0.049x −83.42	0.049	Increasing	3.58	320	0.39	Increasing
March	Maximum	y = 0.019x + 19.19	0.019	Increasing	−0.03	−4	−0.005	Decreasing
	Minimum	y = 0.037x −52.65	0.037	Increasing	1.99	179	0.21	Increasing
April	Maximum	y = 0.018x −1.54	0.018	Increasing	0.66	6	0.073	Increasing
	Minimum	y = 0.025x −25.75	0.025	Increasing	0.86	7.8	0.096	Increasing
May	Maximum	y = 0.037x −39.72	0.037	Increasing	2.08	186	0.22	Increasing
	Minimum	y = 0.035x −44.31	0.035	Increasing	2.66	238	0.29	Increasing
June	Maximum	y = 0.029x −25.93	0.029	Increasing	2.86	256	0.31	Increasing
	Minimum	y = 0.008x + 9.94	0.008	Increasing	1.62	146	0.17	Increasing
July	Maximum	y = 0.035x −37.12	0.035	Increasing	4.01	358	0.43	Increasing
	Minimum	y = 0.012x + 2.74	0.012	Increasing	2.22	199	0.24	Increasing
August	Maximum	y = 0.033x −33.82	0.033	Increasing	3.95	353	0.43	Increasing
	Minimum	y = 0.009x + 6.82	0.009	Increasing	2.16	194	0.23	Increasing
September	Maximum	y = 0.039x −46.75	0.039	Increasing	4.51	403	0.49	Increasing
	Minimum	y = 0.019x −12.41	0.019	Increasing	3.15	282	0.34	Increasing
October	Maximum	y = 0.018x −3.57	0.018	Increasing	1.44	130	0.15	Increasing
	Minimum	y = 0.024x −23.93	0.024	Increasing	2.32	208	0.25	Increasing
November	Maximum	y = 0.009x + 10.65	0.009	Increasing	1.04	94	0.11	Increasing

(continued)

Table 1 (continued)

Month	Monthly average	Linear regression model result			Mann–Kendall test result			
		Linear regression equation	Slope	Trend	z	S	tau	Trend
	Minimum	$y = 0.033x - 47.09$	0.033	Increasing	3.33	298	0.36	Increasing
December	Maximum	$y = -0.025x + 75.85$	−0.025	Decreasing	−1.78	−160	−0.19	Decreasing
	Minimum	$y = 0.053x - 90.55$	0.053	Increasing	3.76	336	0.41	Increasing

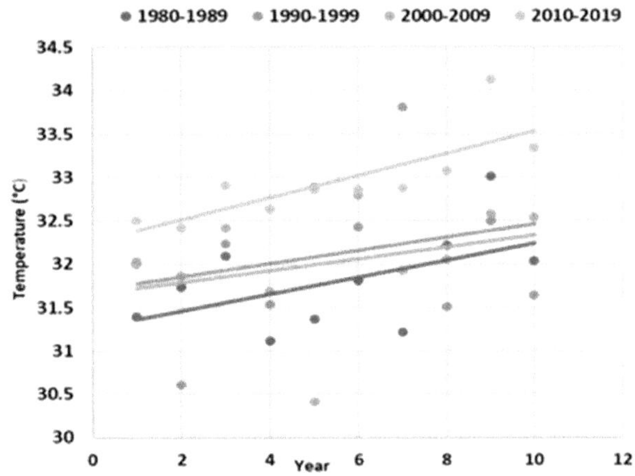

Fig. 6 September's decade-by-decade trend analysis of the average monthly maximum temperature

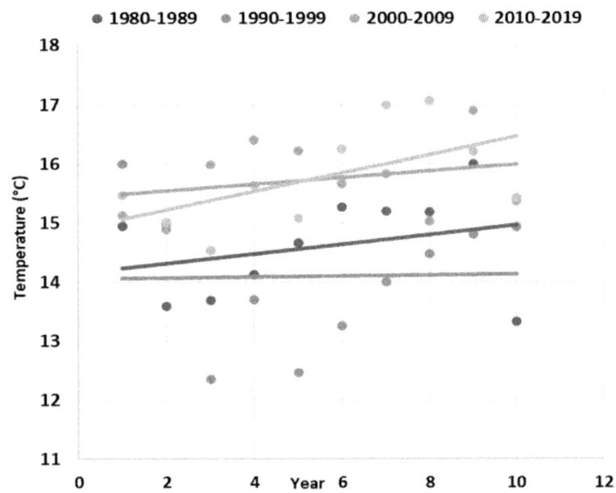

Fig. 7 December's decade-by-decade trend analysis of the average monthly minimum temperature

Table 2 Cold and hot days in Dhaka City by decade from 1980 to 2019

Decades	1980–1989	1990–1999	2000–2009	2010–2019
Number of hot days with maximum temperature exceeding 30 °C	2277	2325	2316	2443
Number of hot days with maximum temperature exceeding 35 °C	297	292	258	386
Number of cold days with minimum temperature below 15 °C	557	573	1246	402
Number of cold days with minimum temperature below 10 °C	31	44	7	14

4 Conclusion

The major conclusions drawn from this study are:

- Over our study period, both the average annual maximum and minimum temperatures of Dhaka City increased by about 0.15 and 0.28 °C each decade, respectively.
- The rate of increase in average minimum temperature in winter (0.45 °C/decade) is more than that of the average maximum temperature in summer (0.15 °C/decade).
- The trend of rise in temperature has been higher in the recent past decade (2010–2019) than in the previous decades.
- The number of hot days (highest temperatures > 30 and 35 °C) is increasing, and the number of cold days (lowest temperatures < 15 and 10 °C) is decreasing. The number of cold days in winter is decreasing at an alarming rate (54.8%).

The present study shows that climate change will result in a constant rise in the temperature of Dhaka. In winter, the highest rise in temperature and reduction in cold days are observed. The results of this study are expected to further knowledge of current climate changes, likely future climatic scenarios, and the climate change consequence in various sectors of Dhaka. This paper also highlights the urgent need for global action to limit temperature rise. Allowing climate change to continue unchecked will have devastating effects, but by acting now, we can avert the worst impact on lives and economies.

References

1. Aerts JCJH, Droogers P (2004) Climate change in contrasting river basins : adaptation strategies for water, food, and environment, p 264
2. Riahi K, Roehrl RA (2000) Greenhouse gas emissions in a dynamics-as-usual scenario of economic and energy development. Technol Forecast Soc Change 63:175–205. https://doi.org/10.1016/S0040-1625(99)00111-0
3. Ledley TS, Sundquist ET, Schwartz SE, Hall DK, Fellows JD, Killeen TL (1999) Climate change and greenhouse gases. Eos (Washington DC) 80. https://doi.org/10.1029/99EO00325
4. AR6 Climate Change (2021) The Physical Science Basis—IPCC. https://www.ipcc.ch/report/sixth-assessment-report-working-group-i/. Accessed 16 Aug 2022
5. Avoiding Dangerous Climate Change - Google Books. https://books.google.com.bd/books?hl=en&lr=&id=7Ui8fBmNlm4C&oi=fnd&pg=PA10&dq=climate+change+&ots=YAs0Gn8PCK&sig=VquvagMx362SYNMDhlgtwGQnlpo&redir_esc=y#v=onepage&q=climate change&f=false. Accessed 16 Aug 2022
6. Alam M, Rabbani GMD (2016) Vulnerabilities and responses to climate change for Dhaka: http://dx.doi.org/101177/0956247807076911 19:81–97. https://doi.org/10.1177/0956247807076911
7. Adger WN, Huq S, Brown K, Declan C, Mike H (2016) Adaptation to climate change in the developing world 3:179–195. https://doi.org/10.1191/1464993403PS060OA
8. Mega-Stress for Mega-Cities: A Climate Vulnerability Ranking of Major Coastal Cities in Asia | ALNAP. https://www.alnap.org/help-library/mega-stress-for-mega-cities-a-climate-vulnerability-ranking-of-major-coastal-cities-in. Accessed 16 Aug 2022

9. Ahmad QP, Warrick RA (1994) Bangladesh: greenhouse effect and climatic change. Briefing documents. https://doi.org/10.3/JQUERY-UI.JS
10. Caesar J, Janes T, Lindsay A, Bhaskaran B (2015) Temperature and precipitation projections over Bangladesh and the upstream Ganges, Brahmaputra and Meghna systems. Environ Sci Process Impacts 17:1047–1056. https://doi.org/10.1039/C4EM00650J
11. Wahiduzzaman M, Luo JJ (2020) A statistical analysis on the contribution of El Niño–Southern Oscillation to the rainfall and temperature over Bangladesh. Meteorol Atmos Phys 1331 133:55–68. https://doi.org/10.1007/S00703-020-00733-6
12. Ahmed AU, Alam M (1999) Development of climate change scenarios with general circulation models. Vulnerability Adapt to Clim Chang Bangladesh 13–20. https://doi.org/10.1007/978-94-015-9325-0_2
13. Kumar Basak J, Al R, Titumir M, Dey NC (2013) Climate Change in Bangladesh: A Historical Analysis of Temperature and Rainfall Data. J Environ 02:41–46
14. Mohiuddin H, Bhuiya MR, Mahmud M, Mamun A (2014) An analysis of the temperature change of Dhaka city. Proc 5th Int Conf Environ Asp Bangladesh [ICEAB 46–48]
15. Google Earth. https://earth.google.com/web. Accessed 28 Aug 2022
16. Dhaka - Banglapedia. https://en.banglapedia.org/index.php/Dhaka. Accessed 28 Aug 2022
17. Zou K, Tuncali K, Radiology SS (2003) Undefined correlation and simple linear regression. uomustansiriyah.edu.iq. https://doi.org/10.1148/radiol.2273011499
18. Kendall MG (1948) Rank correlation methods. Rank Correl methods
19. Yue S, Pilon P, Cavadias G (2002) Power of the Mann-Kendall and Spearman's rho tests for detecting monotonic trends in hydrological series. J Hydrol 259:254–271. https://doi.org/10.1016/S0022-1694(01)00594-7
20. Climate Change (2021) The physical science basis | Climate change 2021: The physical science basis. https://www.ipcc.ch/report/ar6/wg1/. Accessed 22 Aug 2022

BIM and CFD Based Simulation Approach in Reducing Thermal Energy Demands of a Residential Building: A Case Study in Bangladesh

S. A. Islam, M. T. Hassan, and M. R. Awall

Abstract With rising energy demands and impending climate crises driven by high energy consumption, household energy demand reduction is becoming vital for resilient development. This paper demonstrates how building information modeling (BIM) based thermal demand simulation, and computational fluid dynamics (CFD) analysis can help in sustainable building design decisions. Especially regarding selecting the sets of construction materials and components that help reduce building cooling energy demand while considering natural wind flow. Thus, the designers can forecast occupant thermal satisfaction utilizing a bioclimatic chart. The fourth level of a typical five-story residential building in Bangladesh was selected as the case study since residential building types are responsible for a significant percentage of national electricity demands, especially in the month of June. The case study building was replicated on Autodesk Revit. The energy analysis was done using BIM-centric cooling loads calculations based on building location, design characteristics, and material properties. Autodesk CFD software was used for ventilation simulation. Then material modifications with two alternative designs were conducted in the BIM model to produce different cooling load demands. These results are then compared with the initial design that uses conventionally used materials. The study examines the various building envelope materials designs, the different building components, and their individual leverage in reducing thermal energy demands. This will help orient further research toward selecting materials and building components with the most potential for thermal energy management while considering natural ventilation potential—especially regarding Bangladeshi climatic conditions.

S. A. Islam (✉) · M. T. Hassan · M. R. Awall
Department of Building Engineering and Construction Management, Rajshahi University of Engineering and Technology, Rajshahi, Bangladesh
e-mail: sai.ruet16@gmail.com

M. T. Hassan
e-mail: mthassan.ruet16@gmail.com

M. R. Awall
e-mail: robi95@ce.ruet.ac.bd

S. Arthur et al. (eds.), *Proceedings of the 6th International Conference on Advances in Civil Engineering*, Lecture Notes in Civil Engineering 368,
https://doi.org/10.1007/978-981-99-3826-1_25

Keywords BIM · CFD · Natural ventilation · Building envelops · Thermal energy demand

1 Introduction

Buildings account for 40% of total global direct and indirect carbon emissions and a net total of one-third of global energy demand [1], of which the majority is used for heating and cooling [2] and also ventilation, accounting for no less than 25, 9, and 10% correspondingly [3]. The lack of sustainable engineering regarding building occupancy and the global energy demands of the past few decades has demonstrated the importance of researching building energy efficiency concerning thermal demands [4]. Thus, both commercial industry trends and academia have started focusing on targeting building energy demand analysis in a simulative approach based on BIM [5]. Under the current circumstances, BIM is proving itself as a critical technology in the Architecture, Engineering, and Construction (AEC) industry [6], especially due to its accurate digital modeling of physical building assets with functional characteristics representation [7, 8]. Determining cooling loads from the building BIM model has become a reliable and consistent approach to assessing thermal energy demands [9, 10]. Several studies focusing on various elements and components of buildings and their designs in evaluating their leverage in reducing thermal energy demands have proved that the material properties of windows and external walls and WWR (window to wall ratio) are pivotal in thermal energy efficient building envelop design assessment [11–14]. On the other hand, using natural ventilation is one of the most economical techniques for ensuring thermal comfort in hot seasons and reducing cooling loads. Research on thermo-comfort zones shows that the standard mechanical HVAC system does not provide the range of thermo-comfort levels that buildings designed with natural air circulation parameters can [15]. Moreover, a study focusing on energy reduction by introducing natural air circulation in high-rise public housing in Hong Kong has substantiated the potential for natural ventilation [16]. In the United Kingdom, a study demonstrated that natural ventilation in an office building could effectively reduce cooling energy demand by 14–41 kWh/m^2, saving almost 10% of total annual energy consumption [17].

On the other hand, simulations based on computational fluid dynamics (CFD) are being used more and more to model how natural air flows through buildings [18–20]. The CFD approach is also preferable because of its reliable validation with experimental data and competence in the economy compared to datasets with unnecessary high-resolution levels. Hence, it proves its favorability to assign the natural air circulation perplexity, cost-effectiveness, and real data validation [21–23]. Thus, natural ventilation assessment via CFD simulations and BIM-based thermal energy propagation simulations has proven to be a reliable way to reduce building thermal energy demands. However, an extensive literature review on this topic found no comprehensive studies that merged BIM and CFD modeling simulation in Bangladeshi climatic conditions. Therefore, this paper focuses on simulating a building's thermal energy demands by (a) comparing various design material sets and building components, and (b) assessing the feasibility of CFD-centric natural ventilation simulation in reducing thermal energy demands by ensuring occupant thermal comfort.

2 Materials and Methods

Overview: The case study building is a typical residential housing located in Ishwardi, in the district of Pabna, Bangladesh. The building is under construction and of reinforced concrete structure, consisting of five stories. The fourth level is the focus of the study, as the upper stories hold significant natural ventilation potential. The level has 10-foot walls, and the plan illustrates the building orientation (Fig. 1). The building was modeled as a BIM entity for analysis in Autodesk Revit and put through the workflow illustrated in Fig. 2. The cooling load demands of two alternative design material sets were compared to the initial design based on conventional materials. The CFD simulation of natural ventilation to assess occupant thermal comfort in Bangladeshi climatic conditions was done in parallel.

BIM-based thermal energy load simulation: To derive the cooling loads for the building, the building was replicated according to the initial design plan in Autodesk

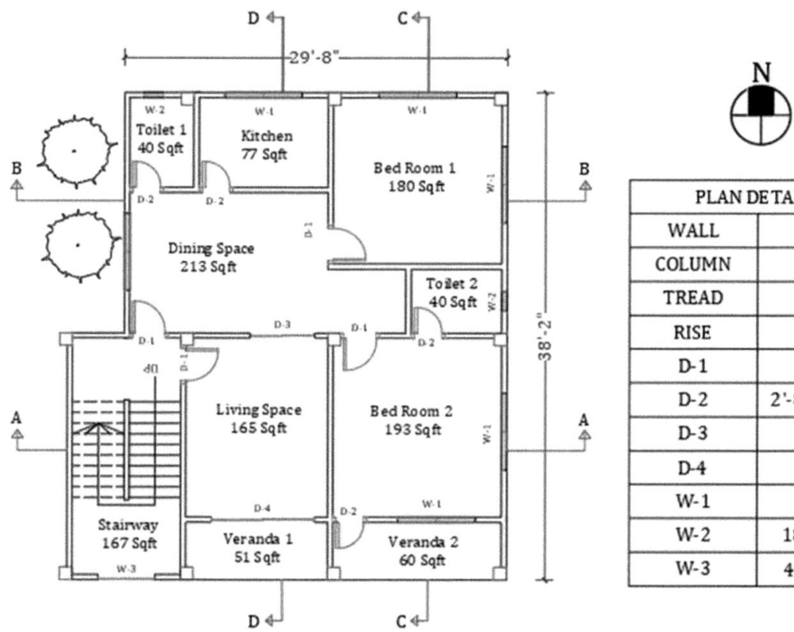

PLAN DETAILS	
WALL	5"
COLUMN	1' X 1'
TREAD	10"
RISE	6"
D-1	3' X 7'
D-2	2'-8" X 6'-6"
D-3	5' X 7'
D-4	8' X 7'
W-1	6' X 4'
W-2	18" X 12"
W-3	4'-5" X 7'

Fig. 1 Typical floor plan of the building

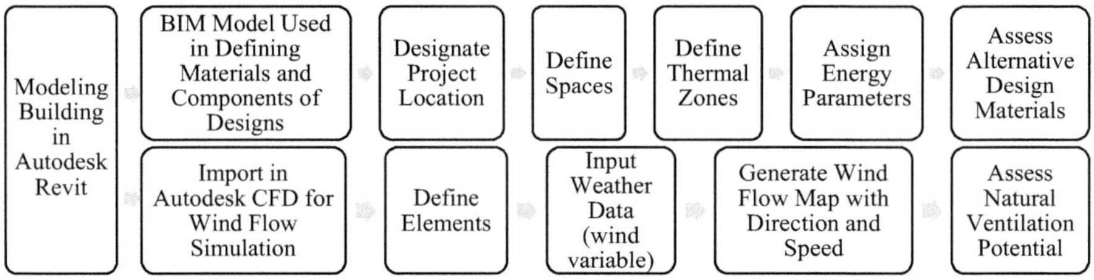

Fig. 2 Workflow methodology for the study

Revit with their respective material properties. However, to assess alternative materials in different components of walls, roofs, and windows, two more separate models (designated as Alternative 1 and 2) were created, having material thermal characteristics described in Table 1.

The material properties and project location (from where Revit determines the climatic circumstances, correlating with material properties to derive desired cooling loads) were set up by tuning Revit's energy settings and location options embedded within its analysis features. After that, spaces were created, assigning each room as an individual space (Fig. 3), from which energy models were created for the whole story. Then the thermal loads were generated from the model considering the general energy parameters in Table 2.

CFD-based natural ventilation simulation: In CFD analysis, the conservation of mass and momentum for modeling the motion of the fluid is shown by the filtered time-dependent Navier–Stokes equation, the governing equations used for eddies of

Table 1 Considered building components and their thermal properties

	Building components	Material properties
Initial design	Roof	4″ Lightweight concrete (U = 0.2245 BTU/(h ft^2 °F))
	Exterior walls	5″ Brick/block wall (U = 0.0774 BTU/(h ft^2 °F))
	Exterior windows	1/8″ Single glass pane in unsealed openable frames (U = 0.5806 BTU/(h ft^2 °F), SHGC = 0.86)
Alternative 1	Roof	6″ lightweight concrete (U = 0.1121 BTU/(h ft^2 °F))
	Exterior walls	5″ Brick cavity full mineral insulation and lightweight plaster (U = 0.0892 BTU/(h ft^2 °F))
	Exterior windows	Double glazing—1/8″ thick—low-E/clear (e = 0.05) glass (U = 0.3500 BTU/(h ft^2 °F), SHGC = 0.41)
Alternative 2	Roof	6″ Heavyweight concrete with 2″ insulation (U = 0.0899 BTU/(h ft^2 °F))
	Exterior walls	5″ Brick cavity with dense plaster (U = 0.2627 BTU/(h ft^2 °F))
	Exterior windows	Double glazing—1/4″ thick—low-E/clear (e = 0.05) glass (U = 0.3500 BTU/(h ft^2 °F), SHGC = 0.37)

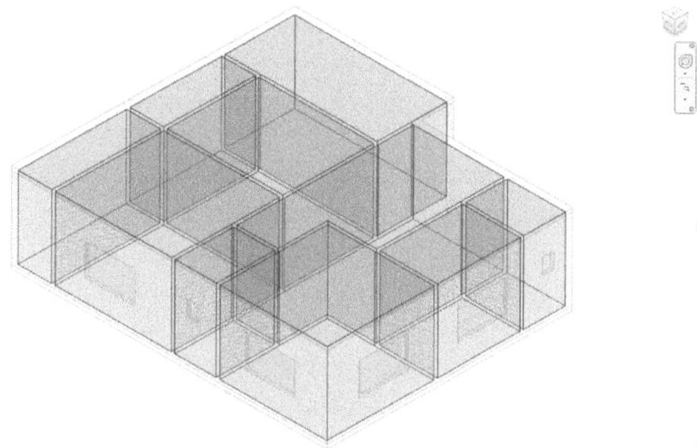

Fig. 3 Energy model representing spaces for each room

Table 2 General energy parameters

Parameter	Value
Building type	Single-family
Location	Ishwardi Upazila, Bangladesh
Ground plane	Level 1
Project phase	At completion
Sliver space tolerance	1′ 0″
Building envelope	Functional parameters
Building service	Single duct
Schematic types	Building
Building infiltration class	None
Report type	Standard
Load credits	Optional

an incompressible flow of a Newtonian fluid being [24]:

$$\frac{\partial \rho}{\partial t} + \frac{\partial(\rho u_i)}{\partial x_i} = 0 \tag{1}$$

$$\frac{\partial \rho u_i}{\partial t} + \frac{\partial(\rho u_i u_j)}{\partial x_j} = -\frac{\partial P}{\partial x_i} + \frac{\partial}{\partial x_j}\left[\mu\left(\frac{\partial u_i}{\partial x_j} + \frac{\partial u_j}{\partial x_i}\right)\right] \tag{2}$$

where i and j = 1, 2, 3 represent the x, y, and z directions respectively; ρ = fluid density; t = time; u = velocity; P = pressure.

However, to assess the thermo-comfort environment in junction to DBT (dry bulb temperature) and RH (relative humidity), CFD simulation is carried out to derive wind speeds inside various portions of the building envelope that specifically meet the requirement for occupant thermal comfort. The 3D building model was replicated in Autodesk Revit, where the inlet and outlet extensions were defined. The inlet extensions were five times the opening width, and the outlet extensions were ten times the opening width, following Autodesk CFD specifics. After defining building and component geometry and computational domain in Autodesk Revit, the model was then exported to Autodesk CFD following the workflow illustrated in Fig. 2. The walls were assigned as brick walls, and the floor and slab were assigned as concrete, and the voids in the model were defined as air (Fig. 4). The mesh sizing was set to auto-sizing (Fig. 5). Finally, the model solution was derived through 300 iterations.

The CFD simulation was set under the climatic parameters of June. The monthly average RH and DBT were calculated to be 71.58% (Fig. 6) and 29.48 °C (Fig. 7). Using these two parameters in Olgyay's bioclimatic chart [25], which has proven to be decisive in assessing the comfort zone for occupants [26, 27], we obtained the minimum wind speed required for the closest achievable, practical comfort zone as 0.1–0.4 m/s (Fig. 8). This is considered the required wind velocity for occupant thermal comfort under the site's DBT and RH parameters. On-site wind speeds of south facing 3.1 m/s and east facing 3.26 m/s were modeled in CFD simulation by analyzing wind rose data obtained from Autodesk Green Building Studio's (GBS)

Fig. 4 Material assigned in CFD model

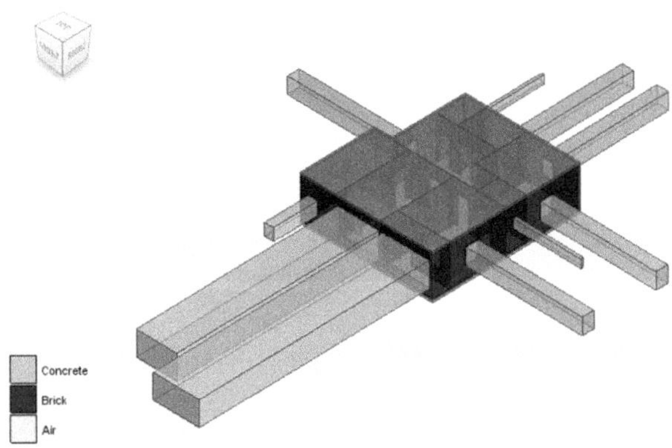

Fig. 5 Mesh sizing of the model

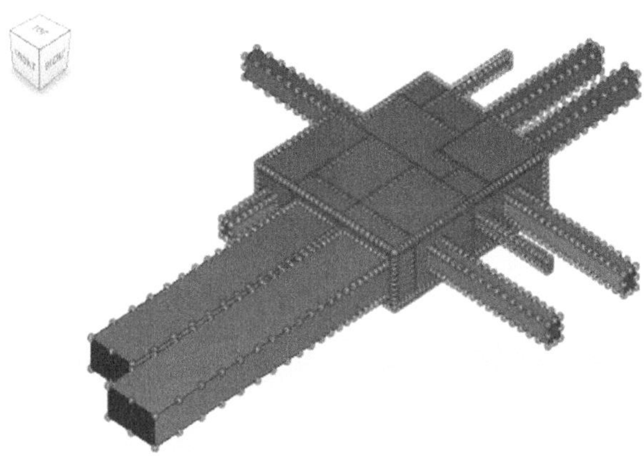

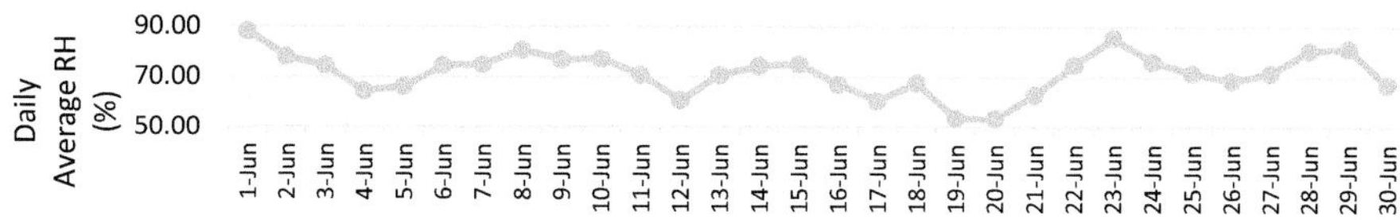

Fig. 6 RH data for June obtained from the GBS weather station

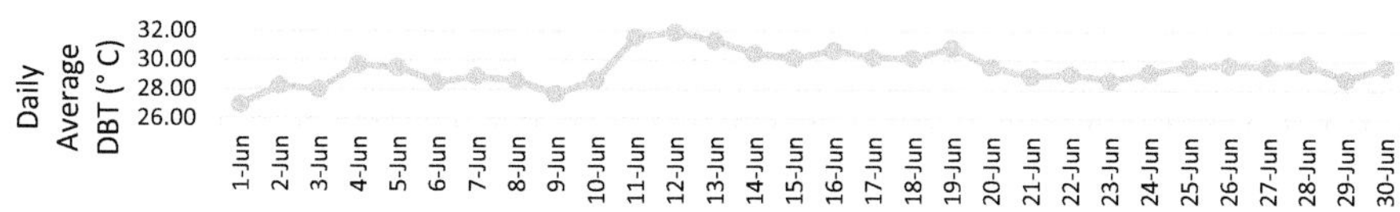

Fig. 7 DBT data for June obtained from the GBS weather station

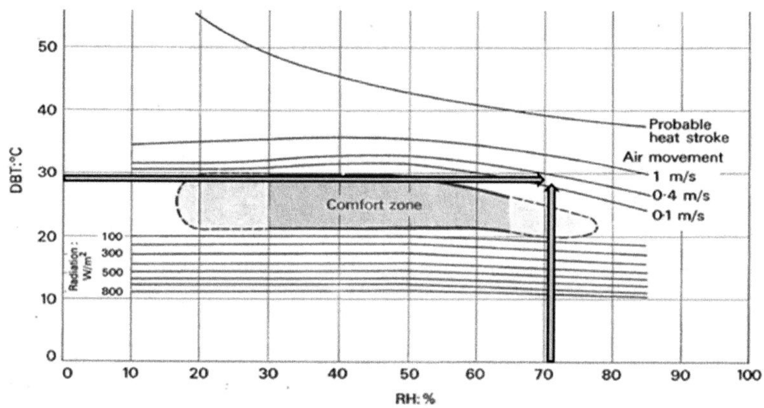

Fig. 8 Olgyay's bioclimatic chart relating to wind speed [25]

access to weather stations closest to the building location (Fig. 9). The CFD simulation will represent the interior wind flow from these constraints and assess how much of the building interior sustains the required wind speed of 0.4 m/s that ensures occupant thermal comfort.

3 Results and Discussions

Results of cooling loads were categorized by three metrics: peak cooling loads of rooms, design types (comparing alternative 1 and 2), and finally, building components of walls, windows, and roofs.

Figure 10 shows the different rooms with their cooling loads compared with the two alternative designs. The dining area requires the most cooling, followed

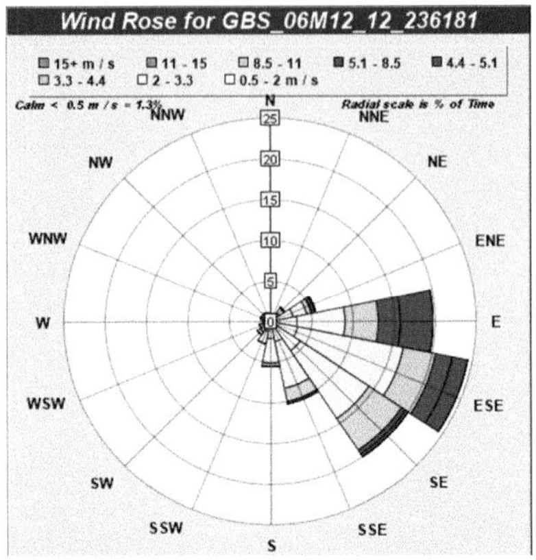

Fig. 9 Summer wind rose of Ishwardi

by Bedroom 1 and Bedroom 2. In measuring peak cooling load reduction from the initial design, Fig. 11 indicates that the alternative design materials can reduce it by a minimum of 34% from the initial design. Moreover, alternative 1 has a more homogeneous cooling load reduction across the rooms.

In Fig. 12, which compares building component sets and their individual potential for reducing cooling load demand, roof materials have been shown to be the most important, except for the Alternative 2 design case, where wall materials have been shown to be the most effective.

Results for CFD analysis were sorted by June's average and the required wind velocity of 0.4 m/s (obtained from Olgyay's bioclimatic chart) for thermal comfort. In Fig. 13a, CFD analysis for south-facing wind indicates that there is high wind velocity (color-coded) and circulation (vector arrows) across the building envelope in the living, two bedrooms, and portions of the dining. But substantial portions of toilets and stairways are relatively less ventilated. We see that Figs. 14a and 15a are the confirmation of the wind circulation speed data assessed via Fig. 13a. Again, in Figs. 16a, 17a, and 18a CFD analysis for east-facing wind indicates that there is a low wind velocity and circulation across the building envelope as the wind circulation

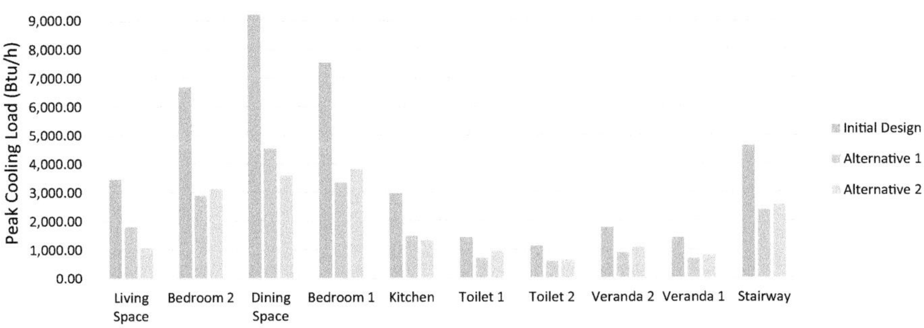

Fig. 10 Peak cooling loads in Btu/h by rooms and design alternatives

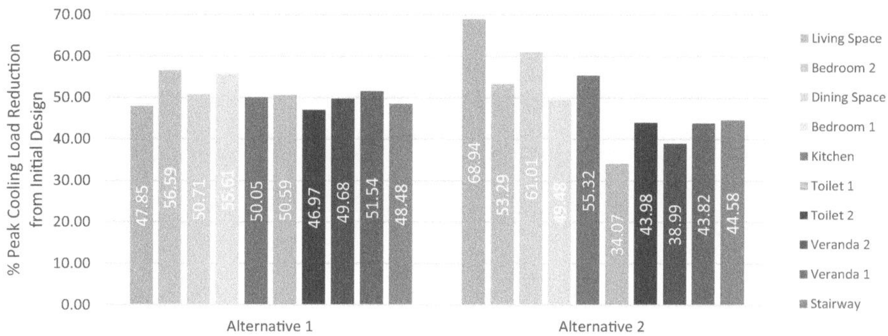

Fig. 11 Alternative design materials and their percentage reduction of peak load by rooms

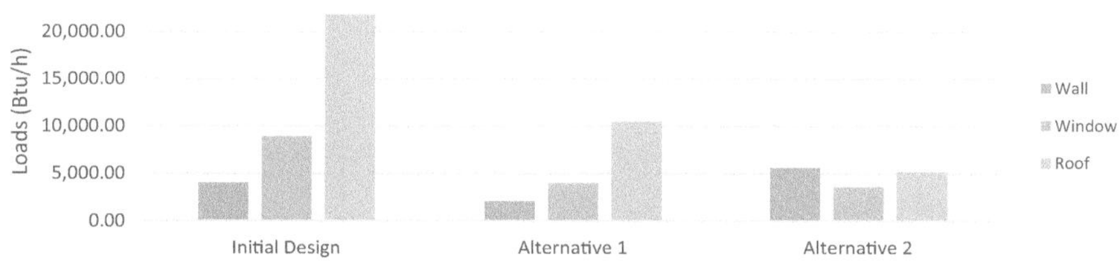

Fig. 12 Building components and their leverage in thermal energy demand reduction in the three design sets

is contained within bedroom 1, bedroom 2, and toilet 2. However, the CFD analysis conducted with the required wind velocity to reach thermal comfort indicates that natural ventilation can achieve relative thermal comfort across the building envelope interior (area color-coded with blue, i.e., wind speed of no less than 0.4 m/s). For south-facing wind, as seen in Figs. 13b and 14b, most of the area covers the required speed. But Fig. 15b shows less speed as it is near toilet 1, which has less opening. In the east-facing wind (Figs. 16b, 17b, and 18b) we see most of the area has less wind speed, but this drawback is mitigated by the dynamic circulation (vector arrow) pattern.

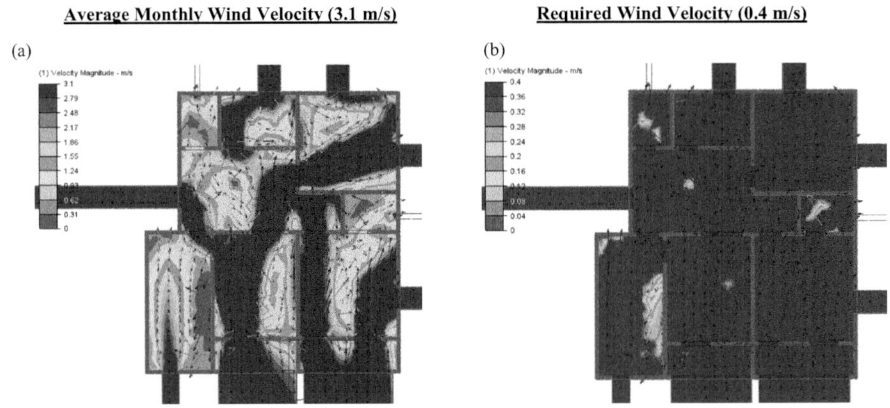

Fig. 13 CFD analysis results of south-facing wind (at 5 ft. elevation from the floor)

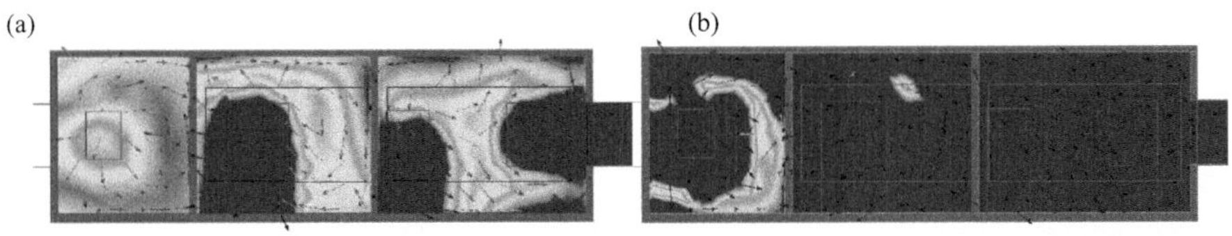

Fig. 14 CFD analysis results of south facing wind (at section A–A)

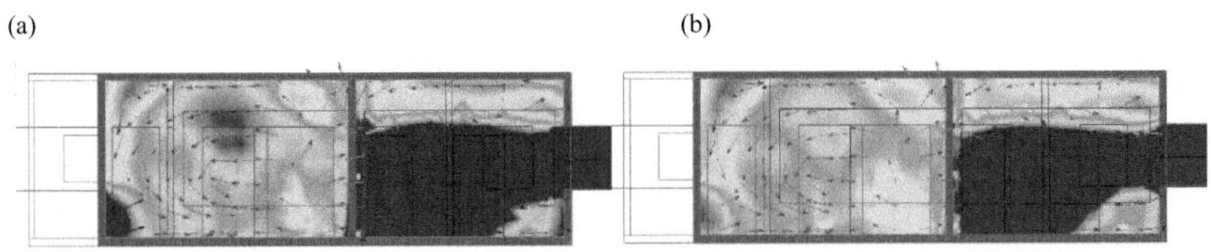

Fig. 15 CFD analysis results of south-facing wind (at section B–B)

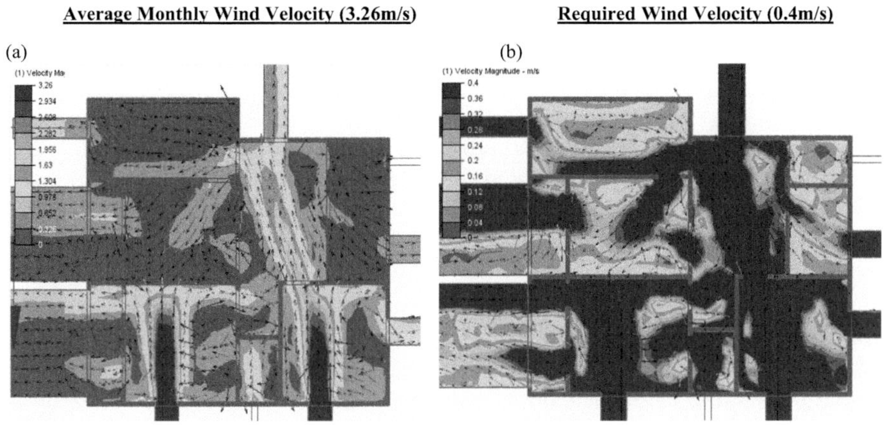

Fig. 16 CFD analysis results of east-facing wind (At 5 ft. elevation from the floor)

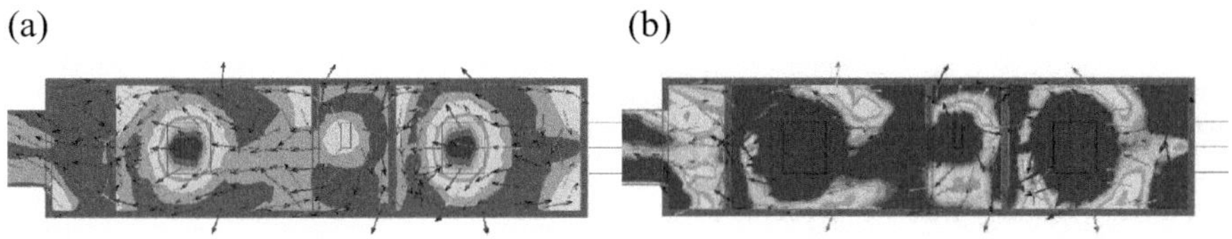

Fig. 17 CFD analysis results of east-facing wind (at section C–C)

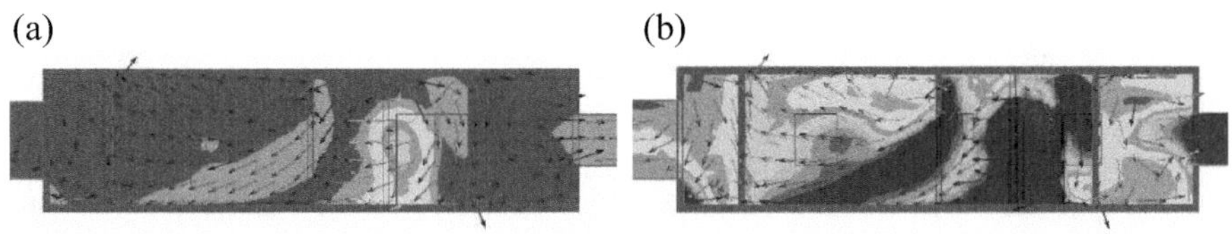

Fig. 18 CFD analysis results of east-facing wind (at section D–D)

4 Conclusion

This study demonstrated that both the CFD-based method for analyzing natural ventilation for occupant thermal comfort and the BIM simulation for assessing the possible decrease of a building's thermal demand by modifying material thermal characteristics through different components hold huge potential.

BIM-based analysis for cooling indicated that the walls and roofs of a building's top story might be leveraged the most, while windows are considerably less feasible.

The study demonstrated that, for ensuring thermal comfort, CFD-centric natural ventilation evaluation helps in visualizing wind flow patterns and velocity concerning exterior and interior openings. This can help with better decision-making in the design phase of the building on how to utilize natural ventilation more. The study also illustrated that in the case study building, the service rooms (toilet, kitchen, and stairway) indicated a general need for mechanical ventilation. But, all the served rooms (bed, living, and dining spaces) can be naturally ventilated for occupant comfort. However, this does not consider the occupant cooling demand when high physical activity is conducted but rather the ventilation required in a resting situation by the human body to achieve thermal comfort. Further studies can be conducted considering this parameter.

The study will help guide future studies toward using a BIM and CFD simulation-based early design process that puts the comfort of building occupants first and reduces the amount of thermal energy the building needs. Thus, maximizing the building's long-term efficiency through strategic placement of openings and material selection for wall, roof, and window components. Particular attention should be paid to the upper floors, which need the most cooling and traditionally leave natural ventilation underutilized.

References

1. International Energy Agency (2021) Buildings: a source of enormous untapped efficiency potential. https://www.iea.org/topics/buildings
2. Comstock M, Garrigan C, Pouffary S, Feraudy TD, Halcomb J, Hartke J (2012) Building design and construction: forging resource efficiency and sustainable development. United National Environmental Program (UNEP) 1
3. CBECS 2012: Energy Usage Summary. https://www.eia.gov/consumption/commercial/reports/2012/energyusage/. Accessed 20 May 2022
4. Chen Z, Jiang C, Xie L (2018) Building occupancy estimation and detection: a review. Energy Build 169:260–270. https://doi.org/10.1016/j.enbuild.2018.03.084
5. Østergård T, Jensen RL, Maagaard SE (2016) Building simulations supporting decision making in early design—a review. Renew Sustain Energy Rev 61:187–201
6. Ilter D, Ergen E (2015) BIM for building refurbishment and maintenance: current status and research directions. Struct Surv 33:228–256. https://doi.org/10.1108/SS-02-2015-0008
7. El-Diraby T, Krijnen T, Papagelis M (2017) BIM-based collaborative design and socio-technical analytics of green buildings. Autom Constr 82:59–74. https://doi.org/10.1016/j.autcon.2017.06.004
8. Zhao X (2017) A scientometric review of global BIM research: analysis and visualization. Autom Constr 80:37–47. https://doi.org/10.1016/j.autcon.2017.04.002
9. Liu J, Xu D, Hyyppä J, Liang Y (2021) A survey of applications with combined BIM and 3D Laser scanning in the life cycle of buildings. IEEE J Sel Topics Appl Earth Obs Remote Sens 14:5627–5637. https://doi.org/10.1109/JSTARS.2021.3068796
10. Tian Z, Zhang X, Jin X, Zhou X, Si B, Shi X (2018) Towards adoption of building energy simulation and optimization for passive building design: a survey and a review. Energy Build 158:1306–1316
11. Bruno R, Bevilacqua P, Cuconati T, Arcuri N (2019) Energy evaluations of an innovative multi-storey wooden near zero energy building designed for Mediterranean areas. Appl Energy 238:929–941. https://doi.org/10.1016/j.apenergy.2018.12.035
12. Albatayneh A, Alterman D, Page A, Moghtaderi B (2020) Alternative method to the replication of wind effects into the buildings thermal simulation. Buildings 10:237

13. Shadram F, Johansson TD, Lu W, Schade J, Olofsson T (2016) An integrated BIM-based framework for minimizing embodied energy during building design. Energy Build 128:592–604. https://doi.org/10.1016/j.enbuild.2016.07.007
14. Gan VJL, Deng M, Tse KT, Chan CM, Lo IMC, Cheng JCP (2018) Holistic BIM framework for sustainable low carbon design of high-rise buildings. J Clean Prod 195:1091–1104. https://doi.org/10.1016/j.jclepro.2018.05.272
15. De Dear RJ, Brager GS (2002) Thermal comfort in naturally ventilated buildings: revisions to ASHRAE Standard 55. Energy Build 34:549–561
16. Yik FW, Lun YF (2010) Energy saving by utilizing natural ventilation in public housing in Hong Kong. Indoor Built Environ 19:73–87
17. Emmerich SJ, Dols WS, Axley JW (2001) Natural ventilation review and plan for design and analysis tools. US Department of Commerce, Technology Administration, National Institute of Standards and Technology
18. You W, Qin M, Ding W (2013) Improving building facade design using integrated simulation of daylighting, thermal performance and natural ventilation. Springer, pp 269–282
19. Porras-Amores C, Mazarrón FR, Cañas I, Sáez PV (2019) Natural ventilation analysis in an underground construction: CFD simulation and experimental validation. Tunn Undergr Space Technol 90:162–173
20. Albuquerque DP, Mateus N, Avantaggiato M, da Graça GC (2020) Full-scale measurement and validated simulation of cooling load reduction due to nighttime natural ventilation of a large atrium. Energy Build 224:110233
21. Hawendi S, Gao S (2017) Impact of an external boundary wall on indoor flow field and natural cross-ventilation in an isolated family house using numerical simulations. J Build Eng 10:109–123
22. Ferrucci M, Brocato M (2019) Parametric analysis of the wind-driven ventilation potential of buildings with rectangular layout. Build Serv Eng Res Technol 40:109–128
23. De Faria L, Cook M, Loveday D, Angelopoulos C, Manu S, Shukla Y (2018) Sizing natural ventilation systems for cooling: the potential of NV systems to deliver thermal comfort while reducing energy demands of multi-storey residential buildings in India
24. Gan VJ, Deng M, Tan Y, Chen W, Cheng JC (2019) BIM-based framework to analyze the effect of natural ventilation on thermal comfort and energy performance in buildings. Energy Procedia 158:3319–3324
25. Olgyay V, Frontado J (1998) Arquitectura y clima: manual de diseño bioclimático para arquitectos y urbanistas. Gustavo Gili Barcelona
26. Matsumoto H, Tsuzuki K, Susanti L (2017) Bioclimatic analysis in pre-design stage of passive house in Indonesia. Buildings 7:24
27. Pontes RH, Najjar MK, Hammad AW, Vazquez E, Haddad A (2022) Adapting the Olgyay bioclimatic chart to assess local thermal comfort levels in urban regions. Clean Technol Environ Policy 24:661–675

Assessing the Fuel Potential of Fecal Sludge and Coal Through Co-combustion

R. P. Saha, T. S. Raaz, and A. B. M. Badruzzaman

Abstract On-site sanitation facilities meet the sanitation needs of a major portion of the population of Bangladesh. However, the generated fecal sludge (FS) is not properly handled and is instead discharged in low-lying areas, posing a significant risk to human health and the environment. There are many examples from the United States, Europe, Japan, and China, where industries use sewage sludge as fuel. This strategy of generating energy from sludge through thermal combustion, if applied in Bangladesh, could be a sustainable solution that provides two simultaneous benefits of energy generation and sludge management. This study aims to evaluate the possibility of recovering energy from fecal sludge for use in energy generation through co-combustion with coal. Hence, for this study, the fecal sludge sample was collected from WSUP (Water and Sanitation for the Urban Poor)'s sludge collection chamber (located in Dhaka) and tested in the laboratory. The samples were subjected to proximate and ultimate analysis, in order to determine the properties of the FS. To determine the effect of sludge addition on calorific value, the sludge samples were mixed with different percent (by weight) of coal and combusted in bomb calorimeter. From the experiment, the gross energy content of the dried sludge sample was found to be 5.9 MJ/kg which had a moisture content in the range of 67–70% as determined through proximate analysis. After deducting the amount of heat required to evaporate the moisture from the sample, the net energy content of the dried FS sample was estimated to be 4.13 MJ/kg. For a sample containing sludge-coal mix in the ratio 1:1 (by weight), the gross and net energy contents were found to be 17.2 and 15.4 MJ/kg respectively, which exhibits acceptable capacity for extraction of energy. Although there are scopes of energy generation through co-combustion of fecal sludge with other fuels, the commercial viability of this study depends on identifying sustainable and cost-effective methods of dewatering and energy recovery technologies.

Keywords Fecal sludge · Co-combustion · Fecal sludge management · Sanitation

R. P. Saha (✉) · T. S. Raaz · A. B. M. Badruzzaman
Department of Civil Engineering, Bangladesh University of Engineering and Technology, Dhaka, Bangladesh
e-mail: rituprogga27@gmail.com

S. Arthur et al. (eds.), *Proceedings of the 6th International Conference on Advances in Civil Engineering*, Lecture Notes in Civil Engineering 368,
https://doi.org/10.1007/978-981-99-3826-1_26

1 Introduction

The sanitation requirement of approximately 2.7 million people around the world are met through onsite (non-sewered) sanitation facilities, with that figure predicted to rise to 5 billion by 2030 [1]. This population consists of both rural and urban inhabitants, especially of low and middle-income countries that lack sewer infrastructures [1]. In Bangladesh, all urban areas are supplied by an on-site sanitation system, except for about 20% of Dhaka's metropolitan area (which is served by an almost non-functional sewerage system) [2]. It is clear that fecal sludge management is a critical issue that must be met, and that it will continue to play an important role in global sanitation management in the future.

Safe reuse or disposal of the generated FS is a crucial part of an effective fecal sludge management system. Traditionally, fecal sludge is disposed of through landfilling. However, due to land constraints and rising sludge production volume, this alternative is no longer as viable as it previously was. The sludge might need to be further treated, or utilized through some type of resource recovery. Typically, the most common type of resource recovery from FS has been as a soil conditioner and organic fertilizer since excreta consist of essential plant nutrients and organic matter that boosts the water holding capacity of soils. In addition, there are several other treatment options that allow for resource recovery. For example, during anaerobic digestion of FS, biogas may be created, and the residual sludge can be utilized as a soil conditioner. Furthermore, energy recovery technologies such as pyrolysis, gasification, incineration, and co-combustion that would result in a more sustainable management of fecal sludge are being developed to recover end products [3].

Energy recovery from fecal sludge can be described as one of the most appealing strategies for utilizing the rising amount of FS and lowering waste volume while also having the potential to recover nutrients and/or metals from it. One of the major factors for effective conversion of sludge to energy is maintaining the moisture content of the sludge. So, reducing the moisture content of the FS simply by drying can significantly enhance its organic content and calorific value [3]. For the removal of fluids present in the primary sludge, different dewatering and drying techniques such as mechanical dewatering, direct drying systems, indirect drying systems, etc. are used [4].

In global context, there are many examples from the United States, Europe, Japan, and China, where industries use sludge (specially sewage sludge obtained as a byproduct of conventional wastewater treatment plants) as fuel [5]. The invention of the idea of using sludge as a fuel has been mainly driven from the lack of disposal options, where landfilling area is limited. A study showed that FS has volatile matter ranged between 39 and 50%, which qualify FS as a fuel [6]. Hence, it can be claimed that dried FS can be used as an alternative to primary fuel in coal fired thermal power plants. Fecal sludge can also be used as an auxiliary fuel in power plants and cement kilns which will decrease the fuel costs of these plants and also provide an economical method of sludge disposal. As a result, the immense pressure on the consumption of primary fuels can be reduced. This strategy of generating energy from sludge through thermal combustion, if applied in Bangladesh, could be a sustainable solution that provides two simultaneous benefits of energy generation and sludge management. Hence, the main objective of this study is to assess whether fecal sludge and sludge-coal combinations can be used as fuel to generate energy.

2 Materials and Methods

The outline of the methodology conducted during this study is shown in Fig. 1 and details are discussed briefly.

For this study, fecal sludge sample was collected from WSUP (Water and Sanitation for the Urban Poor)'s sludge collection chamber, located at Kochukhet, Dhaka (23.7940° N, 90.3901° E). WSUP collects septage from the septic tanks of the adjacent areas using vacutug service. Then the sludge is dewatered mechanically and stored, before being conveyed to the treatment plant. Hence, the sample collected for this study was already dewatered and it existed in a moderately hard semi-solid form.

2.1 *Proximate Analysis*

In this study, proximate analysis was carried out on the fecal sludge sample according to the ASTM D3172-89 guidelines [7]. The analysis involves determination of the moisture content, volatile matter, ash and fixed carbon content of the sample.

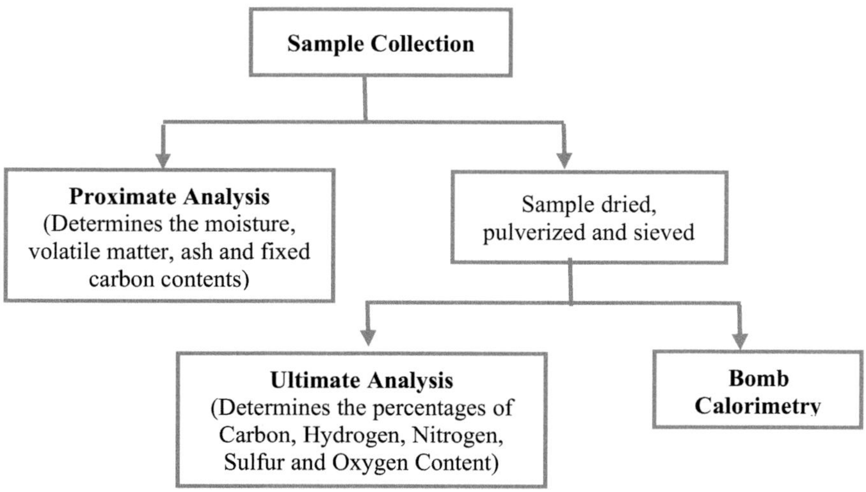

Fig. 1 Outline of methodology

- **Moisture Content**

The moisture content was determined by measuring the weight loss of samples due to heating at 77 °C for 24 h. 2–3 g FS sample was weighed and placed in a clean, dry crucible into the oven for 24 h at a temperature of 77 °C. The crucible containing the dried sample was then placed in a desiccator for 30 min before measuring the dry weight of the sample. Then, the moisture content was calculated using the following equation:

$$\text{Moisture content} = \frac{\text{Ww} - \text{Wd}}{\text{Ww}} \times 100\%$$

where,

Ww wet weight of sample
Wd dry weight of sample

- **Volatile Matter Content**

The dry sample from the moisture content test was heated at 105 °C for 1 h to obtain the initial sample for volatile solid content test. To obtain the percentage of volatile content, the sample was kept in the furnace oven at 550 °C for 1 h in a ceramic crucible. Then, the crucible containing the sample was placed in a desiccator for 30 min before measuring the dry weight again. The percent of volatile solid content was calculated using following formula:

$$\text{Volatile solid content} = \frac{w_1 - w_2}{w_1} \times 100\%$$

where,

w_1 dry weight of sample at 105 °C
w_2 dry weight of sample at 550 °C

- **Ash Residue**

The percent of ash residue, that remained in the crucible after burning the sample at 550 °C, was determined using the following formula:

$$\text{Ash residue} = (100 - \text{volatile solid content})\%$$

2.2 Ultimate Analysis

The ultimate analysis includes an assessment of the levels of carbon, hydrogen, oxygen, nitrogen and sulfur contents present in a sample of biomass. Fine pulverized samples of FS were used for the determination of carbon, nitrogen, sulfur and hydrogen contents. Oxygen content was calculated by the difference of carbon, nitrogen, hydrogen, sulfur and ash from unity.

$$\mathrm{O} = 100 - \mathrm{Ash} - \mathrm{C} - \mathrm{N} - \mathrm{S} - \mathrm{H}$$

where,

C carbon, wt%
H hydrogen, wt%
N Nitrogen, wt%
S Sulphur, wt%
O Oxygen, wt%

2.3 Determination of Calorific Value of FS Sample

Various empirical equations have been developed through mathematical modelling to estimate the energy content of biomass using the percentage composition of the components. The models used in this study are mentioned in Table 1.

Table 1 Mathematical equations for theoretical estimation of energy content

Model	Mathematical equation	Unit	References
Modified Dulong formula	$HHV = 337(\mathrm{C}) + 1419\left(\mathrm{H} - \frac{\mathrm{O}}{8}\right) + 93(\mathrm{S}) + 23.26(\mathrm{N})$	kJ/kg	[8]
Steuer formula	$HHV = 81(\mathrm{C}) - \left(\frac{3\mathrm{O}}{8}\right) + 171\left(\frac{\mathrm{O}}{8}\right) + 345\left(\mathrm{H} - \frac{\mathrm{O}}{10}\right) + 25(\mathrm{S}) - 6\,(9\mathrm{H} + \mathrm{W})$	kcal/kg	[8]
Scheurer–Kestner formula	$HHV = 81(\mathrm{C}) - \left(\frac{3\mathrm{O}}{4}\right) + 171\left(\frac{\mathrm{O}}{4}\right) + 342.5\left(\mathrm{H} - \frac{\mathrm{O}}{10}\right) + 22.5(\mathrm{S}) - 6\,(9\mathrm{H} + \mathrm{W})$	kcal/kg	[8]
Niessen formula	$HHV = 0.2322(\mathrm{C}) + 0.7655(\mathrm{H}) - 0.072(\mathrm{O}) - 0.0419(\mathrm{N}) + 0.0698(\mathrm{S}) + 0.0262(\mathrm{Cl}) + 0.1814(\mathrm{P})$	MJ/kg	[9]

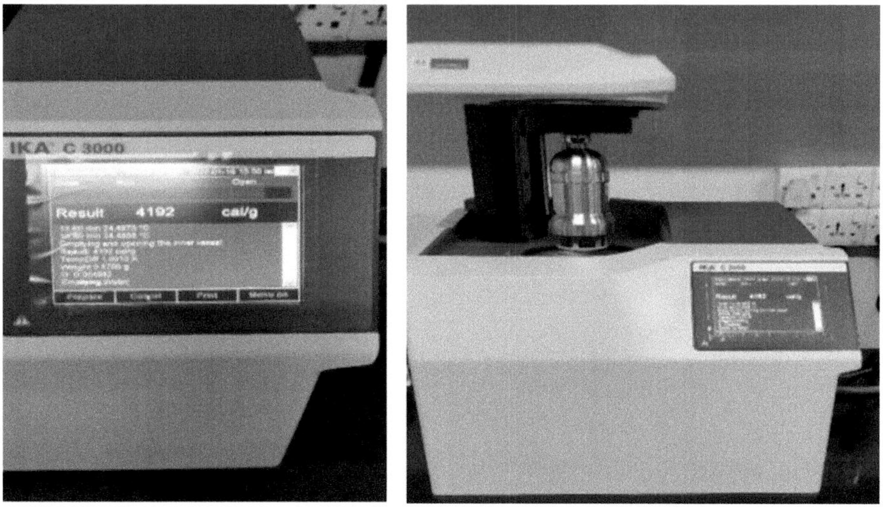

Fig. 2 IKA C3000 bomb calorimeter

2.4 *Bomb Calorimetry*

The bomb calorimeter, shown in Fig. 2, is a device that determines the heating value of a sample. The heat produced by the combustion of the platinum fuse wire and sample, as well as any benzoic acid employed as a combustion aid, is absorbed by the bomb and the surrounding water jacket when the ignition button on the calorimeter is pressed. The calorific value of samples is determined by the amount of heat created during the combustion process.

3 Results and Discussions

3.1 *Proximate Analysis*

The results of the proximate analysis of raw sludge samples are shown in Table 2. The analysis was performed in dry basis.

Table 2 Proximate analysis of fecal sludge samples

Physical parameters	Moisture content (in dry basis) %	Volatile solids %	Ash residue %
Sample 1	67.23	40.4	59.6
Sample 2	70.24	41.1	58.9

The analysis shows that the sample had moisture content ranged between 67 and 70%, which seems to be very high for the raw sludge to be used as fuel. The sample has to be dried well to remove the excess moisture, so that its energy content can be determined. The volatile solids ranged between 40 and 41%. The volatile solids content should be greater than 40% of the total mass for it to qualify as a fuel [6]. The percentage of residual ash ranged between 58 and 60%.

3.2 *Ultimate Analysis*

The elemental composition of the sludge samples has been determined through ultimate analysis as shown in Table 3 and is represented diagrammatically in Fig. 3.

The analysis shows that the elemental composition of the samples of FS does not show much variation. This is because the samples belong to the same origin, as it was collected from a single site.

Table 3 Elemental composition of samples obtained from ultimate analysis

Parameters	Carbon (%)	Hydrogen (%)	Nitrogen (%)	Sulfur (%)	Oxygen calculated (%)
Sample 1	15.2	2.15	1.62	0.5	20.93
Sample 2	15.1	2.1	1.64	0.48	21.78
Sample 3	15.2	2.11	1.65	0.51	21.13

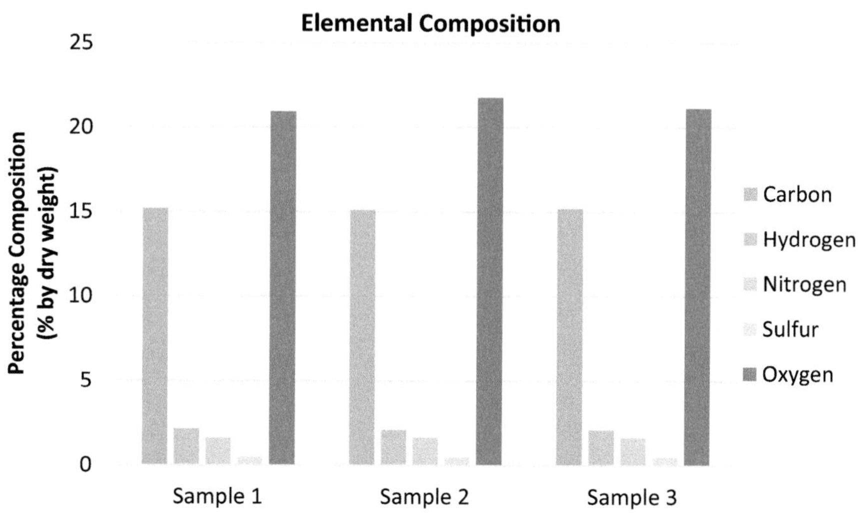

Fig. 3 Elemental composition of samples

Table 4 Theoretical calorific values of sludge samples

	Modified Dulong formula		Steuer formula		Scheurer-Kestner formula		Niessen formula
	kJ/kg	MJ/kg	Kcal/kg	MJ/kg	Kcal/kg	MJ/kg	MJ/kg
Sample 1	4544.972	4.545	933.515	3.906	743.753	3.112	3.635
Sample 2	4288.159	4.288	877.350	3.671	680.325	2.846	3.510
Sample 3	4454.365	4.454	915.465	3.830	724.028	3.029	3.590

3.3 Theoretical Calorific Values

The theoretical calorific values of the sludge samples were calculated using the mathematical models shown in Table 1 and the results are summarized in Table 4.

The calorific value of sludge obtained from mathematical models is found to be too low for FS to be considered for power generation. When compared to the calorific values of other low-quality fuel like lignite coal (14.7–19.3 MJ/kg, ASTM D388-99) and sugar bagasse (18.61 MJ/kg) [10], it can be concluded that the calorific value of fecal sludge is quite low.

3.4 Experimental Calorific Value

From the Bomb Calorimeter, Gross Heating Value was obtained and to calculate the Net Heating Value, energy of vaporization by the moisture content was subtracted. The experimental results from the bomb calorimeter are shown in Table 5 and represented graphically in Fig. 4.

It can be observed in Fig. 4, that as the coal-to-FS mixing ratio increases, the experimental heating value increases. The Net Heating Value from the FS alone (Sample A) is found to be 4.13 MJ/kg without any coal mixing. When the FS is mixed with coal at a 1:1 ratio (Sample C), the maximal extraction of Net Heating Value is almost four times (15.4 MJ/kg) than that of the FS alone.

Table 5 Heating value of FS-coal sample mixes from experimental analysis

Sample	Component		Gross heating value FS-coal mixes	Net heating value FS-coal mixes
	Coal (%)	Fecal sludge (%)	MJ/kg	MJ/kg
A	0	100	5.89	4.13
B	25	75	11.54	9.78
C	50	50	17.16	15.4
D	100	0	27.66	26

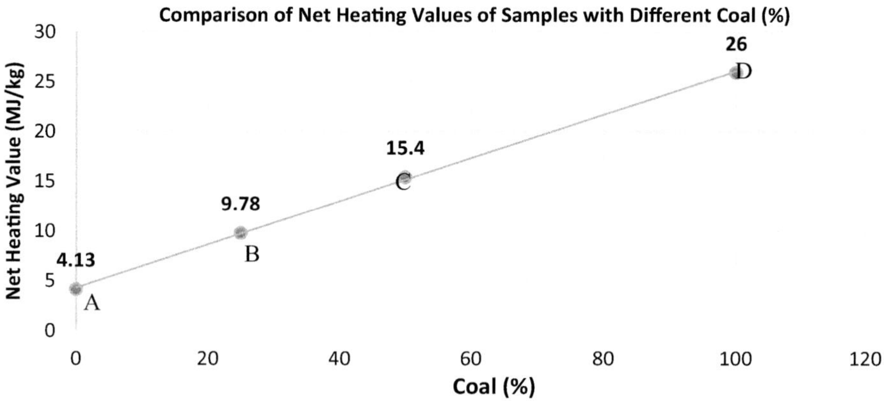

Fig. 4 Comparison of heating values between samples

3.5 *Comparison Between Experimental and Theoretical Results for FS*

From Table 6, it is evident that each of the four theoretical methods produce quite close values for all the three samples due to having same composition. Taking average of theoretical energy values for the three samples in each method, Table 7 was prepared in order to compare between theoretical values and experimental value and analyze the variation.

It can be observed that Modified Dulong Formula produced maximum energy and shows least variation when compared to the experimental heating value, among the other four mathematical models. Also, experimental method using Bomb Calorimeter exhibits more heating value than the theoretical models. This variation between the theoretical and experimental heating values might be because the mathematical models used in this study were not developed for fecal sludge.

Table 6 Theoretical and experimental calorific values from FS (only) samples

Samples	Theoretical models				Experimental (net value)
	Modified Dulong formula	Steuer formula	Scheurer-Kestner formula	Niessen formula	Bomb calorimeter
Unit	(MJ/kg)				
Sample 1	4.545	3.906	3.112	3.635	4.13
Sample 2	4.288	3.671	2.846	3.51	
Sample 3	4.454	3.83	3.029	3.59	

Table 7 Theoretical (Average) and experimental calorific values of FS (only) samples

Value	Modified Dulong formula	Steuer formula	Scheurer-Kestner formula	Niessen formula
	(MJ/kg)			
Gross experimental	5.89			
Net experimental	4.13			
Theoretical	4.43	3.802	2.996	3.578
(%) Variation with net experimental value	7.26	7.94	27.46	13.36

3.6 *Comparison of Energy Density Between FS-Coal Mixes and Common Fuels*

The energy density of coal sludge mixes as found from the experimental analyses was compared with the energy density of the commonly used fuels in Bangladesh. The results are displayed in Fig. 5.

From the energy density plot in Fig. 5, it can be concluded that the gravimetric energy densities of the coal-sludge Mixes are comparable to the gravimetric energy density of the commonly used fuels, such as wood, lignite coal and sugarcane bagasse. It can be noticed that the gravimetric energy densities of sample B (25% Coal and 75% FS) and sample C (50% Coal and 50% FS) are very close to the values of lignite coal. This is an interesting finding since lignite coal is widely used in Bangladesh.

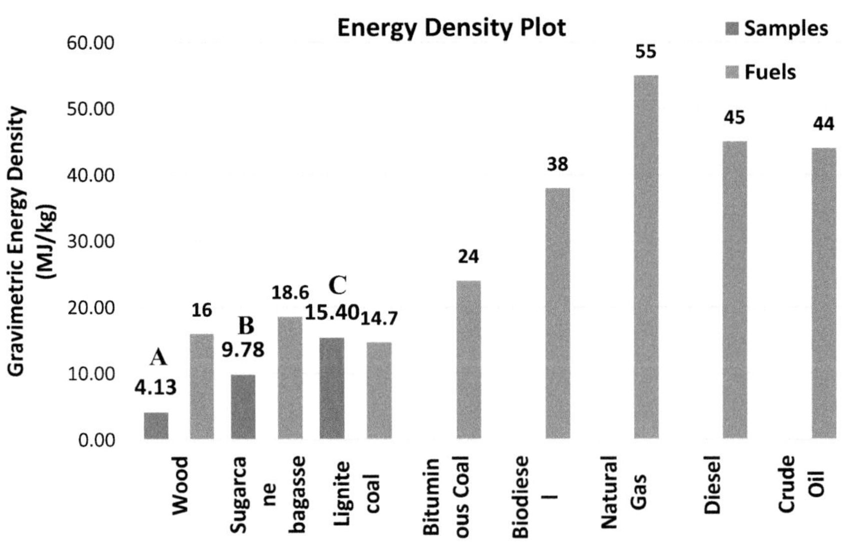

Fig. 5 Energy density plot between FS-coal mixes and common fuels [11]

4 Conclusion

The main objective of this study was to assess whether fecal sludge can be used as a fuel for power generation. For this, characteristic properties such as moisture content, volatile solids content and proportion of ash was determined, in addition to determining the elemental composition of the FS samples. Then, different portions of coal were mixed with FS sample to determine the fuel potential of FS-coal Mixes.

Based on the findings of the investigations, the following conclusions can be drawn:

- The volatile organic content was found to be in the range of 40–42% which ensures good amount of energy extraction from the organic samples as stated in different literature [6].
- The Gross Calorific Value and Net Calorific Value of FS sample was found to be 5.89 and 4.13 MJ/kg respectively from experimental process, which is quite low when compared to other existing fuels. So, it can be concluded that the combustibility of FS alone is not enough for it to be used as fuel for power generation.
- The calorific value of FS obtained experimentally was cross-checked and compared with the heating values obtained theoretically from different empirical equations. Modified Dulong Formula yielded the closest value (4.43 MJ/kg) to the experimental result of all the empirical equations used. As it turned out, the Modified Dulong Formula proved to be a better formula for relating the Bomb Calorimeter results.

Since the calorific value of fecal sludge alone is not enough to be used as fuel, integrating FS with coal proved to be a useful idea for getting a fuel with a higher energy content.

- With increasing coal content, the calorific value of the coal-sludge mixture increases. When the FS is mixed with coal at a ratio of 1:1 (Sample C), the heating value is found to increase to almost three times than that of the FS sample alone. When compared with other fuels, the calorific value of coal-sludge mixture with coal content of 50% (Sample C), is found to be greater than the calorific values of lignite coal. Hence, it appears from Fig. 4, that coal-sludge mixes containing about 50% coal, having a calorific value of approximately 15 MJ/kg, can be used as an alternative to lignite coal.
- The gravimetric energy densities of some commonly used fuels were compared with that of the sludge and coal Mixes. It was observed that the gravimetric energy densities of sample B (25% Coal and 75% FS) and sample C (50% Coal and 50% FS) are very close to the values of lignite coal, sugarcane bagasse and wood.

Acknowledgements The authors of this paper would like to thank the WSUP (Water and Sanitation for the Urban Poor), from where the sample to be studied was collected. We are also thankful to the Institute of National Analytical Research and Service (INARS) at Bangladesh Council of Scientific and Industrial Research (BCSIR) for providing the facilities and equipment for Elemental analysis.

References

1. Strande L (2014) Faecal sludge management—systems approach for implementation and operation
2. Rahman MM, Ali MA, Choudhury MR, Rahman MA, Redwan AM, Noor NF, Sohan AI (2016) Fecal sludge management (FSM) scenario in urban areas of Bangladesh. In: 6th South Asian Conf Sanit, pp 1–6. https://doi.org/10.13140/RG.2.2.25766.73280
3. Oladejo J, Shi K, Luo X, Yang G, Wu T (2019) A review of sludge-to-energy recovery methods. Energies 12:1–38. https://doi.org/10.3390/en12010060
4. Kamizela T, Kowalczyk M (2019) Sludge dewatering: processes for enhanced performance. In: Ind Munic Sludge Emerg Concerns Scope Resour Recover, pp 399–423. https://doi.org/10.1016/B978-0-12-815907-1.00018-0
5. Kuleape R, Cobbina SJ, Dampare SB, Duwiejuah AB, Amoako EE, Asare W (2014) Assessment of the energy recovery potentials of solid waste generated in Akosombo, Ghana. Afr J Environ Sci Technol 8:297–305. https://doi.org/10.5897/ajest2014.1663
6. Chowdhury Ankan M, Murad Hasan M, Jobaer Howlader M (2020) Assessment of energy recovery potential of Faecal sludge. Landsc Archit Reg Plan 5:21. https://doi.org/10.11648/j.larp.20200502.11
7. Standard practice for proximate analysis of coal and coke. https://www.astm.org/d3172-89r02.html. Accessed 8 May 2022
8. Liu J-I, Paode RD, Holsen TM (1996) Modeling the energy content of municipal solid waste using multiple regression analysis. J Air Waste Manage Assoc 46:650–656. https://doi.org/10.1080/10473289.1996.10467499
9. Tchobanoglous G, Theisen H, Vigil SA (1993) Integrated solid waste management engineering principle and management issue. McGraw Hill Inc., New York—References—Scientific Research Publishing
10. Abd S, Rafie A, Haider N (2016) Assessing the energy values of sewage sludge from Pagla sewage treatment plant. In: BUET-ANWAR ISPAT 1st Bangladesh Civ Eng SUMMIT 2016
11. Energy density–energy education. https://energyeducation.ca/encyclopedia/Energy_density. Accessed 8 May 2022

On the Method to Evaluate the Post-earthquake State of Collapsed Structures and Its Applications to Numerical Analyses

Zongmu Liu and M. Saitoh

Abstract Devices to control the direction of structural collapse during earthquakes have been proposed in recent years where a rigid block or a chain is placed in between the column and the beam of a structure to asymmetrically accumulate the inelastic response, leading the collapse in the designated direction. The effectiveness of such devices has been verified by numerical and experimental studies, where symmetric simple structures were targeted. To assess the effectiveness of such devices on asymmetric or eccentric structures, on the other hand, an appropriate method to capture their collapsed state is desired. This study proposes a new method to capture the direction of collapse and the amount of deformation of target structures. Moreover, by employing Open Dynamic Engine (ODE), a three-dimensional time-history analysis that can simulate the collapse of structures consisting of rigid bodies, the proposed method is verified. In this method, the structural members are virtually divided into elements based on the unit weight and the movement of each element is measured as the vector before and after earthquakes. The origin of the vectors is then moved to concentrate at the origin to integrate their direction and displacement. The information of the direction is expressed as the percentage of the accumulated masses of the elements normalized by the total mass of the structure while the information of the displacement is shown as the color map. This graphical mapping can render the state of collapse, expressing the direction and displacement in a simple manner. The proposed method is applied to the results of the time-history analysis in case of structures with and without the collapse direction control devices, which successfully shows the quantitative differences between them. This method is a promising tool to evaluate the state of the collapse of various types of structures.

Keywords Resilience · Collapse · Collapse direction · Graphical mapping

---

Z. Liu (✉)
Department of Civil and Environmental Engineering, Saitama University, Saitama, Japan
e-mail: liuzongmu@gmail.com

M. Saitoh
Graduate School of Science and Engineering, Saitama University, Saitama, Japan
e-mail: saity@mail.saitama-u.ac.jp

S. Arthur et al. (eds.), *Proceedings of the 6th International Conference on Advances in Civil Engineering*, Lecture Notes in Civil Engineering 368,
https://doi.org/10.1007/978-981-99-3826-1_27

1 Introduction

In recent years, a number of severe damages to structures have occurred due to unprecedented earthquake ground motions, which were not considered in the design. Resilient structures have been recognized as a marked paradigm shift from the ordinary safety and reliability engineering approaches to mitigate unwanted outcomes due to uncertainties. Hollnagel et al. [1, 2] defines resilience as the intrinsic ability of a system to adjust its functioning prior to, during, or following changes and disturbances, so that it can sustain required operations under both expected and unexpected conditions. There is a high possibility that emergency roads might be blocked as a result of collapsed structures in an event of an earthquake, exacerbating the recovery of cities and evacuation of people to safety as shown in Fig. 1. Such unwanted situations, in fact, have been experienced in the past earthquakes. It should be primarily admitted, therefore, that the collapse of structures could occur due to unexpected large ground motions despite possessing sophisticated knowledge from historical earthquake events pertinent to the seismic design. Thus, it is desired to consider ways to avoid blocked roads for quick recovery after an earthquake. In recent studies, devices to control the direction of structural collapse have been proposed by Saitoh et al. [3], where a rigid block or a chain is placed in between the column and the beam of a structure to asymmetrically accumulate the inelastic response, leading the collapse in the designated direction. The effectiveness of the devices has been verified by recent experimental studies [4], where symmetric simple structures were targeted. In general, not all buildings simply deform in one direction. At least, there is no realistically designated direction. Mostly, the post-collapse condition is complicated. For example, in a two-story wooden house, there are cases where the first and second stories collapse in different directions. The two-story gate of Aso Shrine collapsed during the 2016 Kumamoto Earthquake, where the first and second floors collapsed in different ways, making it difficult to determine the direction of collapse. To assess the effectiveness of such devices on asymmetric or eccentric structures, an appropriate method to capture their collapsed state is desired.

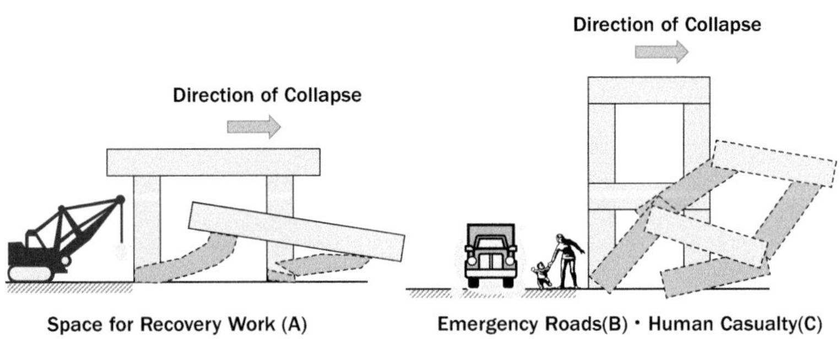

Fig. 1 Resilient structures controlling the direction of collapse [3] (**a–c** are possible status where collapse of structures should be avoided)

This study proposes a new method to capture the direction of collapse and the amount of deformation of target structures. Moreover, by employing Open Dynamic Engine (ODE), a three-dimensional time-history analysis that can simulate the collapse of structures consisting of rigid bodies, the proposed method is verified.

2 Method Proposed

In this method, the structural members are virtually divided into elements based on the unit weight and the movement of each element is measured as the vector before and after earthquakes. The origin of the vectors is then moved to concentrate at the origin to integrate their direction and displacement. The information of the direction is expressed as the percentage of the accumulated masses of the elements normalized by the total mass of the structure while the information of the displacement is shown as the color map. This graphical mapping can render the state of the collapse, expressing the direction and displacement in a simple manner.

2.1 Discretization of Structural Members

Figure 2 shows a column standing in the Cartesian coordinate system. The column is divided into small segments with the total number of n, where they are sufficiently small and have the same mass. The reason for the use of the consistent mass for the division is to be discussed later. These elements can be expressed in the three-dimensional coordinates system; the location of each element can be expressed in a vector form as:

$$[n_x\, n_y\, n_z] \tag{1}$$

where, n_x, n_y, and n_z are the x, y, and z-axis coordinates, respectively. This is a position vector of each element. Therefore, the location of all elements of the column can be written in the following matrix, referred as position matrix.

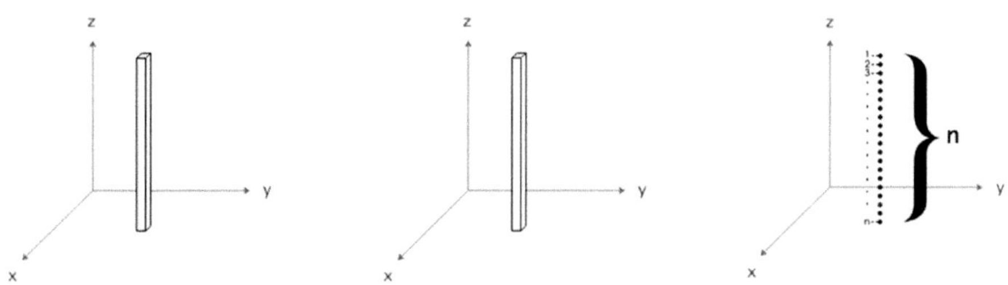

Fig. 2 Discretization of columns

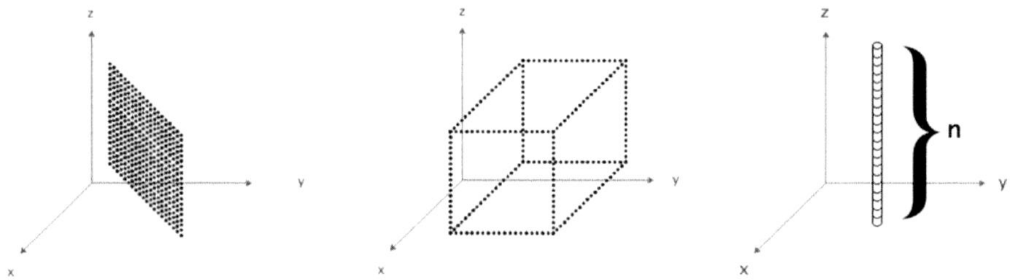

Fig. 3 Discretization of other structural members

$$\begin{bmatrix} n_{1x} & n_{1y} & n_{1z} \\ \vdots & \vdots & \vdots \\ n_{nx} & n_{ny} & n_{nz} \end{bmatrix} \tag{2}$$

This expression can be applied not only to the standing column but also laterally located beams and inclined ones. Moreover, to extend this expression to plates and solid members such as a wall (a board) and a block shown in Fig. 3, the straightforward division into elements with a consistent mass can be applied, where they are also expressed as the matrix form.

There is a primal reason for the use of consistent mass in the division of structural members. It is considered that the mass can be an indicator of how much energy or work should be provided to reinstate damaged structures into their original positions. This study will present a graphical mapping to express the status of damages to structures in the latter section. An appropriate combination of the graphical mapping and the mass information will provide the information of easiness and difficulties in the recovery after earthquakes, which is a promising tool to measure the "Resilience" of the structures.

2.2 Vector Expressions of Elemental Movement

The matrix form defined above is available to calculate various physical quantities representing the status of movement and collapse of structures. The displacement of structural members $\vec{p}$ is expressed in the following matrix form by using Eq. 2, where the position matrix after the earthquake is subtracted from that before the earthquake.

$$\vec{p} = P' - P \tag{3}$$

$$P = \begin{bmatrix} n_{1x} & n_{1y} & n_{1z} \\ \vdots & \vdots & \vdots \\ n_{nx} & n_{ny} & n_{nz} \end{bmatrix} \tag{4}$$

$$P' = \begin{bmatrix} n'_{1x} & n'_{1y} & n'_{1z} \\ \vdots & \vdots & \vdots \\ n'_{nx} & n'_{ny} & n'_{nz} \end{bmatrix} \tag{5}$$

$$\vec{P} = \begin{bmatrix} n'_{1x} - n_{1x} & n'_{1y} - n_{1y} & n'_{1z} - n_{1z} \\ \vdots & \vdots & \vdots \\ n'_{nx} - n_{nx} & n'_{ny} - n_{ny} & n'_{nz} - n_{nz} \end{bmatrix} \tag{6}$$

where, P and P' are the matrices before and after the event, respectively.

2.3 *Graphical Mapping*

The purpose of graphical mapping is to visualize the location of the structural members, the direction of their displacements, and the amount of mass movement. Equation 2 can be used for a graphical mapping to capture the location of the members and their direction from the origin of the coordinate after the event of earthquakes while Eq. 6 can be used to exhibit the direction of the displacements and the amount of mass movement. The former mapping has a benefit to capture the location of members and structures at a glance whereas the disadvantage is that the mapped values depend on where the origin of the coordinate is set. The latter mapping is independent of the location of the origin so that a more comprehensive observation of the mass movement can be performed. In this study, the former graphical mapping is defined as "absolute system mapping" while the latter is as "relative system mapping". To capture the movement of structures from the top view, the matrix of either Eq. 2 or Eq. 6 is projected on the horizontal X–Y plane, that is, just an extraction of the X and Y components from the original matrix such that Eq. 7 represents the matrix and Eq. 8 is the projected matrix.

$$\begin{bmatrix} n_{1x} & n_{1y} & n_{1z} \\ \vdots & \vdots & \vdots \\ n_{nx} & n_{ny} & n_{nz} \end{bmatrix} \tag{7}$$

$$\begin{bmatrix} n_{1x} & n_{1y} \\ \vdots & \vdots \\ n_{nx} & n_{ny} \end{bmatrix} \tag{8}$$

Equation 8 is converted into the polar coordinate according to the following general geometrical relations:

$$\begin{bmatrix} \alpha_1 & l_1 \\ \vdots & \vdots \\ \alpha_n & l_n \end{bmatrix} \tag{9}$$

where,

$$\alpha_n = \arctan \frac{n_{ny}}{n_{nx}} \tag{10}$$

$$l_n = \sqrt{n_{nx}^2 + n_{ny}^2} \tag{11}$$

Figure 4 shows a graphical mapping based on Eqs. 10 and 11. In the mapping, the total circumferential angle is equivalently divided into angular sections while the range of the displacement is discretized into segments. All elements are classified in terms of angular sections and displacement segments. In the mapping procedure, the number of elements within each angular section is accumulated in each displacement segment. The number of elements is expressed as the percentage per the total number of elements in the mapping; range of the displacement is mapped in the form of color differences as shown in the figure.

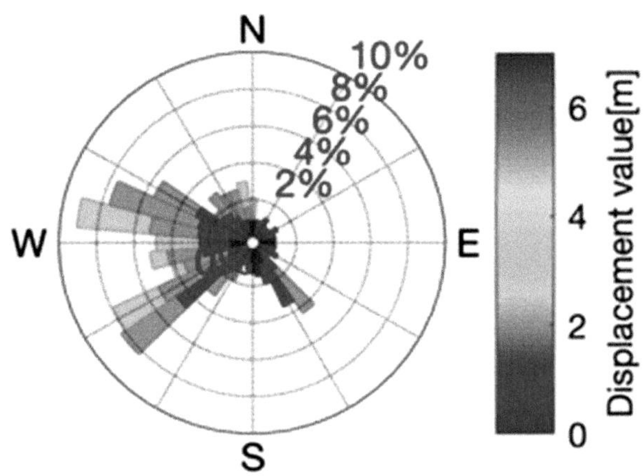

Fig. 4 Proposed graphical mapping

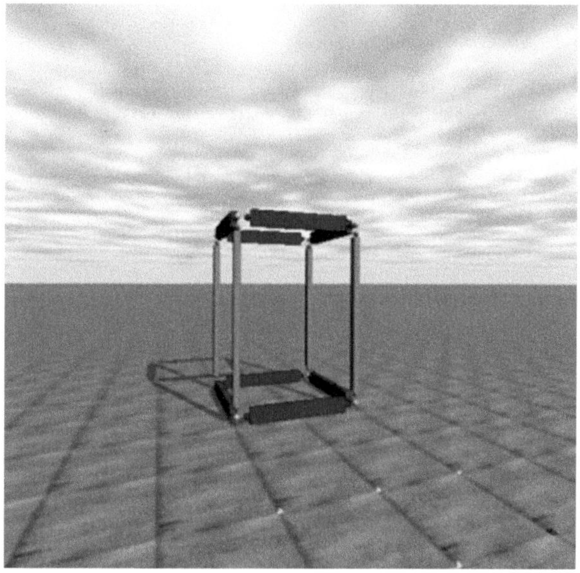

Fig. 5 Model used for analysis in ODE

3 Applications of Proposed Graphical Mapping

3.1 Target Structural Model

The proposed method is applied to the results of the time-history analysis in case of structures with and without the collapse direction control devices. Open Dynamic Engine (ODE) was used for the analysis targeting a frame model consisting of four rigid columns connected to four rigid beams with elastic–plastic rotational joints as shown in Fig. 5. The size of each column was 105 mm in width, 105 mm in depth and 2500 mm in height, and weight 262 kg. The size of each beam was 105 mm in width, 240 mm in depth and 1520 mm in height, and weight 261 kg. The size of model was 2025 mm in width, 2025 mm in depth and 3410 mm in height. Herein, restitution coefficient to define the energy loss by contact was set by 0 in ODE. The load-deformation relationship reproduced by the two-column frame wall in the lateral direction by using the joints is shown in Fig. 6.

3.2 Collapse Direction Control Device

Collapse direction control device is briefly introduced here; details are described in Saitoh [3]. A rigid block is connected to the beam while it maintains a small gap from the column as shown in Fig. 7. The block acts as a restraint when the column is displaced in the undesired direction. It has, however, no effect on the movement

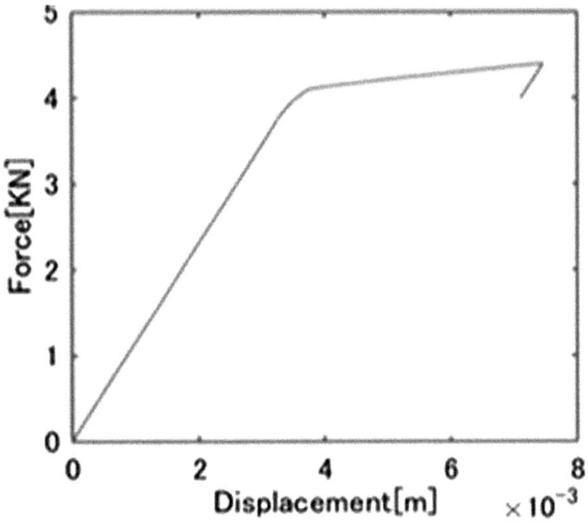

Fig. 6 Load-deformation relationships of two-column frame wall in ODE model

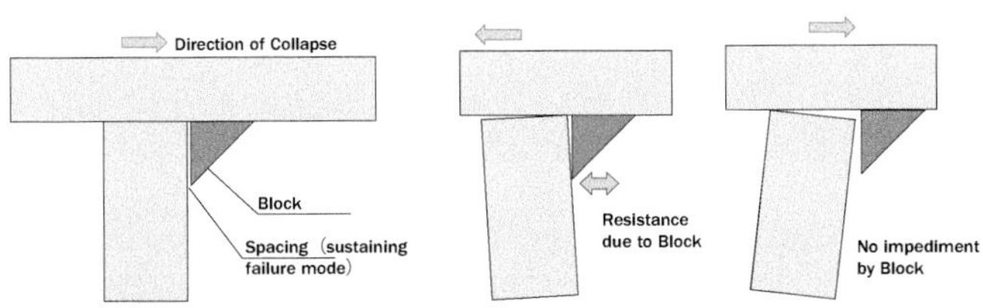

Fig. 7 Device with block for controlling the direction of collapse [3]

of the column in the opposite direction. Consequently, the residual displacement is accumulated in the desired direction.

In this study, two devices were modeled as a rigid body and placed at the top of the beams to allow the collapse in the right-hand side as shown in Fig. 8.

3.3 Time-History Analysis in ODE

Figure 9 shows the simulated earthquake ground motion (displacement) integrating the acceleration record at the station 2004 K-NET Ojiya EW in time domain. The time interval was set 0.0001 s in the analysis. This ground motion was applied to the base of the structure. In the analysis, the direction of the excitation was arbitrarily varied to reproduce different status of collapse. Figure 10 shows the results of the analysis selected where a various type of collapse can be observed.

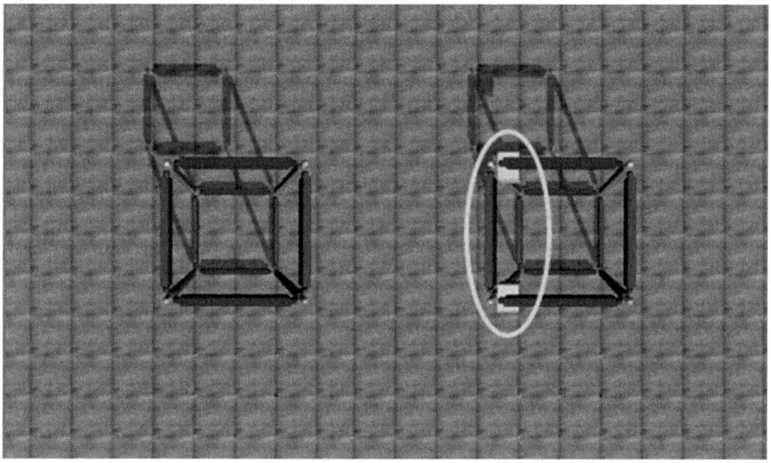

Fig. 8 Location of collapse direction control devices in ODE model

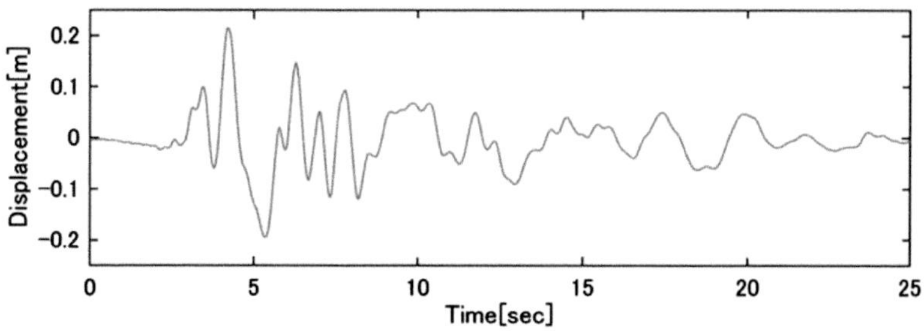

Fig. 9 Incident force seismic waves

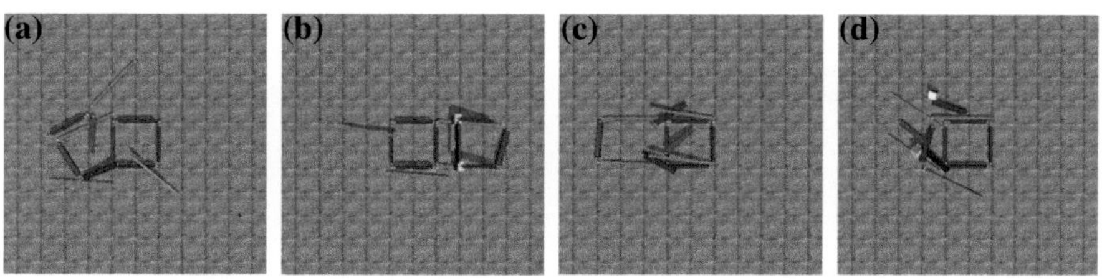

Fig. 10 Top view of ODE results from left to right (**a** EW excitation without device; **b** EW excitation with device; **c** SW excitation without device; **d** SW excitation with device)

3.4 Results of Graphical Mapping

The results of graphical mapping are presented in Fig. 11. The origin of the coordinate is set at the center of the edge line of the beam (south side). Figure 11a shows the graphical mappings of the absolute system and the relative system, respectively, in case of EW excitation without device. The absolute system mapping shows that more than 50% of the elements are located in between the north-west and south-west directions from the origin. It is conceivable that the distribution of the elements after the event of earthquakes can be grasped smoothly from the mapping. On the other hand, the relative system mapping clearly shows that most of the elements displace to the west direction: the relative system mapping may be suitable to evaluate the effect of collapse direction control devices. It is noted that one of the benefits of the relative system mapping is that the visualized quantities are independent of the location of the origin. Therefore, a consistent evaluation can be performed by using the relative system mapping.

Comparisons of Fig. 11a, b exhibit the effect of the collapse direction control device. From the absolute system mapping, a larger number of elements in the case with device are in the east direction than those in the case without device after the earthquake. The relative system mapping with the device indicates that the collapse direction control accomplishes perfectly: all elements displace to the east direction, whereas most elements move to around west direction in the case without device. In contrast, comparisons of Fig. 11c, d indicates that the device is not effective to control the collapse direction as both structures collapse to almost the same west direction.

As shown above, the absolute system mapping and the relative system mapping have their own advantage of grasping the status of structures after the event of earthquakes.

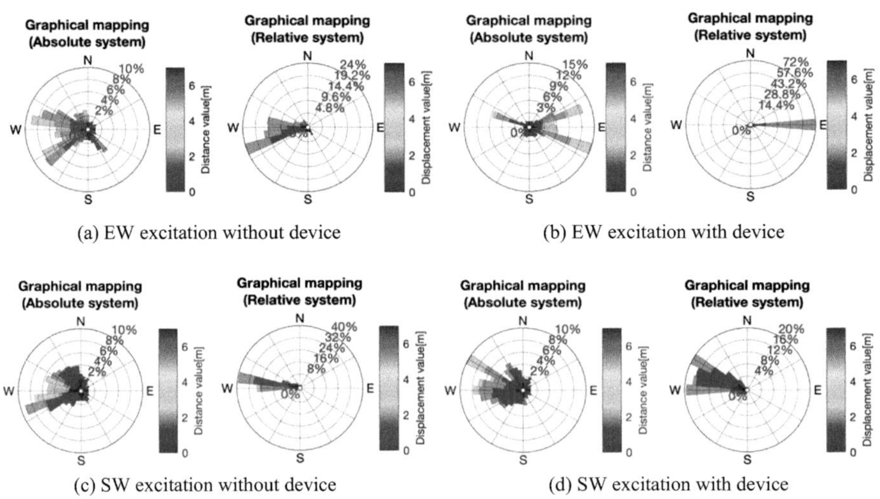

(a) EW excitation without device (b) EW excitation with device

(c) SW excitation without device (d) SW excitation with device

Fig. 11 Graphical mapping (absolute system and relative system)

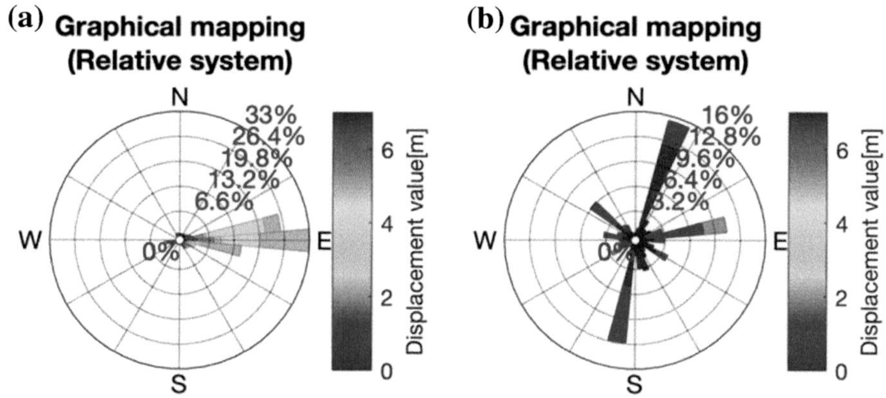

Fig. 12 Graphical mapping using impact matrix (**a** EW excitation; **b** SW excitation)

4 Extended Applications

4.1 Effectiveness of Collapse Control Devices (Impact Matrix)

The effectiveness of the device is better evaluated by taking differences between the post-collapse matrices with and without the device. The equation can be written as follows,

$$\vec{d} = B' - A' \tag{12}$$

where, $\vec{d}$ is Impact Matrix; B' is relative matrix with device; and A' is relative matrix without device.

Figure 12 shows the graphical mappings based on the impact matrices. The left-hand side of the figure shows a dominant effect of the collapse direction control device, where all elements are controlled to move to the desired direction. The right-hand side of the figure implies that, although the moving direction of the elements tends to disperse, the device seems to avoid the movement to the undesired direction (the west direction).

4.2 Anisotropy of Collapse Direction (Anisotropic Matrix)

Grasping the anisotropy of the collapse status in structures subjected to ground motions may illustrate how to spread the elements and pitfall of the device. To visualize the anisotropy, anisotropy matrix is calculated by applying ground motions from different angles. The matrix in each angle is superimposed into a matrix, so

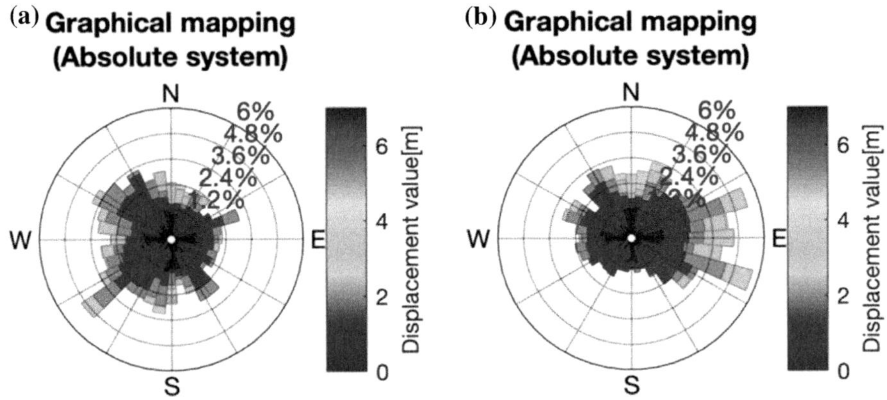

Fig. 13 Graphical mapping using anisotropic matrix (**a** No device; **b** With device)

called "Anisotropic matrix" defined by the following form.

$$P_T = \begin{bmatrix} P_1 \\ P_2 \\ \vdots \\ P_n \end{bmatrix} \tag{13}$$

where, P_T is Anisotropic matrix; $P_1 P_2 P_n$ is the absolute system matrix when subjected to ground motions in different directions.

Figure 12 shows the graphical mappings based on the anisotropic matrix in cases with and without device. It is conceivable that the device works well to control the direction of collapse: it avoids the undesired direction (west) while there is a trend that the structure collapses in the east region. Therefore, this anisotropic mapping is effective to verify the performance of the device against the diversity of the input angle.

5 Conclusions

This study proposes a new graphical mapping method to visualize the direction of collapse, the location of the structural members, and the amount of deformation of target structures. The absolute system mapping and the relative system mapping are a powerful tool to capture the quantitative status of collapse. Moreover, by employing Open Dynamic Engine (ODE), a three-dimensional time-history analysis that can simulate the collapse of structures consisting of rigid bodies, the proposed method is verified for the case of collapse direction control device. This study also proposes extended applications of proposed graphical mappings for detailed verification of the effectiveness of the device. This method is a promising tool to evaluate the state of the collapse of various types of structures.

References

1. Hollnagel E, Woods DD, Leveson N (2006) Resilience engineering—concepts and precepts. CRC Press
2. Hollnagel E, Paries J, Woods DD, Wreathall J (2010) Resilience engineering in practice: a guidebook. Ashgate Publishing
3. Saitoh M, Murono Y, Motoyama H (2015) Resilient structural systems for earthquake disaster mitigation using collapse direction control device. In: Proceedings of the 4th international symposium on engineering, energy and environment, 8–10 November 2015. Thammasat University, Pattaya Campus, Thailand
4. Toyooka A, Murono Y, Saitoh M (2018) Development of the collapse direction control device to improve anti-catastrophe performance of a viaduct

Relationship Between Rate of Corrosion of TMT Bars and Quality of Electrolyte in Electrochemical Corrosion Process

M. N. Bari and M. Mohshin

Abstract A large number of coastal and offshore infrastructures are subjected to deterioration due to continuous effect of physical, chemical and electro-chemical process when exposed to high salinity either directly or indirectly. Reinforcement corrosion has been identified as the predominant mechanism of deterioration of reinforced concrete structures which seriously affects the serviceability, safety and durability of the structures. The researchers are interested to explore the marine effect on concrete as well as reinforcement through the laboratory experiment in artificial marine environment. However, the change in marine environment might affect the corrosion mechanism as well as rate of corrosion. This study focused to establish the relationship of marine environment and rate of corrosion of TMT steel bars. In this regard the change in electrical conductivity (EC), total dissolved solids (TDS), total suspended solids (TSS) and pH value of electrolyte solution as well as rate of corrosion of TMT steel bars were determined at an interval of 15 min during the corrosion process. The experimental results show that the electrical conductivity (EC) and total dissolved solids (TDS) decreased with the progress of corrosion while total suspended solids (TSS) and pH value increased. It indicates that salinity of electrolyte reduced by producing corrosion product and low concentration of salinity hampered the uniformity of corrosion. Therefore, the quality of electrolyte must be maintained to determine the effect of marine environment in electro-chemical corrosion process.

Keywords TMT bars · Marine environment · Electrolyte · Corrosion

M. N. Bari (✉)
Department of Civil Engineering, Rajshahi University of Engineering and Technology, Rajshahi, Bangladesh
e-mail: niamulbari@gmail.com

M. Mohshin
Department of Civil Engineering, Bangladesh Army University of Engineering and Technology, Qadirabad, Bangladesh

S. Arthur et al. (eds.), *Proceedings of the 6th International Conference on Advances in Civil Engineering*, Lecture Notes in Civil Engineering 368,
https://doi.org/10.1007/978-981-99-3826-1_28

1 Introduction

The electrochemical corrosion process is a complicated series of reactions between the electrolyte and metal surfaces. The corrosion process primarily depends on various interdependent parameters and combines mechanical, biological and chemical factors such as electrolyte environment and its compositions, corrosion products and corrosion test period. The electrical conductivity (EC), total dissolved solids (TDS), total suspended solids (TSS), type of ions and pH value are the several dominant factors influencing the rate of corrosion as well as the quality of electrolyte [1]. The electrical conductivity (EC) that comes from the ions in the seawater [1, 2] are greatly affected by the formation of iron oxide (rust) on rebar surfaces [3]. The first solid phase to precipitate from the dissolved ions produced by the corrosion of the steel surfaces is the green rusts (GRs) that are mixed-valence Fe(II,III) layered double hydroxides (LDH) mainly containing Fe(II) cations [4]. The various GRs products, such as GR(Cl^-), GR(SO_4^{2-}) and GR(CO_3^{2-}) are produced depending on the anions present in the electrolyte environment as well as its chemical compositions [5–7]. Among them sulphate green rust is predominant as revealed by the previous results accumulated in the past years [8, 9]. Hence, the understanding of the influence of those parameters as well as their impacts on the variation of electrolyte properties is the key to the optimization of the corrosion and corrosion test period. A focus is made on the formation of the rust compounds and its effect on the quality of electrolyte solution. The majority of the earlier researchers are mainly concerned with corrosion strategies [10], effects of chemical compositions of metal body on the corrosion [11] and therefore the corrosion effects on the mechanical properties of the metal body [12]. Studies were also conducted on the regression analysis of electrode reaction time [13], influence of suspended solids on pitting corrosion of carbon steel [14] and effect of pH on the corrosion and re-passivation behavior within the test solution [2]. Hence, there's a paucity of research on electrochemical corrosion process and consequent behavioral changes of electrolyte properties. Therefore, this study aims to investigate the quality of electrolyte with the corrosion of TMT steel bars in electrochemical corrosion process. The study also presents the numerical correlation between percent corrosion and factors influencing the rate of corrosion.

2 Materials and Methods

2.1 Selection of TMT Steel Bars

Locally produced TMT 500 W high strength steel bars of 10 mm have been investigated in this research work. These bars were collected from the local market.

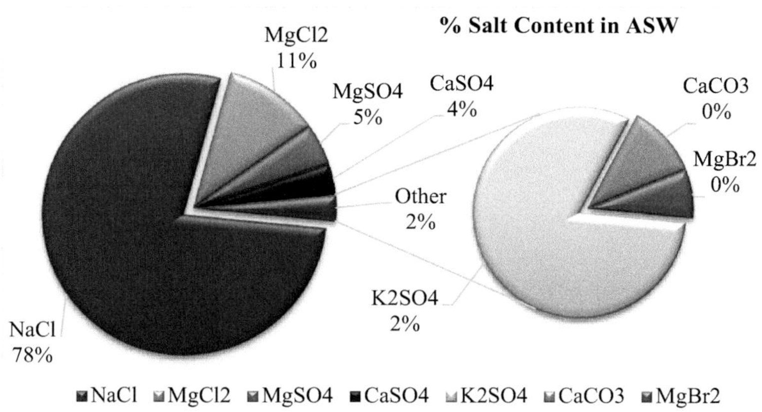

Fig. 1 Composition of artificial sea water (Islam 2016; Mayers 1969)

2.2 Preparation of Electrolyte

In this study artificial sea water (ASW) was used as electrolyte environment. ASW was simulated in the laboratory by mixing the exact amount and ratio of various chemical compounds found in natural seawater with distilled water. The most commonly dissolved ions in seawater are sodium, chloride, magnesium, sulfate and calcium. The chemical composition of artificial sea water is shown in Fig. 1.

2.3 Electrochemical Corrosion Process

In this research work the electrochemical corrosion technique was adopted to accelerate the corrosion process. Impressed current method was used in an integrated system incorporating an adjustable DC power supply with an output of 15 V and 5 amps as shown in Fig. 2.

2.4 Measurement of Water Quality Parameters

Since the electrical conductivity (EC), total dissolved solids (TDS), total suspended solids (TSS), type of ions and pH value are the dominant factors influencing the rate of corrosion as well as the quality of electrolyte as investigated by the previous researchers, these parameters were measured while performing corrosion test to

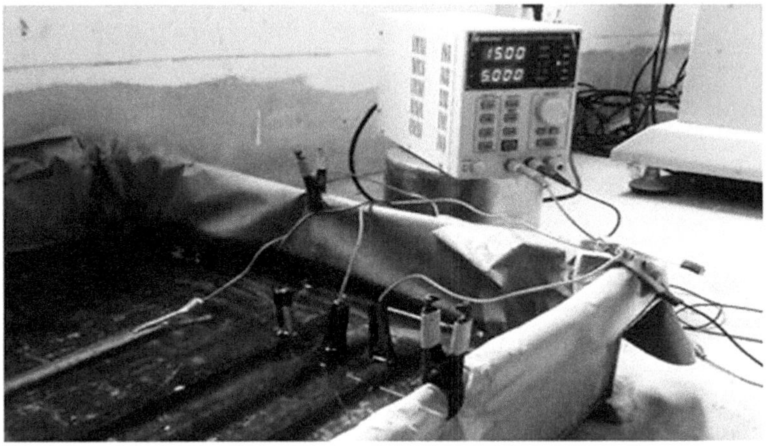

Fig. 2 Electrochemical corrosion process

describe the variation of the quality of seawater as well as to develop numerical correlation between the influencing factors and the rate of corrosion. Electrical conductivity and TDS were measured using calibrated EC meter. A pH meter was also used to determine the pH of the seawater as shown in Fig. 3. The water quality parameters were measured for six samples taking after every fifteen minutes while conducting corrosion test.

Fig. 3 Measurement of pH, EC, TDS and TSS of electrolyte solution

2.5 Experimental Procedure

The experimental steel samples were corroded to six corrosion levels (0.5, 1.0, 1.5, 2.0, 2.5 and 3.0%) at an interval of 15 min. After achieving the desired degree of corrosion, electrolyte samples were collected and tested the quality parameters of the electrolyte solution to find the actual concentration. The experimental program was designed in a marine environment and can be summarized in the following steps:

- Marine environment by preparing artificial sea water (ASW) in laboratory was used as electrolyte.
- Electrochemical corrosion cell was adopted to accelerate the corrosion as well as to achieve the desired degree of corrosion of TMT steel bars.
- The electrolyte quality parameters such as EC, pH, TDS and TSS were measured to develop numerical correlation between the influencing factors and the percent corrosion.

3 Results and Discussions

The variation of electrical conductivity (EC) of electrolyte with percent corrosion is presented in Fig. 4. It was observed from the results that the electrical conductivity decreased linearly with the increase of degree of corrosion. The electrical conductivity comes from the ions in the seawater. The formation and deposition of dissolving and non-dissolving corrosion products on the rebar surfaces hinder both electrode reactions and mass transfer [3]. Since ions are not free to move in solid crystals i.e. iron oxide (rust) accumulates on rebar surfaces, electrical conductivity decreases with the formation of rust on rebar surfaces [15].

Figure 5 represents the variation of pH of electrolyte with percentage of corrosion. The pH value increased with increase of corrosion. This phenomenon increases the

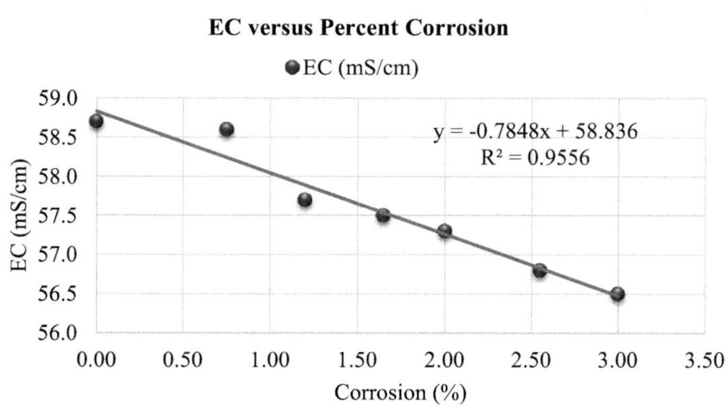

Fig. 4 Variation of EC of electrolyte with percent corrosion

corrosion resistance of reinforcing bars due to the development of passivation layer in moderately alkaline (high pH) solutions [2].

The increased pH levels may be attributed to water electrolysis, which led to the production of hydrogen and (OH) ions. However, the relationship between pH and percentage of corrosion is found to be weak as the coefficient of determination (R^2) is only 87.5% in linear regression model.

The variation of total dissolved solids (TDS) with percent corrosion is illustrated in Fig. 6. It was evident from the experimental results that the TDS value decreased with the increased degree of corrosion. TDS is found to be linearly correlated to percent corrosion in the test solutions. Since the TDS value decreased with the increased experimental time, the EC of the electrolyte also decreased which greatly depends on the TDS.

On the other hand, total suspended solids (TSS) in electrolyte increased with elapse time and increase of corrosion (Fig. 7). During the electrochemical process

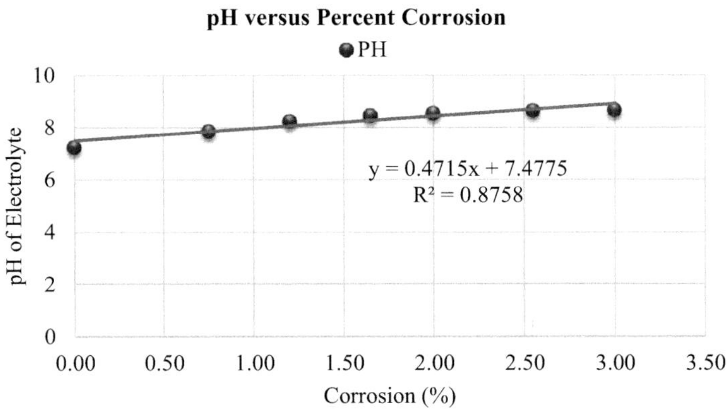

Fig. 5 Variation of pH of electrolyte with percent corrosion

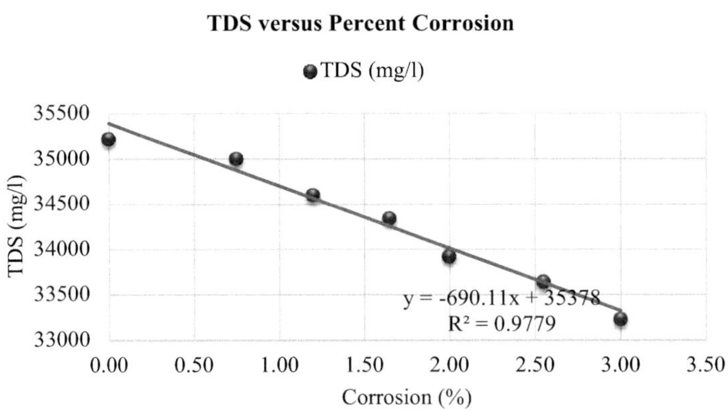

Fig. 6 Variation of TDS of electrolyte with percent corrosion

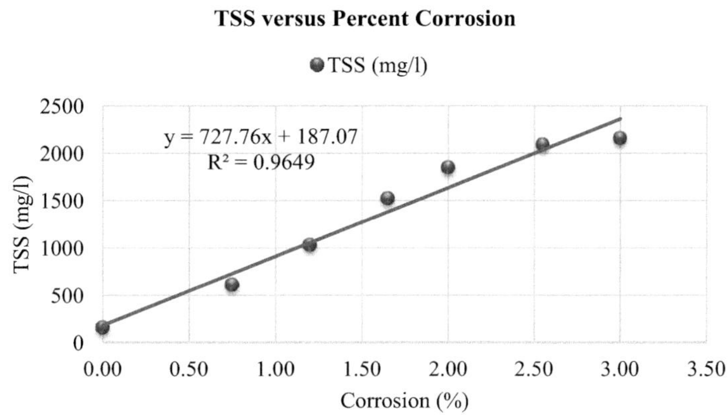

Fig. 7 Variation of TSS of electrolyte with percent corrosion

seawater (electrolyte solution) get darker due to non-dissolving corrosion products formed on the surface of TMT steel bars when exposed to corrosive environment. The deposition of suspended solids was increased with the increase of degree of corrosion. Since the dissolved iron ions from pitting corrosion act as a coagulant, the suspended solids coagulated and deposited [16]. The suspended solids accelerate pit formation and enlarge tubercles by corrosion products, hence TSS increased with the formation of rust on rebar surfaces [17]. TSS also increased with the formation of scaling due to the less soluble calcium and magnesium salts (carbonates or sulphates) presents in the electrolyte solution [18].

The variation of the quality parameters of electrolyte with experimental time is presented in Fig. 8. The experimental results revealed that both the TSS and pH value of electrolyte increased with the increase of experimental duration. The pitting corrosion increases due to the electrochemical reaction which in turn increases the corrosion products as well as the TSS values in electrolyte environment [19]. The electrolysis of electrolyte solution also results in an increase of pH at the cathode due to reduction of hydrogen and generation of hydroxide ions that produces an alkaline (high pH) environment [20]. It was also observed from the results that the TDS and EC value of electrolyte decreased with the increase of experimental time. The chemical compositions present in electrolyte break down due to the electrochemical corrosion process and the cations (M^+) separates from the anions (A^-). The negatively charged chloride ions are attracted to the positive side of the electrical charge where it bonds with oxygen and hydrogen from water and converted to hypo-chlorous acid molecule [21]. The positively charged sodium, magnesium and calcium cations are attracted to the negative charge, where it also bonds with oxygen and hydrogen and is electrochemically converted to hydroxide of cations (NaOH, KOH, $Mg(OH)_2$ etc.) [22]. On the other hand, the suspended solids in electrolyte environment coagulate by the dissolved iron ions resulting from pitting corrosion [16]. Thus electrocoagulation process continues with the increasing experimental time and increases the removal efficiency of TDS and turbidity of electrolyte environment. The EC that comes from

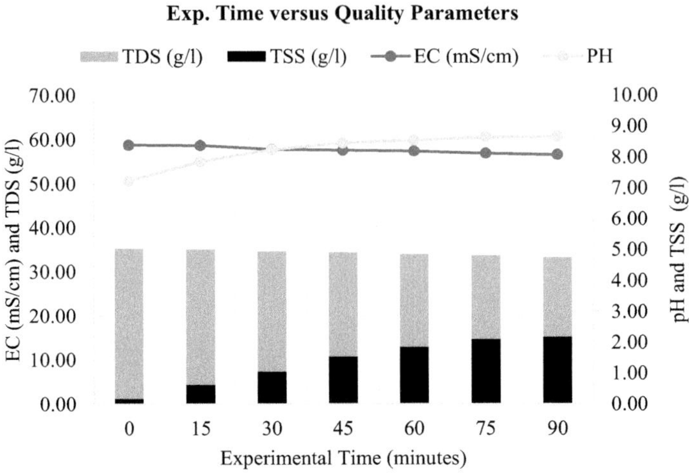

Fig. 8 Variation of the quality parameters of electrolyte with experimental time

the ions present in the electrolyte is also reduced due to the formation and deposition of dissolving and non-dissolving corrosion products [3, 23].

It was also observed from the results that the EC and TDS values decreased by 0.0252 ms/cm-min and 22.31 mg/l-min respectively while the TSS and pH values increased by 23.29 mg/l-min and 0.015 per min of corrosion process. The factors were found to be changed significantly when the experimental time exceeds 15 min.

4 Conclusion

The experimental results revealed that the corrosion reaction as well as the formation of rust (corrosion products) on rebar surfaces greatly affects the electrical conductivity (EC), total dissolved solids (TDS), total suspended solids (TSS) and pH value of electrolyte solution (seawater) which are the important factors influencing the rate of corrosion. These factors also indicate the quality of electrolyte solution as well as the salinity level. The alkalinity level increased highly and the electrolyte solution got darker due to non-dissolving corrosion products formed on the surfaces of TMT steel bar. From the obtained results it can be mentioned that the electrolyte solution should be changed at an interval of 15 min during the corrosion process to maintain the actual concentration of the salinity of the electrolyte solution in order to find out the actual effect of the corrosive environment. Finally the developed numerical correlation between influencing factors and percent corrosion may be used to predict the quality of the electrolyte environment at desired degree of corrosion.

Acknowledgements The author would like to express his gratitude to Department of Civil Engineering, RUET for providing laboratory facilities and technical supports to conduct this research work.

References

1. Assiry AA, Gaily MH, Alsamee M, Sarifudin A (2015) Electrical conductivity of seawater during ohmic heating. Desalination. https://doi.org/10.1016/j.desal.2010.05.015
2. Zhang Y, Yan T, Fan L, Liu Z, Song L, Li X (2021) Effect of pH on the corrosion and repassivation behavior of TA2 in simulated seawater. Materials 14:6764. https://doi.org/10.3390/ma14226764
3. Trethewey KR (1998) Corrosion for students of science and engineering. Longman Scientific and Technical, London, England
4. Refait P, Grolleau AM, Jeannin M, Rémazeilles C, Sabot R (2020) Corrosion of carbon steel in marine environments: role of the corrosion product layer. Corros Mater Degrad 1:198–218. https://doi.org/10.3390/cmd1010010
5. Pineau S, Sabot R, Quillet L, Jeannin M, Caplat CH, Dupont-Morral I, Refait PH (2008) Formation of the Fe(II-III) hydroxysulphate green rust during marine corrosion of steel associated to molecular detection of dissimilatory sulphite-reductase. Corros Sci 50:1099–1111
6. Lanneluc I, Langumier M, Sabot R, Jeannin M, Refait P, Sablé S (2015) On the bacterial communities associated with the corrosion product layer during the early stages of marine corrosion of carbon steel. Int Biodeter Biodegrad 99:55–65
7. Refait P, Nguyen DD, Jeannin M, Sablé S, Langumier M, Sabot R (2011) Electrochemical formation of green rusts in de-aerated seawater-like solutions. Electrochim Acta 56:6481–6488
8. Refait P, Grolleau A-M, Jeannin M, François E, Sabot R (2016) Localized corrosion of carbon steel in marine media: galvanic coupling and heterogeneity of the corrosion product layer. Corros Sci 111:583–595
9. Refait P, Jeannin M, François E, Sabot R, Grolleau AM (2019) Galvanic corrosion in marine environments: effects associated with the inversion of polarity of Zn/carbon steel couples. Mater Corros 70:950–961
10. Apostolopoulos CA, Papadopoulos MP, Pantelakis SG (2006) Tensile behavior of corroded reinforcing steel bars BSt 500s. Constr Build Mater 20(9):782–789
11. Durgam SK, Das G, Singh AK (2017) Corrosion effects of Cr and Ni in thermo-mechanical treated steel bar in marine environments. Int Res J Eng Technol (IRJET) 4(7):870–875
12. Islam MA (2015) Corrosion behaviors of high strength TMT steel bars for reinforcing cement concrete structures. Procedia Eng 125:623–630
13. Rugebregt MJ, Nurhat IS (2020) Preliminary study of ocean acidification: relationship of pH, temperature, and salinity in Ohoililir, Southeast Maluku. In: IOP conference series: earth and environmental science.https://doi.org/10.1088/1755-1315/618/1/012004
14. Uchida T (1993) Influence of suspended solids on pitting corrosion of carbon steel in cooling water. J Soc Mater Sci Jpn 42(4):912–916
15. Koleva DA, De Wit JHW, Van Breugel K, Lodhi ZF, Ye G (2007) Investigation of corrosion and cathodic protection in reinforced concrete: II. Properties of steel surface layers. J Electrochem Soc 154(5):C261
16. Garcia-Segura S, Eiband MMS, de Melo JV, Martínez-Huitle CA (2017) Electrocoagulation and advanced electrocoagulation processes: a general review about the fundamentals, emerging applications and its association with other technologies. J Electroanal Chem 801:267–299
17. Lister DH, Cook WG (2014) Nuclear plant materials and corrosion. The essential CANDU—a textbook on the CANDU nuclear power plant technology, 14

18. Salman MA, Safar M, Al-Nuwaibit G (2015) The effect of magnetic treatment on retarding scaling deposition. Tojsat 5(3):62–77
19. Ewis D, Talkhan AGG, Benamor A, Qiblawey H, Nasser M, Ba-Abbad MM, El-Naas M (2020) Corrosion behavior of API-X120 carbon steel alloy in a GTL FT process water environment at low COD concentration. Metals 10(6):707
20. Yuzer BURAK, Selcuk HÜSEYİN, Chehade G, Demir ME, Dincer I (2020) Evaluation of hydrogen production via electrolysis with ion exchange membranes. Energy 190:116420
21. Eaves DR (2013) A study of novel filiform corrosion phenomena on hot dip organically coated Zn-Al-Mg steel. Swansea University, United Kingdom
22. Speight JG (2017) Lange's handbook of chemistry. McGraw-Hill Education
23. Koushik BG, Van den Steen N, Terryn H, Van Ingelgem Y (2021) Investigation of the importance of heat transfer during thin electrolyte formation in atmospheric corrosion using a novel experimental approach. Corros Sci 189:109542

Performance of Rice Husk Ash Concrete in Acidic Environment

Md. S. Islam, Md. A. Hasan, Md. M. Islam, and S. S. Das

Abstract As a suitable supplementary cementitious material with high pozzolan reactivity, low energy requirements, and minimal greenhouse gas emissions during production, rice husk ash (RHA) concrete has gained recognition. In relation to the silica dissolution from RHA, RHA's pozzolanic reactivity is affected by its amorphous silica content, fineness, mix proportions, alkaline media availability and temperature. This paper discusses mainly the strength performance of RA blended concrete in plan water (PW) and acidic environment. Five different cement replacement levels by RHA (i.e., 0%, 10%, 15%, 20% and 25%) were selected to make blended concrete specimens. A total of 500 nos 100 mm cubical specimens were cast for M28 grade concrete. All the specimens were precured for 7 days in PW and then exposed to PW and acidic environment (5% H_2SO_4 solution) over the period of 14, 28, 60 and 180 days. Different tests including visual examination, compressive strength, split tensile strength, porosity, alkalinity (PH levels) at different depth levels of the specimens were conducted to observe the strength properties and alkaline conditions of the concrete. RHA concrete showed significant resistance against strength deterioration. Among the various RHA concretes, 10–15% cement replacement level is found effective from strength and durability point of view.

Keywords Compressive strength · Split tensile strength · Acidic environment · Rice husk ash (RHA) · Alkalinity

---

Md. S. Islam · S. S. Das
Dept. of Civil Engineering, Stamford University Bangladesh, Dhaka, Bangladesh
e-mail: msislam@cuet.ac.bd

Md. A. Hasan (✉)
Dept. of Civil Engineering, Southern University Bangladesh, Chattogram, Bangladesh
e-mail: hasancuet90@gmail.com

Md. M. Islam
Department of Civil Engineering, CUET, Chattogram, Bangladesh
e-mail: msislam@cuet.ac.bd

S. Arthur et al. (eds.), *Proceedings of the 6th International Conference on Advances in Civil Engineering*, Lecture Notes in Civil Engineering 368,
https://doi.org/10.1007/978-981-99-3826-1_29

1 Introduction

With the increasing global development as well as the ever-raising population, the demand for constructions of infrastructures and buildings is increasing tremendously. Existing construction system require natural raw materials, consume high energy and produce waste during materials processing and construction. The constructions of buildings and infrastructures using cement concrete has become the most popular due to low cost, more durability, high mechanical strength and most convenient and easier to apply [1]. However, the production of cement requires costly limestone as raw materials, consume high energy and emits CO_2 [2]. The manufacturing of regular Portland cement contributes between 5% and 8% of the globe's CO_2 emissions (OPC) [3]. Researchers are looking for alternate sources of cement made from agricultural and mill byproducts due to concerns about the environment, energy use, and the economy. This leads to sustainable, green and environment friendly construction [4]. The key characteristics of supplementary cementitious materials (SCM), like RHA, are good pozzalonic activities and suitable bond-ability with aggregates [5]. RHA has showed that it is a sustainable and eco-friendly SCM in concrete [6]. The United States Department of Agriculture (USDA) estimates that 499.31 million metric tons of rice were produced worldwide in 2019–2020 (World agriculture production, 2020). When rice is milled, 0.28 kg of rice husk is extracted from each kg of rice. As a result, a significant amount of waste is generated annually and used as fuel in many businesses to generate heat energy, such as incineration and burning units. Following complete burning of the rice husk, 20–25% RHA is formed [7]. Unfortunately, most of RHA are dumped to open landfills although a very limited amount of RHA is used as fertilizing agent. RHA can be produced by the burning of rice husk either through open field burning or through incineration conditions in which temperature and duration are controlled. Open field burning is not encouraged due to pollution problems and also for producing poor quality of RHA. It results in a highly crystalline form of ash having high carbon content that adversely affect concrete performance for its lower reactivity [8]. The RHA in the amorphous form of silica which has the potential to be used for concrete is produced through controlled incineration conditions. According to [9] for burning of RH at temperature not exceeding 700 °C, the nature of RHA silica is predominantly amorphous and thus reactive under alkaline conditions like the ones created in the hydrated cement paste. The highest amorphous silica could be obtained by burning the rice husk at the temperature range of 500–700 °C [8]. Depending on the mix proportions, aggregate and cement characteristics, w/c ratio, curing time, and RHA replacement level, the compressive strength of RHA concrete is proportional to its density [10, 11]. Due to their contributions to pozzolanic activity and binder hydration, RHA content and fineness both play a significant role in the strength development of RHA blended concrete. The recommended mean particle size of RHA ranges from 3.6 to 9 μm [12]. The optimal RHA content varies according to RHA characteristics and binder type. For RHA, the ideal cement replacement level is thought to be around 20–30% [13]. Strength increment of about 25–30% upto 56 days were observed upto

15% cement replacement. Similar trend in the compressive and split tensile strength development of concrete with RHA is reported in the existing literature and the typical range being around 15–20%. Durability of concrete in an adverse environment is an important issue to the researchers regarding the effective performance of structural concrete throughout the life span of the structures. It is caused by the penetration and absorption of liquids, ions, and gases into concrete from the surrounding environment. As a result of both chemical and physical loading, the concrete in the core structure deteriorates and degrades. A few literatures are available regarding the performance of RHA blended concrete in aggressive environment including the acidic one. Concrete containing supplementary cementitious materials (SCM's) is reported to have lower permeability that may resist the penetration of objectionable chemical ions into the hardened concrete mass thereby increase its longevity. The paper aims at to observe the performance RHA concrete in acidic environment (H_2SO_4 solution) by conducting some relevant tests.

2 Materials and Methods

2.1 Experimental Program

The experimentations program included the collection of concrete ingredient materials, casting of required number of test specimens, creation of acidic environment, curing and testing of specimen etc. which are stated as follows.

2.1.1 Materials

Cement

Ordinary Portland Cement (OPC) confirming to ASTM C150 was used in concrete mix. Table 1 shows the properties of cement used in the study.

Rice Husk As

The rice husk ash (RHA) was collected from a rice milling industry. The RHA, the residual materials were grounded in loss angel machine to have desired fineness. Table 1 shows the properties of RHA used in the study.

Table 1: Physical and chemical properties of cement and RHA

Relevant properties	OPC	RHA	Relevant properties	OPC	RHA
Physical	3.10	2.06	Chemical (%)		
Fineness (Blaine, cm^2/gm)	4000	6000	SiO_2	21.20	87.2
Setting times(mins)			Al_2O_3	4.65	0.15
Initial	145		Fe_2O_3	2.27	0.16
Final	275		CaO	63.55	0.58
Compressive strength (MPa)			MgO	3.27	0.36
3 days	15.3	–	MnO	–	–
7 day	24.8	–	Na_2O	0.11	1.11
28 day	35.2	–	K_2O	1.04	3.60
			SO_3	2.19	0.32
			LOI	2.30	6.58

Aggregate

Sylhet sand having FM of 2.58 and specific gravity (SG) of 2.60 was used as fine aggregate. Coarse aggregate comprises of well graded crushed stone chips of 20 mm nominal size and having SG of 2.78 was used in making concrete specimens.

Acid

Laboratory grade sulphuric acid (H_2SO_4) was used in creating artificial environment in this program.

2.1.2 Mix Proportions

Concrete mix proportions for a particular grade of concrete (M28) were found after several trial following ACI mix design procedures. Concrete mixes having various cement replacement level by RHA (0%, 10%, 15%, 20% and 25% by weight) were used to cast the test specimens keeping the proportions of fine and coarse aggregate fixed. 0% replacement indicate the control specimens having 100% OPC concrete. Table 2 shows the details of mix design.

2.1.3 Casting of Test Specimens

OPC, rice husk ash and required amount of fine aggregates were mixed thoroughly in dry condition for 1 min water was then added and the materials were mixed for 3 min. 100 mm cubical size molds were made ready to cast the specimens and compaction

Table 2 M_{28} RHA blended concrete mix proportion

Mix no	Cement: RHA	Quantities in kg/m^3					
		Cement	RHA	F.A	C.A	Water	w/c
01	100:0	435	0.0	545	1150	218	0.5
02	90:10	391.5	43.5	545	1150	218	
03	85:15	369.7	65.5	545	1150	218	
04	80:20	348.0	87.0	545	1150	218	
05	75:25	326.2	108.8	545	1150	218	

was done in two equal layers. 16 mm diameter and 0.45 m long bar was used for compaction with 25 blows in each layer.

The cast specimens were demolded after 24 h of casting and precured in plain water (PW) for 7 days before immersion in acidic environment. Total 500 nos specimens were cast of which 250 nos were immersed in PW and the remaining 250 nos for acidic environment.

2.1.4 Exposure Environment

Two types of curing environment i.e., PW environment and Acidic environment was used for the period of 14, 28, 60, 90 and 180 days was used to cure the specimens. Acidic environment was created by mixing 5% H_2SO_4 in PW. Special curing tanks were used for acidic environments in order to avoid accidents.

2.1.5 Test Conducted

Various test including visual examinations and strength tests (compressive strength and split tensile strength), porosity, pH value was carried at different curing period i.e., 14, 28, 60, 90 and 180 days in PW as well as in acidic environment. The identical specimens from each mix combination were tested and average value of test results were considered as the representative data. The compressive and split tensile strength tests were conducted according to BS 1881-166:1983 and IS 5816:1999, respectively. For strength, all the specimens were tested at SSD condition and loads were applied to other than casting faces and at a particular rate. For porosity test, the weight of the same specimen was taken in air at SSD condition, oven dry condition and water under submerged condition. To know the alkalinity level, concrete powder was collected from different depth levels (1 mm, 15 mm & 25 mm) of the specimens by masonry drill. Drilled powder was made finer after grounding to pass through No. 200 sieve and preserved in individual plastic bag in sealed condition to avoid carbonation. The powdered sample was mixed with distilled water with periodic stirring, then filtered with filter paper and finally pH was determined by using pH meter.

(a) PW Environment (b) Acidic Environment

Fig. 1 Concrete specimens exposed to PW and Acidic environment. **a** PW environment, **b** Acidic environment

3 Results and Discussions

3.1 Visual Examinations

He specimens were subjected to visual examination after taking out of acidic environment. No sign for crack or surface damage was detected. However, the specimens showed change in color from off white to brownish as shown in Fig. 1. The color change for RHA concrete was reported as minimum. The specimen in PW showed practically no change in color. In acidic environment, the change in color at exposed surfaces of the specimen may be primarily due to ionic effects. The texture insides the specimens were found in original color.

3.2 Compressive Strength

As per experimental program, compressive strength test of both control and RHA blended concrete specimens exposed to PW and H_2SO_4 acid environment was conducted and the test results are shown in Fig. 2. From Fig. 2, it is observed that for control concrete i.e., 100% OPC concrete specimens in PW the strength increases with curing ages and rate of gain in strength is higher in early ages followed by slower strength gain at later ages. In acidic environment, the strength increases at early ages i.e., upto 60 days and then decreases. The decrease in strength at later period is due to formation of some new compounds including ettringite as a result of reactions of hydrated cement products with acidic ions.

With acidic ions. These products being expansive and leachable in nature cause microcracking that results in strength deterioration. For RHA concrete specimens' strength gaining at early ages is observed to be lower but significant at later ages.

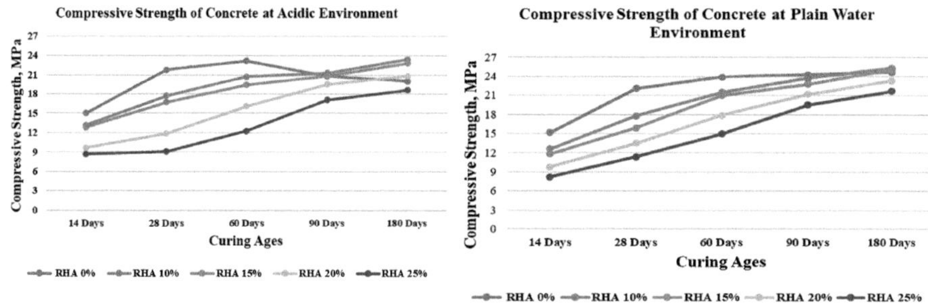

Fig. 2 Concrete specimens' compressive strength in PW and acidic environments

The formation of secondary gel due to reaction of RHA with cement component is reported to be responsible for strength development at later ages. Upto 15% cement replacement level, the strength development of RHA concrete specimens is found significant, even more than control concrete specimens. In H_2SO_4 environment, the strength of RHA concrete specimens also decreases at later ages but at slower rate as compared to control concrete. However, for specimen with higher RHA content, the strength loss is observed are observed to higher at later ages. The overall strength data indicate that concrete specimens upto 15% RHA content attain 2–3% higher strength than control specimens. In acidic environment, after 180 days of curing, the strength loss for control specimens is 19%. Whereas specimen upto 15% RHA content showed 5–8 strength loss as compared to PW cured control specimens. It clearly indicates the better strength performance of RHA concrete in acidic environment.

3.3 Split Tensile Strength

Figure 3 illustrate the split tensile strength development of control and RHA concrete specimens exposed to PW and acidic environment for different curing ages. It is seen that the split tensile strength development of control and RHA concrete specimens follows the similar trend as that of compressive strength both in PW as well as in acidic environment. The strength development of RHA concrete is reported to be lower at early ages and enhanced at later ages.

Also, in case of split tensile strength cement replacement level upto 15% showed significant strength development in PW and acidic environment. From the relevant strength results it is seen that RHA concrete specimens upto 15% RHA content attain slightly higher strength than that of control concrete in PW at 180 days curing period. In acidic environment control concrete specimens show around 25%. Strength loss as compared to PW cured concrete specimens. On the other hand, 10% and 15% RHA blended concrete specimens exhibit 8–12% reduction in strength as compared to PW cured strength.

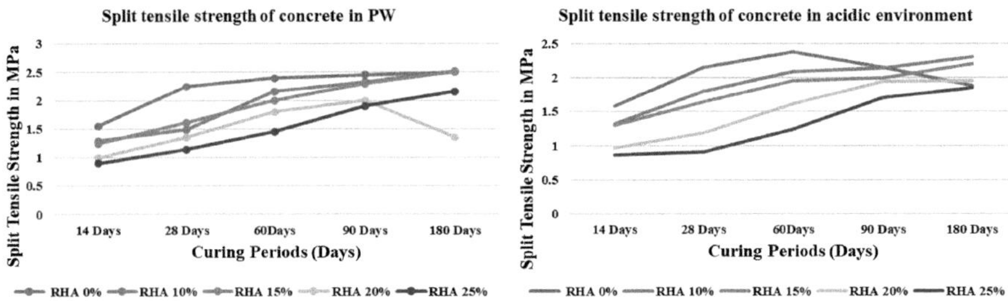

Fig. 3 Concrete specimens' split tensile strength in a PW and acidic environment

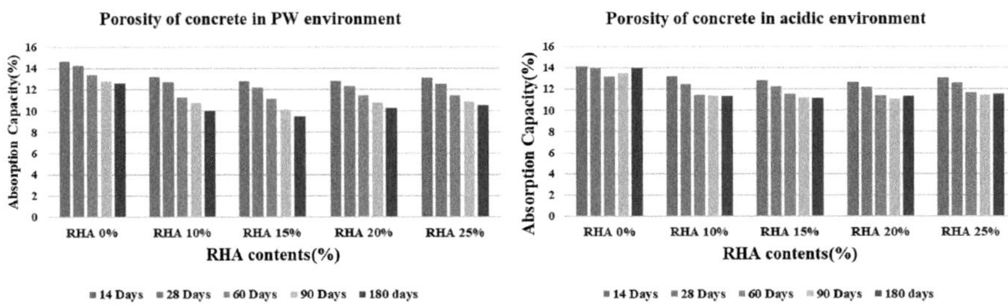

Fig. 4 Porosity of concrete specimens in PW and acidic environment

3.4 *Porosity*

As per program, porosity test of control and RHA concrete specimens exposed to PW and acidic environments were carried out and the test results are shown in Fig. 4. In PW environment, both control and RHA blended concrete specimens indicate the gradual reduction in porosity with time. The porosity of RHA concrete specimens is observed to be less than the control concrete. Among the RHA concrete mixes, 10% and 15% RHA concrete specimens showed lower porosity. In acidic environment, the porosity of control specimens decreases upto 60 days and then increase. This may due to microcracks as a result of formation of expansive compounds and also for leaching action. For RHA blended specimens, 10–15% RHA concrete specimens show gradual reduction in porosity with time whereas 20% and 25% RHA concrete specimens show reduction upto 90 days and then again increase at 180 days. Overall study indicates the better performance of RHA concrete both in PW and acidic environment.

3.5 *pH Values*

To assess the alkaline condition of the hardened concrete mix, the pH values of powder concrete samples taken at different depth level of control and RHA concrete

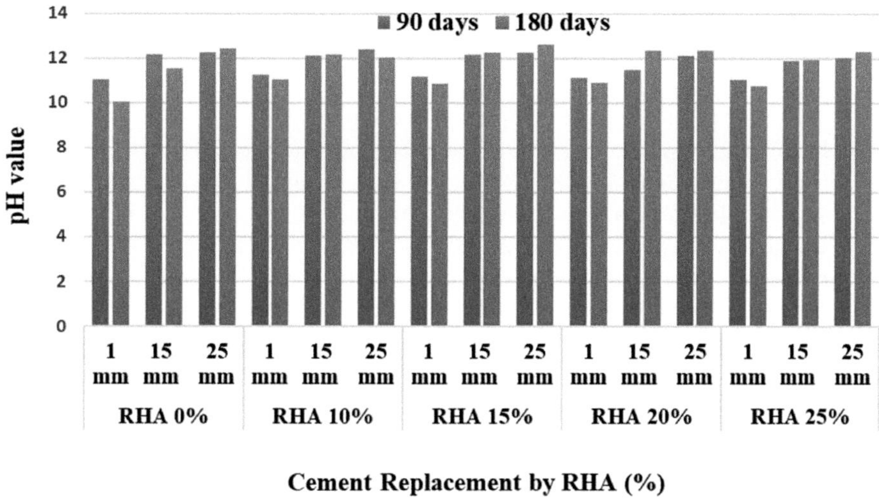

Fig. 5 pH value of concrete specimens at different depth levels in acidic environment

specimens exposed to acidic environments are measured and reported in Fig. 5 From practical point of view, pH of concrete was measured at 3- and 6-month exposure only. The pH value of hardened concrete generally varies in the range of 12–13. However, the observed pH values at different depth levels for control and RHA blended concrete specimens cured in acidic environment are observed to vary from 10.05 to 12.60. At surface levels, the measured pH values were minimum and it increases with depth levels for both control and RHA blended concrete specimens in acidic environment although the change in pH values are marginal. Again, at any depth level, pH values for 10% and 15% RHA concrete specimens are relatively higher than that of control specimens. It may be due to impermeable characteristics of RHA concrete that resist the penetration of acid ions inside the concrete. However, all the measured pH values are well above the limiting value ($\geq$9.5) for initiation of rebar corrosion in concrete. Thus, it is seen that RHA concrete have more capability to preserve alkalinity in concrete pore as compared to identical control (100% OPC) concrete.

4 Conclusion

The stated study presents the performance of RHA blended concretes (0%, 10%, 15%, 20% and 25% RHA) exposed to PW and acidic environment over a period of 6 months. It is a part of durability study although the time frame is too short to predict the various aspects of concrete in an aggressive environment like acidic one. However, based on the limited number of variables studied, the following conclusions can be drawn.

(i) Both control and RHA concretes showed change in color off white to brownish due to acidic reaction.
(ii) The strength gaining of RHA concrete is observed to be lower at early ages and faster at later ages as compared to control concrete.
(iii) In PW environment, upto 15% cement replacement level, RHA concrete specimens attained 2–3% higher compressive strength and slightly around 1% higher split tensile strength than control specimen.
(iv) In acidic environment, both control and RHA concrete losses strength after 6 months. The maximum losses in compressive and split tensile strength were reported around 19% and 25%, respectively for control concrete. The corresponding strength losses for 10–15% RHA concrete varied from 5% to 8% and 8% to 12%, respectively.
(v) In PW environment, the porosity of RHA concrete is observed to be less than the control concrete at any curing ages. In acidic environment, the porosity of control concrete decreases upto 60 days and then increases. 10% and 15% RHA concrete showed gradual reduction in porosity with time.
(vi) RHA concrete exhibited higher PH values than control concrete at any depth level indicating higher capability of RHA concrete in preserving alkaline condition in concrete matrix.

RHA concrete with 10–15% cement replacement levels is found optimum regarding strength, durability and economic point of view.

References

1. Nuaklonga P, Jongvivatsakula P, Pothisirib, T, Sata, V, Chindaprasirt, P (2020) Influence of rice husk ash on mechanical properties and fire resistance of recycled aggregate high-calcium fly ash geopolymer concrete. J Clean Prod 252
2. Rattanachu P, Toolkasikorn P, Tangchirapat W, Chindaprasist P, Jaturpitakkul C (2020) Performance of recycled aggregate concrete with rice husk ash as cement binder. Cem Concr Compos 108
3. Khan MI, Abbas YM, Fares G (2017) Review of high and ultrahigh performance cementitious composites incorporating various combinations of fibers and ultrafibers. J King Saud Univ Eng Sci 29:339–347
4. Siddika A, Al Mamun MA, Alyousef R, Mohammadhosseini H (2021) State-of-the-art-review on rice husk ash: A supplementary cementitious material in concrete. J King Saud Univ Eng Sci 29:294–307
5. Al-kutti W, Saiful Islam ABM, Nasir M (2019) Potential use of data palm ash in cement-based materials. J King Saud Univ Eng Sci 31:26–31
6. Medah MS, Praveenkumar TR, Vijayalakshmi MM, Manigandan S, Arunachalam R (2020) Mechanical and infractural characteristics of rice husk ash and Al_2O_3 nanoparticles modified cement concrete. Concr Build Mater 255:119358
7. Siddika A, Mamun MAA, Arman YHM, Alabduljabbar H (2019) Properties and utilization of waste tire rubber in concrete: A review. Concr Build Mater 224:711–731
8. Hwang CL, Chandra S (2016) The use of rice husk ash in concrete. Book-by Satish-Chandra, 220504

9. Ahmed AE, Adam F (2007) Indian incorporated silica from rice husk and its catalytic activity. Microporous Mesoporous Mater 103(1–3):284–295
10. Wille K, Naaman AE, EI-Tawil, S., Parra- Mantesinos, G, J., (2012) Ultra-high-performance concrete and fiber reinforced concrete: Achieving strength and durability without heat curing. Mater Struct 45:309–324
11. Yang IH, Joh C, Kum BS (2010) Structural behavior of ultra-high-performance concrete been subjected to bending. Eng. Structure. 32:3478–3487
12. Mosaberpanah MA, Umar SA (2019) Using rice husk as supplement to cementitious materials on performance of ultra-high-performance concrete: review. Mater Today Sustain
13. Jamil M, kaish, A.B. M.A., Rahman, S.N., Zain, M.F.M., (2013) Pozzolanic contribution of rice husk in cementitious system. Constr Build Mater 47:558–593

Shape Effects on Shear Behavior of Superdisk Systems Studied by Level Set Discrete Element Method

Haoran Jiang, Reid Kawamoto, and Takashi Matsushima

Abstract This paper investigates the shear behavior of superdisk particles using the 2D level set discrete element method (LS-DEM). Specifically, dense samples of various superdisk shapes (blockiness $n \in [2, 4]$) were prepared through isotropic compression tests, and the prepared samples were then subjected to simple shear tests under different shearing speeds. We find the following. Firstly, it is observed that the final compression state is strongly influenced by the particle shapes. Larger surface blockiness leads to more ordered packing structures, resulting in higher solid fractions and mean coordination numbers in larger values of n. Secondly, the quasi-static shear response shows significant differences for different values of n. The stress ratios q/p and solid fractions ϕ in the peak and critical state increase monotonically with all blockiness values explored. Finally, the study tests the constitutive rheological model of steady flows, and the findings indicate that $\mu(I)$ and $\phi(I)$ relations can be well captured by power-law functions for all samples. Furthermore, the fitted exponents are found to be nearly independent of the shape of the samples. The results of this study provide deeper insights into the particle shape effects on granular materials.

Keywords Non-circular particle · Level set discrete element method · Granular shear behavior · Rheology

H. Jiang
Doctoral Program in Engineering Mechanics and Energy, University of Tsukuba, Tsukuba, Japan

R. Kawamoto
Independent Scholar, Tsukuba, Japan

T. Matsushima (✉)
Department of Engineering Mechanics and Energy, University of Tsukuba, Tsukuba, Japan
e-mail: tmatsu@kz.tsukuba.ac.jp

S. Arthur et al. (eds.), *Proceedings of the 6th International Conference on Advances in Civil Engineering*, Lecture Notes in Civil Engineering 368,
https://doi.org/10.1007/978-981-99-3826-1_30

1 Introduction

Granular materials are widely encountered in industrial and engineering applications, and to gain a better understanding of their complex behavior, numerical modeling using the discrete element method (DEM) [1] has been extensively employed. However, earlier studies [2, 3] have primarily limited to circular or spherical particles due to the difficulties associated with contact detection and the high computational costs of modeling non-spherical particles. In recent times, there has been a growing interest in exploring the impact of particle shape on granular materials, owing to its significant influence on the behavior of such materials. The incorporation of the influence of particle shape into theories and models that are based on circular particles [2, 3], as well as the effects of non-circular shapes on granular flows [4–13], have become topics of interest in recent studies. Moreover, there is a need to develop constitutive laws that can describe the behavior of dense flows of non-circular particles. These issues have been the subject of recent investigations and have been partially addressed in mentioned studies. While most of these studies have predominantly examined ellipses [4, 5], polygons [6, 8], and clump particles [9, 10, 13], the present study aims to investigate the properties of superdisk particles. Although superdisk particles are part of the superellipse family, they have received much less attention compared to their counterparts, the elliptic particles. This can be attributed to the more challenging contact detection involved in modeling superdisks, for which the continuous function representation (CFR) method using an iterative Newton–Raphson approach is often utilized [14, 15]. As the alternate, the discrete function representation (DFR) method [16] is considered more numerically stable for superdisk-related issues [14]. The present study utilizes the level set discrete element method (LS-DEM) [17, 18], which is a type of DFR method, to model the granular systems consisting of superdisks, and aims to fill the gap in the literature by investigating the shear responses of superdisk particles. LS-DEM has been successfully employed in previous studies to predict the behavior of sands [17, 19], understand the fabric effect on shear wave velocity in soils [20], and reproduce particle breakage process [21, 22]. The LS-DEM method incorporates two fundamental components to represent the particle morphology, namely the level set function [23] and the discrete surface nodes, which are described in Sec. 2. By applying this method, all dense packs prepared by isotropic compression are sheared up to a large deformation under different shearing speeds, as outlined in Sec. 3. The obtained results are presented and discussed in Sec. 4. Finally, conclusions and remarks on the displayed results are provided in Sec. 5.

2 Level Set Discrete Element Method

2.1 *Superdisk*

The mathematical formulation of a superdisk surface in the Cartesian coordinate system is

$$|x/a|^n + |y/a|^n = 1, \tag{1}$$

where the positive number n measures the edge sharpness of a given superdisk, and is commonly referred to as blockiness [15]. In this study, we focus on the convex shapes with $n \in [2, 4]$, where $n = 2$ restores a disk and larger values of n correspond to increasingly square-like shapes (Fig. 1). The area of a superdisk is computed via Green formula as

$$A = \frac{4a^2}{n} B\left(\frac{1}{n}, \frac{1}{n} + 1\right), \tag{2}$$

where the term $B(x, y)$ is a beta function related to the gamma function $\Gamma(x, y)$

$$B(x, y) = 2 \int_0^{\frac{\pi}{2}} \sin^{2x-1}\theta \cos^{2y-1}\theta d\theta = \frac{\Gamma(x)\Gamma(y)}{\Gamma(x + y)} \tag{3}$$

Furthermore, the moment of inertia is

$$I_z = \frac{2a^4}{n} B\left(\frac{3}{n}, \frac{1}{n}\right) \tag{4}$$

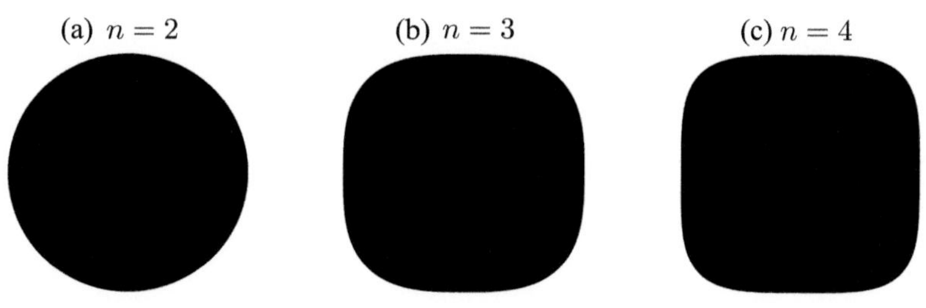

Fig. 1 Superdisk shapes with different surface blockiness. (a)$n = 2$, (b)$n = 3$, and (c)$n = 4$

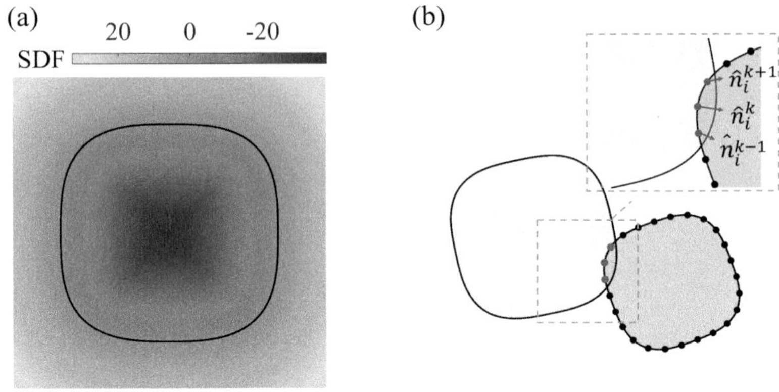

Fig. 2 **a** Signed distance function (SDF) of a superdisk particle ($n = 3$). The value is indicated by color. **b** Illustration of the node-to-surface contact detection algorithm. Overlap is exaggerated for clarity

2.2 *Level Set Function*

For the representation of a particle surface, the level set function $\Phi(\boldsymbol{x})$ should possess the following characteristics: (1) $\Phi(\boldsymbol{x}) < 0$ in the particle interior; (2) $\Phi(\boldsymbol{x}) > 0$ in the particle exterior; (3) $\Phi(\boldsymbol{x}) = 0$ denotes the particle surface. While various mathematical functions can potentially meet these criteria, previous studies [17–19] have mainly utilized a signed distance function (SDF) for its ability to readily assess contact information and other geometric properties, including the center of mass of the particle. Therefore, the SDF was adopted and a specific example of a signed distance function (SDF) describing a superdisk is shown in Fig. 2a.

2.3 *Surface Node Discretization*

In the context of LS-DEM, discrete nodes seeded onto the particle surface are also responsible for the shape representation. In this study, we applied a sufficiently small, uniform node spacing (evenly distributed 150 surface nodes) to all shapes in the simulations. This choice was made in order to meet the requirement previously discussed in Ref. [17]. Then a node-to-surface contact algorithm is considered (see Fig. 2b), which is frequently used in finite element modeling. Specifically, the contact is determined by checking each surface node $\boldsymbol{x}_i^k$ ($k = 1, 2, \ldots, 150$) of a particle i against the interface of another particle j constructed by its signed distance function Φ_j. The penetration sand contact normal on each code can be found by

$$d_i^k = \left|\Phi_j\left(x_i^k\right)\right|, \quad \hat{n}_i^k = \nabla\Phi_j\left(x_i^k\right)/\left|\nabla\Phi_j\left(x_i^k\right)\right| \tag{5}$$

where ∇ is the gradient. Upon obtaining the contact information, a linear spring-dashpot model [1] is used for the contact force computation. For more information on LS-DEM, the readers may refer to the Refs. [17, 19].

3 Numerical Simulation Procedures

3.1 Preparation of Dense Samples

To prepare the dense packing samples for each blockiness value, we first generated 10,000 particles in space with a small dispersity ($d_{max} = 2d_{min}$) uniformly distributed around the unit particle size $\langle d \rangle = 2\sqrt{\overline{A}/\pi}$. This choice was made in order to avoid crystallization [3, 5]. The mean particle area $\overline{A}$ for all shapes was kept constant to ensure equivalent sizes for all samples. Subsequently, the packings were performed by isotropic compression within a rectangular frame with a bi-periodic boundary, with the left and bottom boundaries fixed and the other two boundaries subjected to confining pressure P and allowed to move slowly (as shown in Fig. 3a). The particle stiffness k_n was set to a sufficiently large value to ensure penetrations $\delta/\langle d \rangle \ll 1$ under this loading condition. The jammed mechanically stable state was obtained until the fluctuations of particle positions and contact forces were below 0.01% [3]. During this compression process, the gravity and the inter-particle friction were turned off to obtain uniform, dense packs without the stress gradient [8].

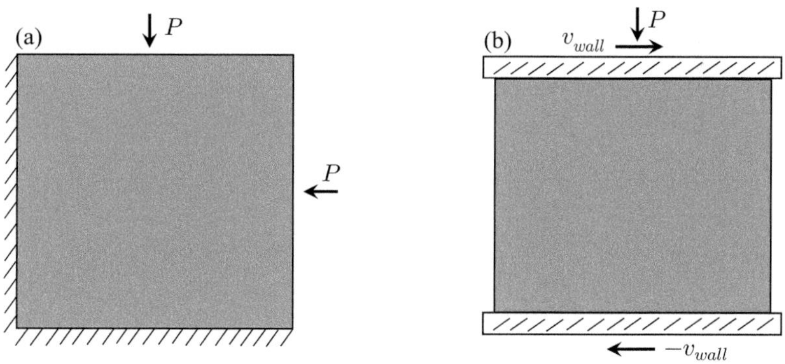

Fig. 3 Simulation setups for **a** isotropic compression and **b** simples shear tests

3.2 Simple Shear Tests

In a subsequent simple shear process, the inter-particle friction was adjusted to 0.5 to resemble most real-life granular materials. Two plates were formed by rigidly connecting particles at the top and bottom of each sample [8, 10]. The top plate was given a constant lateral speed v_x and allowed to move vertically under the same pressure P used in the compression tests, while the bottom plate was fixed vertically with an opposite speed $-v_x$ (see Fig. 3b). The shearing speed was gradually increased before reaching the target speed to avoid initial shock disturbance [24]. The samples were sheared up to a large cumulative shear strain $\gamma = \Delta x/H$, where Δx and H denote the overall wall displacement and sample height. This choice ensures the critical state is reached, where the shear rate $\dot{\gamma} = 2v_x/H$ and normal stress $\sigma_{yy} \approx P$ become steady. Furthermore, an empirical quantity I, known as the inertial number, was used to describe different shear regimes, which is given by

$$I = \dot{\gamma} d \sqrt{\rho_p/\sigma_{yy}}, \tag{6}$$

where $\langle d \rangle$ and ρ_p are the effective size and density of particles. A wide range of I was explored, spanning from approximately 10^{-3} to 1.5×10^{-1}, indicating the transition from a quasi-static state to a dense regime. A summary of simulation parameters is listed in Table 1.

Table 1 LS-DEM simulation parameters used in this study

Parameter	Value
Number of particles, n_p	10,000
Particle blockiness, n	2.0/2.4/3.0/3.4/4.0
Particle surface nodes, s_n	150
Particle size and dispersity, $\langle d \rangle \pm \epsilon$	1 ± 0.33
Particle density, ρ_p	1
Confining pressure, P	100
Coefficient of restitution, e	0.10
Inter-particle friction, μ_p	0.0 (compression)/0.5 (shear)
Normal stiffness, k_n	400 P
Lateral stiffness, k_t	1.0 k_n
Simulation time step, Δt	$0.01 \sqrt{\rho_p \overline{A}/k_n}$
Width of periodic cell, W	$\approx 100\langle d \rangle$
Height of periodic cell, H	$\approx 100\langle d \rangle$

4 Results and Discussion

4.1 Definition of Macroscopic Quantities

In a granular system, the expression of contact stress tensor $\boldsymbol{\sigma}$ [25] reads

$$\sigma_{ij} = \frac{1}{V}\sum_{\alpha \in N_c} f_i^\alpha l_j^\alpha, \tag{7}$$

where $\boldsymbol{f}^\alpha$ is the contact force of the contact α and $\boldsymbol{l}^\alpha$ is the branch vector joining the mass centers of two particles involved. V is the total volume of the sample and N_c is the contact number in total. By computing the eigenvalues σ_1 and σ_2 ($\sigma_1 > \sigma_2$) of stress tensor $\boldsymbol{\sigma}$, an internal friction reflecting the shear strength μ is defined as

$$\mu = q/p, \tag{8}$$

where $q = (\sigma_1 - \sigma_2)/2$ and $p = (\sigma_1 + \sigma_2)/2$ are the deviatoric and mean stresses, respectively. The solid fraction ϕ, which compares the volume occupied by solid particles V^s and total volume V, is adopted to evaluate the system packing density

$$\phi = V^s/V. \tag{9}$$

Given that the shearing plates are not perfectly flat, 100 slices are taken vertically for accurate measurement of ϕ.

Furthermore, the mean coordination number $\overline{z}$ is measured as the mean contact number of non-rattler particles as

$$\overline{z} = 2(N_c - N_r)/(n_p - n_r), \tag{10}$$

with N_r and n_r denoting the number of contacts and particle number of rattlers, respectively.

4.2 Packing Properties After the Isotropic Compression

Two snapshots of our packings after the isotropic compression are displayed in Figs. 4a and b. The particles with zero or one contact, called rattlers, are shown in grey since they do not carry any loads. The solid fraction ϕ and coordination number $\overline{z}$, which are crucial parameters for evaluating the structural stability and geometric properties of the packings, are reported in Fig. 4c. The solid fraction ϕ describes how densely the system is packed, and the result ($\phi = 0.843$) for the circular system ($n = 2$) coincides with random close packing (RCP) value observed in Refs. [8,

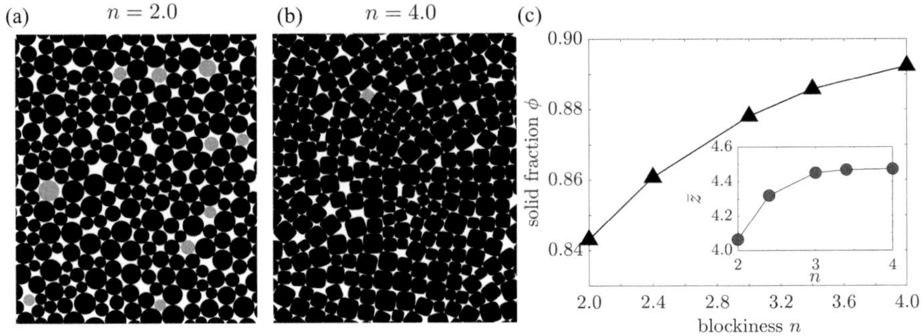

Fig. 4 Prepared packing samples with **a** $n = 2$ and **b** $n = 4$ after the isotropic compression in a particle-scale view. Rattler particles are shown in light gray. **c** Solid fraction ϕ and mean coordination number $\bar{z}$ (inset) as a function of the blockiness n

26]. For non-circular shapes, the solid fraction increases monotonically with the blockiness n we explored. This is due to the ordered alignment of particles in large values of n, thus leading to fewer voids created among particles. A similar trend can also be observed in the variation of coordination number $\bar{z}$, which increases with increasing blockiness n, indicating more contacts are required to constrain a single particle due to additional freedom of rotation. The isostatic theory [3] suggests that in the frictionless limit, two contacts per degree of freedom d_f are needed to constrain a particle, thus leading to $\bar{z} = 2d_f = 4$ for circular shape in two dimensions. However, our result is slightly larger than 4 since the circle is not perfectly represented due to the shape discretization. Nevertheless, such deviation is considered acceptable since all the results obtained for the circular system in the following match very well with previous studies.

4.3 Stress–strain Response in the Quasi-Static Limit

Figure 5 displays the evolution of the stress ratio q/p and solid fraction ϕ as a function of shear strain γ for quasi-static cases with minimum inertial number I we simulated. The shear resistance of all particle shapes increases from 0 to a shape-dependent peak value and then relaxes to a constant value corresponding to the residual state in soil mechanics [27] (Fig. 5a). This behavior is consistent with the observations from [24] for dense samples. Moreover, we find that the peak and residual shear stress ratios have a clear dependency on particle shapes, with larger blockiness n tending to increase them, as shown in the inset of Fig. 5a. It should be noted that the shear strength in the quasi-static state μ_0 is averaged from the stress ratios after the critical state is reached, and the value for the circular system agrees well with the previous study [4] introduced in the following. The peak and residual stress ratios both exhibit similar tendencies, rising rapidly when starting to deviate from the circle, then slowing down and saturating when $n \to \infty$, ultimately reaching the shear strength limit of the square system. To understand this phenomenon, we show two example assemblies near the peak states together with their force chains (Figs. 6a and b). We observe that particles in the assembly with $n = 4$ tend to contact nearby particles through their 'flat sides'. According to the mechanism [28] described in Fig. 6c, the 'side-side' contacts allow the transmission of the force moments through contact pairs and hinders particle rotations, resulting in a larger shear strength. Similar observations are also reported for polygonal systems in previous studies [7, 29].

The evolution of solid fraction (Fig. 5b) exhibits an initial rapid dilation phase that matches the rapid increase in the stress ratio q/p at the beginning. The maximum dilatancy angle (slope of the dilation curve) and the peak stress ratio are reached almost simultaneously. However, unlike the stress ratio q/p, which converges to an almost constant value when $\gamma \approx 0.35$, the sample volume V gradually increases until the sample was sheared to a sufficiently large strain around $\gamma \approx 10$. Therefore,

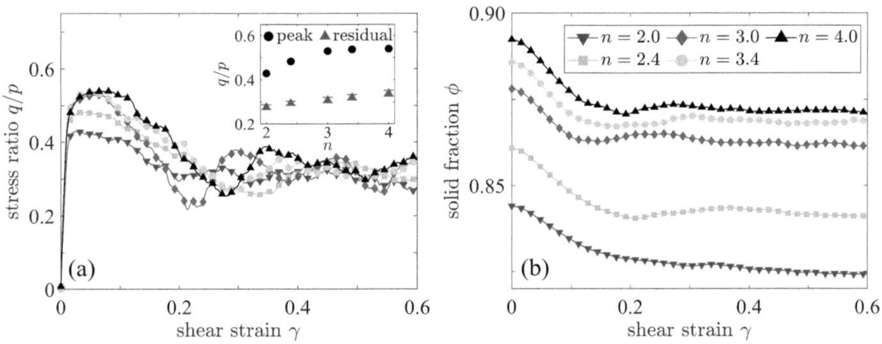

Fig. 5 **a** Stress ratio q/p and **b** solid fraction ϕ of different shapes as a function of shear strain γ in the quasi-static limit. Inset of (a) shows the peak and residual stress ratios as a function of blockiness n. Error bars represent the standard deviation in the residual state

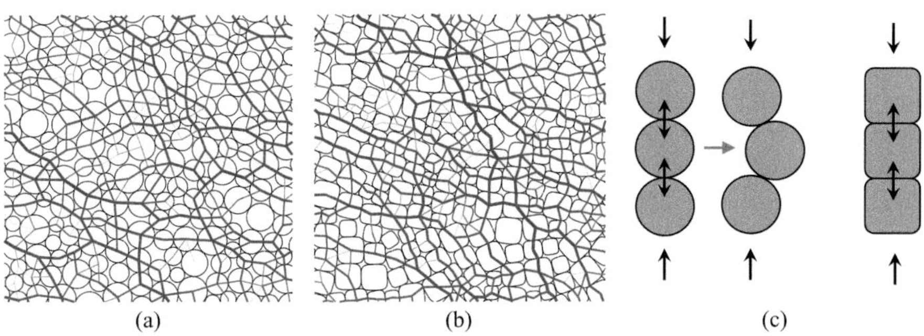

Fig.6 Snapshots of a part of assemblies with **a** $n = 2$ and **b** $n = 4$ in the peak state. Red segments joining the mass centers represent the force chains, and the width is proportional to the magnitude. The difference in shear strength can be explained by the **c** mechanism of particle shape effects [28]

the quasi-static limit ϕ_0 is evaluated when the sample volume no longer increases, which is much lower than the value that the residual state has just achieved (Fig. 5b).

4.4 Rheological Features

As the shearing speed of plates increases, samples tend to flow, indicating the transition from quasi-static to dense regimes. Previous research [2, 30] has established that the shear strength μ and solid fraction ϕ of granular flows, averaged from the steady state rather than just the residual state at small shear strains, can be predicted using the inertial number I. This empirical law, known as $\mu(I)$ rheology model, reveals the non-Newtonian fluid nature of granular flows. It demonstrates a monotonic relationship between the inertial number I and the above two parameters, wherein the shear strength μ increases while the solid fraction ϕ decreases as the inertial number I increases. However, this rheology law was originally developed for circular systems and requires application to non-circular systems. The results of our numerical simulations for all explored inertial numbers I and blockiness n are presented in Fig. 7. Our findings indicate that the numerical data of circular systems ($n = 2$) coincide well with the results obtained from Ref. [4], while non-circular shapes exhibit higher shear strengths (Fig. 7a) or solid fractions (Fig. 7b) for any given inertial number I. However, these differences weaken as I increases, indicating that the system evolves towards a dilute system dominated by binary collisions, and the influence of particle shape gradually diminishes [4, 8]. To capture the exact relations between I and μ or ϕ, power-law functions [31] are considered in this study to fit the data

$$\mu = \mu_0 + AI^{\alpha}, \tag{11}$$

$$\phi = \phi_0 - BI^{\beta}, \tag{12}$$

with μ_0 and ϕ_0 the values in the quasi-static limit introduced in Sec. 4.3, respectively. As previously discussed, the value of ϕ_0 is found to be lower than the critical state value due to additional shear strains. The curves fitted by Eqs. (11) and (12) are also presented in Fig. 7, where all data points appear to align with the obtained curves, indicating the reliability of the power-law functions in describing the $\mu(I)$ and $\phi(I)$ relationships. Intriguingly, we observe that the variations of these two quantities yield similar exponents (α, $\beta \approx 0.8$), regardless of blockiness (see the insets of Fig. 7). This suggests that predicting the shear strength or solid fraction of the superdisk systems may be simpler than previously anticipated. However, further investigation is necessary to verify whether these findings hold for frictionless systems. Moreover, it should be noted that similar observations are made for elliptic systems [5], implying that extending these results to the entire superellipse family may be promising.

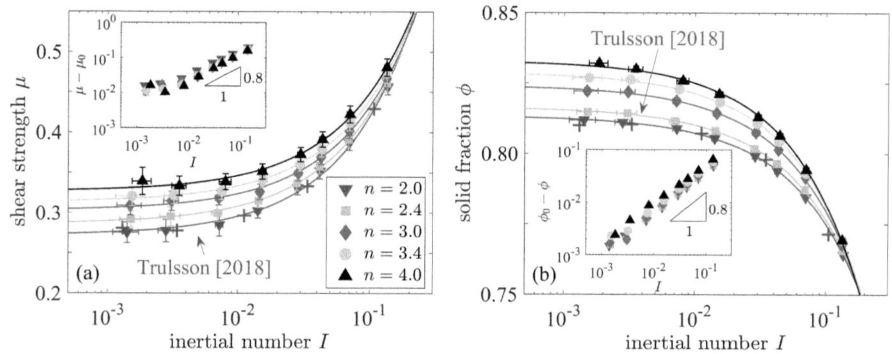

Fig.7 **a** Shear strength μ and **b** solid fraction ϕ as a function of inertial number I for flowing superdisks with different blockiness n in log-linear scale, along with best fit obtained using Eqs. (11) and (12). Error bars represent the standard deviation in the residual state. Data of sheared circles from Ref. [4] is shown in red cross for comparison. Inset: Test of exponents (a) α and (b) β

5 Conclusions

To conclude, we utilized the 2D level set discrete element method (LS-DEM) to investigate the shear responses of superdisk particles. By varying the blockiness n from 2 to 4, we prepared mechanically stable samples composed of various superdisks (from circular to square-like shapes) and carried out a series of simple shear simulations at different shearing speeds. Our findings indicate that: (1) Both the solid fraction ϕ and the mean coordination number $\overline{z}$ of the packs after isotropic compression increase with the blockiness n due to the ordered alignment and one additional freedom of rotation; (2) The difference in peak and residual stress ratios under quasi-static shear between different shapes is significant. We observe a positive correlation between the quasi-static shear resistance and blockiness n, which seems to saturate at large blockiness values. We attribute this enhancement to the rotational resistance provided by shape effects, following the mechanism proposed in previous literature. Additionally, unlike the quick convergence of shear strength, the dilation continues even after the residual state is achieved, leading to a much lower solid fraction in a sufficiently large shear strain; (3) The obtained rheological curves follow the $\mu(I)$ rheology law proposed previously for circular/spherical systems and can be well-described by power-law fitting functions. In the dense regime limit, the larger the blockiness n, the larger the shear strength q/p and solid fraction ϕ still hold. Surprisingly, our analysis of the power-law functions reveal that the exponents describing these variations are similar across all shapes (α, $\beta \approx 0.8$), suggesting that the prediction of superdisk behavior may not be as difficult as previously thought. Our findings highlight the significant influence of particle shape on the behavior of granular assemblies under external loadings and suggest that further investigation of more square-like shapes is needed in future studies.

Acknowledgements H.J. acknowledges the financial support of JST-SPRING program Grant Number JPMJSP2124. T.M. acknowledges the financial support of JSPS KAKENHI Grants 20K20434A, 21H01422, and 21KK0071.

References

1. Cundall PA, Strack OD (1979) A discrete numerical model for granular assemblies. Geotechnique 29:47–65. https://doi.org/10.1680/geot.1979.29.1.47
2. Midi GDR (2004) On dense granular flows. Eur Phys J E 14:341–365. https://doi.org/10.1140/epje/i2003-10153-0
3. Matsushima T, Blumenfeld R (2014) Universal structural characteristics of planar granular packs. Phys Rev Lett 112:1–5. https://doi.org/10.1103/PhysRevLett.112.098003
4. Trulsson M (2018) Rheology and shear jamming of frictional ellipses. J Fluid Mech 849:718–740. https://doi.org/10.1017/jfm.2018.420
5. Jiang X, Matsushima T (2021) Rheological and Microstructural Characteristics in Granular Shear Flow of 2D Elliptical Particles. J Japan Soc Civ Eng Ser A2 (Applied Mech 77:I_297–I_305. https://doi.org/10.2208/jscejam.77.2_i_297
6. Peña AA, García-Rojo R, Herrmann HJ (2007) Influence of particle shape on sheared dense granular media. Granul Matter 9:279–291. https://doi.org/10.1007/s10035-007-0038-2
7. Zhou W, Xu K, Yang L, Ma G (2017) Influence of particle shape on mechanical behavior of granular materials. Springer Proc Phys 188:v–vi. https://doi.org/10.1007/978-981-10-1926-5
8. Azéma É, Radjaï F, Roux JN (2018) Inertial shear flow of assemblies of frictionless polygons: Rheology and microstructure. Eur Phys J E 41https://doi.org/10.1140/epje/i2018-11608-9
9. Mandal S, Khakhar D V. (2016) A study of the rheology of planar granular flow of dumbbells using discrete element method simulations. Phys Fluids 28https://doi.org/10.1063/1.4963310
10. Matsushima T, Chang CS (2011) Quantitative evaluation of the effect of irregularly shaped particles in sheared granular assemblies. Granul Matter 13:269–276. https://doi.org/10.1007/s10035-011-0263-6
11. Nagy DB, Claudin P, Börzsönyi T, Somfai E (2017) Rheology of dense granular flows for elongated particles. Phys Rev E 96:2–6. https://doi.org/10.1103/PhysRevE.96.062903
12. Nagy DB, Claudin P, Börzsönyi T, Somfai E (2020) Flow and rheology of frictional elongated grains. New J Phys 22https://doi.org/10.1088/1367-2630/ab91fe
13. Salerno KM, Bolintineanu DS, Grest GS, Lechman JB, Plimpton SJ, Srivastava I, Silbert LE (2018) Effect of shape and friction on the packing and flow of granular materials. Phys Rev E 98:3–7. https://doi.org/10.1103/PhysRevE.98.050901
14. Lu G, Third JR, Müller CR (2015) Discrete element models for non-spherical particle systems: From theoretical developments to applications. Chem Eng Sci 127:425–465. https://doi.org/10.1016/j.ces.2014.11.050
15. Zhao S, Zhang N, Zhou X, Zhang L (2017) Particle shape effects on fabric of granular random packing. Powder Technol 310:175–186. https://doi.org/10.1016/j.powtec.2016.12.094
16. Williams JR, O'connor R (1995) A linear complexity intersection algorithm for discrete element simulation of arbitrary geometries. Eng Comput 12:185–201. https://doi.org/10.1108/02644409510799550
17. Kawamoto R, Andò E, Viggiani G, Andrade JE (2016) Level set discrete element method for three-dimensional computations with triaxial case study. J Mech Phys Solids 91:1–13. https://doi.org/10.1016/j.jmps.2016.02.021
18. Jerves AX, Kawamoto RY, Andrade JE (2016) Effects of grain morphology on critical state: A computational analysis. Acta Geotech 11:493–503. https://doi.org/10.1007/s11440-015-0422-8

19. Kawamoto R, Andò E, Viggiani G, Andrade JE (2018) All you need is shape: Predicting shear banding in sand with LS-DEM. J Mech Phys Solids 111:375–392. https://doi.org/10.1016/j.jmps.2017.10.003
20. Harmon JM, Arthur D, Andrade JE, Pazmiño SA, Jerves AX, Dijkstra J, Medina DA, Jostad HP, Mital U, Kawamoto R, Andrade JE (2020) Effect of fabric on shear wave velocity in granular soils. Comput Methods Appl Mech Eng 15:112961. https://doi.org/10.1007/s11440-019-00766-1
21. Harmon JM, Arthur D, Andrade JE (2020) Level set splitting in DEM for modeling breakage mechanics. Comput Methods Appl Mech Eng 365:112961. https://doi.org/10.1016/j.cma.2020.112961
22. Pazmiño SA, Jerves AX, Dijkstra J, Medina DA, Jostad HP (2022) A generalized 3DLS-DEM scheme for grain breakage. Comput Methods Appl Mech Eng 399:115383. https://doi.org/10.1016/j.cma.2022.115383
23. Osher S, Fedkiw R, Piechor K (2004) Level set methods and dynamic implicit surfaces. Appl Mech Rev 57:B15–B15. https://doi.org/10.1115/1.1760520
24. Katagiri J, Matsushima T, Yamada Y (2014) Variations in shear behavior among specimens with different packing patterns. Granul Matter 16:891–901. https://doi.org/10.1007/s10035-014-0530-4
25. Rothenburg L, Bathurst RJ (1989) Analytical study of induced anisotropy in idealized granular materials. Geotechnique 39:601–614. https://doi.org/10.1680/geot.1989.39.4.601
26. Vågberg D, Olsson P, Teitel S (2016) Critical scaling of Bagnold rheology at the jamming transition of frictionless two-dimensional disks. Phys Rev E 93:1–20. https://doi.org/10.1103/PhysRevE.93.052902
27. Thurairajah A (1988) Critical state soil mechanics. Engineer 16:53–66
28. Matsushima T (2005) Effect of irregular grain shape on quasi-static shear behavior of granular assembly. Powders Grains 2005—Proc 5th Int Conf Micromechanics Granul Media 2:1319–1323. https://doi.org/10.1201/NOE0415383486
29. Estrada N, Azéma E, Radjai F, Taboada A (2011) Identification of rolling resistance as a shape parameter in sheared granular media. Phys Rev E - Stat Nonlinear, Soft Matter Phys 84:1–5. https://doi.org/10.1103/PhysRevE.84.011306
30. Da Cruz F, Emam S, Prochnow M, Roux JN, Chevoir F (2005) Rheophysics of dense granular materials: Discrete simulation of plane shear flows. Phys Rev E—Stat Nonlinear, Soft Matter Phys 72:1–17. https://doi.org/10.1103/PhysRevE.72.021309
31. Hatano T (2007) Power-law friction in closely packed granular materials. Phys Rev E—Stat Nonlinear, Soft Matter Phys 75:1–4. https://doi.org/10.1103/PhysRevE.75.060301

Morphological Evolution of Andesite Railway Ballast Particle Under Rotating Drum Abrasion Test

O. C. Debanath, T. Matsushima, T. Ijichi, and M. Miwa

Abstract Railway ballast is a natural aggregate widely used in railway infrastructure. The mechanical strength and geometric shape of ballast particles have an important role in the durability and maintenance economy of the ballasted track. This paper aimed to investigate the abrasive behavior of andesite ballast particles using a rotating drum abrasion test and 3D morphological shape analysis. The photogrammetric reconstruction method is adopted here and modified to make it more efficient and user-friendly. The evolution of mass, particle shape, and grain indices were compared for different shape classifications. The experimental results show that the evolution of elongation and flakiness that describe overall particle shape is inconsistent during abrasion, whereas the equivalent ellipsoid volume ratio and rotational resistant angle, which mainly describe particle surface angularity evolve gradually throughout this long-term abrasion process. So, the latter two shape indices describe particle abrasion behavior more accurately.

Keywords Railway ballast · Long-term abrasion · Rotating drum test · Photogrammetric reconstruction · Particle shape analysis

O. C. Debanath
Doctoral Program in Engineering Mechanics and Energy, University of Tsukuba, Tsukuba, Japan

Department of Civil Engineering, CUET, Chittagong, Bangladesh

T. Matsushima (✉)
Department of Engineering Mechanics and Energy, University of Tsukuba, Tsukuba, Japan
e-mail: tmatsu@kz.tsukuba.ac.jp

T. Ijichi · M. Miwa
JR Central, Nagoya, Japan
e-mail: takuya.ijichi@jr-central.co.jp

M. Miwa
e-mail: m.miwa@jr-central.co.jp

S. Arthur et al. (eds.), *Proceedings of the 6th International Conference on Advances in Civil Engineering*, Lecture Notes in Civil Engineering 368,
https://doi.org/10.1007/978-981-99-3826-1_31

1 Introduction

Natural stones are the most common type of material with various engineering applications since prehistoric times, specifically for railway lines; crushed stones are the prime element of the ballast layer. Andesite rock is one of the most common igneous rocks used in railway infrastructure. A major concern about rail tracks is the instability of the track, which mainly depends on the geometric stability and settlement of the underlying ballast layer. Moreover, the stability of the ballast layer depends on the interlocking of particles, gradation, particle shape, hardness, and abrasion resistance. The mechanical stability and economic efficiency of railway tracks greatly depend on the rate of ballast degradation (crushing and abrasion) due to the cyclic traffic load throughout its lifecycle.

Several researchers [1–3] investigated ballast particle degradation by laboratory tests such as Los Angeles abrasion, micro-Deval abrasion, direct shear, and large-scale triaxial test. They suggested that the ballast abrasion rate depends on its rock type, physical properties, external loading condition, etc. The grain shape is another important parameter that significantly affects the ballast particle performance. The relationship between the particle shape and abrasion value was reported by previous research [4–6]; typically, the flaky or elongated particles are more susceptible to breaking. The correlation between the mechanical strength and the physical properties of rock material was also studied by previous researchers [5, 7]. Guo [8] proposed a technique based on abrasion depth and 3D sphericity of particles to quantify the degradation. Similar research findings also revealed the relationship between the micro-Deval loss and the angularity of western Virginia aggregate [9].

Considering the durability phenomena, the long-term abrasive behavior is important, which requires some additional tests beyond the standard test methods. Czinder [10] reported that the short-term micro-Deval abrasion test is insufficient to represent the overall abrasion process. A typical andesite rock requires approximately 2 million abrasion cycles for 80% weight loss. In contrast, relatively soft rocks require up to 300,000 cycles, and the abrasion process is reported as a function of mechanical work done following exponential form. A limited number of research studies was done on 3D shape analysis of particle, which only focused on short-term abrasion. However, it is still important to understand ballast particles' long-term abrasion behavior, including morphological changes.

The present study attempts to observe the abrasive behavior of ballast particles using a rotating drum device to look up the in-depth morphological transformation during long-term abrasion by using 3D particle shape analysis. Particle shape is defined by considering flakiness and elongation. There are many conventional shape descriptors to describe the grain shape, such as true sphericity, form factor, shape factor, angularity, texture, etc. In the current research, we introduced two new parameters ellipsoid volume ratio and rotational resistant angle. The evolution of these parameters was then correlated with volumetric degradation of ballast particles along with physical photographs at different abrasion stages for clear visualization of long-term abrasion phenomena.

2 Materials and Experimental Workflow

The material used in this study was irregularly shaped andesite ballast particles of about 5 cm. Their average density was about 2.62 gm/cm^3. The Photogrammetric reconstruction technique was used to measure the 3D surface morphology of about 50 particles. Among them, nine particles were selected for morphological evolution observation considering the variety of particle shapes. The evolution of particle mass, 3D models, and other shape indices were recorded at various stages of the abrasion process. Details of the experimental setup and workflow are described in the following subsections.

2.1 *Three-Dimensional Shape Extraction*

Recent articles have reported the use of 3D scanning techniques such as laser scanners, X-ray CT scanners, photogrammetric reconstruction, and light-based 3D scanners to extract the three-dimensional shape of ballast particles [11–13]. Laser or X-ray CT scanning is a very popular and primitive technique for digital model reconstruction of any solid object; specifically, laser scanning is an active method of shape reconstruction, where the laser beams reflected from the surface are recorded on the sensor. The suitability of the above-mentioned techniques depends on the object size, shape, surface irregularity, scanning resolution, etc. Several previous researchers used the laser technique to study 3D shapes. For an angular shape ballast particle, it often requires scanning the particle in different orientations and then aligning the meshes by using common surface points to generate the final shape. However, this workflow includes significant time starting from scanning and mesh processing, which is less effective for a large-scale experimental campaign. To overcome this drawback of previous digital reconstruction methods, Paixão [14] introduced the photogrammetry method to make this process more efficient. In this current study, we modified this workflow to make it more user-friendly and time efficient. A detailed description of the workflow is shown in Fig. 1.

For the photographic session, a photographic box was assembled with diffused light source at the top and all the box's inner walls with reflective material to ensure enough light to the object. We used green screen background and a turntable assembled with a tripod needle stand. The tripod stand facilitates the easy placement of particles. In contrast, the previous method [14] needs to drill an orifice at the particle surface to mount on the single pedestal support that may propagate microcracks on the rock surface. The turntable is placed at the center of the assembly and then rotated at a constant speed of 2 rpm. For each particle placed in the turntable, sequential photographs were captured to cover the 360° view at every 10° interval to confirm sufficient overlaps on successive images.

We used a digital camera (Canon EOS Kiss X7 18 MP) equipped with EFS 17–18 mm lens; the camera settings were as follows: manual focus with exposure time 1/

Fig. 1 Photographic arrangement

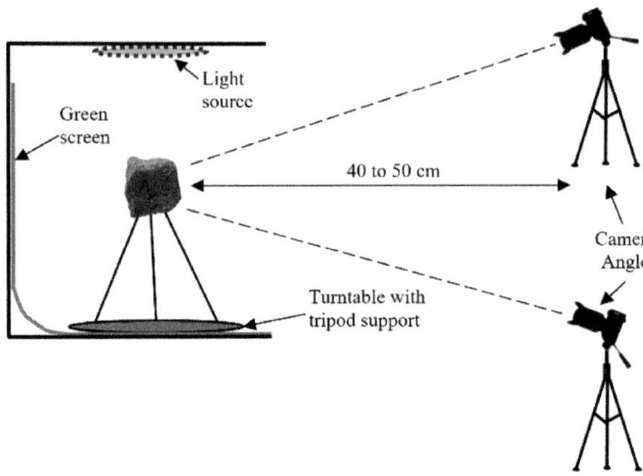

400 s., ISO-6400, and aperture F/8. For each 3D model, 72 photographs were captured from two different vertical positions (approximately 45° with horizontal). The overall photographic session took approximately 3 min per particle, from placement on the tripod stand to the end of the photo session. The next part of the reconstruction is done using Agisoft Metashape (student edition) software [15]. The first step is to align photos sequentially, taking the key point limit per MP as 5000; Afterward, the dense cloud is built at moderate depth filtering; Fig. 2 (a) shows the noise-free dense point cloud. In this workflow, no undesired points are generated, which makes the next step (mesh processing) easier, whereas the VisualSFM and CMVS algorithms used by previous authors build numerous background noises in the dense cloud. Moreover, the Agisoft Metashape software facilitates simultaneous data processing of multiple particles, which makes the workflow more efficient. The rest of the part is done by another software, CloudCompare [16], to erase the tripod support from the dense cloud and Screened Poisson mesh reconstruction. The reconstructed meshes need to be scaled correctly to represent the original particle. In the current study, the geometric scale factor is computed by taking the bouncy volume of the particle and applied for uniform scaling in all directions. Figure 2(b–c) represent the final 3D model and original particle, respectively.

In this modified photogrammetric reconstruction workflow, the average reconstruction time taken for each particle is about 10 min, while the method proposed by Paixão [14] takes around one hr. per particle, and other 3D reconstruction techniques are more time-consuming. At the beginning of this experimental campaign, we scanned the same particle using a laser scanner (Roland LPX-60) with 0.2 mm accuracy; after that, both models were overlapped to check the accuracy of photogrammetric reconstruction. Figure 2(d) illustrates the deviation between the laser-scanned and the photo-reconstructed mesh. Both meshes were closely matched, and for most of the vertices, the deviation was observed between −0.2 mm and 0.3 mm, whereas the accuracy of the scanner was 0.2 mm. Moreover, the laser scanner took around 45 min to scan the particle in one orientation. Multiple orientations were needed for a very angular particle, and the laser beams sometimes failed to capture the extreme

Fig. 2 **a** Reconstructed dense cloud **b** Final 3D mesh **c** Original Ballast particle **d** Deviation with laser scanning

concave surface points. Hence the modified photogrammetric reconstruction method was chosen for the rest of this research.

2.2 Sample Selection

Around 50 representative particles were scanned by photogrammetric reconstruction, and 3D model has generated afterward. In this study, we used the best-fit ellipsoid concept to determine the dimensions of ballast particles. First, the best-fit ellipsoid was defined such that the three principal moments of inertia are identical to those of the original particle, and the maximum, intermediate, and minimum axes, denoted as L, I, and S, respectively, were computed from the ellipsoid. Figure 3(a) illustrates the best-fitted ellipsoid over the original particle.

Therefore, elongation (I/L) and flakiness (S/I) are used to describe the overall shape of the particle. In this study, modified Zingg [17] form terminology was used

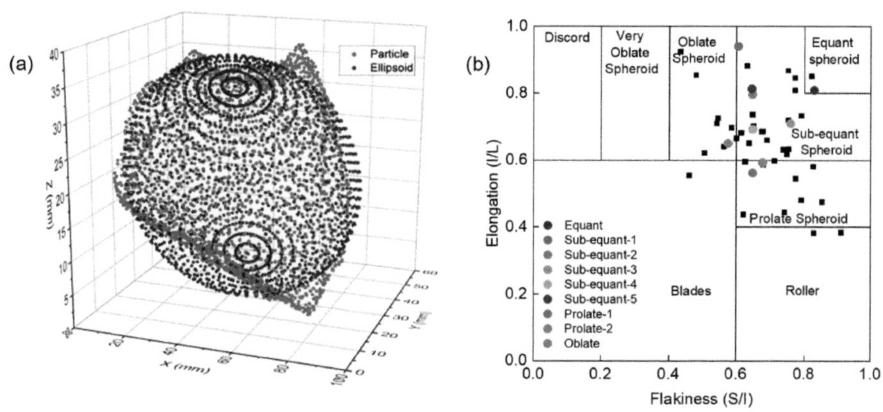

Fig. 3 **a** Illustration of Equivalent ellipsoid **b** Morphological classification of scanned particles

based on the degree of elongation and flatness. Black square dots in Zingg's diagram shows the shape of the measured particles, Fig. 3(b). Among them, five Sub-equant, two prolate, one oblate, and one equant spheroid were chosen for details inspection, marked by colored circles in Zingg's diagram.

2.3 Test Setup

We used a modified micro-Deval setup to accelerate the abrasion process, this modification was made based on available laboratory facilities. Figure 4 shows the setup of the test device. The inner diameter of the drum was 30 cm, and an inner shaft of 5 cm width was fixed to let the particle fall from a certain height and collide with the inner wall of the drum during the drum rotation. The rotational speed was kept constant (70 rpm) throughout the test. The test was conducted for an extended abrasion cycle to observe the long-term degradation until the particle mass loss reached 70–90%. The previous studies [18] reported that the rate of abrasion is very high initially and trending to be constant afterward. The experimental program was set to interrupt the abrasion process at irregular intervals; the test data was initially recorded at intervals of 2000, 14,000, and 50,000 rotations, respectively; afterward, approximately fifty thousand rotations intervals until desired weight loss was achieved. The mass of individual particles was recorded at every incremental stage of abrasion, and the 3D shape extraction was done simultaneously for all selected particles. If any particle was broken during this process, the fine fragments were discarded, and the remaining part of the particle was subjected to the next stage.

Fig. 4 Rotary drum abrasion test setup

3 Results and Discussions

To describe the grain shape evolution, here we considered two new grain descriptors named: the equivalent ellipsoid volume ratio and rotational resistance angle for shape evolution. As mentioned above, the weight loss and 3D model were recorded at each incremental abrasion stage; then, further analysis was carried out based on the extracted virtual shape to obtain various shape indices. The detailed results are described in the following subsections.

3.1 Particle Mass Evolution

The individual particle mass recorded at the incremental stage of abrasion was used to compute the percentage of mass loss compared to the initial mass of the corresponding particle. Figure 5(a) shows the particle mass change during long-term abrasion; each line of this graph represents individual particle mass evolution. The generated abraded materials were mostly fine dust; few particles were broken at a sudden stage of abrasion, and any representable part from broken particles was also further tested; otherwise, the fine fragments were discarded. For example, the particle sub-equant-5 was broken into fine fragments after 50,000 revolutions, and no representing parts remained; hence that particle was discarded from the analysis. On the other hand, the sub-equant-1 particle was divided into two after 142,000 revolutions; these fragments continued to be abraded and analyzed as two independent particles in the next phase. The sudden increase of particle mass loss is due to the particle breaking event at that phase; the sub-equant-3, prolate-2, and oblate particles were subjected to sudden breaking at several test stages. The physical photographs of particles captured at different test stages also confirmed this.

At the initial phase of abrasion, specifically below 10% mass loss, all the shapes showed a similar abrasion rate, which decreased in the later phase. The angularity of particles can describe this behavior; at the early stage, sharp edges and vertices were broken, resulting in higher mass loss, and reduced afterward. Considering long-term abrasion, the degradation of equant and sub-equant spheroid particles shows analogous behavior. Other shapes like oblate or prolate spheroid undergo high degradation, possibly due to continuous edge breaking.

3.2 Particle Shape Evolution

The typical way of representing the particle shape is based on its flakiness and elongation; as defined earlier, the value of flakiness and elongation is less than 1. Several classification methods are available to describe the particle shape, and the modified Zingg's classification was adopted in our study. Figure 5(b) shows the shape

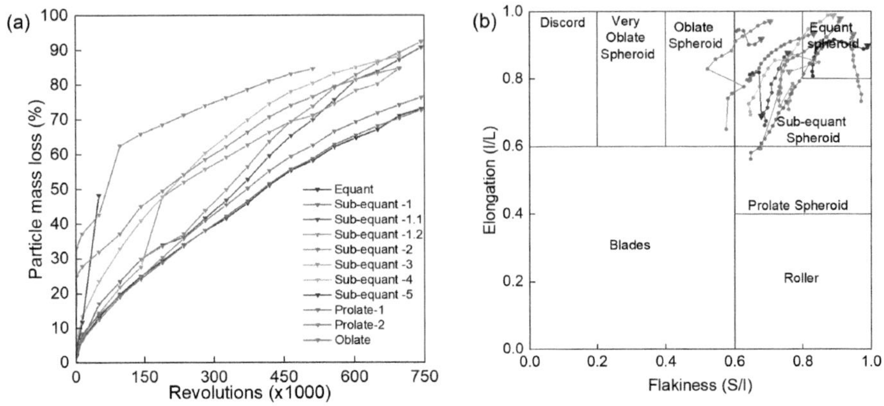

Fig. 5 Evolution of **a** particle mass and **b** shape movement of studied particles in modified Zingg's diagram

evolution of nine particles within modified Zingg's diagram during the abrasion test. The line colors represent the same particle as listed in Fig. 5(a); the circular marker indicates the shape of the particle at different stage of abrasion, whereas the larger circular and triangular markers represent the initial and final shape, respectively.

Significant shape change was recorded during this long-term abrasion process. Figure 6 illustrates the visual change of the studied particles during the test. From visual inspection, it was seen that the sharp edges disappeared quickly, specifically during the first 2000 revolutions. And the smother shapes appeared afterward due to loss of surface roughness by continuous polishing action inside the drum; this characteristic is reported as rolling erosion of rock [19]. At the initial configuration of the test, we started with the particles of different shape classes. However, at the final stage (after 70–90% loss), all particles' classified as an equant spheroid shape. It was also noticed that elongation's evolution was higher than the evolution flakiness. In the conventional micro-Deval abrasion test reported by a previous study [18], no evolution was recorded regarding particle shape classification. But the long-term abrasion test significantly changes the initial form of particle. This shape transformation of the particle during long-term abrasion can be noted as an important finding, as the overall stability of the ballast layer depends on the geometric shape.

3.3 *Morphological Evolution*

Several shape indices were used in previous literature to explain the morphology of angular particles. In this study we considerd two overall indices: (i) flakiness, (ii) Elongation; and two detailed morphological indices: (iii) ellipsoid volume ratio, (iv) rotational resistant angle, respectively. The volume change for a homogeneous rock material is equivalent to the particle mass change. Hence, we considered the volume

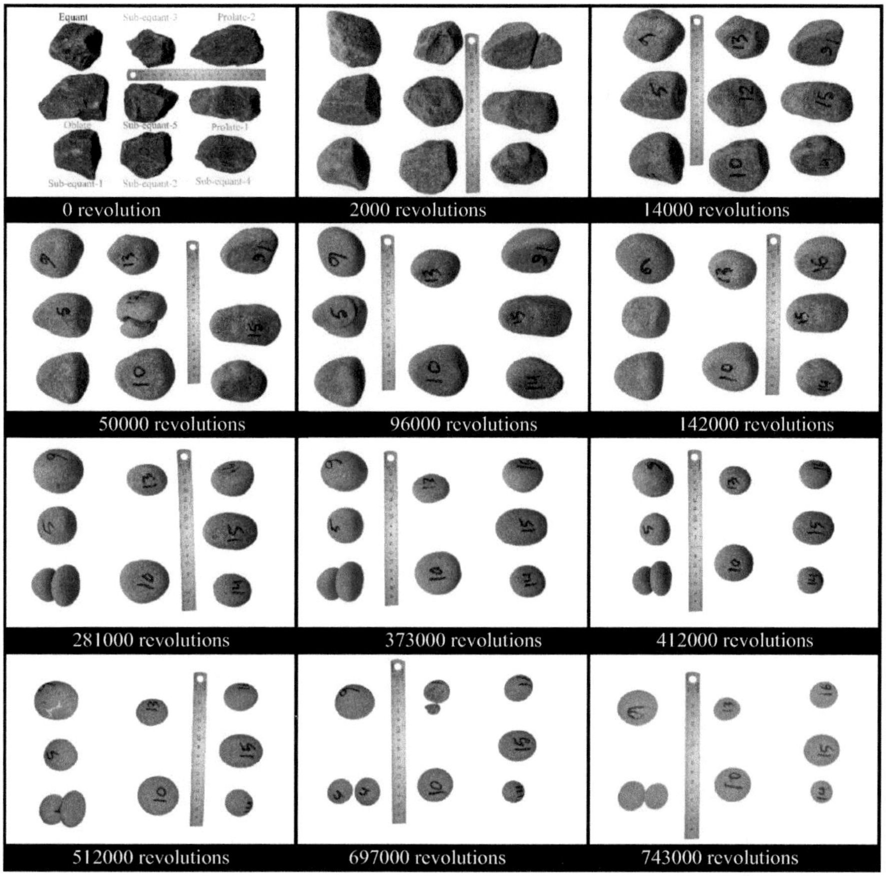

Fig. 6 Changes of particle shape during abrasion test

loss to describe the evolution of these parameters (Fig. 7). Although the particle shape changed significantly, the evolution of flakiness and elongation showed different evolution behavior during this long-term abrasion process. Referring to Fig. 7(a–b), the value of flakiness didn't change much initially, specifically up to 30% volume loss, and after that, this value increased gradually. In contrast, the evolution of elongation happened constantly and merged in the range of 0.8–1; this evolution trend resulted from nonuniform abrasion of particles in all directions.

As described in Sec 2.2, a best-fit ellipsoid is generated for each 3D particle model such that the three principal moments of inertia are identical to those of the original particle. Figure 3(a) illustrates the best fitted ellipsoid over the original particle. Since the volume of the fitted ellipsoid is different from that of the original particle, and this difference is due to the detailed angularity difference, a shape index called, ellipsoid volume ratio (e_v) is defined by the ratio of the original particle volume to the best-fit ellipsoid. The value of the ellipsoid volume ratio is smaller than unity when the

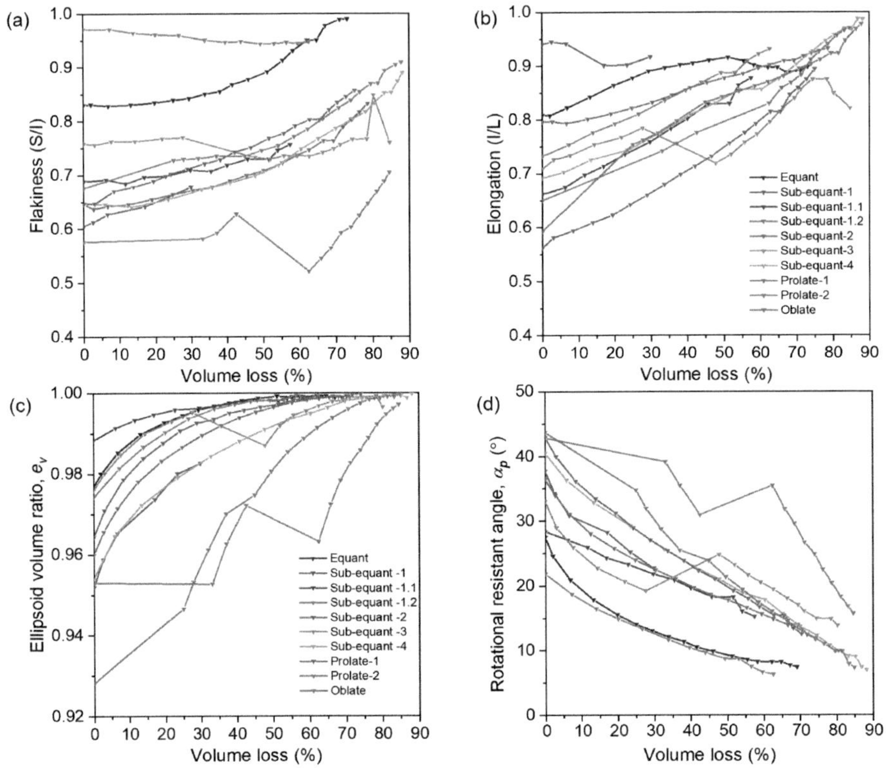

Fig. 7 Evolution of **a** Flakiness, **b** Elongation, **c** Ellipsoid volume ratio and, **d** Rotational resistant angle

particle shape deviates more from the ellipsoid. Figure 7(c) shows the evolution of the ellipsoid volume ratio; at the initial stage of abrasion, the ev value was different for different shapes and typically lies within the range of 0.928–0.98. However, this value moved towards unity for all particles regardless of the shape classification at the final stage of abrasion. This increase of ev was due to the continuous polishing effect inside the drum, which resulted in particles close to an ellipsoid shape.

Another shape parameter is a mechanically based shape index, called the rotational resistant angle, α_p proposed by Kawamoto [20]. For any surface point of 3D mesh, α is the angle between the surface normal of that point and the radial vector of joining the line of particle centroid, and dS is the area of the corresponding triangular face. Finally, α_p is calculated by taking the weighted average α over the entire surface of the particle as described in Eq. (1); here, A is the total surface area of the mesh.

$$\alpha_p = \frac{1}{\mathrm{A}} \int \alpha \, \mathrm{dS} \tag{1}$$

The value of α_p is linked to the rolling resistance acting on two particles in contact, and it increases with increasing particle surface angularity. Regarding the evolution of $\alpha_{p,}$ this parameter varied from 27 to 43 degrees before abrasion; after long-term abrasion, this range was decreased. The Minimum value of α_p was found for the equant shape ballast at the final configuration. For both cases, the abrupt particle breaking was reflected as a sudden transition of shape index, specifically for sub-equant-3 and oblate spheroid. As the shape indices (S/I and I/L) didn't change consistently throughout the abrasion cycle, the other conventional parameters (such as sphericity, roundness, form factor, etc.) based on overall shape are not a good measure to represent the abrasion. On the other hand, the evolution of detailed shape indices (e_v and α_p) was consistent throughout the long-term abrasion and can describe the abrasion of the local corners. Therefore, the latter two indices can be taken as more suitable parameters to track the abrasive behavior of ballast particles.

4 Conclusion

In this paper, the rotating drum test was used to study the morphological evaluation of andesite ballast particles having varying shapes. The test was interrupted at different intervals, and 3D mesh of selected particles was generated by photographic session and 3D reconstruction technique. To observe the effect of accelerated long-term abrasion, the evolution was tracked in terms of weight loss, shape movement, ellipsoid volume ratio, and rotational resistant angle obtained from 3D shape analysis.

The particle mass loss data reveals the high abrasion initially and slower rate afterward, except for some abrupt changes due to fragmentation. Compared to the short-term abrasion test, significant shape change was observed during the long-term abrasion process, where all the particles changed their initial shape classification. The corner breaking and particle fragmentation also occurred throughout the test and were reflected as sudden transitions of mass loss and shape index. The initial ellipsoid volume ratio deviated far from unity and converged significantly because of the rolling action inside the drum. Reasonable change of the shape indices proves the validity of the adopted method. It is suggested that the equivalent ellipsoid volume ratio and rotational resistance angle are good indices compared to overall shape index flakiness and elongation to evaluate the ballast particle abrasion.

Considering the micro-scale morphology, local abrasion parameters such as the local curvature and wear depth are other important parameters that are not analyzed in the current study. However, future study is expected to include these issues to develop an abrasion model that may help to predict ballast particles' abrasive behavior.

References

1. Wu J, Hou Y, Wang L, Guo M, Meng L, Xiong H (2018) Analysis of coarse aggregate performance based on the modified Micro Deval abrasion test. Int J Pavement Res Technol 11:185–194. https://doi.org/10.1016/j.ijprt.2017.10.007
2. Sun Y, Zheng C (2017) Breakage and shape analysis of ballast aggregates with different size distributions. Particuology 35:84–92. https://doi.org/10.1016/j.partic.2017.02.004
3. Qian Y, Tutumluer E, Hashash YMA, Ghaboussi J (2022) Triaxial testing of new and degraded ballast under dry and wet conditions. Transp Geotech 34:100744. https://doi.org/10.1016/j.trgeo.2022.100744
4. Maroof MA, Mahboubi A, Noorzad A, Safi Y (2020) A new approach to particle shape classification of granular materials. Transp Geotech 22:100296. https://doi.org/10.1016/j.trgeo.2019.100296
5. Kim J, Park B-S, Woo SI, Choi Y-T (2020) Evaluation of ballasted-track condition based on aggregate-shape characterization. Constr Build Mater 232:117082. https://doi.org/10.1016/j.conbuildmat.2019.117082
6. Sun Y, Indraratna B, Nimbalkar S (2014) Three-dimensional characterisation of particle size and shape for ballast. Géotechnique Lett 4:197–202. https://doi.org/10.1680/geolett.14.00036
7. Rangaraju PR, Edlinski J (2008) Comparative evaluation of micro-deval abrasion test with other toughness/abrasion resistance and soundness tests. J Mater Civ Eng 20:343–351. https://doi.org/10.1061/(ASCE)0899-1561(2008)20:5(343)
8. Guo Y, Markine V, Song J, Jing G (2018) Ballast degradation: Effect of particle size and shape using Los Angeles Abrasion test and image analysis. Constr Build Mater 169:414–424. https://doi.org/10.1016/j.conbuildmat.2018.02.170
9. Lane DS, Druta C, Wang L, Xue W (2011) Modified micro-deval procedure for evaluating the polishing tendency of coarse aggregates. Transp Res Rec: J Transp Res Board 2232:34–43. https://doi.org/10.3141/2232-04
10. Czinder B, Vásárhelyi B, Török Á (2021) Long-term abrasion of rocks assessed by micro-Deval tests and estimation of the abrasion process of rock types based on strength parameters. Eng Geol 282:105996. https://doi.org/10.1016/j.enggeo.2021.105996
11. Paixão A, Fortunato E (2021) Abrasion evolution of steel furnace slag aggregate for railway ballast: 3D morphology analysis of scanned particles by close-range photogrammetry. Constr Build Mater 267:121225. https://doi.org/10.1016/j.conbuildmat.2020.121225
12. Loz PHF, Angulo SC, Rebmann MS, Tutumluer E (2021) Use of a 3D structured-light scanner to determine volume, surface area, and shape of aggregates. J Mater Civ Eng 33:04021240. https://doi.org/10.1061/(ASCE)MT.1943-5533.0003824
13. Zhihong N, Zhengyu L, Xiang W, Jian G (2018) Evaluation of granular particle roundness using digital image processing and computational geometry. Constr Build Mater 172:319–329. https://doi.org/10.1016/j.conbuildmat.2018.03.246
14. Paixão A, Resende R, Fortunato E (2018) Photogrammetry for digital reconstruction of railway ballast particles—A cost-efficient method. Constr Build Mater 191:963–976. https://doi.org/10.1016/j.conbuildmat.2018.10.048
15. AgiSoft PhotoScan (Version 1.8.6) (2022). Retrieved from http://www.agisoft.com/downloads/installer/
16. CloudCompare (version 2.12) [GPL software] (2022). Retrieved from http://www.cloudcompare.org/
17. Boltt SJ, Pey K (2008) Particle shape: A review and new methods of characterization and classification. Sedimentology 55:31–63. https://doi.org/10.1111/j.1365-3091.2007.00892.x
18. Deiros Quintanilla I, Combe G, Emeriault F, Voivret C, Ferellec J-F (2019) X-ray CT analysis of the evolution of ballast grain morphology along a Micro-Deval test: Key role of the asperity scale. Granul Matter 21:30. https://doi.org/10.1007/s10035-019-0881-y

19. Paixão A, Afonso C, Delgado B, Fortunato E (2022) Evaluation of ballast particle degradation under micro-deval testing using photogrammetry. Lect Notes Civ Eng 165:113–124. https://doi.org/10.1007/978-3-030-77234-5_10
20. Kawamoto R, Andrade J, Matsushima T (2018) A 3-D mechanics-based particle shape index for granular materials. Mech Res Commun 92:67–73. https://doi.org/10.1016/j.mechrescom.2018.07.002

Geopolymer-Based Building Blocks Using Construction and Demolition Waste

Md. Zubair, Md. Humayun Mahmood, and G. M. Sadiqul Islam

Abstract Brick and cement industries are booming with rapid urbanization and infrastructure revolution. Both sectors have a substantial carbon footprint, a significant concern for global warming. Clay brick production requires kiln burning, which causes air pollution. Moreover, traditional brick industries consume a considerable amount of topsoil layer from agricultural land. On the other hand, due to rapid urbanization and the growing housing needs, construction and demolition wastes are increasing each day. This study, therefore, aims to find an alternative solution by producing non-fired geopolymer building blocks using construction and demolition wastes. Fly ash and Ground Granulated Blast Furnace Slag (GGBS) based geopolymer binder and construction, and demolition waste filler are used to produce non-fired building blocks. The physical and mechanical characteristics of geopolymer building blocks are evaluated using various mix proportions. The activator molarity was varied to find out the optimum for geopolymer concrete. The investigated result reveals that all the prepared geopolymer building blocks achieved at least 17.2 MPa compressive strength as per ASTM C62. The produced specimens also satisfied BDS 208: 2009 standard for bricks. The highest compressive strength (36.5 MPa) was achieved with 60% GGBS and a 12 M NaOH activator. In general, the compressive strength of bricks tends to increase as GGBS amount and NaOH activator concentrations increase. These blocks have shown excellent results regarding water absorption capacity and efflorescence. The promising results create a more significant opportunity to establish an industry based on a non-fired, eco-friendly geopolymer building block which will reduce the carbon footprint and the pressure on the fertile topsoil layer of agricultural lands. This study will give a viable indication of the construction and demolition waste disposal issues.

Keywords Geopolymer · Construction and demolition waste · Building block · Fly ash · GGBS · Activator

Md. Zubair · Md. H. Mahmood · G. M. S. Islam (✉)
Department of Civil Engineering, Chittagong University of Engineering and Technology, Chattogram, Bangladesh
e-mail: gmsislam@cuet.ac.bd

S. Arthur et al. (eds.), *Proceedings of the 6th International Conference on Advances in Civil Engineering*, Lecture Notes in Civil Engineering 368,
https://doi.org/10.1007/978-981-99-3826-1_32

1 Introduction

Since the beginning of civilization, bricks have been extensively used as a primary building material. Bangladesh is the 4th largest manufacturer of bricks in the world [1]. A million people are employed in the brick manufacturing sector, which provides around 1% of the national GDP and supports economic expansion [1]. Most brick kilns use highly polluting and inefficient equipment for energy use, despite the overall significance of the sector in national growth. About 6 million tons of CO_2 are produced annually by this outdated technique. The 20,000 crore taka brick industry is harming the environment due to the increased demand for traditional fired bricks [2].

Moreover, the red brickfields are using 284 billion cubic feet of soil which are mostly topsoil; as a result, reducing food production [3]. In addition to the topsoil, this sector consumes more than 5 million tons of coal and 3 million tons of wood per year to produce fuel for burning bricks, which results in 15 million tons of emissions of greenhouse gases every year [3]. The need for housing has grown recently due to rapid urbanization and the increasing population. Construction work has been expanding at a high rate. Therefore, managing waste sustainably generated from Construction and Demolition (C&D) is becoming increasingly essential to protect public health and the natural ecology [4]. C&D waste mainly ends up in landfills, a significant concern for densely populated countries [5].

Every stage of the cement manufacturing process affects the surrounding environment, including releasing airborne dust and gas, generating noise and vibration when machinery is in operation, and quarry blasting. Cement production is estimated to be responsible for approximately 5% of global CO_2 emissions, with the chemical process responsible for 50%, while fuel combustion is responsible for 40%. Every ton of cement produced results in at least 0.9 tons of CO_2 being released into the atmosphere [6]. Therefore, it is critical to identify alternative uses for Ordinary Portland Cement (OPC) to reduce CO_2 emissions. Incorporating a geopolymer binder in place of OPC could be an effective solution. It also provides good durability and physical properties of concrete [7]. Instead of traditional cement paste, geopolymer can be used as a binder to manufacture concrete. Silicon and aluminum in a powder material are activated by sodium hydroxide and sodium silicate to bind coarse and fine aggregates and other unreacted components, resulting in a geopolymer paste [8].

In Bangladesh, the coal-burning power plant will be a significant source shortly. It is assumed that approximately 1 Million Metric tonnes of fly ash will be produced by 2024 once the three large coal-based power plants come into production, which is currently under construction [9]. Fly ash can be an excellent binder if adequately assigned to the concrete [10]. GGBS is found as a by-product material from the steel manufacturing industry. It contains cementitious as well as pozzolanic properties. It is a material with latent hydraulic potential that, in the presence of an alkaline activator, is capable of reacting directly with water [9]. GGBS, with proper alkali activation, can produce high-strength concrete [11]. On the other hand, a practical approach is required to establish alternative brick production.

An eco-friendly building block manufacturing industry is suggested as an alternative to brick industries. Developing geopolymer building blocks as a substitute for traditional fired bricks could be an essential approach to mitigate the negative environmental impact of the brick and cement industries. In the alternative brick industry, topsoil and heat are not used; instead, waste generated from Construction and Demolition (C&D) is used as aggregates. In addition, GGBS and fly ash are utilized as a base for the geopolymer binder. Therefore, this research has been performed using fly ash and GGBS-based geopolymer to transform C&D waste into concrete blocks to reduce carbon footprint and air pollution.

2 Materials

Geopolymer-based concrete building block was produced from GGBS and fly ash-based binder. No Cement was used as the binder. The compressive strength is used as the primary criterion for streamlining the development of these blocks. NaOH (8 M, 10 M, & 12 M) and Na_2SiO_3 are used to activate the geopolymer. Recycled Coarse Aggregate (RCA), i.e., C&D debris from a construction site, was considered aggregates in block production. The physical properties of RCA, such as Los Angeles abrasion value, bulk density, specific gravity, and moisture content, are found to be 33%, 1310 kg/m^3, 2.15, and 8.7%. The recycled coarse aggregates were crushed to produce recycled fine aggregates (RFA). The fineness modulus of RFA was 3.53. The particle size distribution of aggregates is shown in Fig. 1.

Ground granulated blast furnace slag (GGBS) was collected from a local cement industry that imported this for use with cement clinker. The physical properties of GGBS, such as fineness of 3120 cm^2/gm, a specific gravity of 2.3, and moisture content of 0.8%. The chemical composition of GGBS is given in Table 1. Fly ash was also collected from a local Cement Company. Class F fly ash was used for the project and complied with ASTM. The sand used in this project was obtained from the northeastern region of Bangladesh. The unit weight, specific gravity, absorption capacity, moisture content, and fine modulus was 1690 kg/m^3 2.59, 1.0%, 0.53%, and 2.59, respectively. Alkaline Activators (NaOH & Na_2SiO_3) were collected from

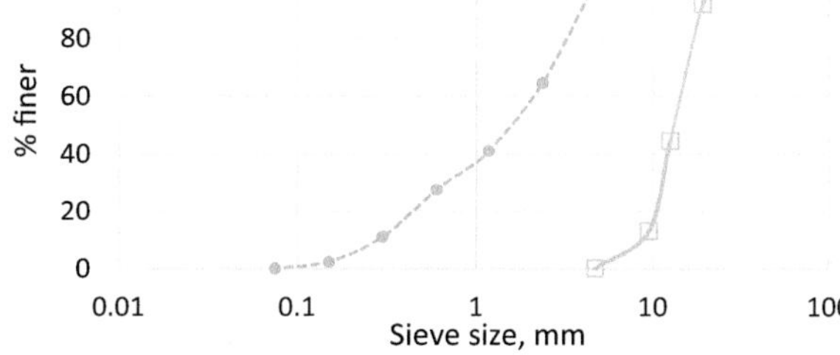

Fig. 1 Particle size distribution of RCA and RFA

Table 1 Chemical composition of GGBS

Oxides	Al_2O_3	CaO	SiO_2	SO_3	Fe_2O_3	MgO	TiO_2	Insoluble residue	Free lime	LOI
Amount, %	18.17	35.38	36.29	0.64	0.72	5.12	–	1.92	–	0.61

Table 2 Proportion of materials

Aggregates			Binder + Activator	
70%			30%	
RCA	RFA	Sand	Binder	Activator (NaOH and Na_2SiO_3)
28%	28%	14%	20.68%	9.32%

a local chemical shop. NaOH was in the form of solid flakes, and it was white. Na_2SiO_3 was in the liquid state, and it was golden brown. Potable water was used for the mixes, which was non-acidic and in the pH range of 6.5–7.5. There was no presence of turbidity.

3 Experimental Methodology

3.1 Mix Design

Six mix designs with GGBS content of 40–60% are set for this study. The remaining portion of the binder was taken with fly ash. The molarity of NaOH was varied to see the influence of activator strength. Three different molarities, viz., 8 M, 10 M, and 12 M, were used. The ratio of Na_2SiO_3 to NaOH was 2.5, and the alkaline activator solution was 45% of the total binder content. The general proportion of materials used for the project is given in Table 2, and the mix proportion to make the geopolymer blocks is shown in Table 3.

3.2 Mixing process

The materials were weighted, mixed, and cast using the brick maker machine, as shown in Fig. 2. The quantity of water used in the mix influenced the strength and shape of the blocks. The brick maker uses vibration to compact the fresh mixture. Therefore, each mix contained a different quantity of water as demanded during the mixing to maintain similar consistency. In the geopolymerization process, the mixed material was activated by NaOH solution and Na_2SiO_3. The materials were then thoroughly remixed after the solutions were added. After that, the materials were

Table 3 Mix proportions for geopolymer bricks

Mix	Aggregates			Binder		Activator		NaOH molarity
	RCA (%)	RFA (%)	Sand (%)	Fly ash (%)	GGBS (%)	NaOH solution (%)	Na_2SiO_3 (%)	
S40M8	28	28	14	12.41	8.27	2.66	6.66	8 M
S40M10	28	28	14	12.41	8.27	2.66	6.66	10 M
S40M12	28	28	14	12.41	8.27	2.66	6.66	12 M
S60M8	28	28	14	8.27	12.41	2.66	6.66	8 M
S60M10	28	28	14	8.27	12.41	2.66	6.66	10 M
S60M12	28	28	14	8.27	12.41	2.66	6.66	12 M

Notes *S40 means GGBS 40% & M8 means the molarity of NaOH is 8 M

(a) Ingredient materials (b) Brick maker machine (c) Freshly prepared blocks

Fig. 2 Geopolymer based block making process

poured into the brick-making machine. After compaction, the fresh blocks were taken out from the device. The blocks were left for air curing for 7 and 14 days before compression and other tests.

3.3 *Laboratory testing*

The bricks were intended to produce as per the PWD prescribed size in Bangladesh. Therefore, the shape, size, and color were evaluated as per BDS 208:2009. The bricks were divided into halves to test under compression. After cutting the bricks, the interface was observed manually. Compressive strength was tested as per ASTM C39. The water absorption and efflorescence of produced blocks were tested as per ASTM C67.

Fig. 3 Colour and microstructure test

Table 4 Dimensions of geopolymer building blocks

Prototype	Length (mm)	Width (mm)	Height (mm)
Geopolymer block	240	115.6	70.2

4 Results and Discussions

4.1 Shape, Size, Colour, and Structure

The form, size, colour, and architecture of bricks considerably influence the aesthetic value of brick-built structures therefore, these play a significant role in the appearance and aesthetic importance of the buildings made with bricks. The compressive strength is also directly related to the shape and microstructure of brick. Smooth and compact microstructure indicates higher compressive strength. Colour indicates the presence of pollutant substances in the brick. Figure 3 shows the colour of the geopolymer building block and its inner structure after cutting it into two-half pieces. The inside surface of the geopolymer bricks was found smooth and uniform. Table 4 shows the mean dimensions of geopolymer building blocks that comply with the PWD recommended size in Bangladesh.

4.2 Compressive Strength

Compressive strength reflects the most crucial characteristic of building components. ASTM C62 suggest a least compressive strength of 17.2 MPa for common bricks in moderate weathering condition. BDS 208:2009 set the limiting value for Grade A brick compressive strength as 15.1 MPa. Figure 4 gives that the mixture S60M12, which contains 12 molar NaOH concentration (activator) and 60% GGBS binder inclusion, has the highest compressive strength of 34.7 MPa following a week of air curing. The lowest compressive strength was found 18.7 MPa for the mix S40M8 with 40% GGBS binder replacement and 8 molar NaOH concentration. However,

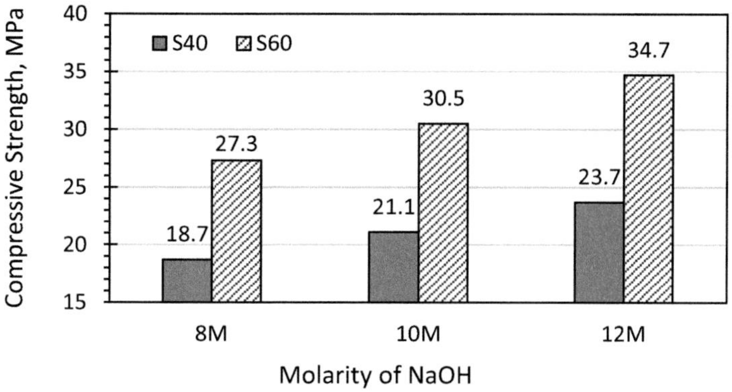

Fig. 4 Compressive strength following seven days of air curing

the lowest value of compressive strength of tested geopolymer brick complied with the requirements of moderate weathering brick (17.2 MPa) as per ASTM C62 and Grade A brick (15.1 MPa) according to BDS 208:2009.

The compressive strength improved significantly with GGBS inclusion in place of the fly ash binder. The improvement in strength is also noticed with the concentration of NaOH. Figure 4 demonstrates a 46%, 44.5%, and 46.4% increase in the compressive strength for 8 M, 10 M, and 12 M NaOH molarity when the percentage of GGBS raised from 40 to 60%. The increase in compressive strength with GGBS binder is consistent with previous research [12, 13]. Nath and Sarker [12] postulated that 10% inclusion of GGBS in place of fly ash binder increases the compressive strength up to 10 MPa. A notable amount of calcium content in GGBS could be the reason behind the increased strength of blocks. The calcium component forms a gel phase (C–A–S–H), improving the microstructure's compactness [14]. Another reason that might be attributed to the strength gain is the presence of a more specific surface area in GGBS compared with fly ash, which intensifies the geopolymerisation process and thus results in rapid strength gain [15]. Also, GGBS content being finer than fly ash could be a triggering factor for more strength gain of geopolymer block including GGBS binder.

The combination of NaOH and Na_2SiO_3 significantly influences the strength properties of geopolymer blocks. NaOH dissolute active alumina-silicates of geopolymer constituents, and Na_2SiO_3 acts as a binding agent [16, 17]. Figure 4 indicates an increase in compressive strength with increasing molarity of the NaOH activator. The 12-molar concentration of NaOH shows the best result for 40% GGBS binder replacement. However, the rate of increase in compressive strength is not as higher as found with 60% GGBS binder inclusion. This increase in compressive strength is consistent with previously carried out research works [18, 19]. Sodium ions are responsible for charge balance and, in that process, arrange an alumino-silicate network [19]. The emergence of gel and strength gain is dependent upon charge balancing and the alumino-silicate network. In case the concentration of NaOH is less, the gel that forms co-precipitate results in less strength gain. However, when the

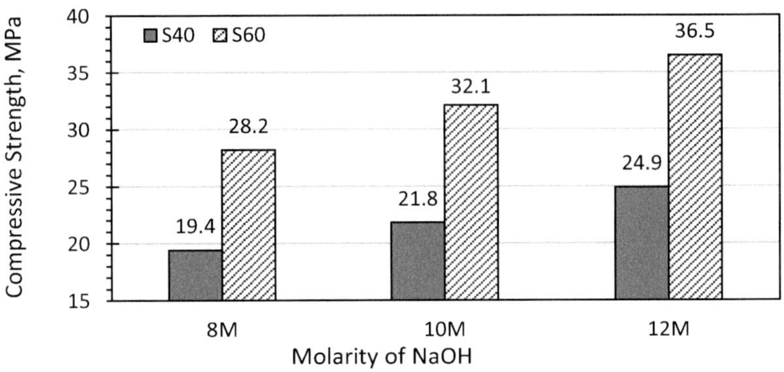

Fig. 5 Compressive Strength following 14 days of air curing

concentration exceeds the prime level, the compressive strength decreases with the increasing molarity of NaOH. The explanation is that greater alkali dissolves more silica, which forms silicic acid and reduces the strength gain of geopolymer [18]. This research found maximum compressive strength at a 12 molar concentration of NaOH, consistent with previous work [19].

Figure 5 shows that the mixture with a 12 molar NaOH activator and 60% GGBS binder inclusion has the highest compressive strength of 36.5 MPa following 14 days of air curing. However, the increase in compressive strength at 14 days was trivial (2–5%) compared to 7 days strength. This finding is harmonious with previously carried out research works. When 100% recycled aggregate was used in place of the original aggregate and 30% GGBS was added, Hu et al. [20] found a maximum increase of 7% compressive strength following 28 days compared with 7 days strength. This study was carried out without any heat-curing process. At low temperatures, GGBS governs the activation process in fly ash-GGBS geopolymers [21]. Fly ash shows pozzolanic activity slowly compared to GGBS [22]. Since GGBS is the governing factor at low temperatures, most strength is developed in the early days after casting. This results in an insignificant variation of compressive strength between 7 and 14 days.

4.3 Water Absorption

The water absorption value indicates the compactness of bricks. In geopoplymer blocks, the water absorption capacity is directly related to compressive strength. Less water absorption capacity means fewer pores in the blocks and a densely packed structure. The Alkali-activated geopolymer blocks generally show a lower water absorption capacity [7]. Figure 6 indicates the highest water absorption capacity (9.21%) for the mix S40M8. This value is well below the limiting water absorption value of severe weathering (Grade SW) brick as per ASTM C62 [23]. BDS 208:2009 permits no more than 10%, 15%, and 20% water absorption capacity for

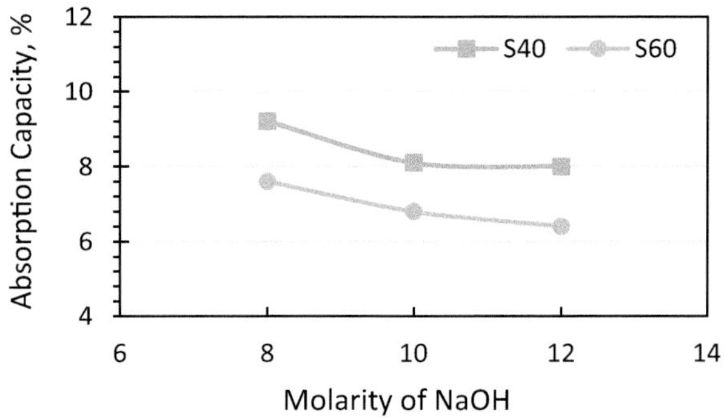

Fig. 6 Water absorption capacity of geopolymer blocks

Grade S, Grade A, and Grade B brick. Each mix design of this research satisfies the Grade S brick requirement as per BDS 208:2009. The lowest value of water absorption capacity was found for the mix S60M12. Figure 6 suggests a decrease in water absorption capacity with an increased amount of GGBS inclusion. The test results found from water absorption capacity for different mix supports the increase of corresponding compressive strength. The findings are consistent with previously carried out research works [7]. The mechanism behind this phenomenon is the presence of a more specific surface area in GGBS, and GGBS is finer than fly ash [15]. Another explanation suggests GGBS form (C–A–S–H) gels which are thicker and more impenetrable, resulting in lower water absorption [20].

The test results indicate a gradual decrease in water absorption with increasing NaOH molarity, similar to the prior research [24]. A greater NaOH molarity enhances the geopolymerization process, resulting in a compact microstructure and less water absorption. The inclusion of recycled construction and demolition wastes in place of natural aggregates negatively affects water absorption. Recycle aggregates are generally more porous than natural aggregates, resulting in more water absorption. However, all the mix specimens gave well below the limiting water absorption value of Grade SW (Severe Weathering) as specified in ASTM C62-12.

4.4 Efflorescence Test

The efflorescence test was carried out by checking any salt deposition on the tested brick surface. A pair of bricks from the same mix was placed together for the efflorescence test. Test results indicate all the specimens can be categorized as "not effloresced" as per ASTM C67. These test values were obtained by visual inspection. No presence of salt deposit was detected on the surface of geopolymer building blocks. Previous studies also indicated less chance of efflorescence for geopolymer-based building blocks [25, 26].

4.5 Cost Analysis of the Blocks

Cost is one of the critical factors when it comes to industrial-level production. Mix S40M8 shows a minimum unit price of 13.6 Tk, slightly higher than the price for 1st class brick in the market, which is 10 Tk. [27]. Na_2SiO_3 is responsible for almost 60% of the cost per unit block. A study suggests that Na_2SiO_3 to NaOH ratios of 1.5 and 2 give good results concerning geopolymer bricks' physical and mechanical properties [7]. Hence further investigation is recommended to reduce the proportion of Na_2SiO_3 to NaOH and study the compressive strength pattern for this ratio. This study didn't apply compaction pressure while casting the geopolymer blocks. The cost for binding material of geopolymer block is lowered by 50% when the compaction pressure is increased by 101% [28]. This will help to reduce the unit price of geopolymer brick to a large extent.

4.6 Practical Implication

Brick is considered one of the essential building materials globally, while Bangladesh is the fourth largest manufacturer of bricks in the world [1]. The industry provides approximately 1% of the nation's GDP and supports over one million individuals. With the current population growth rate, Bangladesh will need to build around 4 million houses yearly [1]. This will create a massive market for the brick industry. However, the problem associated with the traditional fired brick industry is its huge carbon footprint. The study reported manufacturing one brick consumes 2.0 kWh of energy and results in 0.4 kg CO_2 emissions [29]. On the other hand, the production of geopolymer building blocks does not require kiln burning; thus, no CO_2 emission in the environment occurs.

Additionally, the technology proposed in the current study uses construction and demolition waste in place of natural sand or clay. This will significantly reduce the pressure on the fertile topsoil layer of agricultural land. Geopolymer ingredients such as fly ash and GGBS are also by-products from industries and create administration concerns for safe environmental disposal. These waste materials can be used extensively in the building block business. Implementation of cost optimization can result in about 5% savings compared to conventional fired bricks [30]. The incorporation of geopolymer brick technology in place of traditional brick manufacturing will occupy the 2.53 billion USD market in Bangladesh. This study established a course of action to use geopolymer building blocks to replace traditional fired bricks for a better environment and economy in the upcoming future.

5 Conclusion

The alternative building block manufacturing process is described with industrial, construction, and demolition waste products. All the mix designs achieved the least compressive strength of 17.2 MPa as per ASTM C62 for building brick with moderate weathering. The test results also conform to BDS 208:2009. Furthermore, the water absorption complied with Grade SW (Severe Weathering) as specified in ASTM C62 and Grade S brick as per BDS 208:2009. The following inferences can be made from the test results:

The amount of GGBS inclusion noted a gradual increase in the compressive strength of the geopolymer building block. The maximum compressive strength was found for 60% GGBS binder inclusion and 12 M NaOH at 7 and 14 days of air curing. The compressive strength of geopolymer bricks improves with increasing concentration of the activator (NaOH). Therefore, water absorption capacity decreases with increasing GGBS inclusion and activator strength. Thus, the minimum water absorption was found for the mix design S60M12.

With 60% GGBS binder and 12 M NaOH gives the best result considering the compressive strength, water absorption, and efflorescence of geopolymer building blocks. Mix design S40M8 indicates the lowest price for unit bricks. The strength obtained for this mix satisfies the Grade MW (Moderate Weathering) as per ASTM C62-12. The water absorption capacity for this mix is also satisfactory for Grade SW (Severe Weathering). Hence, this mix can be used for industrial-level production of geopolymer building blocks.

Geopolymer bricks are eco-friendly. The blocks were made without any kiln burning and the use of cement. Production of these blocks will reduce the enormous carbon footprint of brick and cement industries. Construction and demolition wastes were used to make these building blocks. Therefore, this study effectively solved the construction waste dumping problem and natural resource depletion. Further analysis can optimize the cost by compacting the fresh mix with certain pressure.

Acknowledgements The Department of Civil Engineering of CUET provided the necessary laboratory facilities and technical assistance, which are gratefully acknowledged. The authors also thank Engr. Pusan Chakraborty, Sub Divisional Engineer, Chattogram PWD Maintenance, for providing a site for collecting C&D waste. The contributions of fly ash from Diamond Cement Company and GGBS from Royal Cement Company Ltd. are highly appreciated.

References

1. DoE G (2017) National Strategy for Sustainable Brick Production in Bangladesh. 41
2. The Daily Star (2022) Eco-friendly blocks made mandatory in govt projects|The Daily Star
3. The Business Standard (2019) Alternative brick production still less than 1%. In: www.tbsnews.net/environment/alternative-brick-production-still-less-1
4. Kovler K, Roussel N (2011) Properties of fresh and hardened concrete. Cem Concr Res 41:775–792. https://doi.org/10.1016/J.CEMCONRES.2011.03.009

5. Chowdhury FH, Raihan MT, Islam GMS, Ramiz F (2016) Construction waste management practice: Bangladesh perception. In: Proceedings of 3rd International Conference on Advances in Civil Engineering. pp 21–23
6. Mahasenan N, Smith S, Humphreys K (2003) The cement industry and global climate change: Current and potential future cement industry CO_2 emissions. Greenh Gas Control Technol - 6th Int Conf 995–1000. https://doi.org/10.1016/B978-008044276-1/50157-4
7. Islam GMS, Shubbar AA, Sarker S, Sadique M (2022) Ternary combined industrial wastes for non-fired brick. 1–14 https://doi.org/10.1080/13287982.2022.2038406
8. Hardjito D, Wallah SE, Sumajouw DMJ, Rangan BV (2005) On the development of fly ash-based geopolymer concrete. 467–472
9. Sarker S, Islam GMS, Sadique M, Shubbar AA (2020) Non-fired building block using industrial wastes. In: Proceedings of the 5th International Conference on Civil Engineering for Sustainable Development (ICCESD 2020), 7~9 February 2020, KUET, Khulna, Bangladesh. KUET, Khulna, Bangladesh
10. McCarthy MJ, Islam GMS, Csetenyi LJ, Jones MR (2012) Refining the foam index test for use with air-entrained fly ash concrete. Mag Concr Res 64https://doi.org/10.1680/macr.11.00152
11. Ayon MS, Author C (2020) Slag and silica fume based geopolymer mortar using. 21–23
12. Nath P, Sarker PK (2014) Effect of GGBFS on setting, workability and early strength properties of fly ash geopolymer concrete cured in ambient condition. Constr Build Mater 66:163–171. https://doi.org/10.1016/j.conbuildmat.2014.05.080
13. Nasimuzzaman M, Ayon MS, Islam GMS (2022) Slag and silica fume-based geopolymer mortar using locally available waste filler materials. Advances in Civil Engineering. Springer, Singapore, pp 165–176
14. Kumar S, Kumar R, Mehrotra SP (2010) Influence of granulated blast furnace slag on the reaction, structure and properties of fly ash based geopolymer. J Mater Sci 45:607–615. https://doi.org/10.1007/s10853-009-3934-5
15. Gunasekara C, Law DW, Setunage S (2016) Long term engineering properties of fly ash geopolymer concrete. In: Fourth International Conference on Sustainable Construction Materials and Technologies. Las Vegas, USA
16. Wang H, Li H, Yan F (2005) Synthesis and mechanical properties of metakaolinite-based geopolymer. Colloids Surf A Physicochem Eng Asp 268:1–6. https://doi.org/10.1016/j.colsurfa.2005.01.016
17. Xu H, Van Deventer JSJ (2002) Geopolymerisation of multiple minerals. Miner Eng 15:1131–1139. https://doi.org/10.1016/S0892-6875(02)00255-8
18. Singh B, Rahman MR, Paswan R, Bhattacharyya SK (2016) Effect of activator concentration on the strength, ITZ and drying shrinkage of fly ash/slag geopolymer concrete. Constr Build Mater 118:171–179. https://doi.org/10.1016/j.conbuildmat.2016.05.008
19. Abdullah MMA, Kamarudin H, Mohammed H, Khairul Nizar I, Rafiza AR, Zarina Y (2011) The relationship of NaOH molarity, Na_2SiO_3/NaOH ratio, fly ash/alkaline activator ratio, and curing temperature to the strength of fly ash-based geopolymer. Adv Mater Res 328–330:1475–1482. https://doi.org/10.4028/www.scientific.net/AMR.328-330.1475
20. Hu Y, Tang Z, Li W, Li Y, Tam VWY (2019) Physical-mechanical properties of fly ash/GGBFS geopolymer composites with recycled aggregates. Constr Build Mater 226:139–151. https://doi.org/10.1016/j.conbuildmat.2019.07.211
21. Xie J, Wang J, Zhang B, Fang C, Li L (2019) Physicochemical properties of alkali activated GGBS and fly ash geopolymeric recycled concrete. Constr Build Mater 204:384–398. https://doi.org/10.1016/j.conbuildmat.2019.01.191
22. Kejkar RB, Wanjari SP (2020) Sustainable production of commercially viable alkali-activated bricks. Proc Inst Civ Eng Eng Sustain 174:109–119. https://doi.org/10.1680/jensu.20.00001
23. ASTM International (2012) ASTM C62-12 : Standard specification for building brick (solid masonry units made from Clay or Shale). ASTM Int 10–13
24. Huseien GF, Mirza J, Ismail M, Hussin MW, Arrifin MAM, Hussein AA (2016) The effect of sodium hydroxide molarity and other parameters on water absorption of geopolymer mortars. Indian J Sci Technol 9. https://doi.org/10.17485/ijst/2016/v9i48/109629

25. Kejkar RB, Wanjari SP, Sharma D, Rajankar R (2018) Experiment investigation and physical performance of geopolymer fly ash bricks. IOP Conf Ser Mater Sci Eng 431https://doi.org/10.1088/1757-899X/431/9/092005
26. Subramani T, Sakthivel P, Ash F (2016) Experimental Investigation On Flyash Based. Int J Appl Innov Eng Manag 5:216–227
27. Public Works Department (PWD) Bangladesh (2022)
28. Ahmed E, Hassan SM-U, Islam GMS (2022) Effect of compaction pressure on the properties of eco-friendly building block produced from industrial by-products. Advances in Civil Engineering. Springer, Singapore, pp 189–198
29. Muñoz Velasco P, Morales Ortíz MP, Mendívil Giró MA, Muñoz Velasco L (2014) Fired clay bricks manufactured by adding wastes as sustainable construction material—A review. Constr Build Mater 63:97–107
30. Youssef N, Lafhaj Z, Chapiseau C (2020) Economic analysis of geopolymer brick manufacturing: A French case study. Sustainability 12:7403. https://doi.org/10.3390/su12187403

Burst Pressure Prediction of Dent-Cracked Steel Pipeline Subjected to Tensile Force

B. C. Mondal, A. Farzana, and C. K. Debnath

Abstract The dent-crack defect is considered to be one of the fatal defects in a pipeline system. The pipelines often face the presence of external loadings like tension during service life. In this study, the burst pressure of an X70 steel pipeline having a dent-crack defect has been analyzed using the Extended Finite Element Method (XFEM) when the pipe was subject to an axial tensile force. The Maximum Principal Strain (Maxpe) criterion was considered for crack initiation, and the Fracture Energy criterion was used for damage evolution. The dent-cracked pipes were modeled considering different crack sizes but one dent size. The pipes were pressurized internally up to the failure subjecting to a specified axial tensile force. The study reveals that the crack size and axial force evidently reduce the burst pressure of the pipe, where crack depth plays the leading role in reducing the pressure compared to the crack length. Finally, four interaction curves between normalized burst pressure and normalized crack area have been developed. The interaction curves might be the useful tools to predict the remaining strength of the defected pipe in terms of internal pressure.

Keywords Burst pressure · Axial tensile force · Dent-crack defect · Extended finite element method (XFEM) · Interaction curve

1 Introduction

With the advancement of industries, pipeline systems have become a more popular and essential transfer system worldwide. Particularly the oil and gas industries cannot sustain themselves without pipelines at all. Therefore, the integrity of the pipelines in the system is the most important criterion to keep the transferring and production process rolling. A pipeline system's integrity depends on various things. In the working environment, almost always, the pipelines are subject to simultaneous

B. C. Mondal (✉) · A. Farzana · C. K. Debnath
Department of Civil Engineering, CUET, Chittagong, Bangladesh
e-mail: bipul@cuet.ac.bd

S. Arthur et al. (eds.), *Proceedings of the 6th International Conference on Advances in Civil Engineering*, Lecture Notes in Civil Engineering 368,
https://doi.org/10.1007/978-981-99-3826-1_33

tension, compression and bending [1]. There are two widely used standards named BS 7910 [2] and R6 [3] those are used to evaluate the fracture resistance in terms of strain rate. They are substantially limited to minor plastic strains and established for an elastic response only, whereas the pipe material shows considerable plastic deformation before failure. Therefore, a precise and more comprehensive method for fracture evaluation of pipelines is needed which will also be compatible with the deformation capacity of the pipeline under those types of external forces.

During the service life, a pipeline may be subjected to different types of geometric and mechanical defects such as corrosion, dent, crack, corrosion with crack, and dent-crack defects. The defect is represented by the defect area which is the product of defect length, L and defect depth, d (i.e., L × d). The presence of these defects reduces the strength of the pipelines considerably. The extend of strength reductions is influenced by the size, shape, location, and orientation of the defects. Over the last few decades, comprehensive investigations were conducted on corroded pipelines [4–14], dented pipelines [15–18], cracked pipelines [5, 7, 9], and corroded pipelines with crack like defects [19–21]. But the investigations on dent-crack defects are limited up to the knowledge of the authors. Using numerical analysis, Okodi [22] studied the effects of dent-crack on the strength of the pipeline subjected to internal pressure only, whereas the pipelines are subjected the combined loads of internal pressure and axial load during operating condition.

In laboratory test, the cracks are depicted as notches and gouges with a minor fatigue crack at the tip, and in numerical analysis, the cracks are simulated as notches with small openings [7, 23]. Moreover, in the existing literature, the cracked pipelines were analyzed numerically using fracture mechanics approach based on stress intensity factors [24]. Besides, in calculating the stress intensity factors, the material separations, by crack propagation, are not allowed that may affect the strength of the cracked pipelines predicted by conventional finite element analysis (FEA). The above-mentioned shortcoming of the conventional finite element method can be resolved by the advanced finite element method named Extended Finite Element Method (XFEM). This new method nearly simulates the experimental scenario. Therefore, for precise strength prediction of a pipeline with a dent-crack defect, the analysis should be conducted using XFEM. In XFEM, the material discontinuity due to cracking is modelled using a displacement jump function, $H(x)$, where the nodal displacement vector is defined using $H(x)$, as shown in the following equation [25].

$$\boldsymbol{u} = \sum_{I=1}^{N} N_I(x)\left[u_I + H(x)a_I + \sum_{\alpha=1}^{4} F_\alpha(x)b_I^\alpha\right] \tag{1}$$

where $N_I(x)$ = nodal shape function

u_I = usual nodal displacement vector.

a_I = nodal enriched degree of freedom vector.

$H(x)$ = displacement jump function across the crack surface

$$= \begin{cases} 1 & \text{if}(x - x^*) \cdot n \geq 0 \\ -1 & \text{Otherwise} \end{cases}$$

$F_\alpha(x)$ = elastic asymptotic crack-tip function

$$= \left[\sqrt{r} sin\frac{\theta}{2}, \sqrt{r} cos\frac{\theta}{2}, \sqrt{r} sin\theta sin\frac{\theta}{2}, \sqrt{r} sin\theta cos\frac{\theta}{2} \right] 0$$

$b^\alpha{}_I$ = nodal enriched degree of freedom vector
x = sample (Gauss) point
x^* = point on the crack closest to x
n = unit outward normal to the crack at x^*
r, θ = polar coordinate system with its origin at the crack tip

In this study, the pipelines with dent-crack defect have been analyzed using XFEM, where the pipelines were subjected to combined loads of axial force and internal pressure. The material non-linearity and geometric non-linearity were employed in the FE models. From the results of analyses, the interaction diagrams between burst pressure and defect area (i.e., a) have been developed that can be used to predict the remaining strength of the defected pipeline pipelines. Here, the defect area, "a" is defined by the area of the crack which is equal to length times width of the crack.

2 FE Analysis

The FE analysis provides a powerful tool for modelling complex problems. Among the commercially available software for FE analysis, Abaqus is most commonly used for analysis of pipelines. Abaqus has the capability of modelling the non-linear deformation during yielding of corroded pipeline under high pressure. Abaqus/Standard module with XFEM is used in this study for calculation of burst pressure of dent-cracked pipelines. The crack initiation and its propagation were defined using maximum principal strain criteria (Maxpe) and fracture energy-based damage evaluation criteria (G_C), respectively, with the linear traction–separation law.

2.1 FE Model

The FE model of the studied pipeline was developed using the geometric and material properties of the pipeline shown in Table 1 and in Fig. 1. The properties were collected from Okodi [22]. To save the computation time, the pipe domain was modeled using four nodded continuum shell elements with reduced integration (Abaqus element name, S4R). To simulate the dent-crack model, a solid strip of length 250 mm, width 20 mm, and depth 8.5 mm was developed using eight nodded brick elements with reduced integration (Abaqus element name, C3D8R) in which the crack was

Table 1 Geometric and material properties of pipeline [22]

Property	Value
Outer diameter, D (mm)	762
Wall thickness, t (mm)	8.5
Pipe Length (mm)	1250
Density, ρ (kg/m^3)	7850
Yield strength, σ_Y (MPa)	540
Ultimate strength, σ_U (MPa)	620
Modulus of elasticity, E (GPa)	204
Poisson's ratio, ν	0.3
Maximum principal strain, Maxpe	0.045
Fracture energy, G_C (kJ/m^2)	2

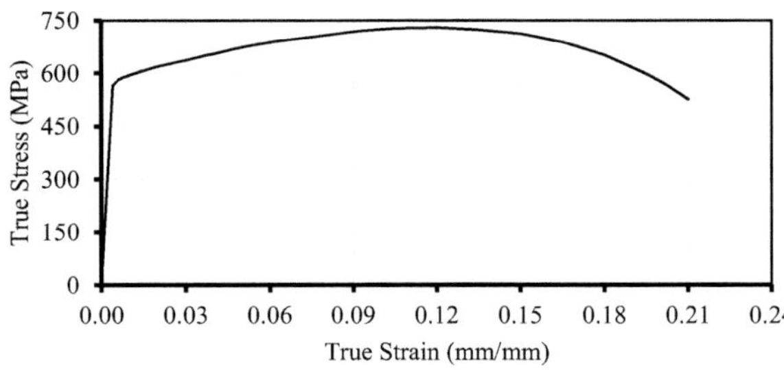

Fig. 1 Stress–Strain Diagram of X70 Steel (After [22])

embedded. A portion of the shell pipe, similar to the solid strip, was removed and the solid strip was placed there. The shell-solid coupling constraints were employed to connect the solid strip to the shell pipe. Considering the advantage of symmetricity, half of the pipe was modeled applying suitable symmetric boundary conditions.

To apply the axial force to the pipe wall, the defect free pipe end was connected to a reference point located at the center of the pipe section using kinematic constraint allowing suitable degree of freedom. To create the required dent, a discrete rigid solid block, known as indenter, located above the center of the crack was moved downward up to the specified distance. The indenter's dimensions were 100 mm × 50 mm. The surface-surface contact was applied between the indenter and the pipe surface. A full analysis was completed following the steps outlined below:

Step-1: The displacement-controlled load was applied to the solid trip using the indenter to create the required dent.
Step-2: The specified axial tensile force (T) was applied to the reference point.
Step-3: The internal pressure was applied to the inner surface of the pipe and the solid trip. The pressure was increased gradually up to the failure of the pipe.

The mesh sensitivity analysis was conducted to determine the optimum mesh size. Fine mesh was applied at the dent-crack zone and coarse mesh was applied away from the dent-crack zone with appropriate bias ratio. Figure 2 shows a typical FE

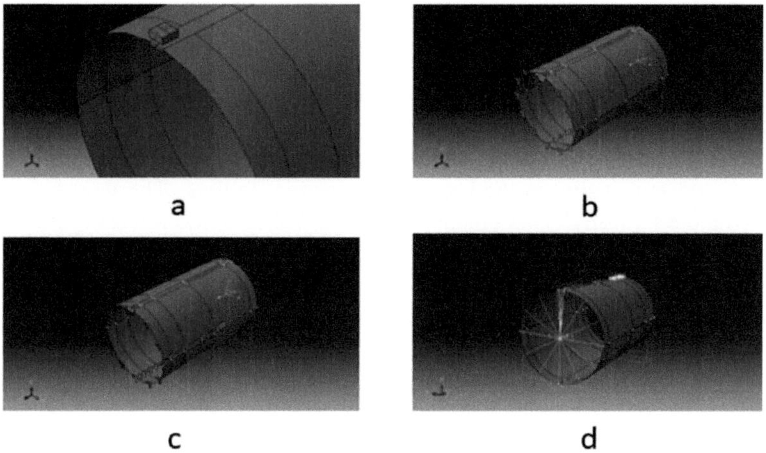

Fig. 2 **a** Assembly of the parts **b** Z-symmetric Boundary Condition **c** Pin Boundary Condition **d** Coupling

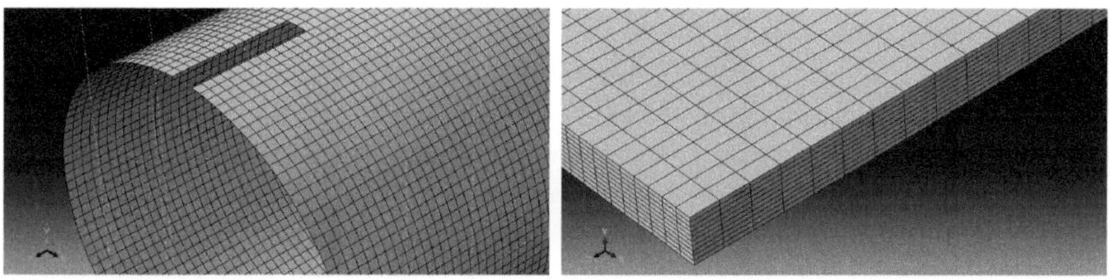

(a) Meshing of Shell Pipe (b) Meshing of Solid Strip

Fig. 3 Typical Meshing of a FE model **a** Meshing of Shell Pipe, **b** Meshing of Solid Strip

model of the pipe with boundary conditions and Fig. 3 shows a typical meshing of the FE model.

2.2 Failure Criteria

The strength determined by a numerical analysis depends on the failure criteria considered in the analysis. The commonly used von Mises failure criteria was employed in this investigation, where the internal pressure producing the average von Mises equivalent stress, through the ligament of the pipe wall, similar to the ultimate tensile strength of the pipe material (i.e., 620 MPa) is known as the burst pressure (*P*) [26].

2.3 Validation of the FE Model

For the validation purpose, a dent-cracked pipe was modelled considering a dent size of 100 mm long, 50 mm wide and 30 mm depth and an embedded crack size of 60 mm long and 4.3 mm depth. The dimensions were collected from Okodi [22]. All other properties using in the model are given in Table 1. The dent was created by applying a 63.5 mm displacement load to the indenter and releasing the indenter again. Figure 4 shows the load–displacement curves obtained by the FE analysis and the experiment as well. The test results were collected from Okodi [22]. The figure indicates that the FE model reasonably simulates the test result.

3 Results and Discussions

To investigate the effects of crack dimensions on the burst pressure of the dented pipe, cracks with different lengths and depths were considered in the study. Table 2 depicts the dimensions of the cracks considered in the study. An intact pipe (i.e., defect free pipe) was also considered in the study to calculate the degradation in strength of the dent-cracked pipes. The ultimate tensile force (T_0) of the intact pipe was calculated by assuming that the full pipe section reaches the ultimate tensile strength (i.e., 620 MPa) and it was found as 12.5×10^6 N. Similarly, the burst pressure of the intact pipe (P_0) was calculated according to Burlow's Equation and it was found as 14 MPa. The tension was applied as a percentage of the ultimate tensile force ($0.0T_0$, $0.2T_0$, $0.4T_0$, $0.6T_0$, $0.8T_0$). For rest of the analysis, the dent size of 100 mm long, 50 mm wide and 30 mm depth was used.

Fig. 4 Load–Displacement Curve

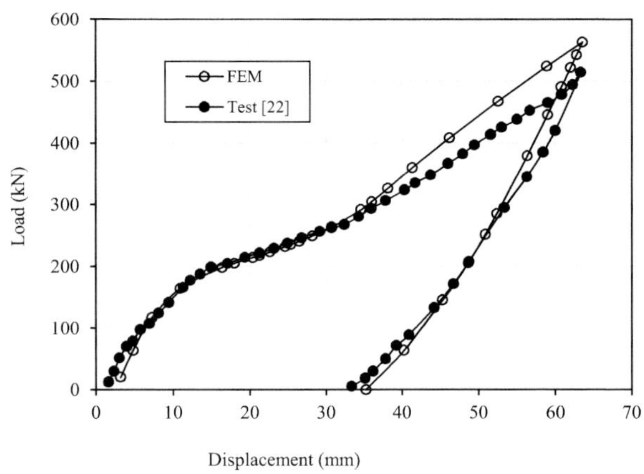

Table 2 Crack dimensions

Model no	Length, L (mm)	Depth, d (mm)
1	20	4.3
2	60	4.3
3	100	4.3
4	140	4.3
5	60	1.7
6	60	6.8
7	100	6.8

3.1 Length of Crack

To study the influence of crack length on the burst pressure of dent-cracked pipes (*P*) subjected to tensile force, four different crack lengths ranging from 20 to 140 mm with a constant crack depth of 4.3 mm were selected. Figure 5 shows the variation of burst pressure with crack length where the pipes were subjected to different tensile forces (*T*) varying from 20% to 80% of T_0. The figure reveals that the crack length reduces the burst pressure with exponential decay for any tensile force. The figure also shows that the tensile force reduces the burst pressure considerably, where the reduction is higher for longer crack length. For the studied magnitude of tensile forces, all curves provide almost similar trendline with high value of coefficient of determination (i.e., $R^2 > 0.9619$).

3.2 Depth of Crack

Similar to the crack length, the influence of crack depth on the burst pressure of dent-cracked pipes (*P*) subjected to tensile force was investigated considering three different crack depths ranging from 1.7 to 6.8 mm with the constant crack length of 60 mm. The magnitudes of tensile forces were similar to that for the investigation of crack lengths. Figure 6 shows the variation of burst pressure with crack depth for the studied tensile forces. The figure indicates that the crack depth reduces the burst pressure almost linearly. Similar to crack length, the reduction is higher for greater crack depth.

3.3 Tensile Force

Although, the effects of tensile forces on the burst pressure of a dent-cracked pipe have already been discussed briefly in the previous sections, but it has been investigated

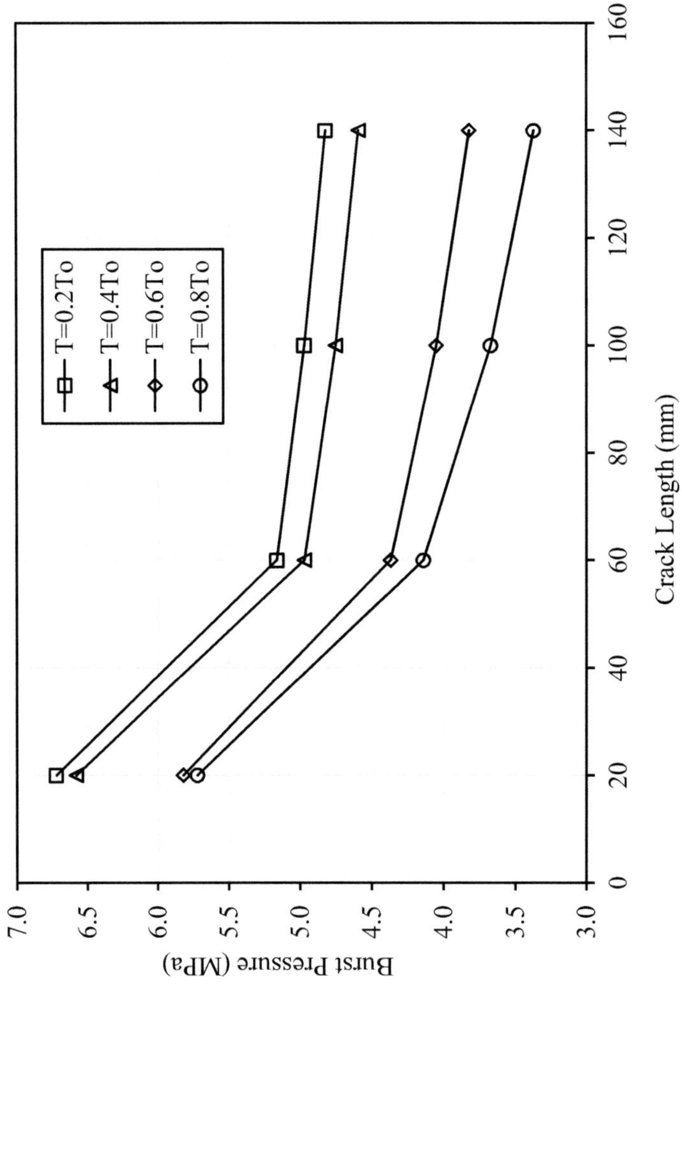

Fig. 5 Effect of crack length in association with tensile force on the burst pressure

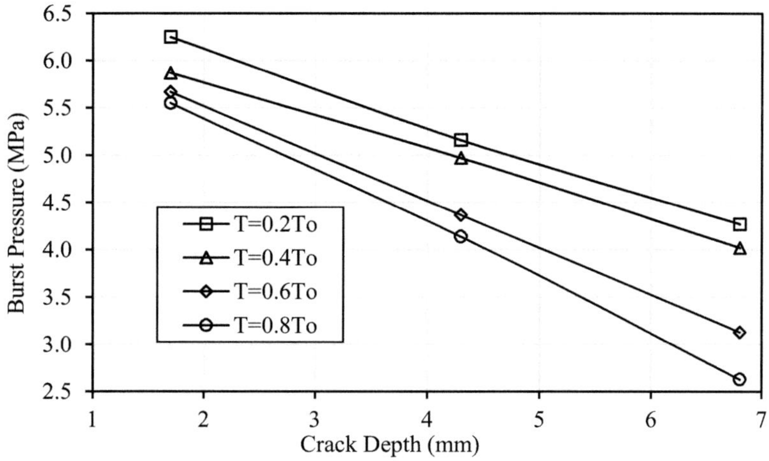

Fig. 6 Effect of crack dept in association with tensile force on the burst pressure

again in details in this section. For this objective, the burst pressures of the dent-cracked pipes with the dimensions given in Table 2 are determined for different magnitude of axial tensile forces. The burst pressures are categorized based on crack depths and crack lengths. The lengths and depths are normalized by the wall thickness (t). Similarly, the burst pressures and axial tensile forces are normalized by P_0 and T_0, respectively.

Figure 7 shows the variations of normalized burst pressure with normalized axial tensile force, where Fig. 7(a) belongs to the category based on crack length and Fig. 7(b) belongs to the category based on crack depth. The figure indicates that the burst pressures reduce drastically up to the normalized tensile force of 0.20 for both categories. The reduction in burst pressure does not vary significantly from the normalized tensile force of 0.20 to 0.80. Figure 7(a) also reveals that the burst pressure remains almost unchanged when the crack length is greater than 7 t, but the crack depth reduces the burst pressure continuously as shown in Fig. 7(b). The crack depth is more harmful the crack length. Therefore, more attention should be paid to the crack depth compared to the crack length to ensure the integrity of the pipe during operation.

4 Interaction Curve

The remaining strength (i.e., burst pressure) of a dent-cracked pipe can be predicted using closed form equations or using interaction curves. In this study, four interaction curves have been developed for four axial tensile forces as shown in Fig. 8. From the above discussion, it is observed that the remaining strength of a dent-cracked pipe depends on crack dimensions (i.e., crack length and crack depth) for the studied

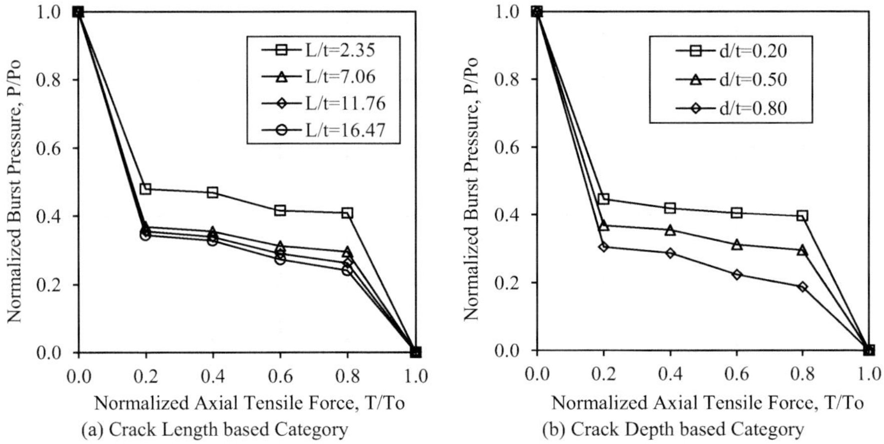

Fig. 7 Effect of axial tensile force on the burst pressure. **a** Crack Length based Category, **b** Crack Depth based Category

pipe with the studied dent size. Therefore, to apply the interaction curve for any crack sizes with the said dent size, the interaction curves have been constructed in terms of normalized burst pressure (i.e., P/Po) and normalized crack area (i.e., a/A) in percentage, where "a" indicates the area of the crack (= L × d)) and "A" indicates the cross-sectional area of the intact pipe ($= \frac{\pi}{4}(D_o{}^2 - D_i{}^2)$), where D_o and D_i indicate the outer and inner diameter of the intact pipe. Applying these interaction curves, the remaining strength of a cracked pipe with the studied dent size can be predicted for any crack size having the known geometric and material properties of the pipe. Further study should be conducted to develop interaction curves for other dent sizes.

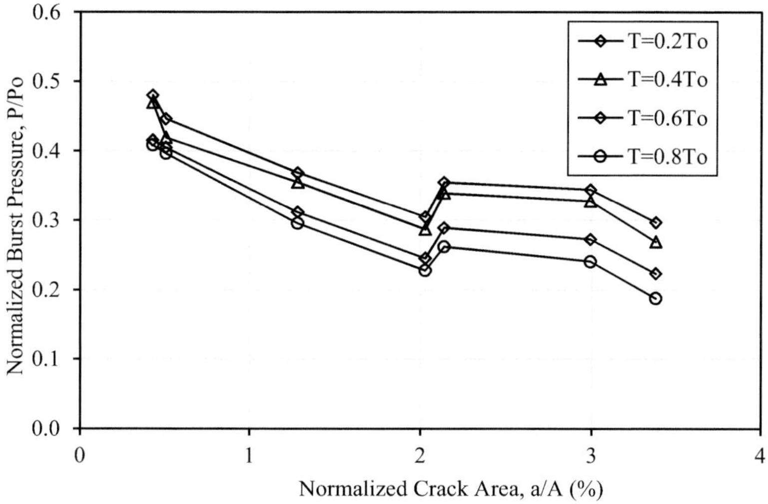

Fig. 8 Failure Loci for axial tensile force

5 Conclusion

This paper presents an investigation on the remaining strength (i.e., burst pressure) of a pipe with a dent-crack defect subjected to axial tensile force using Extended Finite Element Method (XFEM). The findings from this study are summarized below:

- The crack dimensions reduce the burst pressure of the dent-cracked pipes considerably, where the crack depth is more harmful compared to the crack length.
- The remaining strength keeps unchanged when the crack length is greater than 7 t, but it is degraded continuously with crack depth.
- The axial tensile force also reduces the burst pressure of pipe significantly.
- Since, the remaining strength depends on crack lengths as well as crack depths, so four interactions curves have been constructed for four axial tensile force in terms of normalized crack area. The curves are useful tools to predict the remaining strength of the dent-cracked pipe with the studied dent size.

References

1. Saneian M, Han P, Jin S, Bai Y (2020) 'Fracture response of steel pipelines under combined tension and bending'. Thin-Walled Struct 155: 106987. https://doi.org/10.1016/j.tws.2020.106987
2. 'Guide on methods for assessing the acceptability of flaws in metallic structures'(1999) British Standard Institution London, UK
3. 'Assessment of the integrity of structures containing defects' (2001) R6 Revision

4. Kiefner JF, Vieth PH, Pipeline Research Council International I, Institute BM (1989) A modified criterion for evaluating the remaining strength of corroded pipe. Pipeline Research Committee, American Gas Association
5. Chiodo MSG, Ruggieri C (2009) 'Failure assessments of corroded pipelines with axial defects using stress-based criteria: Numerical studies and verification analyses'. Int J Press Vessel Pip 86(2–3): 164–176.https://doi.org/10.1016/J.IJPVP.2008.11.011
6. Mondal BC, Dhar AS, Hafiz HI (2022) 'Burst pressure assessment of pipe bend/elbow for transmission pipelines'. Thin-Walled Struct 174: 109148. https://doi.org/10.1016/J.TWS.2022.109148
7. Fekete G, Varga L (2012) 'The effect of the width to length ratios of corrosion defects on the burst pressures of transmission pipelines'. Eng Fail Anal 21: 21–30.https://doi.org/10.1016/J.ENGFAILANAL.2011.12.002
8. Star N, North H, Avenue S, Sn S, Tel UE (1965) 'Revisiting burst pressure models for corroded pipelines'. 1(709): 1–41
9. Mondal BC, Dhar AS (2017) 'Improved folias factor and burst pressure models for corroded pipelines'. J Press Vessel Technol 140(1). https://doi.org/10.1115/1.4038720
10. Mondal BC, Dhar AS (2017) 'Improved folias factor and burst pressure models for corroded pipelines'. J Press Vessel Technol 140(1https://doi.org/10.1016/J.ENGSTRUCT.2019.02.010
11. Mondal BC, Dhar AS (2019) Burst pressure assessment of corroded pipelines using fracture mechanics criterion. Eng Fail Anal 104:139–153. https://doi.org/10.1016/J.ENGFAILANAL.2019.05.033
12. Zhang S, Zhou W, Zhang S (2020) 'Development of a burst capacity model for corroded pipelines under internal pressure and axial compression using artificial neural network'. https://doi.org/10.1115/IPC2020-9631
13. Zhu XK (2021) A comparative study of burst failure models for assessing remaining strength of corroded pipelines. J Pipeline Sci Eng 1(1):36–50. https://doi.org/10.1016/J.JPSE.2021.01.008
14. Zhang S, Zhou W (2020) Assessment of effects of idealized defect shape and width on the burst capacity of corroded pipelines. Thin-Walled Struct 154:106806. https://doi.org/10.1016/J.TWS.2020.106806
15. Jajo JO (2014) 'Scholarship at UWindsor Dent behaviour of steel pipes under pressure load'
16. Tian X, Zhang H (2017) Failure criterion of buried pipelines with dent and scratch defects. Eng Fail Anal 80:278–289. https://doi.org/10.1016/J.ENGFAILANAL.2017.06.049
17. Shuai Y, et al. (2020) 'Local buckling failure analysis of high strength pipelines containing a plain dent under bending moment'. J Nat Gas Sci Eng 77 https://doi.org/10.1016/j.jngse.2020.103266
18. Cosham A, Hopkins P (2004) The effect of dents in pipelines—Guidance in the pipeline defect assessment manual. Int J Press Vessels Pip 81(2):127–139. https://doi.org/10.1016/j.ijpvp.2003.11.004
19. Saneian M, Han P, Jin S, Bai Y (2020) 'Fracture response of steel pipelines under combined tension and torsion'. Thin-Walled Struct 154. https://doi.org/10.1016/j.tws.2020.106870
20. Seif T (2014) 'Ductile fracture analysis in a steel plate by cohesive zone modeling'. Memorial University of Newfoundland
21. Pluvinage G, Capelle J, Schmitt C (2016) 'Chapter 3—Methods for assessing defects leading to gas pipe failure'. In: Makhlouf ASH, M. B. T.-H. of M. F. A. with C. S. from the O. and Aliofkhazraei GI (Eds.) Butterworth-Heinemann. pp 55–89
22. Okodi AR (2021) 'Burst pressure of pipeline with longitudinal crack in dent defects using extended finite element method'. University of Alberta
23. Bai Y, Song R (1997) Fracture assessment of dented pipes with cracks and reliability-based calibration of safety factor. Int J Press Vessels Pip 74(3):221–229. https://doi.org/10.1016/S0308-0161(97)00113-0
24. Orynyak IV, Bogdan AV, Rozgonyuk VV (2004) 'Ductile fracture model for a pipe with a dent'. In: Denys R (Ed.) Proc. 4th Int. Conf. on Pipeline Technology, Ostend, Belgium, pp 949–960

25. Belytschko T, Black T (1999) Elastic crack growth in finite elements with minimal remeshing. Int J Numer Meth Eng 45:601–620
26. Mondal BC, Dhar AS (2019) Burst pressure of corroded pipelines considering combined axial forces and bending moments. Eng Struct 186(February):43–51. https://doi.org/10.1016/j.engstruct.2019.02.010

Heavy Metal Pollution Assesment in Urban Road Network; a Case Study of Chattogram

S. Deb, R. Prodhan, and S. K. Pal

Abstract Road surfaces generate a significant level of pollution derived from the road traffic environment influenced by the surrounding land uses. Among may different pollutants associated with road deposited sediment, selected heavy metals are commonly found in elevated concentrations and found to be varied with site-specific characteristics, and hence, it's assessment towards human and aquatic health is deemed necessary. The present investigation was conducted at five different locations with different land uses pattern in Chattogram city to assess pollution potential of road dust at present time frame. Upon collection of road dust from selected sites, the concentrations of Cu, Cd, Pb, Ni and Cr were determined using atomic absorption spectroscopy (AAS) following strong nitric acid digestion of sediment for heavy metal extraction from solid to liquid form. To assess the pollution potentials, degree of contamination, ecological risk index, integrated pollution index, the geo-accumulation index, were computed and evaluated. Among the five study locations, named as Bahaddarhat, Patenga, Bayazid, Didar Market and Kaptai Rastar Matha, the maximum concentration of Cu, Pb, Cd, and Cr are at Patenga sampling site with 121, 164, 2, 92 mg/kg, respectively, while the maximum concentration of Ni was found as 54 mg/kg at Didar Market site. The Pb concentration was found as high as 13 times of its background concentration, indicating strong anthropogenic inputs. Among sites, Didar Market and Patenga sampling sites standing at top exhibiting the highest levels of contamination, whereas Bahaddarhat and Kaptai Rastar Matha indicating moderate pollution and Bayazid site was found with bottom standing levels. The high level of metal contamination in urban soil enriched with the pollution indices illustrates that a significant human and aquatic health hazards are associated with road dust that needs appropriate control measures.

S. Deb (✉) · R. Prodhan · S. K. Pal
Department of Civil Engineering, CUET, Chittagong, Bangladesh
e-mail: u1601098@student.cuet.ac.bd; soniadeb59@gmail.com

R. Prodhan
e-mail: u1601117@student.cuet.ac.bd; rashedprodhancuetce16@gmail.com

S. K. Pal
e-mail: sudip@cuet.ac.bd

S. Arthur et al. (eds.), *Proceedings of the 6th International Conference on Advances in Civil Engineering*, Lecture Notes in Civil Engineering 368,
https://doi.org/10.1007/978-981-99-3826-1_34

Keywords Heavy metal contamination · Road dust · Risk index · Pollution index · Geo-accumulation index

1 Introduction

Environmental pollution is a prevalent issue nowadays, and it is becoming worse every day. Due to increased urbanization, industrialization, population growth, land usage, and food production techniques, a significant issue for developing nations is the environmental damage brought on by heavy metals [1]. Every day, the number of automobiles increases in a city with rapid growth like Chattogram. The most significant causes are traffic patterns, road surfaces, and air deposition [2]. The vehicular emissions are a major source of pollution that is seriously harming the ecosystems. The configuration of the roads, the speed of the vehicles, and the nearby infrastructure all have an impact on the pollution load. Pollutant build-up is influenced by road shape and traffic circulation patterns [3]. The street, which is caused by traffic or industrial activities, is the largest source of pollution in metropolitan areas. It has a detrimental effect on both our health and the environment. Numerous hazardous compounds found in street dust raise the risk to human health. Street dirt and soils have extremely high concentrations of Cr, Ni, Cu, Pb, Zn, Cd, and other metals. The majority of the metals found in soils and road dirt, such as Cd, Cr, Ni, Cu, Pb, and Zn, have evolved through industrial and vehicular emissions. Pollutants are primarily produced as a result of human activity anthropogenic activities include trade, construction labour, traffic, and metal erosive processes [4].

In most circumstances, city wastewater and car exhaust can also have an impact on the sources of metals in agricultural soils. It has a detrimental effect on both our health and the environment. It is a problem to be concerned about when heavy metals like cadmium, lead, chromium, copper, etc. contaminate the soil. Heavy traffic is a significant cause of infection in the majority of the roadside soils, even though heavy metals are naturally present in the soil. Typically, low levels of heavy metals can be found in soil, vegetation, and living things. However, at large concentrations, these are poisonous to living things. The root system is harmed as a result of heavy metal poisoning, which also impacts plant growth. Traffic density, one of the primary causes of heavy metal contamination, is typically the cause of the high levels of heavy metals in the soil and plants beside metropolitan roads. Due to its unsettling effects, recently, there has been a lot of interest in determining the level of contamination and risk that heavy metals in road dust represent. The effects of heavy metal poisoning on Chattogram city's inhabitants and environment in Bangladesh are also a major issue. Chattogram, the port metropolis and second-largest city in Bangladesh, is one of the most important commercial hubs in South Asia. This population and its current growth have increased the need for services in all areas of civilization. A significant amount of development work, such as road construction, flyover or elevated road construction, building construction, industrial infrastructure, etc., has been completed in both the public and private sectors to meet their needs and provide them with the amenities they require. In addition, this city's transportation and traffic systems are becoming increasingly congested due to its growing urbanization and commercial activity. It accelerates the emission of particulate matter, industrial waste, oil lubricants, and automobile parts. As a result, the amount of road dust in Chattogram is drastically increasing, and it is also thought that the city's heavy metal contamination is increasing at the same time. This could have a negative impact on city residents, pedestrians, tourists, and other groups. The assessment of environmental pollution with relation to the risk of heavy metals in this city's road dust is unavoidable in this situation.

Although several studies exist in relation to heavy metal assessment and pollution aspects, according to earlier studies the concentration of heavy metals varies greatly depending on traffic volume, urban development, traffic maneuvering pattern, road alignment etc., that are too site specific. The present study conducted different types of indices to assess the hazard associated with road sediment with heavy metals in different areas of Chattogram city. The purpose of this study is to assess the pollution potential of road sediment towards probable ecological risk and human health caused by heavy metals in Chattogram urban area. It is hope that the outcomes of this study might be helpful for city council to adopt appropriate measure to control urban point and nonpoint pollution associated with urban road sediment. Furthermore, it might be helpful for building awareness among road users to abate health hazard.

2 Materials and Methods

2.1 Study Area

Five different locations were selected in Chattogram city. These areas are Bahaddarhat (Bh), Patenga (Pt), Bayazid (Bz), Didar Market (D.M) and Kaptai Rastar Matha (K.R.M). Those locations were chosen because heavy metals are approaching and there are moving buses, trucks, and other motor vehicles in the region. All the samples were collected near the junction of the road. Location details are as follows: Patenga is an industrial area such as electronic and electric industries, refineries, Karnafully EPZ, shoe factories, oil and agro chemicals plants, truck parks etc. The busiest roads of Patenga is overcrowded with trucks, lorries and buses. Bahaddarhat is the partially commercial and marketplace busy with heavy vehicles and public transports. Didar Market is the marketplace. Here also pass the significant amount of vehicles. Kaptai Rastar Matha is partially commercial and industrial area such as textiles factories, shoes factories and metal welding shops. Here the road sites also busy with heavy vehicles which contributes the heavy metal pollutions. In comparison to the other study site, Bayazid has fewer heavy vehicles and low vehicle congestion because it is a residential and institutional area (Fig. 1).

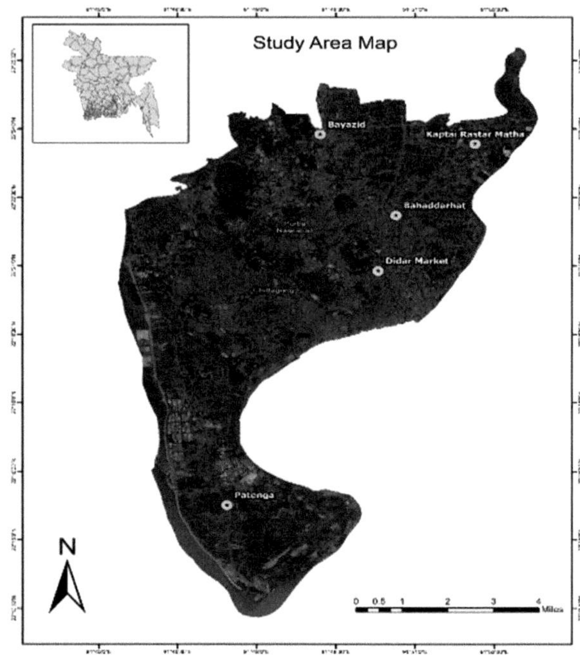

Fig. 1 Sampling location marked in green dot at Chattogram City, Bangladesh

2.2 *Collection of Samples*

A brush was used to gather sample of the dry dust that had accumulated on the road surfaces. This is the most widely used sampling method that has frequently been observed in other studies [5–7]. We marked 1 m^2 area on the road surface with spray paint. The marked area was cleaned to take sample. The number of samples collected from each site was about 90 bags consists of at least 2 to 3 bags after 24 consecutive hours for 30 days. A 1 m^2 collection area was set up and the particulate materials were gathered from the enclosure, which was located close to the curb, at each of the 5 study locations. Near the curb, there is a significant portion of sediments. With the aid of a fine brush, the marked area was properly brushed several times to collect as much road dust as possible. During sampling, care was taken to minimize sweeping pressure so that detachment of road material particles could be avoided. The sample was kept in a sealed bag and then transported back to the Environmental Engineering Laboratory at CUET. Point to be noted that each bag has been marked or levelled properly for identification.

2.3 List of Apparatus and Equipment Used

The following apparatus and equipment are used for sample collection, preservation and analytical determination: (a) Sample collecting plastic bags, (b) Plastic brush, (c) Painting spray, (d) Flask, (e) Filter paper, (f) Volumetric flask, (g) Nitric Acid & Hydrochloric Acid.

2.4 Sample Preparation and Laboratory Testing

After successfully collecting the data from five sites the sample was carried out to the laboratory. Then the sample was sieved by No. 200 sieve. Then we weighted the sample which is passed No. 200 sieve also retained samples are weighted. Then we store the sample which is passed No. 200 sieve. The sample which is passed through the No. 200 sieve had been digested with strong HCl and HNO_3 acid using the process identified in this research [8]. The ratio of HCl and HNO_3 was 3:1 for acid digestion, and 5 gm of road sediment was digested with 30 mL of acid mixer. The sample was taken into flask and heated with 150–180 °C in electric hot plate until white fumes appeared from the sample. Then the sample was cooled in room temperature for filtration through Whatman No. 42 filter. After filtration sample was ready to test. The concentration of Cu, Pb, Ni, Cd and Cr was tested by Atomic Absorption Spectroscopy method employing accepted analytical procedure [8, 9].

2.5 Assessment of Pollution Level of Heavy Metal

There is a wealth of material in the literature about pollution caused by heavy metals in relation to sediment deposited on roads or in any sediment [10, 11]. To determine the contamination of suspended or re-suspended road dust collected from roadside vegetation, obstacles, and islands, the several indices, such as geo-accumulation index (Igeo), contamination factor (Cf), degree of contamination (Cdeg), ecological risk factor (Er), risk index (RI), pollution index (PI) and integrated pollution index (IPI) were estimated and evaluated.

Geo-accumulation index (Igeo) is used to measure anthropogenic rates of toxicity and contrast various soil and sediment metals developed by Muller proposed the following formula [12].

$$\text{Igeo} = \log_2\left(\frac{C_n}{1.5 \times B_n}\right) \tag{1}$$

where C_n represents the concentration of the element in the enriched samples and B_n represents the background concentration of the element. Igeo was divided into seven categories, ranging from exceedingly unpolluted to heavily polluted, as values increased numerically.

According to Hakanson [13], Pal [8], and Suryawanshi et al. [11], the degree of contamination (Cdeg), shown in Eq. (1), is the total of the contamination factors for each element under consideration.

$$\mathrm{C\,deg} = \sum_{i=1}^{n} \mathrm{Cf}^{i} \tag{2}$$

The contamination factor Cf was described in Eq. (2)

$$\mathrm{Cf}^{i} = \frac{C_i}{C_n} \tag{3}$$

where C_i is the mean concentration of metal and C_n is the concentration of a reference or background value for each metal. The Indian natural soil background was used to generate C_n for soil in this study by Kuhad et al. [14] and Gowd et al. [15]. According to Suryawanshi et al. [11], who suggested revising the categories of degree of contamination linked with six heavy metals in road sand tested here, as necessary based on the concept derived by Hakanson [13].

The determination of the level of heavy metal pollution is carried out using the Risk Index (RI), developed by Hakanson and the following formula [13].

$$\mathrm{RI} = \sum_{i=1}^{n} \mathrm{Er}^{i} \tag{4}$$

$$\mathrm{Er}^{i} = \sum_{i=1}^{n} \mathrm{Tr}^{i} \cdot \mathrm{Cf}^{i} \tag{5}$$

where RI is the requested potential ecological Risk Index, Er^{i} is the Potential ecological risk factor and Tr^{i} is Toxic-response factor. Eight pollutants formed the basis of RI classification recommended by Hakanson [13]. In order to evaluate the RI by the analyzed six heavy metals Table, a modified RI categorization given by Zhang et al. [10] was used in this study.

The following equations are used to estimate a revised pollution index (PI) for each metal and an integrated pollution index (IPI) of the six metals for each sampling location in order to determine the degree of metal contamination derived by Huang [16].

$$\mathrm{PI} = \tfrac{C}{X_a} \text{ where } C \le X_a \tag{6}$$

$$\mathrm{PI} = 1 + \tfrac{(C - X_a)}{(X_b - X_a)} \text{ where } X_a < C < X_b \tag{7}$$

$$\text{PI} = 2 + \frac{(C - X_b)}{(X_c - X_b)} \text{ where } X_b < C < X_c \tag{8}$$

$$\text{PI} = 1 + \frac{(C - X_c)}{(X_c - X_b)} \text{ where } C > X_c \tag{9}$$

where C represents the estimated concentration of a particular metal. X_a is the threshold concentration for metal accumulation, X_b is the threshold concentration for low level pollution and X_c is the threshold concentration for high level of pollution were derived by Huang [16] and Bai et al. [17]. The difference between each element's PI and IPI is added together to form the IPI of all measured elements for each sample [17].

$$\text{IPI} = \sum_{i=1}^{n} (\text{PI} - 1) \tag{10}$$

According to Bai et al. [17], who proposed revising the categories of pollution based on the number of pollutants used based on PI and IPI is seen more relevant and appropriate for this investigation and hence used.

3 Results and Discussions

3.1 Heavy Metal Concentrations

The samples were collected from five different locations of Chattogram City for 30 days after 24 consecutive hours. From 30 days samples, only the 10, 15 and 20th days samples were used for measuring the concentration of five metals like Cu, Pb, Cd, Ni, Cr etc. The maximum and minimum concentrations vary for Cu, Pb, Cd, Ni & Cr are 121 to 7, 164 to 49, 2 to 0.5, 54 to 17 & 92 to 25 mg/kg, respectively. The experimented concentration was compared with the background value taken from the Indian natural soil [9, 10]. With respect to background value for Cu, Bahaddarhat, Didar Market and Patenga exceeds the background value, while for Pb all five sites were exhibiting the background value, and for Cd only Didar Market and Patenga site are with higher than background. And for the concentration of Cr all the five sites were found below the background value. The maximum concentration of Cu, Pb, Cd and Cr was found as 121, 164, 2 mg/kg at Patenga site following on Didar Market and Kaptai Rastar Matha sites stand based on similar fashion with respective concentrations, as seen in Table 1. The results indicate significant enrichment of heavy metals in road sediment derived from the road dust is primarily linked with road traffic sector, for example, Cu with brake wearing/brake lining, Pb with fuel and road marking, Cd with fuel combustion and Cr with road infrastructures [9, 10].

The average quantities of all five metals discovered in this investigation were much greater than the background levels observed in Indian natural soil. Table 2 compares

Table 1 Experimented value of heavy metal concentrations in road dust samples in Chattogram

Parameters (mg/kg)	Bahaddarhat	Didar market	Bayazid	Patenga	Kaptai Rastar Matha
Cu	61.38	108.91	7.44	120.88	40.76
Pb	49.19	138.05	50.98	164.55	82.74
Cd	0.71	1.04	0.66	2.1	0.51
Ni	20.26	53.82	16.59	24.98	18.29
Cr	25.13	37.37	39.39	92.33	26.12

Table 2 Road Dust Heavy Metal Concentrations (mg/kg) in Chattogram and other cities around the globe

City	Cu	Pb	Cd	Ni	Cr	RDS size (μm)	References
Chattogram	121	164	2	54	92	<75	This study
Dhaka	46	74		26	104	<1000	[12]
Calcutta	44	536	3.12	42	54	<600	[18]
Delhi	191.7	120.7	2.65	36.4	148.8	<75	[11]
Kuala Lumpur	35.5	2466	2.9			<63	[19]
Shanghai	196.8	294.9	1.23	83.9	159.3	<125	[20]
Seoul	101	245	3			<2000	[21]
Hong Kong	173	181	3.77			<2000	[22]
Indian natural soil background	56.5	13.1	0.9	27.7	114		[14, 15]

research conducted elsewhere with the current study's average heavy metal concentrations in road silt. The results are determined to be within the heavy metal content range that was reported in past investigations. The variation exists are well understood with the variability of site-specific factors and road traffic management. Point to be noted that the sediment size fraction also matters for heavy metal concentrations, as seen in Table 2. The higher the sediment size, the lower the heavy metal concentrations.

3.2 *Assessment of Heavy Metal Contamination*

3.2.1 Geo-Accumulation Index (Igeo)

Figure 2 depicts the geo-accumulation index values estimated for individual metals, indicating how these metals enriched with its geological background. It appears that Pb accumulates at every sampling site with the highest value of above 3 in Patenga, while that of 2.8 in Didar Market and 2 with Kaptai Rastar Matha, illustrating these site sediments are getting anthropogenic input from the road traffic, as Pb is mostly

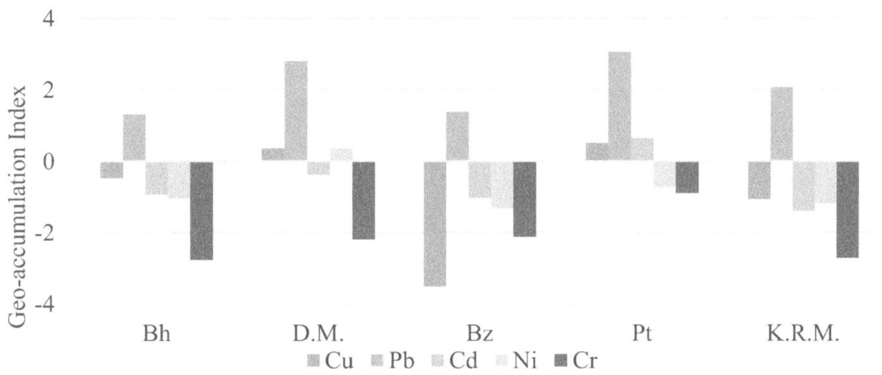

Fig. 2 Geo-accumulation index of road dust by heavy metals in various regions in Chattogram

used as fuel in heavy vehicles along with yellow road marking, as reported by other researchers [1, 8]. The negative Igeo of Cr and Ni at all the sites illustrate that no enrichment of these from road sediments than their respective background values.

3.2.2 Contamination Factor and Contamination Degree

The degree of contamination in road dust from Bahaddarhat, Patenga, Bayazid, Didar Market and Kaptai Rastar Matha areas are found as 6.58, 18.75, 5.70, 15.89, 8.49, respectably, illustrating the road sediments are with moderate to considerable contamination. The considerable degree of contamination is found in Didar Market and Patenga sites, while Bahaddarhat and Kaptai Rastar Matha are with moderate degree of contamination and the lowest degree of contamination is at Bayazid. The degree of contamination follows the similar trend, such as Patenga site exhibit approximately 19 and ranked top, followed by 16 at Didar Market, 9 at Kaptai Rastar Matha and 6.6 at Bahaddarhat and 5.7 at Bayazid sampling site. The indices illustrate road sediment contaminated primarily with anthropogenic input from the traffic in addition to surrounding land uses activities. Such examples include manufacturing processes in the industrial sector, automotive and industry emissions over time. Emissions from fossil fuels, brick kilns, and industrial processes are all closely related to Pb. Meanwhile, Pb was frequently employed in coating, paint, and pigment compounds [19, 23]. A similar investigation was carried out at 32 locations around Chattogram city [1]. Due to the presence of factories, refineries, and a large number of gatherings of vehicles, Patenga has exceptionally high levels of industrial and vehicular pollution. Refer to other researchers who have conducted similar research in other cities throughout the world for additional details [24–26] with this study as presented in Fig. 3.

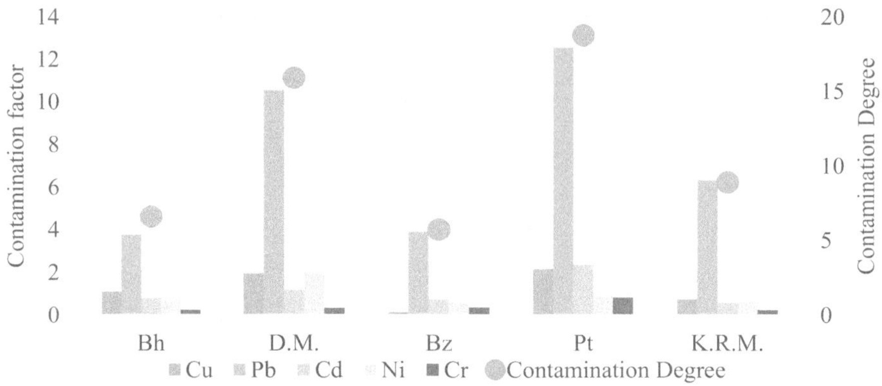

Fig. 3 Contamination factor and degree of contamination of road dust by heavy metals in various regions in Chattogram

3.2.3 Ecological Risk Factor and Risk Index

Figure 4 illustrate ecological risk and risk indices associated with road sediment at studied sites. As discussed earlier, Patenga site road sediment enriched with toxic Pb, Cd poses greater ecological risk and in overall risk index of 150, illustrating significantly higher risk sites. Following the similar fashion, Didar Market site scored 110 in risk index and other sites risk index are varied between 40 and 60. The pollution indices revealed that an immediate attention at Patenga and Didar Market sites are required for managing road sediment as this may pose a greater risk once found its way to the nearby water bodies. An enormous impact on both human and environmental health. Because they mix with biomolecules in the body, heavy metals are known to have major negative impacts on both human health and the environment [27] specially Pb [28].

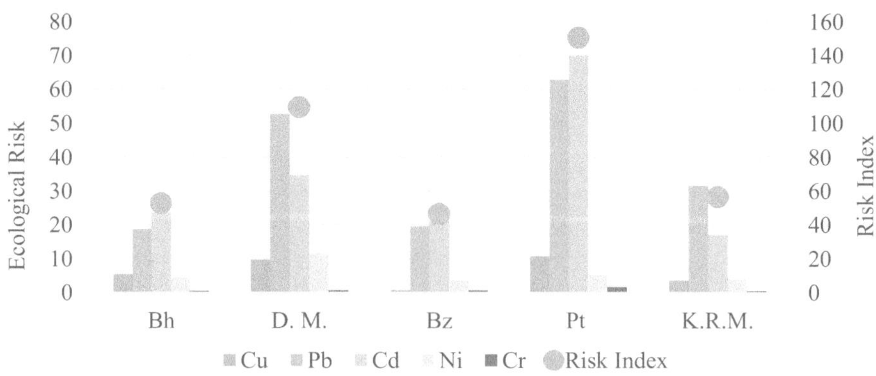

Fig. 4 Ecological risk factor and risk index of road dust by heavy metals in various regions in Chattogram

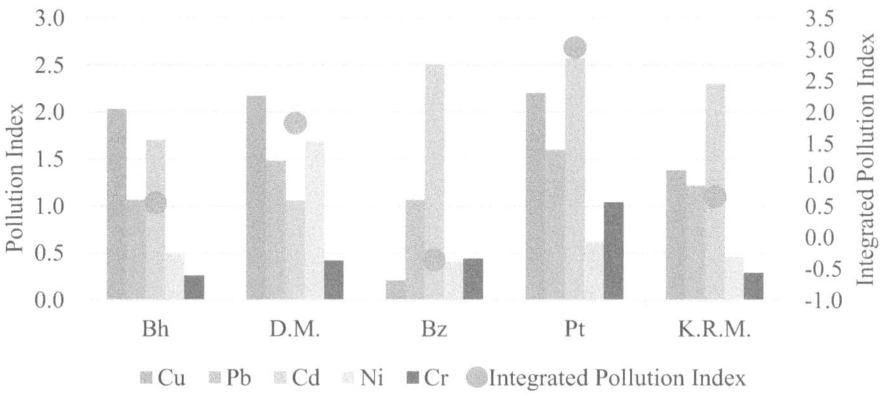

Fig. 5 Pollution index and integrated pollution index of road dust by heavy metals in various regions in Chattogram

3.2.4 Pollution Index and Integrated Pollution Index

The integrated pollution index in road dust from Bahaddarhat, Patenga, Bayazid, Didar Market and Kaptai Rastar Matha areas derived the values of 0.55, 3.03, 0.36, 1.82, 0.65, respectably. The scores are found with low to no pollution hazard at all other sites except Patenga and Didar Market sites. As discussed in preceding sections, these two sites need some immediate steps to abate pollution potential associated with road sediment that may threat human and aquatic health once aquatic lives and human beings are exposed to the road sediment. Consideration should also be given to the huge rise in Pb and Zn concentration in surface dust brought on by industrial activity and high traffic volumes in comparison to studies elsewhere with this investigation as presented in Fig. 5 [11, 29].

4 Conclusion

The study was conducted in five different locations in Chattogram metropolis named as Bahaddarhat, Didar Market, Patenga, Bayazid, Kaptai Rastar Matha collecting suspended and deposited road dust sampling towards assessing heavy metal concentrations that may impact upon human health and ecology of the environment. This study found that Pb is the criteria pollutant for each of the site, and the increasing rate of Cr, Cd, Ni and Cu in road sediment indicate road sediment is sink of pollutants derived from the road site land use pattern along with road traffic design. Among the site, there exist a significant variability where the Didar Market and Patenga sites exhibit the highest levels of contamination, whereas Bahaddarhat and Kaptai Rastar Matha showed intermediate levels and Bayazid had low levels. The spatial variability of heavy metals studied among sampling sites indicates pollutants enrichment is influenced by site-specific characteristics other than traffic volume, road condition and vehicles types. The road sediment enriched with heavy metals are found to pose a moderate to high ecological and human health risk. This study identified site influencing factors such as industrial activities, battery manufacturers, automobile mechanics shops, and construction activities. It will also help to enrich the knowledge gap about heavy metal pollution potential of road dust in Chattogram city.

References

1. Newaz KK, Pal SK, Hossain S, Karim A (2021) Evaluation of heavy metal pollution risk associated with road sediment. Environ Eng Res 26(3). https://doi.org/10.4491/eer.2020.239
2. Irish LB, Lesso WG, Barrett ME, Malina JF, Charbeneau RJ, Ward GH, An evaluation of the factors affecting the quality of highway runoff in the Austin, Texas area. https://repositories.lib.utexas.edu/handle/2152/6741. Last Accessed 18 July 2022
3. Brinkmann WLF (1985) Urban stormwater pollutants: sources and loadings. GeoJournal 11(3):277–283. https://doi.org/10.1007/BF00186341
4. Göbela P, Dierkesb C, Coldeweya WG (2007) Storm water runoff concentration matrix for urban areas. Contam Hydrol 91(1–2):26–42
5. Pal SK, Wallis SG, Arthur S (2008) Assessment of heavy metals emission on road surfaces
6. Kim KW, Myung JH, Ahn JS, Chon HT (1998) Heavy metal contamination in dusts and stream sediments in the Taejon area, Korea. J Geochemical Explor 64(1-3–3) pt 1:409–419. https://doi.org/10.1016/S0375-6742(98)00045-4
7. Robertson DJ, Taylor KG (2007) Temporal variability of metal contamination in urban road-deposited sediment in Manchester, UK: implications for urban pollution monitoring. Water Air Soil Pollut 186(1–4):209–220. https://doi.org/10.1007/S11270-007-9478-X
8. Kumar S (2012) On heavy metal pollution from a suburban road network
9. Pal SK, Wallis SG, Arthur S (2018) Spatial variability of heavy metal pollution potential from an urban road network. Environ Eng Manage J 17(9):2097–2102. https://doi.org/10.30638/eemj.2018.208
10. Zhang J, Deng H, Wang D, Chen Z, Xu S (2013) Toxic heavy metal contamination and risk assessment of street dust in small towns of Shanghai suburban area, China. Environ Sci Pollut Res 20(1):323–332. https://doi.org/10.1007/S11356-012-0908-Y
11. Suryawanshi PV, Rajaram BS, Bhanarkar AD, Chalapati Rao CV (2016) Determining heavy metal contamination of road dust in Delhi, India. Atmosfera 29(3):221–234. https://doi.org/10.20937/ATM.2016.29.03.04
12. Muller G (1981) The heavy metal pollution of the sediments of Neckars and its tributary: a stocktaking, vol 150. In: Chemische Zeit, pp 157–164. References-Scientific Research Publishing. https://www.scirp.org/(S(czeh2tfqw2orz553k1w0r45))/reference/referencespapers.aspx?referenceid=647127. Last Accessed 27 July 2022
13. Hakanson L (1980) An ecological risk index for aquatic pollution control. A sedimentological approach. Water Res 14(8):975–1001. https://doi.org/10.1016/0043-1354(80)90143-8
14. Kuhad M, Malik R, Singh A, Dahiya I (1989) Background levels of heavy metals in agricultural soils of Indo-Gangetic Plains of Haryana. J Indian Soc Soil Sci 37:700–705
15. Srinivasa Gowd S, Ramakrishna Reddy M, Govil PK (2010) Assessment of heavy metal contamination in soils at Jajmau (Kanpur) and Unnao industrial areas of the Ganga Plain, Uttar Pradesh, India. J Hazard Mater 174(1–3):113–121. https://doi.org/10.1016/J.JHAZMAT.2009.09.024
16. Huang R (1987) Environmental pedology. Chain High Educ Beijing 8

17. Bai J, Cui B, Wang Q, Gao H, Ding Q (2009) Assessment of heavy metal contamination of roadside soils in Southwest China. Stoch Environ Res Risk Assess 23(3):341–347. https://doi.org/10.1007/S00477-008-0219-5
18. Ahmed F, Ishiga H (2006) Trace metal concentrations in street dusts of Dhaka city, Bangladesh. Atmos Environ 40(21):3835–3844. https://doi.org/10.1016/J.ATMOSENV.2006.03.004
19. Chatterjee A, Banerjee RN (1999) Determination of lead and other metals in a residential area of greater Calcutta. Sci Total Environ 227(2–3):175–185. https://doi.org/10.1016/S0048-9697(99)00026-1
20. Ramlan MN, Badri MA (1989) Heavy metals in tropical city street dust and roadside soils: a case of Kuala Lumpur, Malaysia. Environ Technol Lett 10(4):435–444. https://doi.org/10.1080/09593338909384759
21. Shi G et al (2008) Potentially toxic metal contamination of urban soils and roadside dust in Shanghai, China. Environ Pollut 156(2):251–260. https://doi.org/10.1016/J.ENVPOL.2008.02.027
22. Chon HT, Kim KW, Kim JY (1995) Metal contamination of soils and dusts in Seoul metropolitan city, Korea. Environ Geochem Health 17(3):139–146. https://doi.org/10.1007/BF00126082
23. Duan Z, Wang J, Zhang Y, Xuan B (2017) Assessment of heavy metals contamination in road dust from different functional areas in Guiyang, Southwest, China. Int J Environ Sci Educ. https://doi.org/10.12973/IJESE.2017.01238A
24. Zhang J, Hua P, Krebs P (2017) Influences of land use and antecedent dry-weather period on pollution level and ecological risk of heavy metals in road-deposited sediment. Environ Pollut 228:158–168. https://doi.org/10.1016/J.ENVPOL.2017.05.029
25. Tokalıog S, Kartal S (2006) Multivariate analysis of the data and speciation of heavy metals in street dust samples from the Organized Industrial District in Kayseri (Turkey) synthesis of silicon compounds view project Serife Tokalioglu Erciyes Üniversitesi Multivariate analysis of the data and speciation of heavy metals in street dust samples from the Organized Industrial District in Kayseri (Turkey). Atmos Environ 40:2797–2805. https://doi.org/10.1016/j.atmosenv.2006.01.019
26. Rawat M, Ramanathan A, Subramanian V (2009) Quantification and distribution of heavy metals from small-scale industrial areas of Kanpur city, India. J Hazard Mater 172(2–3):1145–1149. https://doi.org/10.1016/J.JHAZMAT.2009.07.115
27. Sharma V, Singh P (2015) Heavy metals pollution and it's effect on environment and human health. Int J Recent Sci Res 6(12):7752–7755. https://www.academia.edu/20116419/Heavy_metals_pollution_and_it_s_effect_on_environment_and_human_health. Last Accessed 15 Mar 2023. (Online)
28. Zhang R, Wilson VL, Hou A, Meng G (2015) Source of lead pollution, its influence on public. Int J Heal Anim Sci Food Saf 2(1):18–31. https://doi.org/10.13130/2283-3927/4785
29. Chen Y, Hu Z, Bai H, Shen W (2022) Variation in road dust heavy metal concentration, pollution, and health risk with distance from the factories in a city–industry integration area, China. Int J Environ Res Public Health 19(21). https://doi.org/10.3390/ijerph192114562

Effect of Loading Directionality on the Horizontal Stiffness of Unbonded Scrap Tire Rubber Pad Isolator

M. B. Zisan and A. Igarashi

Abstract The horizontal stiffness of a low-cost base isolator made with scrap tire rubber pad (STRP) is investigated. A finite element approach is used for modeling, verification, and analyses for the STRP isolator. The isolator is assumed to be utilized without any mechanical fastening with the structure, which transfers lateral load through friction between the isolator and the structure. The effect of the loading directionalities and the coupling between the orthogonal lateral displacements are taken into consideration. It was found that the directionality of the input displacement reduces the horizontal stiffness, and its effect is high for a high length-to-width ratio of the isolator. The lateral stiffness is overestimated when coupling between lateral displacements is ignored, and this overestimation is high for a high length-to-width ratio of the isolator.

Keywords STRP isolator · Loading directionality · Coupling effect · Horizontal stiffness · Lateral displacement

1 Introduction

Base-isolation is widely used to protect structures from the damage caused by an earthquake. For this purpose, steel-reinforced elastomeric isolators (SREIs) are widely used. The high cost and heavy weight of SREI limit its use in masonry structures and public buildings. Therefore, a low-cost isolator that can be utilized by following simple design principles is necessary. As an alternative, a low-cost isolator made with scrap tires, called a scrap tire rubber pad (STRP) isolator, has been

M. B. Zisan (✉)
Department of Civil Engineering, Chittagong University of Engineering and Technology, Chittagong, Bangladesh
e-mail: basirzisan@cuet.ac.bd

A. Igarashi
Disaster Prevention Research Institute, Kyoto University, Kyoto, Japan
e-mail: igarashi.akira.7m@kyoto-u.ac.jp

S. Arthur et al. (eds.), *Proceedings of the 6th International Conference on Advances in Civil Engineering*, Lecture Notes in Civil Engineering 368,
https://doi.org/10.1007/978-981-99-3826-1_35

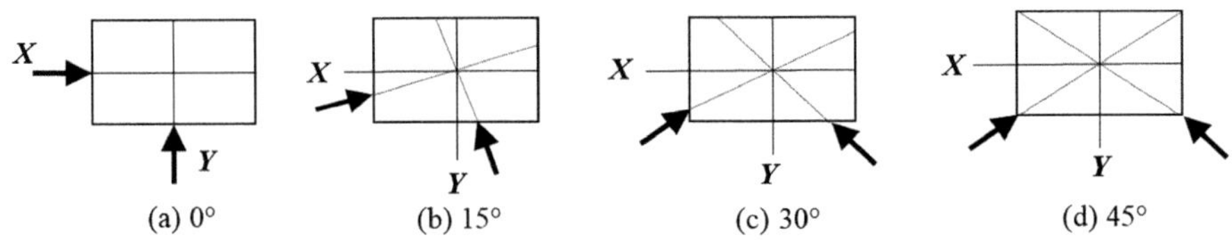

Fig. 1 Different orientations of the lateral load

proposed [1, 2]. Since tires are manufactured by vulcanizing rubber with embedded steel cords, it is expected that the STRP isolator functions in a manner similar to SREIs. An experiment shows that an STRP isolator has an effective damping ratio of 10–22% and a vertical-to-horizontal stiffness ratio in the range of 450–600 [3]. Also, the shear modulus of the tire rubber is around 1.0 MPa [2, 3], which is in the range of natural rubber, 0.55–1.20 MPa, as specified by the AASHTO-LRFD specification.

It is claimed that an STRP isolator without mechanical bonding with the structural elements is more advantageous. An unbonded STRP isolator displays a rollover deformation, which is helpful in bypassing a large amount of tension within the elastomer. In addition, the rollover deformation improves the efficiency of base isolation by lowering the lateral stiffness [4–6]. The previous studies [5–8] assessed the performance of STRP isolators and STRP base-isolated buildings under lateral load. The stress-strain behavior of an unbonded STRP isolator has been assessed under lateral load by Zisan and Igarashi [9]. A hysteresis force model has been developed for horizontal stiffness and assessed under uniaxial lateral load [6]. These studies were done under uniaxial lateral load along the principal axes of the isolator. Since lateral load is bidirectional in nature, it is important to investigate the unbonded STRP isolator subjected to different orientations of the lateral load. When an isolator will be used without any mechanical fastening, the effect of the loading directionality on the lateral load performance needs to be investigated. Therefore, this study concentrated on the effect of the loading direction on the lateral load performance of the unbonded STRP isolator. Four cases of loading orientation, such as 0°, 15°, 30°, and 45°, are considered. Figure 1 shows the orientation of the load with respect to the principal axis of the isolator. Both square and strip isolators are considered in this study. The size of the square-shaped isolator is 100 × 100 × 24 mm and that of the strip-shaped isolator is 400 × 100 × 24 mm. The length-to-width ratio of the strip-shaped isolator is four times that of the square-shaped isolator.

2 Finite Element Modeling of the STRP Isolator

An STRP isolator is made by stacking one STRP layer on top of another and bonding them together with an adhesive. An STRP layer is 12 mm high and contains five layers of steel cord. The equivalent thickness of the elastomer layers and that of the reinforcement layers in an STRP layer are assumed to be 2.4 mm and 0.4 mm, respectively. The properties of the reinforcing steel cords and that of the rubber material of the scrap tire obtained from Bridgestone 385/65R22 are shown in Tables 1 and 2. In Table 2, G_e and G_{eff} indicate the shear modulus of the elastomer at undeformed and 25% shear strain. The hyperelasticity of the rubber material is modeled using the Mooney–Rivlin material model. The hyperelasticity of the rubber material is expressed by C_{10}, C_{01} and C_{11}, called Mooney–Rivlin material constants. The viscoelasticity of the rubber is modeled using the Prony series viscoelastic shear response parameters, where δ is the scalar multiplier of the strain energy function and λ is the relaxation time. The elastomer softening is assumed using the discontinuous phenomenological damage model, representing the so-called Mullin effect. The damage parameters η and m indicate the scale factor and relaxation factor, respectively. A detailed description of the STRP isolator and the procedure for deriving the rubber material constant can be found in Zisan and Igarashi [5, 6].

The finite element modeling and analysis of the STRP isolators was carried out by MSC Marc-Mentat [10]. The bottom and top contact surfaces, which are shown in Fig. 2, represent the foundation and superstructure, respectively, and are modelled as a rigid plane. The bilinear Coulomb friction model with a friction coefficient of 0.80 is assumed between the isolator and the contact surface. Axial force and lateral displacement are applied at the top rigid surface. The vertical compression on the isolator is assumed to be 5.0 MPa. Zisan and Igarashi [5, 6] provide a detailed description of finite element modeling, contact modeling, and FE model verification.

In the present study, the lateral displacement is assumed to have directionality 0°, 15°, 30°, and 45°. Figure 3 shows the lateral displacement time-history of the two components of the input displacements. The phase difference between these components is assumed to be 90°. The input displacement consists of eight reverse cycles of displacement whose amplitude is equal to 25, 50, 75, 100, 125, 150, 175, 200, 225, and 250% shear strain. The isolators subjected to the non-simultaneous action of the displacement component with different loading directions are analyzed first. To consider the coupling effect of the input displacement, each isolator is then analyzed for the simultaneous application of horizontal displacement components. In both cases, the response of the isolator is calculated along two principal axes, called the X and Y-axes.

3 Result and Discussion

The restoring force-displacement relationship of the STRP 400 × 100 × 24 mm isolator subjected to non-simultaneous and simultaneous action of lateral displacement and working at 45° direction with principal axes of the isolator is depicted in Figs. 4 and 5. It is clear that when simultaneous effects are ignored, the isolator exhibits hardening behavior over a large displacement range. This hardening is due to the rollover deformation where the vertical face of the isolator touches the horizontal rigid surfaces, as shown in Fig. 6i. The hardening behavior of an isolator is beneficial because it restrains the large displacement of the isolator under a large

Table 1 Properties of reinforcing cord

Layer	Layer no	No. of filaments	Filament dia (mm)	Single cord area (mm^2)	Orientation	Equivalent thickness t_f (mm)	Yield strength (MPa)	Spacing (mm)	Young modulus E (GPa)	Poisson ratio υ
Carcass	1	5	0.2	0.44	0°	0.40	2800	2.5	200	0.3
Belt	4	14	0.4	0.63	$\mp$ 72°					

Table 2 Properties of elastomers

Mooney–Rivlin constant			Shear modulus (MPa)			Prony viscoelastic parameters				Mullin-damage parameters			
C_{10}	C_{01}	C_{11}	G_{eff}	G_e	υ	δ^1	λ^1	δ^2	λ^2	η_1	m_1	η_2	m_2
0.40	1.22315	0.18759	1.10	1.31	0.49995	0.30	0.2	0.30	0.55	0.01	5	0.05	10

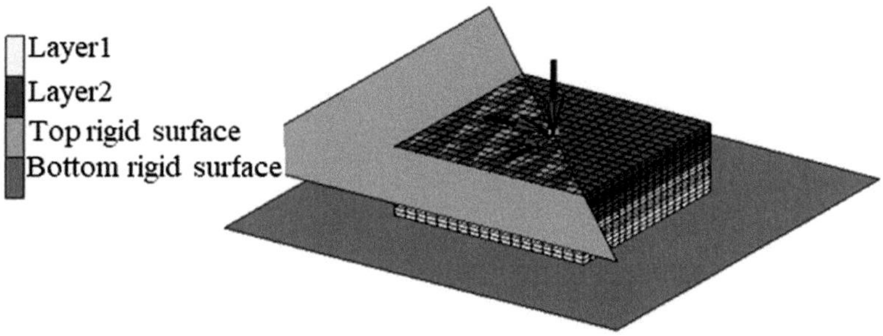

Fig. 2 FE model, support conditions, and lateral loading condition of the STRP isolator

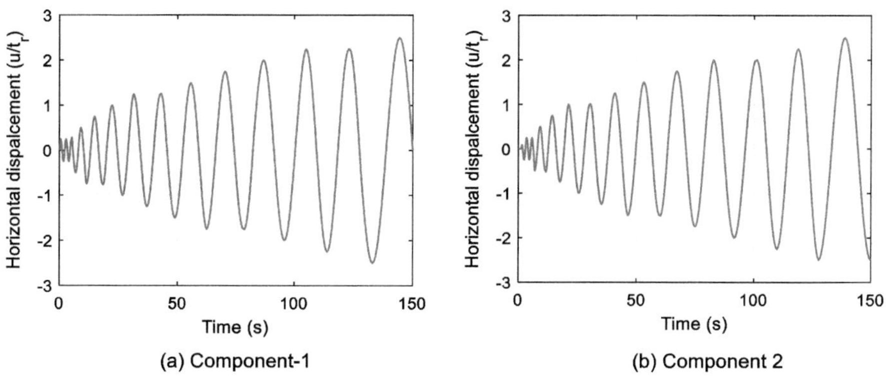

Fig. 3 Time history of the input lateral displacement

ground motion. However, when the coupling effect is taken into consideration, the rollover deformation as well as the hardening effect are negligible. The hardening effect reduces with an increase in loading directionality. Figure 6ii, iii, and iv show that the contact area between the vertical faces of the isolator and the horizontal rigid surfaces reduces when loading directionality increases. The horizontal stiffness of the isolator is defined by the slope of the hysteresis curve. The average slope of the hysteresis curve is calculated by the least squares method, which is indicated by the straight line in Figs. 4 and 5. The slope of the equation is the lateral stiffness of the STRP isolator. Due to the effect of rollover deformation, the slope of the hysteresis curve when loading directionality is 0° is higher than that of the slope when loading directionality is greater than 0°.

Tables 3 and 4 show the lateral stiffness of STRP 100 × 100 × 24 mm and STRP 400 × 100 × 24 mm isolators, respectively, at various orientations of the input displacement. In Table 3, K indicates the horizontal stiffness. K_s is the root sum

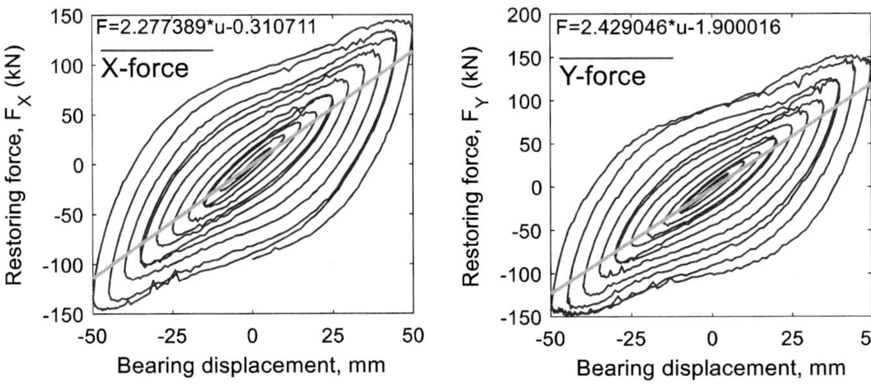

Fig. 4 Hysteresis force of STRP 400 × 100 × 24 mm under the action of displacements at 45°

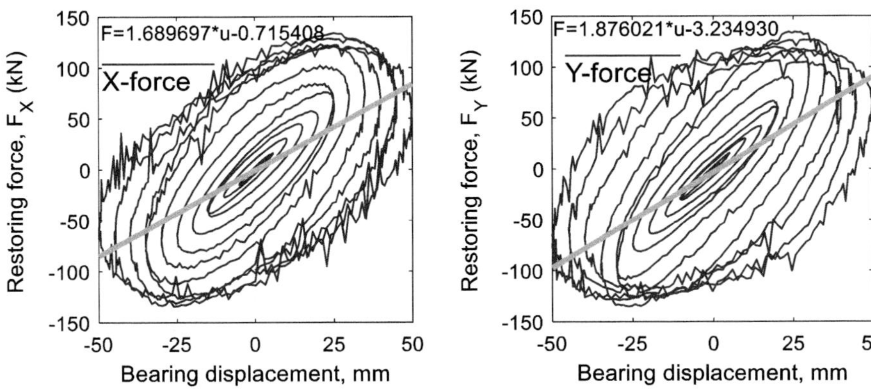

Fig. 5 Hysteresis force of STRP 400 × 100 × 24 mm under simultaneous action of displacements at 45°

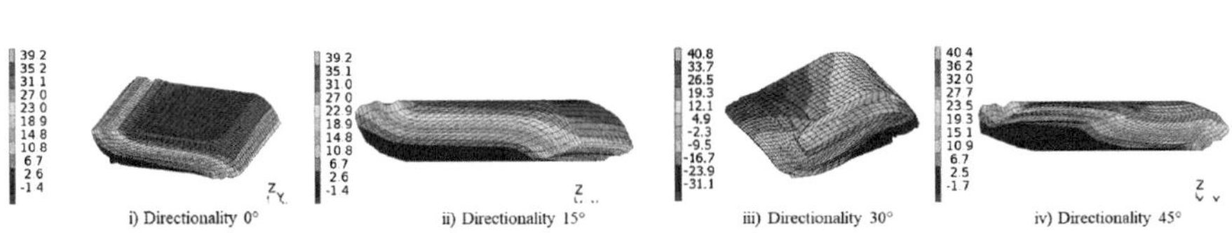

Fig. 6 Rollover deformation at 200% shear strain at various load directionality of the STRP 100 × 100 × 24 isolator

square of the two horizontal stiffnesses, K_{sx} and K_{sy}. Similarly, K_i is the root sum square of the two horizontal stiffnesses, K_{ix} and K_{iy}. The suffixes s and i indicate the conditions of simultaneous and non-simultaneous action of the displacements, respectively. In the case of STRP 100 × 100 × 24 mm, the stiffness in the Y-direction is higher than that in the X-direction due to the orientation of the steel cord. The same is true for the STRP 400 × 100 × 24 mm too when coupling effect is considered. But the stiffness in the X-direction is greater than that in the Y-direction when no coupling is considered in the non-simultaneous application of lateral load. It is due to the fact that the major axis of the isolator is parallel to the X direction. When simultaneous action is considered, the horizontal stiffness is lower than that of the non-simultaneous action of the lateral displacement. It indicates that the structural response will be underestimated when an isolator is designed based on the non-simultaneous application of the lateral load.

Tables 3 and 4 as well as Fig. 7 show that the horizontal stiffness of the isolator decreases by about 5% in the case of the STRP 100 × 100 × 24 mm and 10% in the case of the STRP 400 × 100 × 24 mm isolator due to an increase in loading directionality. However, in the case of the Y direction, it is improved due to the orientation of the steel cord within the isolator. Therefore, it is important to consider the loading directionality effect in the design of a base-isolated structure.

Figure 8 shows the ratio of horizontal stiffness between simultaneous and non-simultaneous applications of lateral load at different levels of directionality. In the case of STRP 400 × 100 × 24 mm, the stiffness ratio changes from 0.85 to 0.75,

Table 3 Stiffness (kN/m) of the square-shaped bearing (STRP 100 × 100 × 24 mm)

Loading direction	Simultaneous action of applied displacement components			Non-simultaneous action of displacement components			K_s/K_i
	K_{sx}	K_{sy}	K_s	K_{ix}	K_{iy}	K_i	
00°	493	509	709	513	561	760	0.93
15°	489	512	708	522	574	776	0.91
30°	480	518	707	531	577	784	0.90
45°	465	504	686	554	596	814	0.84

Table 4 Stiffness (kN/m) of the strip-shaped isolator (400 × 100 × 24 mm)

Loading direction	Simultaneous action of applied displacement components			Non-simultaneous action of displacement component			K_s/K_i
	K_{sx}	K_{sy}	K_s	K_{ix}	K_{iy}	K_i	
00°	1903	2049	2796	2528	2067	3266	0.86
15°	1702	2086	2692	2390	2124	3197	0.84
30°	1616	2120	2666	2404	2192	3253	0.82
45°	1690	1876	2525	2277	2429	3330	0.76

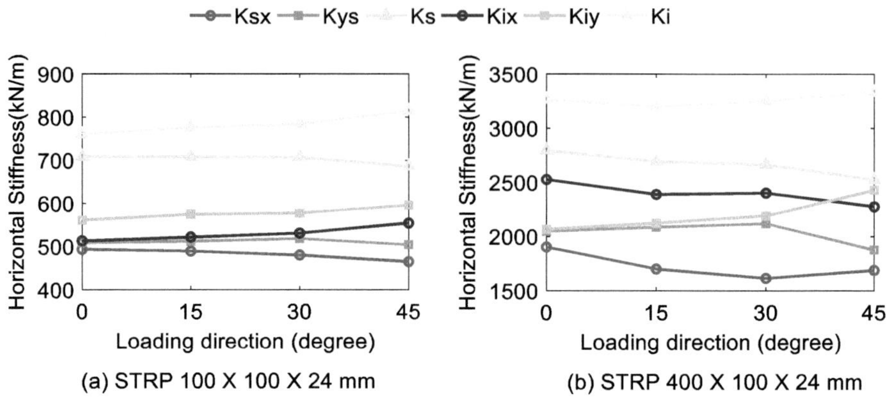

Fig. 7 Stiffness due to simultaneous and non-simultaneous action of the displacement components at various loading directions

whereas it is 0.92–0.84 in the case of the STRP 400 × 100 × 24 mm isolator. It indicates that both directionality and simultaneous loading effects reduce the horizontal stiffness of the unbonded STRP isolator, and this reduction is high when the length-to-width ratio of the isolator is high.

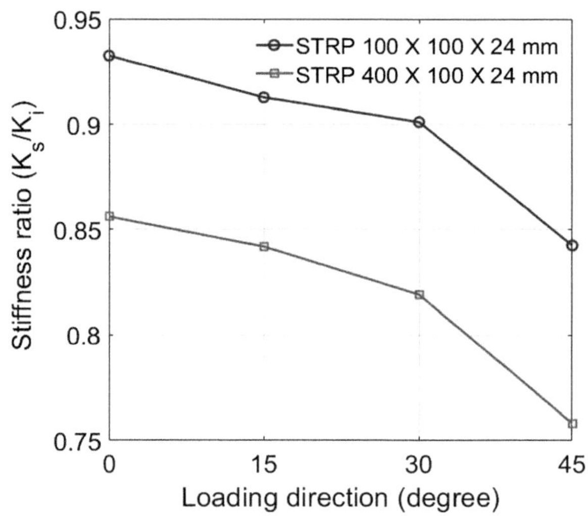

Fig. 8 Effect of simultaneous application of input displacement at different loading directionality

4 Conclusion

This paper describes the effect of loading directionality and the coupling effect of the lateral displacement on the horizontal stiffness of an unbonded STRP isolator. The square and strip-shaped STRP isolators are subject to orthogonal lateral displacements and friction-based unbonded boundary conditions that are analyzed through a finite element program. The finite element model of the STRP isolator is validated by the experimental result. Two cyclic displacement time histories, which have a fixed phase difference of 90°. The findings of the analysis are summarized below:

- When coupling effect is ignored, which mean input displacements are non-simultaneously work, the lateral stiffness is overestimated. This overestimation is high for a high length-to-width ratio of the isolator.
- When the coupling effect is ignored, an increase in the loading directionality increases the lateral stiffness.
- The horizontal stiffness of the isolator decreases with an increase in loading directionality when coupling effect on the input displacement is considered. The effect of loading directionality is high when the high length-to-width ratio of the isolator is high.

Since the horizontal stiffness of the unbonded STRP isolator is influenced by the loading directionality and coupling effect of the input loading, it is necessary to properly estimate the lateral load performance of the unbonded STRP isolator, including the effect of the loading directionality and the simultaneous action of the horizontal load.

References

1. Turer A, Özden B (2007) Seismic base isolation using low-cost scrap tire pads (STP). Mater Struct 41(5):891–908
2. Mishra HK, Igarashi A, Matsushima H (2013) Finite element analysis and experimental verification of the scrap tire rubber pad isolator. Bull Earthq Eng 11(2):687–707
3. Mishra HK (2012) Experimental and analytical studies on scrap tire rubber pads for application to seismic isolation of structures. PhD thesis, Kyoto University, Japan
4. Zisan MB, Igarashi A (2020) Lateral load performance evaluation of unbonded strip-STRP base isolator. In: 17th World conference on earthquake engineering (17WCEE), September 13th to 18th, 2020. Sendai, Japan
5. Zisan MB, Igarashi A (2021) Lateral load performance and seismic demand of unbonded scrap tire rubber pad base isolators. Earthq Eng Eng Vib 20:803–821
6. Zisan MB, Igarashi A (2021) Evaluation of unbonded Strip-STRP bearing based on current design guidelines. In: IABSE congress–resilient technologies for sustainable infrastructure, February 3–5, 2021. Christchurch, New Zealand, pp 1247–1256
7. Zisan MB, Haque MN, Hasan MA (2022a) Seismic vulnerability assessment of masonry building supported by STRP isolators. Asian J Civ Eng. https://doi.org/10.1007/s42107-022-00492-2

8. Zisan MB, Hasan M, Haque M (2022b) Performance assessment of buildings seismically isolated with scrap tire rubber pad isolators. Asian J Civ Eng. https://doi.org/10.1007/s42107-022-00503-2
9. Zisan MB, Igarashi A (2022) A hysteresis force model for unbonded scrap tire rubber pad isolators. In: 3rd International conference on natural hazards & infrastructure, 5–7 July, 2022. Athens, Greece
10. Marc MSC (2018) Theory and user information, vol A. MSC Software Corporation, Santa Ana, CA

Optimization of Particle Packing Density of Industrial Wastes Incorporated Concrete

A. R. Ziad and G. M. Sadiqul Islam

Abstract Growing urbanization and industrialization increase the demand for concrete. A massive amount of natural stone is extracted to meet this demand, rapidly depleting natural resources. One of the steelmaking byproducts, electric arc furnace (EAF) steel slag, could be a sustainable substitute for natural stone. This study aimed to replace natural stone 100% with EAF slag. The concrete was proportioned according to the Modified Andreasen model to ensure the optimal gradation of the particles. Moreover, a wet particle packing density optimization was performed to identify the optimum water content at which the packing density maximizes. Two more industrial wastes/byproducts, viz., fly ash and silica fume, were utilized as supplemental cementitious materials to substitute a portion of cement to investigate their influence on particle packing density and concrete properties. The replacement levels for fly ash were 15 and 30%, while for silica fume, they were 5 and 10%. The workability, strength, water absorption, voids and density tests of the produced concrete revealed that using EAF slag as a coarse aggregate gives promising concrete characteristics. The incorporation of fly ash and silica fume as partial cement substitutes increased the particle packing density of the concrete mixture. However, the silica-fume-based concrete showed better strength characteristics, while the fly-ash-based concrete gave lower reactivity at 28 days.

Keywords Electric arc furnace (EAF) steel slag · Particle packing density · Modified Andreasen model · Fly ash · Silica fume

---

A. R. Ziad (✉) · G. M. S. Islam
Department of Civil Engineering, Chittagong University of Engineering and Technology, Chattogram, Bangladesh
e-mail: u1610018@student.cuet.ac.bd

G. M. S. Islam
e-mail: gmsislam@cuet.ac.bd

S. Arthur et al. (eds.), *Proceedings of the 6th International Conference on Advances in Civil Engineering*, Lecture Notes in Civil Engineering 368,
https://doi.org/10.1007/978-981-99-3826-1_36

1 Introduction

Millions of years ago, when humans first realized that they could use natural resources to make something beneficial to themselves, the journey of extraction and construction started. Concrete has been used to build and design infrastructures for thousands of years. However, this concrete requires a significant amount of high-quality aggregate since it accounts for nearly 70–75% of its overall volume and directly impacts its toughened qualities [1]. Therefore, a massive amount of natural stone is extracted yearly, depleting the world's natural resources reserve. On the other hand, increasing industrialization produces significant amounts of waste each year. Steel slag is one of these produced during the steel refining process. The material could be an alternative to traditional stone coarse aggregate that is sufficiently dense [2]. It is usually composed of metal oxides. Using this slag as an aggregate can reduce the need for natural aggregates and ensure sustainable industrial waste management [3].

To make sure that the concrete meets the requirements in its fresh and hardened state in a cost-effective and long-lasting way, proper proportioning of the concrete mix is essential [4]. The aggregate gradation has long been known to impact the substantial functionality of concrete [5]. Therefore, concrete efficacy may be increased by modifying the gradation of the entire spectrum of the solid particles, ensuring an optimum packing density [5]. The particle packing density describes how firmly the solid particles are packed together in a mix and corresponds to the ratio of solid particle volume to the total volume (including pore volume) filled by the particles. The particle selection should be such that the matrix fills up the spaces among big particles with comparatively smaller particles and further fills up the spaces among small particles with relatively finer particles. This process is expected to lower the volume of voids and maximize the particle packing density to achieve a densely packed firm particle structure [6]. An increased aggregate packing density may either enhance the workability with a fixed paste volume of constant water-to-binder ratio or increase the strength while keeping the workability constant [7].

Since the 19th century, researchers have examined the idea of aggregate packing in relation to concrete [8]. A group of early researchers, Fuller and Thompson, introduced the gradation curve in 1907, which is renowned as Fuller's ideal curve [9]. Andreasen attempted to enhance this curve and suggested employing an exponent 'q' in Fuller's equation with a value falling between 0.33 and 0.50 [10]. Dinger and Funk further adjusted the Andreassen model to account for the smallest particles in actual materials that are limited in size [11]. Therefore, in contradiction to the Fuller curve, their model, which is well-known as the Modified Andreasen model, not only takes into consideration the grain that would be the largest but also the grain that is the smallest. The gradation that yields the highest densities of the aggregates independently may not always yield the highest density while blended with cement and water due to the cement particles' ability to be incorporated into microscopic gaps. The curve should have indicated the gradation with the highest density [12]. Keeping this in mind, 'Kwan' introduced a wet packing method that determines the optimum water content at which the packing density of a mix is maximized [13].

Table 1 Physical properties of EAF slag and sand

Properties	EAF slag	Sand
Specific gravity	3.56	2.63
Unit weight (kg/m^3)	2190	1715
Absorption capacity (%)	0.77	1.9
Aggregate crushing (%)	13.6	–
Los Angeles abrasion (%)	16.8	–

Table 2 Physical properties of cement, fly ash and silica fume

Properties	Cement	Fly ash	Silica fume
Specific gravity	3.15	2.06	2.23
Unit weight (kg/m^3)	1440	870	650
Surface area (m^2/kg)	300–320	300–500	15,000–35,000
Color	Gray	Light gray	Dark gray

*The surface area of fly ash and silica fume as per suppliers

The study aims to fully replace the natural stone aggregate with potential industrial waste EAF slag and design the concrete mix according to particle packing optimization, maintaining a well-graded distribution of the particles. Investigations were also conducted on how adding fly ash and silica fume affected the concrete's characteristics and packing density. Tests included workability, particle packing density, compressive strength, tensile strength, density, water absorption, and permeable voids in hardened concrete.

2 Materials and Methods

2.1 *Materials*

Natural coarse aggregate was replaced entirely by EAF steel slag, while the fine aggregate was coarse sand, and the primary binding material was Ordinary Portland Cement (OPC). In addition, as a partial cement substitute, fly ash (15 and 30%) and silica fume (5 and 10%) were used. Tables 1 and 2 lists the material properties.

2.2 *Mix Design*

Particle packing optimization was employed to design the concrete mixture. This optimization was carried out in two stages. Initially, the Modified Andreassen model was adapted to find the volumetric content of coarse aggregate, fine aggregate, and binder, ensuring an optimized particle gradation. In the next stage, the concrete mix's optimum packing density for varying water/binder ratios was produced to evaluate the optimum w/b ratio according to the wet packing method suggested in [14]. Seven mix proportions and eleven batches of the concrete mix were made.

2.2.1 Particle Grading Optimization

The particle grading optimization was carried out following the Modified Andreassen model, which maintains a well-gradation of the particles and ensures an optimum packing density. According to the model, Eq. 1 estimates the total percentage (volume) of the particles finer than a size class "D". In this investigation, the maximum particle size (D_{max}) was 19 mm. In contrast, the minimum particle size (D_{min}) was taken as 0.075 mm. From the literature, the suggested value of the exponent "q" was 0.25–0.3 for making standard quality concrete [15]. Here, the value of the exponent was selected to be 0.30 for all the batches of concrete. Figure 1 represents the particle size distribution (PSD) curve using the Modified Andreassen model to figure out the volumetric percentage of each particle size class.

In this investigation, particles smaller than 0.15 mm was considered a binder. As a result, coarse aggregate, fine aggregate, and volumetric binder contents were 34.5%, 46.5%, and 19%, respectively. The coarse and fine aggregate size and amount are given in Tables 3 and 4, respectively.

$$P(D) = \left(\frac{D - D_{min}}{D_{max} - D_{min}}\right)^q \tag{1}$$

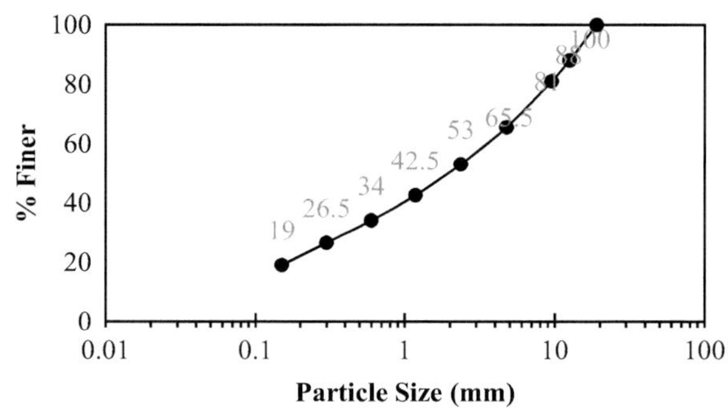

Fig. 1 PSD curve using Modified Andreassen model

Table 3 Coarse aggregate size classes and content

EAF slag coarse aggregate			
Particle size (mm)	12.5	9.5	4.75
Amount (%)	34.8	20.3	44.9

Table 4 Fine aggregate size classes and content

Fine aggregate					
Particle size (mm)	2.36	1.18	0.6	0.3	0.15
Amount (%)	26.9	22.6	18.3	16.1	16.1

2.2.2 Particle Packing Density Optimization

After determining the proportions of the materials, a wet packing density [14] test was performed to find the optimum w/b ratio that maximizes concrete's packing density. Equation 2 was used to compute the particle packing density.

$$\varphi = \frac{{}^{M}/_{V}}{W_w\rho_w + V_c\rho_c + V_f\rho_f + V_b\rho_b} \quad (2)$$

Here, φ denotes packing density, and M/V is the bulk density of the mix. W_w and ρ_w Indicate the water-to-total solid ratio and the water density. V_c, V_f, *and* V_b are the volumetric content and ρ_c, ρ_f, and ρ_b are the solid densities of coarse aggregate, fine aggregate, and binder, respectively. First, the optimum w/b ratio was investigated for the control concrete (100% cement) mix. An initial w/b ratio of 0.3 was used to initiate the control concrete mixing. After mixing, the particle packing density was determined, the w/b ratio was increased by adding water, and the associated particle packing density was measured again. In this process, an optimum w/b of 0.38 was found for the control concrete mix. Similarly, for the other mixes, the optimum w/b ratios were investigated and reported in Table 5.

2.2.3 Concrete Mix Proportions

A different optimum w/b ratio was noted for the various concrete mixtures in which the cement was partially substituted by fly ash, silica fume, or both. This diversity occurred due to the difference in water requirements of these three binders. As the concrete's hardened properties (especially compressive strength) greatly depend on the w/b ratio, comparing the concretes having a similar w/b ratio is required. For that reason, seven concrete mixes (M1 to M7) were prepared while keeping the w/b ratio at 0.38, which was the optimum value of the control concrete mix. Furthermore, to compare all the concrete mixes based on their corresponding optimum particle packing density, four more concrete mixes (M3a, M5a, M6a, and M7a) were produced as these mix proportions' optimum w/b ratio wasn't 0.38.

Table 5 Concrete compositions and mixture proportions

Batch	Concrete ID	w/b	Unit (kg/m^3)						
			Cement	Fly ash	Silica fume	Coarse aggregate (EAF)	Fine aggregate	Water	SP
M1	Control	0.38	477	–	–	982	977	181	3.81
M2	F15_0.38	0.38	405	72	–	982	977	181	3.81
M3	F30_0.38	0.38	334	143	–	982	977	181	3.81
M4	S5_0.38	0.38	453	–	24	982	977	181	3.81
M5	S10_0.38	0.38	429	–	48	982	977	181	3.81
M6	F15S5_ 0.38	0.38	381	72	24	982	977	181	3.81
M7	F30-S10_ 0.38	0.38	286	143	48	982	977	181	3.81
M3a	F30_0.42	0.42	327	140	–	963	959	196	3.74
M5a	S10_0.40	0.40	425	–	47	973	968	187	3.78
M6a	F15S5_ 0.40	0.40	378	71	24	973	968	187	3.78
M7a	F30S10_ 0.44	0.44	278	139	46	954	950	204	3.71

3 Results and Discussions

3.1 Particle Packing Density

Figure 2 represents the particle packing densities and voids ratios of the control mix at different w/b ratios. At the dry condition, the dry packing density was found to be 0.831, and the corresponding voids ratio was 0.204. The graphs show that with the rise in w/b ratio from zero (dry condition), the particle packing density initially decreases, and the voids ratio increases. This is because air became trapped between the particles due to water films forming on their surfaces, which in turn caused the solid particles to become farther apart. This phenomenon is known as the "bulking effect" [14]. The water films eventually converged as the w/b ratio continued to rise and the excess water filled in the gaps, which reduced the amount of air trapped inside the gaps. Consequently, the solid concentration and particle packing density went up while the voids ratio went down. The voids ratio, however, kept decreasing as particle packing density rose until the w/b ratio reached its optimal level. At that point, the particle packing density reached its highest point, and the voids ratio reached its lowest point. Suppose the w/b ratio is increased above the optimal level. In that case, the solid particles will be disseminated throughout the water as a suspension since there will be more water than is required to fill the voids. Consequently, the packing density would decline while the voids ratio would rise [16].

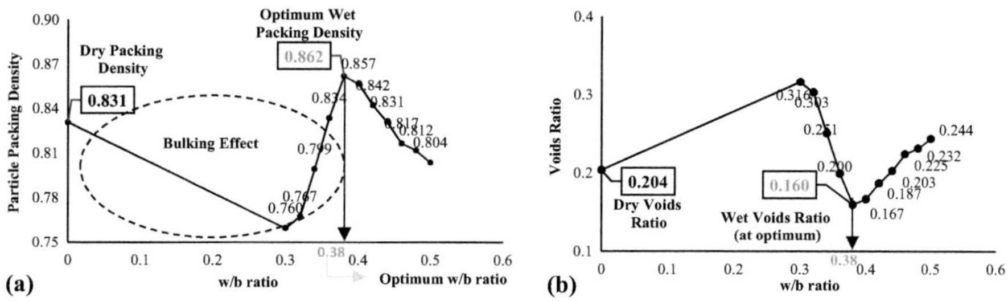

Fig. 2 **a** Particle packing densities, and **b** voids ratios of control mix at different w/b

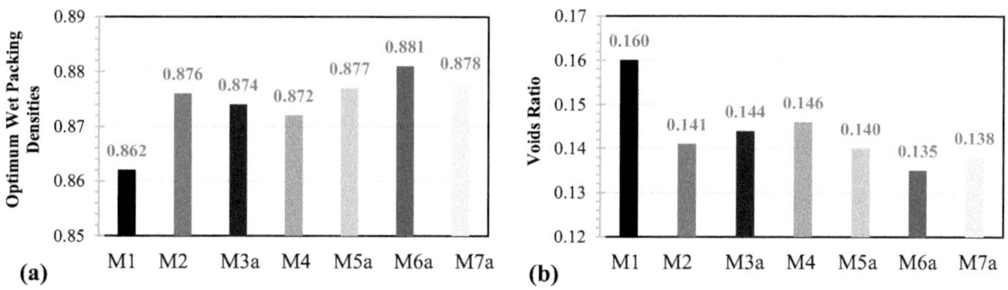

Fig. 3 **a** Optimum wet particle packing densities, and **b** voids ratios at the optimum w/b ratios of mixes

Figure 3 summarizes the optimum wet packing densities and voids ratios for optimum w/b ratios of all the mixes. The optimum wet packing density of the mixes increased in all cases where the cement was partially substituted by fly ash, silica fume, or both. The one with the highest optimum wet packing density was mix M-6a, where fly ash and silica fume partially substitute the cement by 15% and 5%, respectively. Similarly, the voids ratio of this mix M-6a shows the lowest value among all the mixes. When measured against the control, the voids ratio for the M-6a mix was reduced by 15%.

The partial cement substitution results in different optimum w/b ratios of the concrete mixes. It's because of the variation in water requirements of the mixes to reach the optimum level of consistency. As the w/b ratio strongly affects the hardened characteristics of the concrete, working within a certain w/b ratio is necessary to assess the concrete mix's reactivity against the control. Therefore, the packing densities for a w/b ratio of 0.38 (optimum for the control concrete) were also determined. Figure 4 illustrates the concrete mixes' wet packing densities and void ratios for a w/b ratio of 0.38. According to the results, M-2 and M-4 showed a higher wet packing density and a lower voids ratio than the rest of the mixes. It is because those two mixes had the same optimum w/b of 0.38. Though the other four mixes, M-3, M-5, M-6, and M-7, did not have this optimum w/b ratio, their wet packing densities were still higher, and the voids ratios were lower than the control.

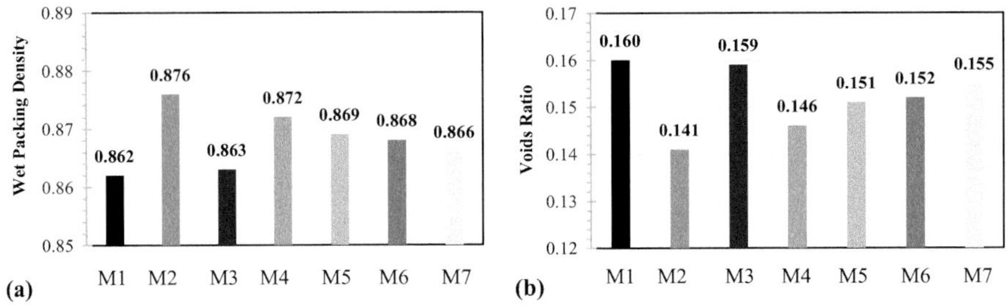

Fig. 4 **a** Wet particle packing densities, and **b** voids ratios of the concrete mixes at the w/b of 0.38

3.2 *Workability*

All concrete mixtures were evaluated for their workability by employing the conventional slump test. Table 6 presents the outcomes of the test. The w/b ratio substantially influences the slump value of the mix. Therefore, the workability is proportional to the w/b ratio. Another major factor is the use of admixture, which also controls the workability of the concrete. A polycarboxylic ether (PCE) based superplasticizer was incorporated in all the concrete batches. From the table, the slump is noticeably reduced with the rise in the incorporation percentage of fly ash and silica fume, even though all the mixes conform to the least target slump of 5 cm. Being less dense, as fly ash or silica fume concentrations rise, more fine particles are present in the mix. Silica fume mainly contains reactive silica and alumina particles with high surface areas and water absorption rates [17]. Since fly ash and silica fume have a smaller mean particle size than cement, the surface area of the mixture increases, leading to an increase in water requirement. Therefore, increasing fly ash and silica fume reduces workability.

Table 6 Workability test results

Batch	Concrete ID	w/b	Slump (cm)
M1	Control	0.38	12.2
M2	F15_0.38	0.38	9.7
M3	F30_0.38	0.38	5.4
M4	S5_0.38	0.38	10.6
M5	S10_0.38	0.38	9.2
M6	F15S5_0.38	0.38	8.1
M7	F30-S10_0.38	0.38	5.3
M3a	F30_0.42	0.42	7.5
M5a	S10_0.40	0.40	9.6
M6a	F15S5_0.40	0.40	8.3
M7a	F30S10_0.44	0.44	9.2

3.3 Compressive Strength

The compression test was conducted using cube specimens produced with a similar w/b ratio. The results are given in Fig. 5. All the samples were subjected to a curing process until 7 and 28 days. The sample with fully substituted natural coarse aggregate by EAF steel slag, i.e., the control concrete, gave 34.4 MPa and 51.5 MPa of compressive strength after 7 and 28 days, respectively (based on particle packing density). While working with a certain w/b ratio, the incorporation of fly ash (M-2 and M-3) reduces compressive strength from the control concrete. The concrete loses its strength with increasing fly ash, though fly ash incorporation showed better wet packing density. The compressive strength at 28 days decreases by 12% for M-2 and 19% for M-3.

Concrete yielded better strength results for M-4 and M-5, where silica fume partially substituted the cement by 5% and 10%, respectively. As a result, the 28-day compressive strength increases by about 7.2% for M-4 and about 10.9% for M-5. The inclusion of silica fume also enhanced the wet packing density for these two mixes relative to the control mix. The compressive strength in cases M-6 and M-7, where both fly ash and silica fume were used as partial cement substitutions, was lower than the control. However, both mixes showed similar compressive strength to M-2 and M-3, respectively, where fly ash was substituted for cement. The 28-day compressive strength decreases by about 6.6% for M-6 and 12.8% for M-7 from the control. Again, with a 5% silica fume incorporation with this 15% fly ash, the 28-day compressive strength increased by about 6%. In the same way, there was a 7.5% increment in compressive strength after 28 days when 10% silica fume was combined with 30% fly ash.

The compressive strength based on optimum wet packing density is shown in Fig. 6. As of the same w/b case, M-4 and M-5 mixes gave the maximum strength. Here, the three mixes (M-1, M-2, and M-4) are with w/b of 0.38, and the other four mixes (M3a, M5a, M6a, and M7a), whose optimum w/b differed from the control, are designed with the optimum w/b of those mixes. The strength of the M-3a mix is 10.5% lower than the previous M-3, though the wet packing density was higher for

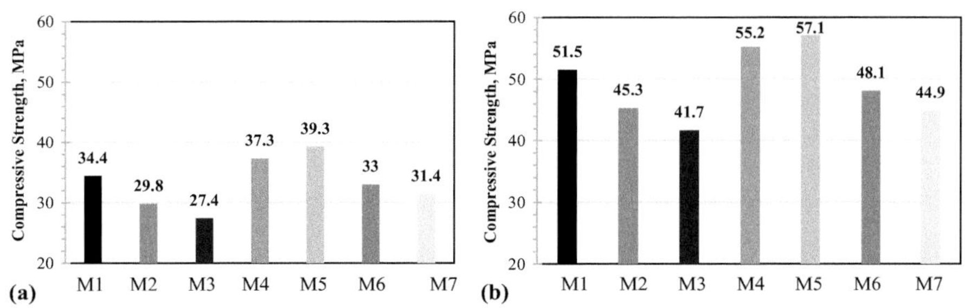

Fig. 5 Compressive strength of the concrete mixes at the w/b of 0.38 at **a** 7 days and **b** 28 days

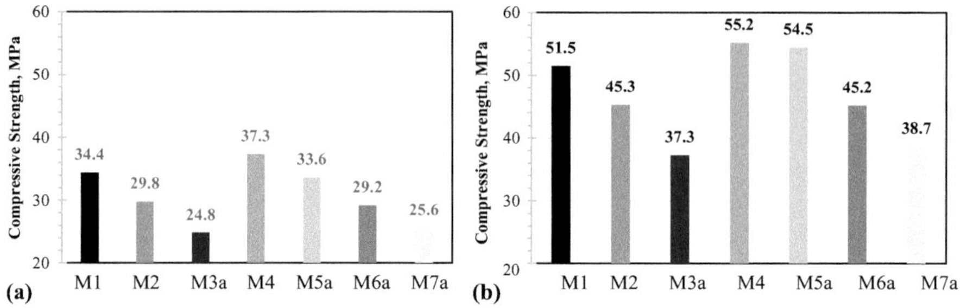

Fig. 6 Compressive strength of the concrete mixes at the optimum w/b at **a** 7 and **b** 28 days

M-3a than for M-3. An increase of about 10.53% in the w/b ratio in M-3a than for M-3 resulted in a strength decrease.

The similarity can also be found in the other mixes. Therefore, the test outcomes show that though the mixes are at their optimum w/b ratio, the strength depends on the w/b at which the mixes reach their optimum consistency and the cementitious material used. For example, in all the cases, the silica fume replacement shows better results than the fly ash replacement due to its better reactivity [17].

3.4 Splitting Tensile Strength

A 28-day splitting tensile test was performed on the cube samples. As shown in Fig. 7**a**, the control concrete gave a 5.24 MPa of tensile strength. While keeping the w/b ratio constant at 0.38, the fly ash inclusion reduces the concrete's tensile strength for both M-2 and M-3 mixes. Similar to the compressive strength, concrete also loses its tensile strength with the rise in fly ash replacement, though the inclusion exhibits better wet packing density. The splitting tensile strength decreases by about 21.9% for M-2 and about 31.7% for M-3 from the control mix. On the other hand, the tensile strengths of M-4 and M-5 mixes (with 5% and 10% silica fume) increase by 12% and 19.3%, respectively. Therefore, as the percentage of silica fume in the mix rises, so does the concrete's tensile strength. [17]. Additionally, it should be emphasized that the silica fume replacement also increased the wet packing density for M-4 and M-5 mixes compared to the control mix.

The tensile strength was observed to be less than the control mix while using both fly ash and silica fume (M-6 and M-7) to replace the cement partially, though both mixes resulted in better tensile strength relative to M-2 and M-3, where only fly ash was replaced. The tensile strength decreases by about 19.5% for M-6 and 17% for M-7 from the control mix. An increase of about 3.2% in the tensile strength was noticed with a 5% silica fume incorporation into the concrete mix in which a 15% replacement was made using fly ash. In the same way, the concrete mix using 30% fly ash would yield about a 6.4% improvement in tensile strength with the addition of

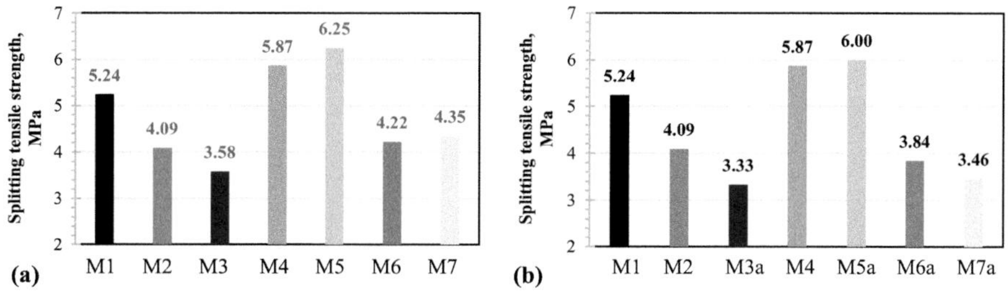

Fig. 7 **a** Splitting tensile strength of the concrete mixes at the w/b of 0.38 and **b** at the optimum w/b

10% silica fume. However, when comparing based on optimum wet packing density (Fig. 7b), as with compressive strength, M-4 and M-5 mixes gave the maximum tensile strength. Here, the three mixes (M-1, M-2, and M-4) are with w/b of 0.38, and the other four mixes (M3a, M5a, M6a, and M7a), whose optimum w/b differed from the control, are designed with the optimum w/b of those mixes. The tensile strength of the M3a mix is 6.9% lower than the previous M3, though the wet packing density was higher. This was due to an increase of 10.5% in the w/b ratio in M3a. Similarly, the strength of the M-6a mix is 9% lower than the previous M-6, and the strength of the M7a mix is 20.5% lower than the previous M7.

3.5 *Density, Absorption, and Voids in Concrete*

The test was conducted on a 2-inch-thick core of a concrete cylinder cured for 28 days following ASTM C 642. The results are given in Table 7. It is noticeable that the control concrete shows the lowest water absorption. As the replacement percentage of fly ash and silica fume increases, so does the absorption rate. The concrete where the binder replacement was with fly ash shows more absorption than with the replacement with silica fume. Compared to the optimum w/b ratio, the absorption rate increases in all cases with an increasing w/b ratio. A similar correlation can be found in the case of the voids in hardened concrete.

On the other hand, in the case of the density (oven-dry and saturated surface dry conditions) of the concrete, the control mix and the mixes in which the cement was partially substituted by silica fume, M-4, and M-5, show almost similar maximum values, while the mixes M-2 and M-3, where the partial replacement was done with fly ash, show less density than the above-mentioned mixes. Compared with the optimum w/b ratio, these densities decrease with the rising w/b. In the case of apparent density, in which the permeable pores are not counted, the control, M-2, M-4, and M-5, show almost similar values, while M-3 shows a higher value as the mix had a higher level of voids. However, the strength of concretes with supplementary cementitious materials improves at later ages [18].

Table 7 Density, absorption, and voids in hardened concrete

Batch	Concrete ID	Water absorption[1] (%)	Water absorption[2] (%)	Voids* (%)	Bulk density, (oven dry) (g/cm^3)	Bulk density[1] (g/cm^3)	Bulk density[2] (g/cm^3)	Apparent density (g/cm^3)
M1	Control	4.1	4.9	12.3	2.53	2.63	2.65	2.88
M2	F15_0.38	4.8	5.5	13.5	2.47	2.59	2.61	2.86
M3	F30_0.38	5.2	5.9	14.6	2.49	2.62	2.63	2.91
M4	S5_0.38	4.2	5.0	12.7	2.53	2.63	2.65	2.89
M5	S10_0.38	4.6	5.3	13.4	2.50	2.61	2.63	2.88
M6	F15-S5_0.38	5.1	5.7	13.8	2.43	2.55	2.57	2.82
M7	F30-S10_0.38	5.6	6.6	15.9	2.40	2.53	2.56	2.85
M3a	F30_0.42	5.3	5.8	14.6	2.51	2.64	2.66	2.94
M5a	S10_0.40	4.7	5.3	13.2	2.49	2.6	2.62	2.87
M6a	F15-S5_0.40	5.3	5.7	13.8	2.42	2.55	2.56	2.81
M7a	F30-S10_0.44	6.2	6.6	15.8	2.37	2.52	2.53	2.82

[1] After immersion
[2] After immersion and boiling
* Volume of permeable pore space (voids)

4 Conclusion

Electric arc furnace steel slag was utilized as a 100% replacement for natural coarse aggregate while designing the concrete based on particle packing optimization. The w/b ratio at which the packing density optimizes differs among the mixes and depends on the cementitious materials used and their replacement level. Fly ash and silica fume incorporation as partial substitutes for cement improved the particle packing density of fresh concrete. However, their influence on the hardened characteristics of concrete was different. Employing the Modified Andreasen model, a high-strength concrete of compressive strength up to 57.1 MPa was achieved. The higher particle packing density (0.881) was obtained by replacing the binder with 5% silica fume and 15% fly ash. Consequently, this combination gave the lowest voids ratio of 0.135.

Although a concrete mix achieves maximum packing density with its optimum w/b ratio, the strength characteristics also depend on the w/b ratio and the combination of the cementitious material used. In all the cases, the use of silica fume showed better strength characteristics due to its better reactivity than the fly ash at an early age, though both mixes showed better packing densities than the control mix. The water absorption increased in all cases when fly ash, silica fume, or both were employed as partial cement substitutes. The bulk density (OD) remains almost identical to the control mix in the case of silica fume replacement. However, for fly ash replacement, it decreases. This study indicates industrial waste EAF slag can be used as aggregate to make high-strength concrete.

Acknowledgements The author appreciates the necessary laboratory facilities and technical assistance provided by the Department of Civil Engineering, CUET. The author would also like to express gratitude and thanks to Abul Khair Group for providing EAF steel slag, without which this research would not be possible.

References

1. Neville AM (2011) Properties of concrete, 5th edn. Pearson, Edinburgh, UK. ISBN 978-0-273-75580-7
2. Arribas I, Santamaría A, Ruiz E, Ortega-López V, Manso JM (2015) Electric arc furnace slag and its use in hydraulic concrete. Constr Build Mater 90:68–79. https://doi.org/10.1016/J.CONBUILDMAT.2015.05.003
3. Islam GMS, Akter S, Reza TB (2022) Sustainable high-performance, self-compacting concrete using ladle slag. Clean Eng Technol 7:100439. https://doi.org/10.1016/J.CLET.2022.100439
4. Newman J, Choo BS (2003) Advanced concrete technology: concrete properties
5. Fennis S, Walraven JC (2012) Using particle packing technology for sustainable concrete mixture design. Heron 57:73–101
6. De Larrard F (1999) Concrete mixture proportioning: a scientific approach. CRC Press
7. Wong HHC, Kwan AKH (2005) Packing density: a key concept for mix design of high performance concrete. Mater Sci Technol Eng Conf Hong Kong 1–15
8. Feret R (1892) Sur la compacité des mortiers hydrauliques. C. Dunod, Paris
9. Fuller WB, Thompson SE (1907) Closure: laws of proportioning concrete. Trans Am Soc Civ Eng 59:169–172. https://doi.org/10.1061/TACEAT.0001978

10. Andreasen AHM (1930) Ueber die Beziehung zwischen Kornabstufung und Zwischenraum in Produkten aus losen Körnern (mit einigen Experimenten). Kolloid-Zeitschrift 1930 503 50:217–228. https://doi.org/10.1007/BF01422986
11. Dinger DR, Funk JE (1997) Particle-packing phenomena and their application in materials processing. MRS Bull 22:19–23. https://doi.org/10.1557/S0883769400034692
12. Fennis WJC, Den Uijl JA (2013) Defined-performance design of ecological concrete. Mater Struct Constr 46:639–650. https://doi.org/10.1617/s11527-012-9922-2
13. Wong HHC, Kwan AKH (2008) Packing density of cementitious materials: part 1-measurement using a wet packing method. Mater Struct Constr 41:689–701. https://doi.org/10.1617/s11527-007-9274-5
14. Li KAKH (2014) Packing density of concrete mix under dry and wet conditions. Powder Technol 253:514–521. https://doi.org/10.1016/j.powtec.2013.12.020
15. Kumar SV, Santhanam M (2003) Particle packing theories and their application in concrete mixture proportioning: a review. Indian Concr J 77:1324–1331
16. Kwan AKH, Wong HHC (2008) Packing density of cementitious materials: part 2-packing and flow of OPC + PFA + CSF. Mater Struct Constr 41:773–784. https://doi.org/10.1617/s11527-007-9281-6
17. Joshaghani A, Moeini MA, Balapour M, Moazenian A (2018) Effects of supplementary cementitious materials on mechanical and durability properties of high-performance non-shrinking grout (HPNSG). J Sustain Cem Mater 7:38–56. https://doi.org/10.1080/21650373.2017.1372318
18. Islam GMS (2012) Evaluating reactivity and sorptivity of fly ash for use in concrete construction. University of Dundee

Comparative Study on Riverbank Erosion and Accretion Using HEC-RAS and Landsat Imagery

B. K. Nath and A. Akter

Abstract In Bangladesh, riverbank erosion is a chronic and persistent problem. Due to erosion and deposition, the Padma river course changes from meandering to braided. Between Pangsha and Goalanda, a 48 km reach was selected for this study. This research aims to assess the riverbank erosion and accretion in the Rajbari area using HEC-RAS and satellite image processing. The historical dataset indicates that the Padma River is highly dynamic, with its position, planform, and morphological features often changing due to flow and sedimentation variations. About 75.59 and 20.83 km^2 were degraded between 2014 and 2022, respectively, due to heavy discharge in the monsoon. Satellite images showed that during the last nine years (2014–2022), erosion on the right bank of the River Padma in Rajbari district had increased, i.e., 75.59 km^2 and 20.83 km^2, respectively. The simulation of sediment transport using a 1-D HEC-RAS model in the study area comprising d_{50} as 0.14 mm and Manning's n = 0.025 showed reasonable responses. Thus, this assessment could be applied to the decision support system for these river stakeholders.

Keywords Padma river · Erosion · Accretion · Remote sensing · Landsat imagery · Morphological change · HEC-RAS

1 Introduction

The slope, the volume of water, the frequency of water, and the river category influence the size and shape of rivers. Most of Bangladesh's primary riverbank line shifts their courses from 60 to 1,600 m annually. Nearly 2,400 km of the bank line annually suffer massive erosion [10]. The Padma is a significant river in Bangladesh, having

B. K. Nath · A. Akter (✉)
Department of Civil Engineering, Chittagong University of Engineering and Technology, Chittagong 4349, Bangladesh
e-mail: aysha_akter@cuet.ac.bd

B. K. Nath
e-mail: u18mce022p@student.cuet.ac.bd

S. Arthur et al. (eds.), *Proceedings of the 6th International Conference on Advances in Civil Engineering*, Lecture Notes in Civil Engineering 368,
https://doi.org/10.1007/978-981-99-3826-1_37

originated as a tributary of the Ganges before that river's bifurcation. Bangladesh's River Padma is the world's third-largest sand-bed river regarding average annual discharge [7]. A deltaic plain produced by the confluence of the Ganges, Brahmaputra, and Meghna and their tributaries creates the low-lying riverine country of Bangladesh. Significantly shaping the characteristics of the Bengal delta was the Ganges (Padma), Brahmaputra (Jamuna), and Meghna, as well as their tributaries and distributaries [12]. The area of Bangladesh's rivers and inland waterways is roughly 7% of its total territory [15]. With a baseline flow of over 30,000 m^3/s, overflows of over 75,000 m^3/s, and a 100-year flood flow of about 130,000 m^3/s, it is one of the world's greatest rivers in terms of flow [8]. Halder [5] studied using Geographic information system (GIS) and remote sensing methods that a total of 17.26 km^2 area experienced erosion, compared to the 18.47 km^2 deposition in the Rajbari district along Padma's right bank between 1979 and 2019. Researchers [6, 11] discovered that the average overland discharge is 95000 m^3/s, whereas the moderate discharge is 4800 m^3/s. The depth of water assessed by the Public Works Department is a horizontal datum believed initially to have zero at the determined Mean Sea Level (m PWD). As a result, the maximum average depth of water is 5.5 m PWD, the minimum is 1.5 m PWD, and the total annual sediment movement is 900 M tons. From 1975–2015, a time series of Landsat satellite imagery MSS, TM, OLI, and TIRS images were analyzed to detect river bank erosion-accretion and bank line shifting. The study also exposes that the total area of islands has increased significantly, with about 50,967 ha of island area in 2015, up from 20,533 ha in 1975. The results show uniformity of sediment deposition in the river bed [1]. Geographic Information System (GIS) is used to evaluate erosion and accretion's spatial and temporal dynamics for varying stages along the Jamuna River in Bangladesh using images for Landsat 1995, 2005, and 2015. The erosion and accretion levels in Jamuna River varied from site to site. The largest eroded area was 3.82 km^2 from 1995 to 2005, and the maximum accreted area was 6.15 km^2 from 1995 to 2015 [14]. The sediment transport simulation for the Kobadak River uses a 1-D HEC-RAS model. Various processes and environmental conditions, including the bed and banks' composition and erosion prospect, influence a river channel's morphology. The changing trends of hydrological and morphological parameters in the Surma and Kushiyara river systems are investigated in this research [13]. This study aims to observe morphological changes in the 48 km of Padma River throughout the year. Thus, the rates and amounts of erosion and accretion in the studied river section in the study area could be determined.

2 Study Area

Bangladesh's Rajbari district and the Padma River are both prone to erosion. The Padma river spans around 48 km across the Rajbari district (23°35' and 23°55' N Latitudes and 89°09'–89°55' E Longitude). The Padma River expanded by almost 3.5 km, and the bar area became more than 40 km. Throughout periods of significant monsoon flow, the Padma River transported massive sediment, causing extensive bank erosion and deposition. The Padma River has been subjected to numerous studies demonstrating the widespread erosion and subsequent deposition. There were 355 km^2 of degraded land and 157 km^2 of accreted land between 1973 and 2001, respectively [2] (Fig. 1).

Bank erosion is a natural disturbance of the river system that is crucial to both the long-term morphologic evolution of river systems and the health of the river ecosystem [3]. The River Padma has been the subject of numerous studies demonstrating the widespread erosion and subsequent deposition. There were 355 km^2 of degraded land and 157 km^2 of accreted land between 1973 and 2001 [2]. Because of its spatial and temporal context, remote sensing research is superior to other methods for assessing river bank erosion and its impacts.

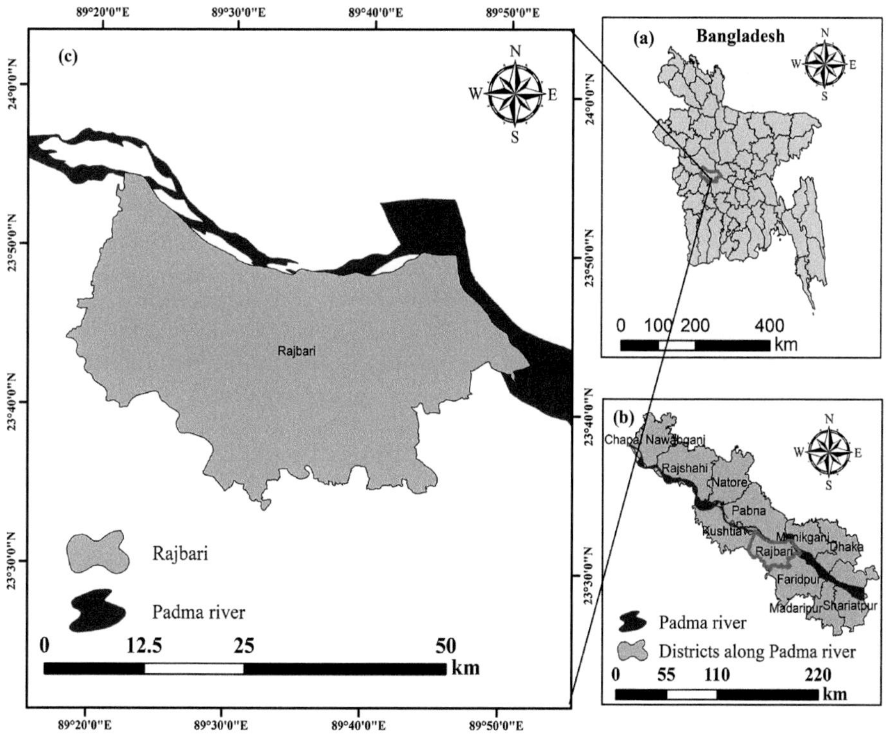

Fig. 1 **a** Bangladesh map, **b** districts along Padma River, **c** Study area (Rajbari district)

2.1 Data Collection

Landsat 8 monitors several frequency ranges over the electromagnetic spectrum—a color that is not necessarily visible to the human eye. Each range is known as a band, and Landsat 8 contains eleven bands. Different acquisition date has been selected to minimize error due to cloud cover of Landsat images. To characterize the erosion and accretion pattern, we used high-resolution Landsat satellite imageries from Band 1 to Band 7, encompassing 2014–2022. In addition, relevant Google Earth images from 2014 and 2022 have been compared to understand better how the right bank of the Padma River in the Rajbari area has transformed over time. Water level, discharge, and sediment flow were collected from the Bangladesh Water Development Board (BWDB) to establish a 1 D HEC-RAS model.

2.2 Methodology

The hydrodynamic model simulates the movement of water using mathematical equations in a given area. Sediment model development refers to the process of developing mathematical models to simulate the flow and settling of sediment in various environments. This includes sediment transport in rivers, lakes, and oceans and material deposition in soil and other porous materials. Computer models are used to simulate the hydrodynamic and sediment models to forecast how the system will behave under various conditions. This can be used to investigate the effects of environmental factors on the design and to predict its future behavior. The study area was selected through a comprehensive literature review. Satellite images were collected during 2014–2022 for image processing (Table 1). The amount of land lost and gained along the Padma River was assessed using ArcMap 10.6.1 (Fig. 2). The changes were obtained from the HEC RAS simulation and compared with the image processing. Figure 3 illustrates the adopted methodology.

An erosion and accretion estimate for the study was calculated using ArcMap 10.6.1. The water ratio index (WRI) method was used to calculate the frequencies of erosion and accretion in this area. A critical component of the WRI approach is extracting a particular water body from the Padma River. Using Landsat-8 (OLI/

Table 1 Details of the Landsat 8(OLI/TIRS) data Path

Path	Row	Acquisition date	Size of the processing	Cloud cover	Resolution (m)
138	043	29-09-2014	L1TP	5.72	30
138	043	23-02-2016	L2SP	2.17	30
138	043	29-12-2018	L1TP	0.01	30
138	043	15-10-2020	L1TP	5.45	30
138	043	26-08-2022	L1TP	8.31	30

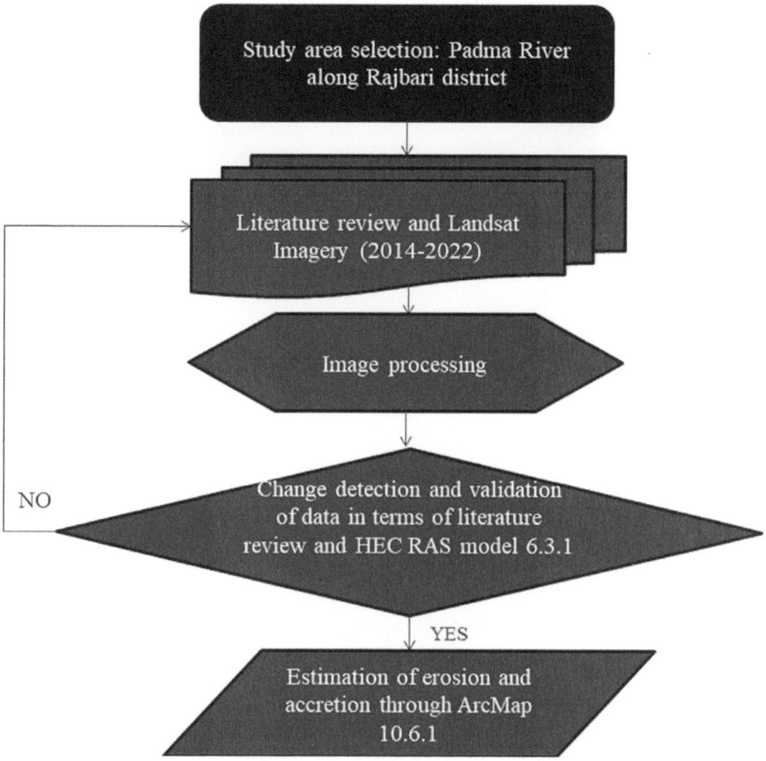

Fig. 2 Annual variations in the Padma River (2014–2022)

TIRES) satellite imagery, WRI performed post-processing. The green (Band 2) and red (Band 3) spectral reflectance of water are much stronger than the near-infrared (Band 4) and medium-infrared (Band 6) spectral reflectance for apparent reasons (Band 5).

WRI displays water values that are often larger than 1 Gautam et al. [4], WRI is described as:

$$\text{WRI} = \frac{\text{Green} + \text{Red}}{\text{NIR} + \text{MIR}} \quad (1)$$

The WGS84 UTM (zone 45N) coordinate system L1T data in GeoTIFF format is made available by the United States Geological Survey (USGS) for each band. A composite was carried out by combining the specifics of each band's separate photos.

For the study, cross-sectional data of the river reach, stage, and hydrograph data were necessary. The cross-sectional data were collected from the BWDB. The boundary conditions were: the Mohendrapur as the upstream boundary (Discharge data) and Goalanda transit as the downstream boundary (Water Level data). The tidal phase hydrograph and rating curve were created using data from the BWDB

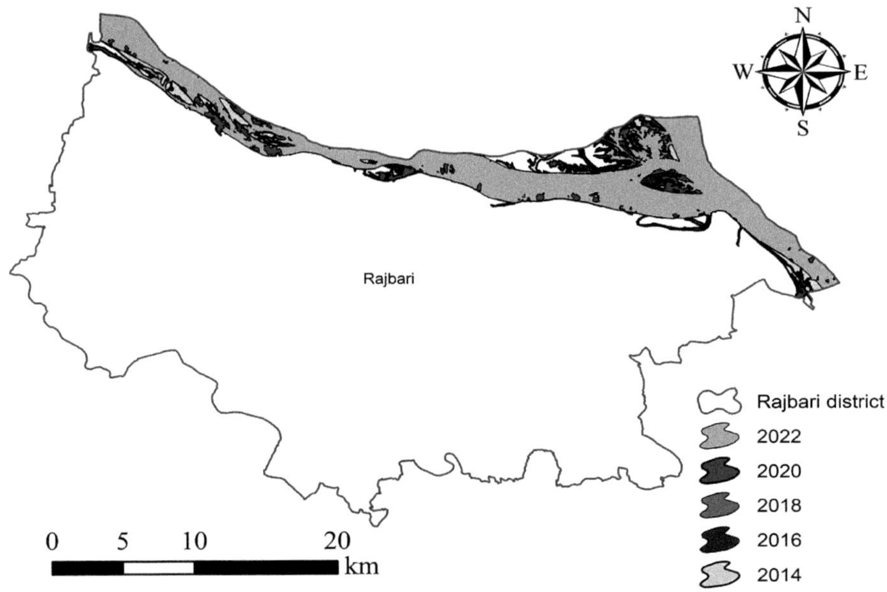

Fig. 3 Shows a schematic representation of the methodology's overall steps

raw data considering six-hour intervals, and the sediment data was used as the calibrating parameter of the model. The Padma River's bed gradation was classified using Fig. 4. The maximum depth for possible erosion at any section was set at 10 m, as shown in Fig. 4.

This study uses Landsat images (30 m resolution) to estimate erosion and accretion from 2014 to 2022. The different dates of satellite images were chosen due to cloud cover, which should be less than 10%. The Padma River flows through the Rajbari district and has an annual water path of around 48 km. In 2014, the water area of the Padma River was around 135.90 km^2, but by 2016, it had reduced to about

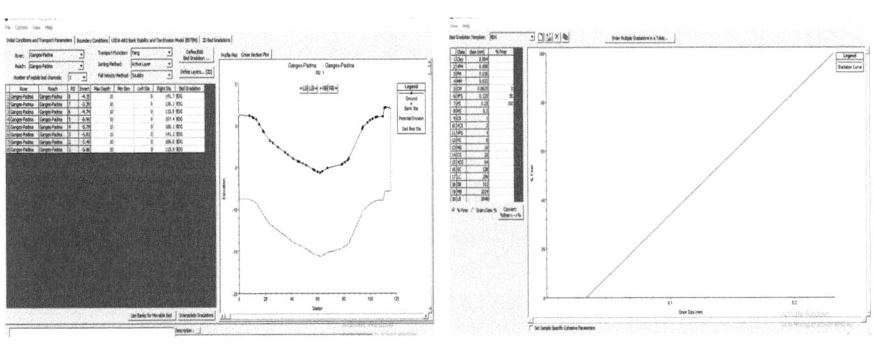

Fig. 4 Hydrodynamic parameter and sediment data input in HEC RAS (6.3.1)

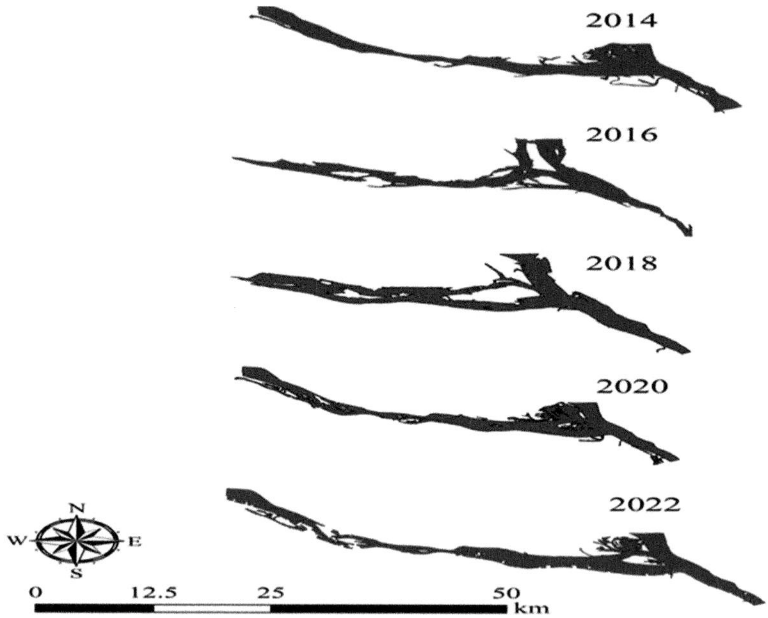

Fig. 5 Changes to the Padma River every year from 2014 through 2022

109.35 km^2. In addition, the change between 2016 and 2018 corresponds to about 107.51 km^2. Between 2020 and 2022, the water area for any hydraulic infrastructure increased from 112.5 to 118.0 km^2 (Fig. 5).

The empirical formula used to compute the erosion and accretion are-

$$\text{Eroison} = (\text{Previous year's area}) - (\text{unchanged area}) \tag{2}$$

$$\text{Accretion} = (\text{Next year's area}) - (\text{unchanged area}) \tag{3}$$

3 Result and Discussions

The outcomes from this research have been described in two phases, i.e., (a) comparison of satellite image processing data and literature, and (b) comparison of HEC RAS simulated and satellite image processing data. Ophra et al. [9] studied that the Padma River's banks started to shift several times in 1988 and 2017 due to land erosion and accumulation. The total land loss for this 29-year time frame was 9.72 km^2. Table 2 shows Padma River's erosion and accretion from 2014 to 2022. The accredited area decreased from 74.64 km^2 to only 4.13 km^2 between 2018 and 2020 and increased again between 2020 and 2022. A sharp increase has been observed between 2014–2016 and 2016–2018. On the other hand, in the case of erosion, a negative tendency was from 2014 to 2016 to 2018 to 2020, resulting in 75.59–3.39 km^2. An increased value of 20.83 km^2 in 2020–22 was found, and similar findings were recorded by [5] and some discontinuity values due to time frame variation. From 2016 to 2018, the erosion area decreased from 13.49 to 3.39 km^2, and this trend continued through 2018 and 2020 (Table 2).

In this study, analysis has been completed using satellite image processing (Fig. 6). The results of ArcGIS also validate with HEC-RAS 6.3.1 sediment transport analysis in bed level for unsteady flow. The Padma river reach is eroded in its downstream end by an amount of 0.05–0.40 m and, in a few locations, deposition of 0.01–0.50 m (Fig. 7).

Table 2 Eroded and accumulated areas during 2014–2022

Year	Erosion (km^2)		Accretion (km^2)	
	This study	Halder [5]	This study	Halder [5]
2014–2016	75.59	–	2.76	–
2016–2018	13.49	–	55.58	–
2018–2020	3.39	–	74.64	–
2020–2022	20.83	17.26 km^2	4.13	18.47 km^2

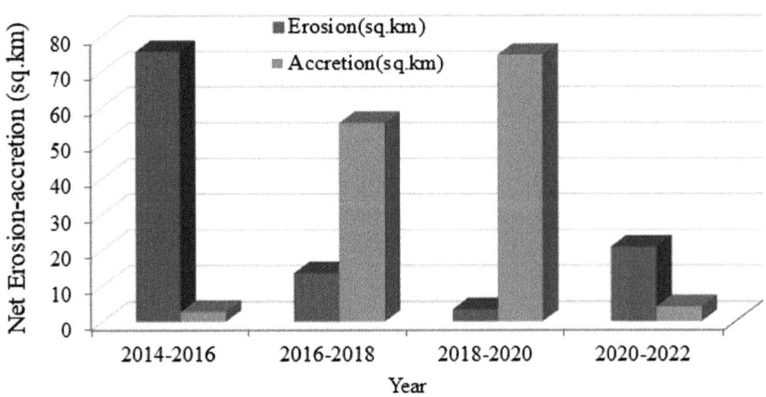

Fig. 6 Net erosion and accretion amounts

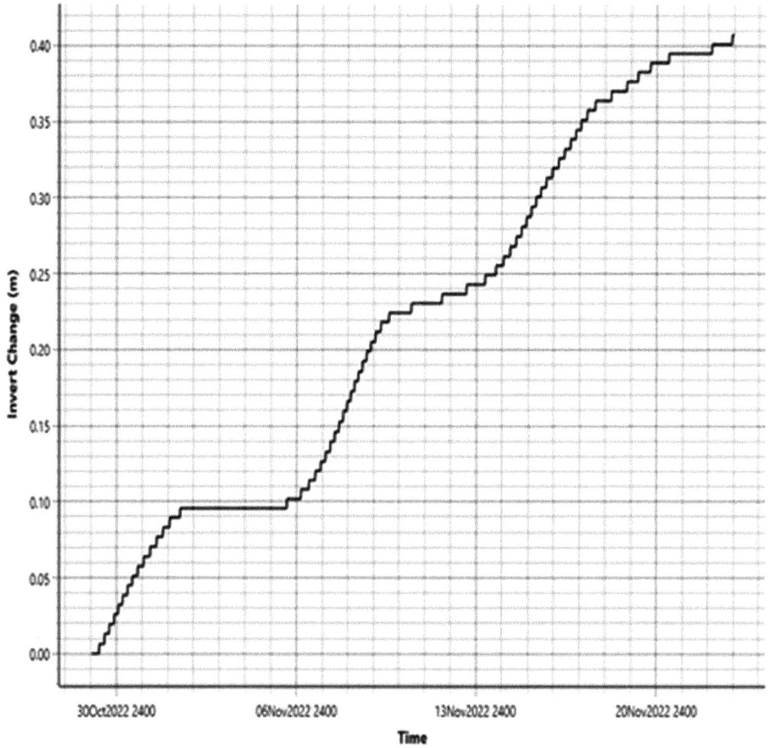

Fig. 7 Profile of invert change at the downstream

4 Conclusion

This study used geographic information system (GIS) and remote sensing techniques with multi-temporal satellite imageries to feature changes in Padma River morphology. The findings demonstrated that the altered fluvial flow system had resulted in a considerable morphological change. Historical data shows erosion and accretion of river channels along the Padma River in the Rajbari district in the last four decades. Rapid changes led to significant land development and loss in this study area. Right-bank erosion areas dominated during this period, with more erosion than accretion. This study comprises satellite image processing and geographical and temporal data assessment. For validating the output of satellite images, the simulation of sediment transport in the Padma River using a 1-D HEC-RAS model determines the bed level changes. Despite its methodological and digital data constraints, this study is helpful since it helps save time and money, especially in places where field survey methods are impractical.

Acknowledgements The authors acknowledge the ongoing research project, "Hydrodynamic Modeling of Selected Riverbank Protections on Bangladesh (CUET/DRE/2021-22/WRE/004)," of the Chittagong University of Engineering and Technology (CUET) in Bangladesh for funding this research work.

References

1. Billah MM (2018) Mapping and monitoring erosion-accretion in an alluvial river using satellite imagery-the river bank changes of the Padma River in Bangladesh. Quaest Geogr 37(3):87–95. https://doi.org/10.2478/quageo-2018-0027
2. Dewan A, Corner R, Saleem A, Rahman MM, Haider MR, Rahman MM, Sarker MH (2017) Assessing channel changes of the Ganges-Padma River system in Bangladesh using Landsat and hydrological data. Geomorphology 276:257–279. https://doi.org/10.1016/j.geomorph.2016.10.017
3. Florsheim JL, Mount JF, Chin A (2008) Bank erosion as a desirable attribute of rivers. Bioscience 58(6):519–529. https://doi.org/10.1641/B580608
4. Gautam VK, Gaurav PK, Murugan P, Annadurai M (2015) Assessment of surface water dynamicsin Bangalore using WRI, NDWI, MNDWI, supervised classification and K-T transformation. Aquat Procedia 4(Icwrcoe):739–746. https://doi.org/10.1016/j.aqpro.2015.02.095
5. Halder (2021) Evaluation of the river Padma morphological transition in the central Bangladesh using GIS and remote sensing techniques. Int J River Basin Manag. https://doi.org/10.1080/15715124.2021.1879095
6. Hassan S, Akhtaruzzaman AFM (2010) Environmental change detection of the Padma River in the North-Western part of Bangladesh using multi-date Landsat data. In: Proceedings of international conference on environmental aspects of Bangladesh (ICEAB10), pp 193–195
7. Kamal MR, Rumman R, Shushmi KW, Hasan KA (2019) River bank protective work of Bangladesh: a case study on the River Padma. Int J Struct Civ Eng Res 171–177. https://doi.org/10.18178/ijscer.8.2.171-177
8. Neill C, Hotopp D, Hunter B (2013) Some hydrotechnical features of Padma River, Bangladesh
9. Ophra SJ, Begum S, Islam R, Islam MN (2018) Assessment of bank erosion and channel shifting of Padma River in Bangladesh using RS and GIS techniques. Spat Inf Res 26(6):599–605. https://doi.org/10.1007/s41324-018-0202-2
10. Rahman MA (2010) Comparative analysis of design and performance of bank protection works of Jamuna River at Titporol and Debdanga. Submitted by Md. Anisur Rahman
11. Rahman M, Ishigaki T, Rahman MM, Nakagawa H, Khaleduzzaman A (2003) Channel stabilization using bandalling. In: Ecosystem services for poverty alleviation (ESPA delta) view project assessing health, livelihoods, ecosystem services and poverty alleviation in populous deltas view project channel stabilization using bandalling. https://www.researchgate.net/publication/47528102
12. Rana M (2017) Impact of riverbank erosion on population migration and resettlement of Bangladesh. Sci J Appl Math Stat 5(2):60. https://doi.org/10.11648/j.sjams.20170502.11
13. Sabbir M, Khan M, Das P (2018) Assessment of hydro-morphological change of Surma-Kushiyara River system. J Mod Sci Technol 6(1):113–123
14. Sarker MH, Sarker MH, Ferdaous MR (2011) River bank protection measures in the Brahmaputra-Jamuna River: Bangladesh experience. https://www.researchgate.net/publication/263125674
15. Whitehead PG, Jin L, Macadam I, Janes T, Sarkar S, Rodda HJE, Sinha R, Nicholls RJ (2018) Modelling impacts of climate change and socio-economic change on the Ganga, Brahmaputra, Meghna, Hooghly and Mahanadi River systems in India and Bangladesh. Sci Total Environ 636:1362–1372. https://doi.org/10.1016/j.scitotenv.2018.04.362

Bangladeshi Drivers' Perception on Vehicle Safety Features

A. Huq, S. Iqra, and Z. Ikram

Abstract The purpose of this study is to understand drivers' perceptions on vehicle safety features such as anti lock braking system (ABS), adaptive cruise control (ACC), and electronic stability control (ESC). Furthermore, this study aims to determine the impact of demographic factors of drivers on their opinion of these safety measures. Structural equation modeling (SEM) was used to assess the correlation between the four latent variables that were generated from multiple exogenous factors acquired from survey responses. It is estimated that drivers' demographic traits have a positive influence on their understanding of all safety elements. Furthermore, drivers' perceptions about the benefits of ACC, ESC, and ABS were positively associated among each other. If a driver has sufficient understanding about any of the safety elements, he is more likely to share a positive perception for other safety elements as well. It is found that exogenous variables connected to ACC and ESC are positively correlated with the latent variable of these groups, but aggressive driving tendency was adversely related to advantages of ABS. The outcomes of these studies can be utilized to make policies on the installation of safety measures in future automobiles on urban roads. This study demonstrates that the aforementioned safety measures are significantly useful for driving in a critical condition, and so they can be implemented in vehicles to prevent crashes in a critical circumstance. For this reason, before granting a driving license, the necessary examination should be implemented

A. Huq (✉)
Accident Research Institute (ARI), Bangladesh University of Engineering and Technology (BUET), Dhaka 1000, Bangladesh
e-mail: ahuq002@fiu.edu; ashuq@ari.buet.ac.bd

S. Iqra
Department of Civil Engineering, Bangladesh University of Engineering and Technology (BUET), Dhaka 1000, Bangladesh
e-mail: 1604135@ce.buet.ac.bd

Z. Ikram
Department of Computer Science and Engineering, Bangladesh University of Engineering and Technology (BUET), Dhaka 1000, Bangladesh
e-mail: 1905111@cse.buet.ac.bd

S. Arthur et al. (eds.), *Proceedings of the 6th International Conference on Advances in Civil Engineering*, Lecture Notes in Civil Engineering 368,
https://doi.org/10.1007/978-981-99-3826-1_38

to assess drivers' awareness of these safety measures. This study will also help future studies on people's perceptions on autonomous vehicles.

Keywords Structural equation modeling (SEM) · ABS · ACC · ESC · Vehicle safety feature

1 Introduction

Advanced driving assistance systems (ADAS) can be utilized to assist humans for preventing accidents [1]. An ADAS is a vehicle control system that uses embedded sensors (such as radar, laser, and vision) to help drivers recognize and respond to potentially hazardous traffic conditions thus increasing driving comfort and traffic safety It is expected that the adoption of smart technologies like ADAS in vehicles might save more than half of all fatalities among people [2]. There are several ADAS that improves safety in vehicles by helping in avoiding collision or increasing vehicle stability such as anti lock braking system (ABS), electronic stability control (ESC), adaptive cruise control (ACC), airbag, seat belt etc. which are commonly installed in modern vehicles [3]. Among these safety features ABS, ESC & ACC are called active safety features as they can activate automatically by sensing the environment. The in-vehicle information flow can be regulated as a final application of this assessment to lighten the driver's workload as much as possible [4].

ABS was one of the first active assistance systems based on proprioceptive sensors; the addition of dynamic driving controls, such as ESC, marked another milestone in the development of the driving assistance system [5]. ABS is a vehicle safety feature that enables a car's wheels to stay in touch with the pavement during braking, preventing the wheels from locking up (ceasing rotation) and preventing uncontrollable sliding [6]. It is interesting to note that the majority of researchers have come to the conclusion that the widespread use of vehicles with ABS has not resulted in the anticipated decrease in collisions [7]. Research on the effects of ABS has shown that it impacts driver behavior by encouraging quicker and more aggressive driving. Moreover, ABS can adversely affect fatal accidents, overturning accidents, single accidents without overturning, and collisions with fixed objects [8].

When turning a corner, braking suddenly, or performing a quick maneuver, ESC helps keeping a car from sliding and the driver from losing control of the car. The vehicle's brakes are immediately engaged using ESC technology to assist with steering [9]. Except for accidents involving pedestrians, bicyclists and animals, ESC appears to lower the number of accidents of all types. It may appear paradoxical that ESC appears to have such unambiguously positive effects on accidents, yet ABS did not [8].

ACC allows car to maintain a safe distance behind a forward vehicle. It is a vehicle function that allows the cruise control system to adjust the vehicle's speed based on the traffic situation. A radar device attached to the front of the vehicle detects whether slower moving cars are in the path of the ACC vehicle. When a slower moving vehicle is recognized, the ACC system slows the vehicle and controls the clearance, or temporal gap, between the ACC vehicle and the advancing vehicle [9]. If the system determines that the front vehicle is no longer in the path of the ACC vehicle, the system will accelerate the car back to its cruise control speed. This action enables the ACC vehicle to autonomously slow down and accelerate in response to traffic without driver's activity. The ACC vehicle's speed is regulated through engine throttle control and limited brake usage [1, 10, 11].

So, the above mentioned ADASs can improve active and integrated safety, however, more market penetration is required to maximize their contribution to overall traffic safety [11]. This can be accomplished by increasing public awareness of ADAS and their benefits, or by mandating their implementation through legislation. Little is known, however, about driver comprehension and knowledge of these technologies. These ADAS can not only warn the driver of hazards and lapses in concentration, but they can even intervene to prevent or decrease the severity of a crash in some cases. If drivers do not know the purpose, function, and limitations of a device, the full safety advantage may not be realized and communicated to the roads [12]. Most of the research in this field focuses either on the technical specification and functions or the commercial advancement of these above-mentioned safety features. There has been little research, particularly in developing countries like Bangladesh, done on the driver's understanding of these ADASs, and much less is known about their feedback. However, it is crucial to know the acceptance of these technologies among drivers in terms of their ability to reduce the risk of accidents.

As a result, the goal of this study is to get a thorough grasp of the driver's perspective of vehicle safety features such as ABS, ESC, and ACC. Finally, this study focuses on the relationship between driver demographic characteristics (age, gender, driving experience, accident involvement, monthly income, etc.) and their perception of these advanced safety features in order to assess driver's knowledge of these advanced safety features across different socioeconomic backgrounds.

2 Materials and Methods

The Survey Procedure

A manual pen-and-paper survey was used in this investigation. The survey was carried out in eleven places around Dhaka city (Khilgaon, Basabo, Dhanmondi, Gulshan, Mirpur, Rupnagar, Vatara, Basundhara, Malibagh, Rampura, Banani). The survey was conducted across two time periods. The first takes place during the Covid-19 time (April 2021—May 2021), while the second takes place following the Covid-19 period (Jun 2022—July 2022). 750 people were questioned in selected places on both weekdays and weekends. Survey respondents were chosen at random from the car repairing shops, rent-a-car shops, garages, streets of the survey areas based on their responses having various characteristics (gender, age, income level, vehicle ownership, driving experience, accident involvement and perception about advanced vehicle safety features like ABS, ESC and ACC). Because Bangladesh is a developing country, many people are unaware of advanced technologies such as ADAS. The same phenomenon was noticed in this study's survey. As a result, the incomplete and invalid responses were omitted out of the analysis. Lastly, 607 valid replies were identified to be included into the Structured Equation Modeling (SEM) for further analysis.

Structural Equation Modeling (SEM)

SEM comprises two components known as Measurement model and Structure model. Measurement model examines relation between measured item and latent items while Structure model estimates internal relationship between latent variables. Measurement equation in SEM approach is:

$$x_i = \Lambda \omega_i + \varepsilon_i \tag{1}$$

where, x_i is a vector of indicators describing the random vector of latent variables ω_i, Λ is matrices of the loading coefficients as obtained from the regressions of x_i on ω_i and ε_i is random vectors of the measurement errors.

The structural equation of SEM is expressed by:

$$\eta_i = \Gamma \xi_i + \delta_i \tag{2}$$

where, η_i is a vector of latent variables, Γ is a regression parameter matrix for relating the endogenous latent variables and exogenous latent variables, ξ_i is a vector of latent variables and δ_i is a vector of disturbances.

3 Results and Discussions

According to Table 1, the majority of drivers hold a neutral stance towards the advantages of ESC. While approximately 40% of drivers feel that ESC aids driving in numerous ways, just 10% of drivers strongly agree that ESC assists driving. However, approximately 20% of drivers believe that this safety function does not actually improve driving.

Table 2 reveals a similar trend; however, the proportion of drivers who strongly agree with the probable benefits of ACC rises marginally. Table 3 demonstrates that the majority of drivers do not have the correct understanding of how the ABS is activated. ABS is activated automatically when skidding is detected, however, only roughly 17% of drivers have the correct understanding of ABS activation. 53% of drivers whose vehicles are equipped with ABS are more likely to engage in aggressive driving.

Table 4 illustrates that around 65% of survey respondents are professional drivers, while approximately 56% of drivers have been involved in accidents. 60% of drivers

Table 1 Perception of drivers about benefits of ESC

Question no	Question	1 = Strongly disagree	2 = Disagree	3 = Fair	4 = Agree	5 = Strongly agree
esc1	ESC helps in driving on a wet road	24.01	11.84	38.16	14.64	11.35
esc2	ESC helps when overtaking on a slippery road	17.43	12.99	37.99	19.57	12.01
esc3	ESC helps in the sudden hard brake	16.61	5.76	41.45	20.89	15.30
esc4	ESC helps while speeding while climbing a steep road	17.43	9.21	39.97	22.04	11.35
esc5	ESC helps while speeding on a flat road	14.64	10.69	41.61	20.89	12.17
esc6	ESC helps to maneuver at regular speed	15.13	10.69	40.79	20.72	12.66
esc7	There's a lower risk of skidding with a car equipped with ESC than a car without ESC	16.61	10.20	41.78	20.72	10.69
esc8	here's a better chance to correct a slide with a car equipped with ESC than a car without ESC	15.63	10.86	43.59	19.57	10.36
esc9	A car with ESC can accelerate faster than a car without it	16.12	9.54	42.11	19.90	12.34
esc10	A car with ESC can be driven faster on slippery roads than a car without it	15.79	11.02	41.28	18.26	13.65
esc11	A car with ESC can be steered and braked simultaneously better than a car without it	18.91	9.54	39.14	19.08	13.32

are between the ages of 18 and 40 and have a range of incomes. Only 9% of these drivers have never performed a safety inspection on their vehicle. The majority of these drivers perform safety inspections at varied intervals, with 18% performing them only when a problem is discovered.

Table 5(a) represents the path value from the observed to the latent variables for SEM calculated using the Maximum Likelihood (ML) method by the lavaan package for SEM and blavaan package in R. The third and fourth column represents the Estimate value (to evaluate effect of a given observed variable on the respective latent variable) and the p-value (to evaluate parameter significance) of SEM. An attribute is considered significant at 95% confidence interval when p-value <0.05 for that Estimate value in SEM. It can be seen from Table 1 that the path values for all

Table 2 Perception of drivers about benefits of ACC

Question no	Question	1 = Strongly disagree	2 = Disagree	3 = Fair	4 = Agree	5 = Strongly agree
acc1	ACC helps in driving on a wet road	20.56	10.69	41.45	15.46	11.84
acc2	ACC helps when overtaking on a slippery road	15.13	13.32	41.12	17.60	12.83
acc3	ACC helps in the sudden hard brake	13.49	9.70	41.78	20.07	14.97
acc4	ACC helps while speeding while climbing a steep road	13.16	11.02	43.09	18.26	14.47
acc5	ACC helps while speeding on a flat road	14.31	12.01	38.16	20.89	14.64
acc6	ACC helps to maneuver at regular speed	14.14	10.69	39.14	21.88	14.14
acc7	There's a lower risk of skidding with a car equipped with ACC than a car without ACC	14.80	8.72	42.43	20.23	13.82
acc8	here's a better chance to correct a slide with a car equipped with ACC than a car without ACC	16.45	12.34	40.95	17.27	12.99
acc9	A car with ACC can accelerate faster than a car without it	14.80	10.53	39.97	19.74	14.97
acc10	A car with ACC can be driven faster on slippery roads than a car without it	14.31	10.53	40.63	20.07	14.47
acc11	A car with ACC can be steered and braked simultaneously better than a car without it	15.46	11.18	38.49	20.72	14.14

latent variables are significant in SEM that indicates positive and negative influence of the observed variable on the latent variable.

Table 5(b) represents latent variable relations of the proposed SEM model. All the variables are statistically significant which indicates the strong relation between latent variables. All of the hypotheses were accepted as they were statistically significant. Figure 1 clearly demonstrates the path analyses for the SEM.

Six hypotheses were developed for determining the correlation among four endogenous variables, ACC, ESC, ABS and demo, after that all of the hypotheses were accepted as they were found to be statistically significant at 1% level of significance.

H1: Primary Perception Regarding ACC has a positive impact on perception regarding ESC- In this study, primary perception regarding ACC refers to the respondent's primary or basic idea, thinking, and knowledge about different aspects of ACC.

Table 3 Perception of drivers about benefits and attitudes about ABS

Question no	Question	Description	
		Category	%
abs1	How to activate ABS	ABS activate by Pressing the Brake at a Certain Level	16.28
		ABS activate when hard brake	13.16
		No idea about ABS activation	22.20
		ABS activate with normal brake	26.64
		ABS will be Activated Automatically if it Detects Skidding	16.78
		Multiple response	4.93
abs2	How ABS helps driving	ABS features better avoids crash	39.31
		ABS features better control	11.84
		ABS helps to follow tailgating vehicle more closely and efficiently	21.71
		ABS features no idea	20.23
		Multiple response	6.91
abs3	Does having ABS in your car encourage you to drive more aggressively	Yes	52.30
		No	47.70

Table 4 Demographic information of the drivers

Question no.	Question	%	Question no.	Question	%
Involved in accident	Yes	55.76	Income	35,000–50,000	16.94
	No	44.24		50,000–75,000	9.38
Professional driver	Yes	65.13		75,000–100,000	6.58
	No	34.87		More than 100,000	7.40
Gender	Female	31.25	Driving experience	Less than 1 year	29.77
	Male	68.75		1–3	18.26
Age	Below 18	20.39		3–5	24.18
	18–25	29.44		5–10	17.93
	26–40	28.62		More than 10 years	9.87
	41–65	14.80	Safety checkup frequency	Never	9.21
	Above 65	6.74		Less than 2 months	31.91
Income	Below 15,000	27.14		2–6 month	23.36
	15,000–25,000	16.94		More than 6 months	17.43
	25,000–35,000	15.63		When issue is found	18.09

Table 5 Structural equation modeling results

(a) Regression weight of structural relationship

Latent variables	Observed variables	Estimates	p-value	Latent variables	Observed variables	Estimate	P-value
acc	acc1	1.027	0.000	esc	esc6	1.234	0.000
	acc2	1.224	0.000		esc7	1.279	0.000
	acc3	1.254	0.000		esc8	1.248	0.000
	acc4	1.321	0.000		esc9	1.145	0.000
	acc5	1.327	0.000		esc10	1.212	0.000
	acc6	1.369	0.000		esc11	1.225	0.000
	acc7	1.347	0.000	abs	abs1	1.224	0.000
	acc8	1.308	0.000		abs2	0.072	0.435
	acc9	1.286	0.000		bs3	−0.446	0.000
	acc10	1.233	0.000	demo	Checkup frequency	1.495	0.000
	acc11	1.154	0.000		Gender	−0.139	0.139
esc	esc1	1.087	0.000		Age	2.738	0.000
	esc2	1.190	0.000		Income range	4.581	0.000
	esc3	1.268	0.000		Driving experience	2.581	0.000
	esc4	1.175	0.000		Involvement with accident	−0.601	0.000
	esc5	1.226	0.000		Professional driver	−0.276	0.000

(b) Regression weight of structural relationship of latent variable

Latent variables		Estimates	P-value
Acc	esc	0.385	0.000
	abs	0.218	0.000
	demo	0.072	0.000
Esc	abs	0.270	0.000
	demo	0.077	0.000‘
abs	demo	0.039	0.018

Particularly in this study, “acc” refers to thirteen aspects of perception regarding ACC i.e., weather ACC assists in driving on a wet road; overtaking on a slippery road; sudden hard brake; while speeding while climbing a steep road; while speeding on a flat road, in maneuvering at regular speed. Also, the latent variable ACC compares the performance of a car having ACC with a car not equipped with ACC, in lowering the risk of skidding, to correct a slide, in faster accelerating, possibility of faster driving on slippery roads and better steering and braking opportunity. The perception regarding ESC, latent variable “esc” also has the similar thirteen aspects including

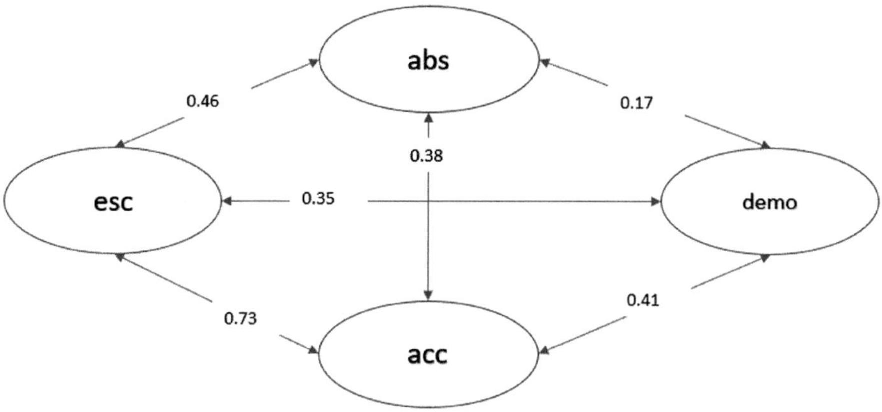

Fig. 1 Path analyses for structural equation modeling

respondent's basic idea, thinking and knowledge about different aspects of ESC. It is assumed that, if a person has a positive perception towards the benefits of ACC he might have the same towards the benefits of ESC.

H2: Primary Perception Regarding ACC has a positive impact on perception regarding ABS- The latent variable "abs" includes driver's knowledge about activation of ABS, whether ABS helps in driving and whether a driver with an ABS equipped car is more prone to aggressive driving than a non-ABS user. The rationale of this hypothesis is that if a person has a positive behavior towards ACC, he is more likely to have a better knowledge and attitude towards ABS, as ABS is comparatively older technology in the sense of vehicle safety than ACC (Table 6).

Table 6 Hypothesis results

Hypothesis	P-value	Results
H1: Primary Perception Regarding ACC has a positive impact on perception regarding ESC	0.001	Accept
H2: Primary Perception Regarding ACC has a positive impact on perception regarding ABS	0.000	Accept
H3: Primary Perception Regarding ACC has a positive impact on perception regarding Driver's demographic characteristics	0.000	Accept
H4: Driver's demographic characteristics has a positive impact on Primary Perception regarding ACC	0.001	Accept
H5: Driver's demographic characteristics has a positive impact on Primary Perception regarding ABS	0.000	Accept
H6: Driver's demographic characteristics has a positive impact on Primary Perception regarding ESC	0.001	Accept

H3: Primary Perception Regarding ACC has a positive impact on perception regarding driver's demographic characteristics- Demographic characteristics constitutes age, gender, income, driving experience, safety checkup frequency, involvement in accidents and whether the driver is a professional driver. ACC might have a direct relationship with demographic characteristics.

H4: Driver's demographic characteristics have a positive impact on Primary Perception regarding ACC- A professional driver might have sufficient knowledge about the benefits of ACC, whereas occasional driver might not have proper knowledge about how ACC can be used as a driving assistance system. Moreover, as young drivers are more lenient to have positive perception towards newer technology, they are more likely to have positive attitude towards the use of ACC in a critical driving condition. However, drivers who have active involvement with accidents might have a negative perspective about ACC as they are less likely to use it at a critical driving condition, or might not find it useful to use it against a crash.

H5: Driver's demographic characteristics has a positive impact on Primary Perception regarding ABS- From previous literatures, it is evident that most of the drivers have wrong idea about activation of ABS, they think ABS might be activated by pressing the brake at a certain level or ABS is activated with hard braking; whereas ABS system is activated automatically if a skidding is detected. Furthermore, it is observed that drivers become aggressive as they feel that they can overcome a skidding or a potential crash as they have ABS to assist them while driving. So, a driver's characteristics may influence their perception towards ABS.

H6: Driver's demographic characteristics have a positive impact on Primary Perception regarding ESC- Driver's knowledge about the benefits of ESC might depend on their demographic characteristics.

The correlation matrix was examined to determine whether the correlations between the variables are particularly large in absolute magnitude. Results for each of the latent variables are shown in Fig. 2. The variables exhibit moderate to weak relationships, consequently, our choice of variables was sound.

Some model fit indices were calculated to check the model fit like Log-Likelihood, Chi-square (p-value), RMSEA, CFI, TLI. AIC, and BIC values for SEM and DIC and PPP for BSEM were analyzed and are shown in Table 7, which indicate an acceptable fit of the model.

4 Conclusion

This study seeks to determine how drivers view vehicle safety features such as ABS, ACC, and ESC. In addition, this study aims to determine how demographic characteristics of drivers influence their opinion of these safety elements. As a result, a comprehensive study of Dhaka's drivers in eleven locations was conducted. The survey contained questions regarding four primary criteria: drivers' perceptions of the

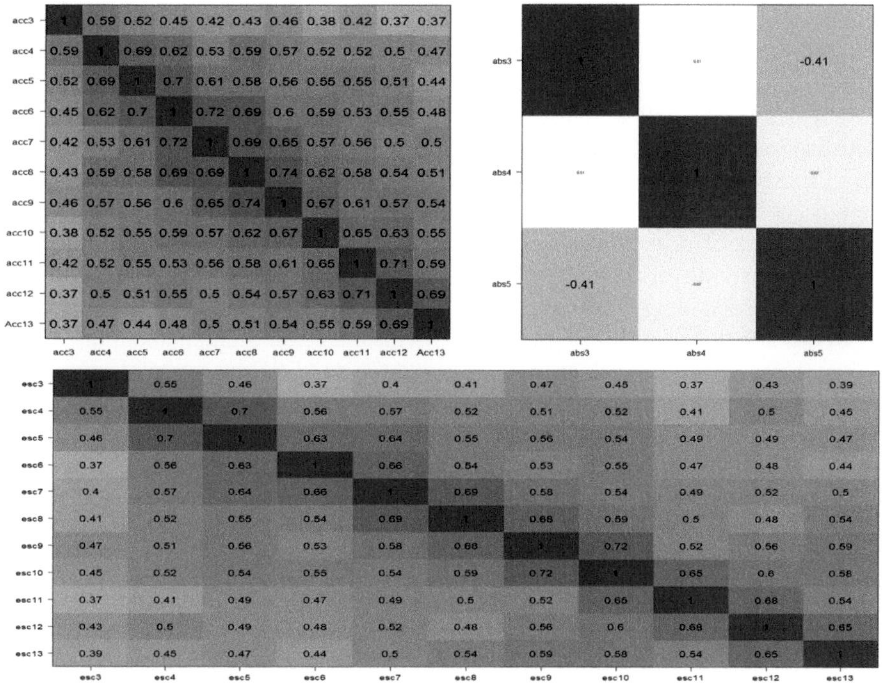

Fig. 2 Correlation matrix for latent variables

Table 7 Model fitness

Test	Values	Standards
Log likelihood	−25360.564	
Chi-squared (p-value)	0.000	
RMSEA	0.083	<0.08
CFI	0.847	>0.7
TLI	0.834	>0.7
AIC	50861.129	
BIC	51169.841	

benefits of ACC and ESC, drivers' attitudes and knowledge about ABS, and drivers' demographic factors. During the survey's two phases, 750 responses in total were collected. After removing the incomplete and invalid responses, structural equation modeling (SEM) was applied to 607 responses to assess the correlation between the latent variables. Numerous exogenous variables collected from survey responses were used to generate four latent variables: ACC, ESC, ABS, and demographic characteristics. According to the descriptive analysis, the majority of respondents held a moderate stance about the advantages of these safety elements. Approximately 20% of drivers strongly agreed that these safety features provided them with major benefits, such as driving on a wet surface, passing on a slick road, speeding when climbing a steep road, speeding on a flat road, accelerating faster, preventing sliding and skidding, etc. The demographic study revealed that the majority of drivers who participated in this poll were professional drivers, and a considerable proportion of them had been involved in accidents. Following the development of six hypotheses for establishing the connection between four endogenous variables, all of the hypotheses were accepted as they were statistically significant. It is claimed that demographic characteristics of drivers have a positive effect on their awareness over all safety elements. Also, drivers' opinions about the benefits of ACC, ESC, and ABS were all linked to each other in a positive way. This means that if a driver knows enough about one safety feature, he is more likely to have a better opinion about the others. Evidently, exogenous variables pertaining to ACC and ESC are favorably correlated with the latent variable of these groups, however aggressive driving tendency was negatively correlated with ABS. The outcomes of this study can influence policy decisions on the installation of safety systems in future automobiles. Also, there is a tendency among customers to prioritize the cost, comfort, luxury, and convenience of a vehicle over its safety features when purchasing one while this research can improvise this attitude. This study demonstrates that the aforementioned safety measures are particularly useful for driving in hazardous conditions; hence, they can be installed in vehicles to prevent accidents. As a consequence, before granting a driving license, the necessary examination should be implemented to assess drivers' awareness of these safety measures. Additionally, this study will aid future research on how people perceive autonomous vehicles.

References

1. Ziebinski A, Cupek R, Grzechca D (n.d.) Review of advanced driver assistance systems (ADAS), vol 5
2. Moravčík E, Jaśkiewicz M (2018) Boosting car safety in the EU. In: 2018 XI international science-technical conference automotive safety, pp 1–5. https://doi.org/10.1109/AUTOSAFE.2018.8373307
3. Bloecher HL, Dickmann J, Andres M (n.d.) Automotive active safety & comfort functions using radar, vol 6
4. Sathyanarayana A, Boyraz P, Hansen JHL (2011) Information fusion for robust 'context and driver aware' active vehicle safety systems. Inf Fusion 12(4):293–303. https://doi.org/10.1016/j.inffus.2010.06.004
5. Bengler K, Dietmayer K, Farber B, Maurer M, Stiller C, Winner H (2014) Three decades of driver assistance systems: review and future perspectives. IEEE Intell Transp Syst Mag 6(4):6–22. https://doi.org/10.1109/MITS.2014.2336271
6. Day TD, Roberts SG (2002) A simulation model for vehicle braking systems fitted with ABS. SAE Trans 111:821–839
7. Garrott WR, Mazzae EN (1999) An overview of the national highway traffic safety administration's light vehicle antilock brake systems research program (SAE technical paper no. 1999-01-1286). SAE International. https://doi.org/10.4271/1999-01-1286
8. Vaa T, Assum T, Elvik R (n.d.) Driver support systems and road safety outcomes: potential effects on fatal accidents, vol 11

9. Headley P (n.d.) ESC as a baseline for active safety, vol 5
10. Gietelink OJ, Ploeg J, De Schutter B, Verhaegen M (2009) Development of a driver information and warning system with vehicle hardware-in-the-loop simulations. Mechatronics 19(7):1091–1104. https://doi.org/10.1016/j.mechatronics.2009.04.012
11. Lu M, Wevers K, Van Der Heijden R (2005) Technical feasibility of advanced driver assistance systems (ADAS) for road traffic safety. Transp Plan Technol 28(3):167–187. https://doi.org/10.1080/03081060500120282
12. McDonald A, Reyes M, Roe C, McGehee D (n.d.) Driver understanding of ADAS and evolving consumer education, vol 10